Space Robotics

Xiu Tian Yan · Gianfranco Visentin
Editors

Space Robotics

The State of the Art and Future Trends

Springer

Editors
Xiu Tian Yan
Department of Design, Manufacturing
and Engineering Management
The University of Strathclyde
Glasgow, UK

Gianfranco Visentin
Automation and Robotics Section,
Mechanical Engineering Department
European Space Agency
Noordwijk, The Netherlands

ISBN 978-3-031-39213-9 ISBN 978-3-031-39214-6 (eBook)
https://doi.org/10.1007/978-3-031-39214-6

This Springer imprint is published by the registered company Springer Nature Switzerland AG
The registered company address is: Gewerbestrasse 11, 6330 Cham, Switzerland

Foreword

Anybody looking at the sky is fascinated by the unknown of space, and scientists and engineers have considered space as a source of inspiration for their creativity.

At the same time, the field of robotics is probably the most exciting scientific and engineering field which, thanks to the current advancements of computer technology, will make things possible which were considered science fiction until a few decades ago. The possibility to discover the unknown of space with robot technology is what this book is about and it is a broad exposition of many different aspects of this endeavour about where we are and where we will be going.

In this short foreword, I will not describe the 21 chapters book structure and its content because this is broadly introduced in Chap. 1 written by the authors, but I would rather like to describe shortly philosophy on the importance of the field that this extensive book covers.

It is in the essence of being humans, that curiosity has brought us to progress and to discover new lands in the last millennium. Such curiosity feeds our intellect, and once we realized that the surface of our planet was finite and totally mapped, there was a need to look outside our world. Any new visionary mission often requires major disruptive steps to implement, and there is probably no other field which is as challenging as space exploration.

As humans, we evolved until now to live in our world, but our intellect has brought us to conceive engineering systems, robots, which can do what we as humans cannot do, and to enable our species to explore what humankind thought unreachable only a hundred years ago.

Robotics is a very special field of science and engineering because at its core it tries to develop machines which can move and act in the physical space but do not have the restrictions that humans have and can, if properly engineered, operate in environments and in places which are impossible for humans. Thanks to robotics humanity is able to extend its presence in space and to explore the beauty of the universe of which we are part of.

The engineering challenges are nevertheless enormous and span all fields of science and engineering like tribology, computer science, and artificial intelligence just to name a few. Robotics is the science of integration and undoubtedly the field which better embodies the concept of multidisciplinarity.

Space is undoubtedly the most challenging environment in which any engineering machine, and therefore robots, has to operate and that is why space robotics is such a fascinating field.

In this book, many of the exciting topics related to this wonderful field are treated and will hopefully not only function as a base of knowledge for those people who intend to gain an insight to contribute to the field, but also as a source of inspiration to those who will continue humankind endeavour to find ways to colonize those parts of the universe which are still unknown.

Enschede, The Netherlands

Prof.dr.ir. Stefano Stramigioli, (M.Sc., Ph.D.)
Vice President Research of euRobotics asbl
Full Professor for Advanced Robotics
IEEE Fellow, ERC Laureate
Member of the Royal Holland Society of Science and Humanities (KHMW)
Scientific Director I-Botics, Vice President MAB of IEEE Robotics and Automation

Contents

1 **Overview of Space Robotics for Sustainability** 1
Xiu Tian Yan and Gianfranco Visentin

2 **Fundamentals of Space Robotics** 13
Xiu Tian Yan and Gianfranco Visentin

3 **Space Robotics Technologies: Perception for Autonomy** 49
Xiu Tian Yan, Jeremi Gancet, Shashank Govindaraj, Raúl Domínguez, Raphael Viards, Clément Bazerque, Thierry Germa, Simon Lacroix, and Michal Smisek

4 **Space Robotics Technologies: Small Dexterous Manipulators** 97
Andrea Rusconi and Piergiovanni Magnani

5 **Locomotion Subsystem Design** 105
Stéphane Michaud, Gerard Kruse, and Michel van Winnendael

6 **ESROCOS: Development and Validation of a Space Robotics Framework** 121
Miguel Muñoz Arancón, Malte Wirkus, Kilian Hoeflinger, Nikolaos Tsiogkas, Saddek Bensalem, Olli Rantanen, Daniel Silveira, Jérôme Hugues, Mark Shilton, and Herman Bruyninckx

7 **Legged Systems for Exploration** 135
Hendrik Kolvenbach, Philip Arm, Giorgio Valsecchi, Nikita Rudin, and Marco Hutter

8 **Air-Bearing Microgravity Simulators for Space Robotics** 157
Tomasz Rybus, Hendrik Kolvenbach, and Xiu Tian Yan

9 **Modular Reconfigurable Multi-robot Systems for Flexible Space Exploration** 191
S. Bartsch, F. Cordes, T. M. Roehr, R. U. Sonsalla, and F. Kirchner

10 **SIROM an Enabler for Reconfigurable Space Robotics** 209
Javier Vinals, Xiu Tian Yan, Sebastian Bartsch, and Manfred Doermer

11 **Coupled Path and Motion Planning for a Rover-Manipulator System** 231
G. J. Paz-Delgado, J. R. Sánchez-Ibáñez, P. Romeo-Manrique, C. J. Pérez-del-Pulgar, and Martin Azkarate

12 **Rover Autonomy Testing in Simulation and on Breadboard Rovers** 249
Matthias Winter

13 **Design and Validation of Orbital Robotic Missions** 281
Marco De Stefano, Máximo A. Roa, Ribin Balachandran, Alessandro M.Giordano, Hrishk Mishra, Nassir Oumer, Bernhard Brunner, Martin Lingenauber, Martin Stelzer, and Roberto Lampariello

14 **Hybrid Testbeds with a Focus on Sensors in Space Robotics Applications** 305
Jörn Thieling and Jürgen Roßmann

15 **In-Space Assembly Operations** 331
Maximo A. Roa, Korbinian Nottensteiner, Ismael Rodriguez, and Gerhard Grunwald

16 **PULSAR: Verification of Technologies for In-Orbit Assembly of a Large Telescope** 355
Maximo A. Roa, Christian Ernst Siegfried Koch, Mathieu Rognant, Antoine Ummel, Pierre Letier, Alessio Turetta, Pablo Lopez Negro, Souriya Trinh, Ismael Rodríguez, Korbinian Nottensteiner, Julien Rouvinet, Vincent Bissonnette, Gerhard Grunwald, and Thierry Germa

17 **An Overview of the Space Servicing Requirements in a Sustainable Space Age** 375
Pablo Lopez Negro, Daniele Filippetto, Claudio Tanzariello, Pierre Letier, and Xiu Tian Yan

18 **Building a Lunar Infrastructure with the Help of a Heterogeneous (Semi)Autonomous Multi-robot-Team** 395
Shashank Govindaraj, Wiebke Brinkmann, Francisco Javier Colmenero, Irene Sanz Nieto, Alexandru But, Matteo De Benedetti, Leon C. Danter, Mercedes Alonso, Enrique Heredia, Simon Lacroix, Dominik Kleszczynski, Joseph Purnell, Kevin Picton, Nabil Aouf, and Luis Lopes

19 The AgriRover: A Space Robotics Inspired Mechatronic Platform Technology for Precision Farming 433
Xiu Tian Yan, Mark A. Post, Alessandro Bianco, Roberto Palazzetti, Ying Lu, Youhua Li, Aron Kisdi, and Wayne Tubby

20 Space Robotics Standard Interface: SIROM Evolution 455
Javier Vinals

21 Robotics Concepts for Future Planetary Exploration Missions 483
Thomas Vögele, Roland Sonsalla, Alexander Dettmann, Florian Cordes, Michael Maurus, Raul Dominguez, and Frank Kirchner

22 HOTDOCK and the MOSAR Walking Manipulator, Technologies for Modular Spacecraft Assembly and Reconfiguration 515
Pierre Letier, Mathieu Deremetz, Jeremi Gancet, Gerhard Grunwald, Maximo A. Roa, Bernhard Brunner, and Xiu Tian Yan

Author Index 539

Topics Index 541

List of Figures

Fig. 1.1 A structure of space robotics themes, topics, future needs and trends and applications fields . . . 5
Fig. 2.1 Reproduction of the original Chess Player (The Turk) [1] (*Photo courtesy of* Marcin Wichary from San Francisco, U.S.A.) . . . 14
Fig. 2.2 Gradual approaching of Comet 67P/Churyumov–Gerasimenko (*Photo courtesy of* DLR) . . . 18
Fig. 2.3 Philae landing on Comet 67P/Churyumov–Gerasimenko (*Photo courtesy of* DLR) . . . 19
Fig. 2.4 Earth orbits and Van Allen belt boundaries . . . 22
Fig. 2.5 A breakdown of subsystems causing satellite failures . . . 33
Fig. 2.6 NASA's robotic refuelling mission 3 [16] (*Photo courtesy of* NASA) . . . 35
Fig. 2.7 A generalised scheme of a planetary sample return mission [10] . . . 36
Fig. 2.8 A&R autonomy modes with regard to delays and control frequency . . . 43
Fig. 3.1 Phase detection principle (*credits* Melexis) . . . 59
Fig. 3.2 Different pattern types for camera calibration . . . 64
Fig. 3.3 Rock feature detection . . . 69
Fig. 3.4 A rover and its coordinate systems using Envire [57]—an open-source robotics library that allows to generate graphs connecting coordinate systems . . . 71
Fig. 3.5 A generic flowchart of the target state estimation for perception . . . 73
Fig. 3.6 Generic flow chart of 3D target reconstruction and tracking using 3D model-based point cloud recognition . . . 74
Fig. 3.7 Illustration of InFuse CDFF-Dev, CDFF-Core, and modular approach to data fusion processes . . . 77

Fig. 3.8 Illustration of Digital Terrain Maps built with a sequence of stereovision data. A luminance value has been associated to each cell (colours encode a notion of traversability), **a** an overview of the terrain of interest for long-distance planning purpose and **b** a local terrain map for detailed planning . . . 80
Fig. 3.9 A flow chart of the processes that generate the digital terrain map and the soil type map . . . 82
Fig. 3.10 Generic flow chart of the satellite state estimation . . . 83
Fig. 3.11 A multiple robot cooperation using different perception sensors mounted on each robot (*credits* CNRS-LAAS) . . . 88
Fig. 3.12 A multiple robot cooperation using perception sensors mounted on a unmanned aerial vehicle which is deployed to enhance the perception of the area terrain of interest, in particular covering a larger area in terrain perception (credit: ESA) . . . 88
Fig. 3.13 An example SLAM used for navigation, guidance and control . . . 91
Fig. 4.1 DEXARM in launch and operational configuration (courtesy of ESA/Leonardo) . . . 99
Fig. 4.2 DEXARM tight integration of joints, electronics and harness (courtesy of ESA/Leonardo) . . . 100
Fig. 4.3 Joint position control with inner motor velocity loop . . . 100
Fig. 4.4 Opportunity MER robotic arm (*credit* NASA/JPL/Caltech) . . . 101
Fig. 4.5 Beagle 2 arm (*credit* Beagle2 team / ESA[8]) . . . 102
Fig. 4.6 DELIAN arm (courtesy of ESA/Leonardo) . . . 103
Fig. 5.1 Mechanics of wheel-soil interaction . . . 107
Fig. 5.2 Phase B Concept of ExoMars chassis kinematics . . . 109
Fig. 5.3 Schematics of a six-wheeled rover with double Ackermann steering geometry. Credit: ETHZ . . . 110
Fig. 5.4 ESA ExoMars rover locomotion verification model in stowed configuration during deployment test at Beyond Gravity. Credit: ESA . . . 112
Fig. 5.5 Hub displacement versus slip for a rigid wheel of diameter 250 mm and width 50 mm at 180 N wheel load on martian soil simulant ES4-SS3590+MS6 . . . 113
Fig. 5.6 Drawbar pull versus slip for flexible BB2 and similar rigid wheel at 180 N wheel load on soils ES1-D6DD and ES3-OMR . . . 113
Fig. 5.7 Unpowered BB2 flexible wheel simulation on ES3-OMR soil and wheel load 180N. Contour shading denotes the pre-consolidation pressure. Credit: DELTARES . . . 115

Fig. 5.8 3D visualisation of the simulated ExoMars model after lander egress on Oxia Planum (Credit: Beyond Gravity) ... 117
Fig. 6.1 The ESROCOS framework ... 124
Fig. 6.2 Testing in the planetary exploration reference scenario: BRIDGET rover in the Mars Yard facility ... 130
Fig. 6.3 Testing in the on-orbit servicing reference scenario: UR-5 manipulator and satellite mock-up in GMV's robotics laboratory ... 131
Fig. 6.4 Testing in the nuclear robotics scenario: CMM robot in the DTP2 facility at VTT ... 132
Fig. 7.1 Several high-priority targets on the Moon and Mars are difficult to reach with traditional wheel-based locomotion. Exemplary shown are on Mars: **a** Edge of the Victoria crater at Meridiani Planum. **b** Cliffs at Meridiani Planum. **c** Aeolis Mons "Mount Sharp" in the Gale crater. **d** South pole ice cap. **e** Visible ice on cliff faces (interior structure is shown in false color). **f** Korolev crater at Planum Boreum (northern polar plain); and on the Moon: **g** Aristarchus crater. **h** Shackleton crater (The crater's interior structure is shown in false color based on data from NASA's LRO probe). **i** Lunar pit located in the Mare Tranquilatis (*Photo courtesy of* NASA/JPL-Caltech/UA / Cornell/Ohio State/GSFC/ESA/DLR/FU) ... 137
Fig. 7.2 Rovers on Mars experienced several high-slip and sinkage events. Depicted are: **a** Spirit. **b** Opportunity. **c** Curiosity (*Photo courtesy of* NASA/JPL-Caltech/MSSS) ... 138
Fig. 7.3 Selection of space legged robot prototypes: **a** CMU Dante II. **b** DFKI Scorpion. **c** DFKI SpaceClimber. **d** NASA JPL LEMUR. **e** DLR Crawler. **f** DLR Bert. **g** FZI Lauron V. **h** ETH Zurich ALoF. **i** ETH Zurich SpaceBok (*Photo courtesy of* CMU/DFKI/NASA/JPL/DLR/FZI/ETHZ) ... 139
Fig. 7.4 Selection of leg-wheel hybrids for planetary exploration: **a** DFKI Asguard v4. **b** NASA JPL ATHLETE. **c** NASA JPL RoboSimian (*Photo courtesy of* DFKI/NASA/JPL) ... 140
Fig. 7.5 Selection of dynamically walking legged robots: **a** Boston Dynamics Spot. **b** IIT HyQ. **c** IIT HyQReal. **d** Unitree Robotics Laikago. **e** Unitree Robotics Aliengo. **f** MIT Mini Cheetah. **g** MIT Cheetah 3. **h** ETH Zurich/ ANYbotics ANYmal B. **i** ANYbotics ANYmal C (*Photo courtesy of* Boston Dynamics/IIT/Unitree/MIT/ETHZ) ... 142
Fig. 7.6 Models of a legged robot with three different levels of simplification [66] ... 146
Fig. 7.7 Thousands of robots simulated in parallel [73] ... 148

Fig. 7.8 SpaceBok (**a**) and ANYmal (**b**) at the robot testing facility of RUAG Space ... 151
Fig. 8.1 Scheme of planar air-bearing (left) and photo of a 50 mm diameter planar round air-bearing (right) ... 159
Fig. 8.2 Test-bed with a 2 m × 3 m granite surface at the Space Research Centre of the Polish Academy of Sciences (left) and 5 m × 9 m epoxy floor at the Orbital Robot test Bench for Integrated Technology (ORBIT) in ESTEC, ESA [54] (right) ... 162
Fig. 8.3 Robotic Space Emulator "Cepheus" operated at the National Technical University of Athens [106] (left) and 3 DOF fixed-base manipulator used for demonstration of redundancy resolution schemes aimed at locally minimizing the reaction torque transferred to the satellite [37] (right) ... 167
Fig. 8.4 Scheme of the satellite-manipulator system used in experiments performed on the planar air-bearing microgravity simulator at the Space Research Centre of the Polish Academy of Sciences ... 168
Fig. 8.5 Photo of the satellite-manipulator system operated at the Space Research Centre of the Polish Academy of Sciences ... 168
Fig. 8.6 Test set-up used in experiments with flexible-joint manipulator [108] (left) and view (acquired by the external vision system mounted above the test-bed surface) of the PINOCCHIO platform operated at the Sapienza Università di Roma [109] (right) ... 168
Fig. 8.7 A 6 DOF air-bearing microgravity simulator used for demonstration of a capture operation performed with a novel hybrid manipulator [110] ... 169
Fig. 8.8 Two vehicles used in approach and capture experiments performed at the Nanjing University of Aeronautics and Astronautics [134] ... 172
Fig. 8.9 Vehicles used in the formation flight experiments at the Spacecraft Robotics Laboratory of the Naval Postgraduate School [149] (left) and vehicle operated at the 5 DOF test-bed at Astrodynamics and Control Laboratory, Yonsei University [48] (right) ... 174
Fig. 8.10 An overview of the spherical air-bearing platform (left) and charging and control interface (right) ... 176
Fig. 8.11 Electrical and hardware system diagram ... 177
Fig. 8.12 Control system overview (left) and schematic overview (right) ... 178
Fig. 9.1 Example for an extensive heterogeneous modular reconfigurable MRS in planetary exploration ... 194

Fig. 9.2 The electromechanical interface (EMI) as central connection of various systems and functional units 197
Fig. 9.3 Two examples of heterogeneous modular reconfigurable MRS 198
Fig. 9.4 Modular communication bus example: Seven paths for sending command openlatch from GCS to bottom EMI of Sherpa with one connected PLI to a payload bay 204
Fig. 10.1 Orbital Scenario. The copyright and credit: ESA – D. Ducros 212
Fig. 10.2 ExoMars rover protptype is shown when it is exploring a potenital planetary area (left), and the concpet that ExoMars is collaborating with an orbitor for a common task is also shown (right) 213
Fig. 10.3 General SIROM/APM architecture 218
Fig. 10.4 SIROM Main parts 220
Fig. 10.5 SIROM coordinate system. *Left:* Front view. *Right:* Top view 221
Fig. 10.6 SIROM Male and Female sides wrt Androgyny axis 222
Fig. 10.7 SIROM thermal IF views 223
Fig. 10.8 CL-FHEM concept 224
Fig. 10.9 SIROM I/F Controller Architecture 225
Fig. 10.10 Termination resistors 226
Fig. 10.11 SIROM Modular IF Product tree 226
Fig. 11.1 Real scale mock-ups of rovers sent by nasa to Mars. From smaller to bigger size: Sojourner (left below), Opportunity (left above) and Curiosity (right). Credit: NASA / JPL-Caltech 233
Fig. 11.2 The Sample Fetch Rover (SFR). Credit: ESA 234
Fig. 11.3 3D model of ExoTeR carrying a 5 DoF manipulator 237
Fig. 11.4 Overview of the experimental terrain 3d model 238
Fig. 11.5 Schematic of the proposed simulation environment 239
Fig. 11.6 Example of an Obstacles Map, differentiating safe areas from non-traversable ones 241
Fig. 11.7 Example of a Cost Map to be used by the path planner, where the higher costs correspond to non-traversable areas ... 241
Fig. 11.8 Example of a planned path for the mobile base of the rover, allowing it to safely reach the sample position avoiding all the intermediate hazards 242
Fig. 11.9 Planned motion for the manipulator, whose movement is constrained by the already drawn rover path, since it generates a tunnel-like volume in its surroundings that encloses all the possible places for the arm to be placed 243

Fig. 11.10 Overview of all sampling cases used for the simulations, including the paths created for the rover base and the end effector 244
Fig. 11.11 Close-up screen-shots of the virtual rover taking a sample during the simulation cases 245
Fig. 12.1 Example of the terrain assessment and path selection by MER ([4], courtesy NASA/JPL-Caltech) 250
Fig. 12.2 ExoMars Rover ([7], courtesy Airbus) 251
Fig. 12.3 ROAMS components of obstacle detection by Sojourner Rover ([11], courtesy NASA/JPL-Caltech) 253
Fig. 12.4 PANGU rendering of Martian terrain and Rover Dynamics Model (from [7], courtesy Airbus) 254
Fig. 12.5 Comparison of slippage in ROAMS simulator to slippage in hardware tests ([14], courtesy NASA/JPL-Caltech) 255
Fig. 12.6 Position estimate without (top) and with (bottom) slippage correction while laterally traversing a slope ([6], courtesy Airbus) 256
Fig. 12.7 Example of the controller error while laterally traversing a sandy slope of 15 degrees ([6], courtesy Airbus) 257
Fig. 12.8 Example of the controller action while laterally traversing a sandy slope of 15 degrees ([6], courtesy Airbus) 258
Fig. 12.9 DEM (Digital Elevation Map) from a simulation on the autonomy test terrain. The DEM is approximately 15 m wide and 7 m deep ([7], courtesy Airbus) 258
Fig. 12.10 Discontinuity and slope map generated by the AutoNav algorithms for the traversability analysis ([7], courtesy Airbus) 259
Fig. 12.11 Nominal placement of a virtual Rover model by the AutoNav algorithms for the clearance check, visualised in PANGU ([7], courtesy Airbus) 259
Fig. 12.12 Persistent map generated by ExoMars AutoNav with currently planned path and complete History of previously driven paths, projected on the simulated terrain. Total traverse length at this stage: 43 m ([7], courtesy Airbus) 260
Fig. 12.13 JPL Mars Yard and test rovers. Front: Sojourner flight spare; Left: MER Project Test Rover; Right: MSL Test Rover ([16] courtesy NASA/JPL-Caltech) 262
Fig. 12.14 Mars Yard at Airbus in Stevenage with MDM Breadboard Rover ([7], courtesy Airbus) 262
Fig. 12.15 Autonomy test scenario in the JPL Mars Yard ([19], courtesy NASA/JPL-Caltech) 263
Fig. 12.16 GESTALT diagnostic output ([19], courtesy NASA/JPL-Caltech) 263

Fig. 12.17 GESTALT autonomous 17.5 m traverse ([19], courtesy NASA/JPL-Caltech) ... 265
Fig. 12.18 Traction control test setup with Scarecrow test rover ([20], courtesy NASA/JPL-Caltech) ... 266
Fig. 12.19 Resulting loads when traversing the test setup shown in Fig. 12.18 ([20], courtesy NASA/JPL-Caltech) ... 266
Fig. 12.20 Test setup for the heading test ([20], courtesy NASA/JPL-Caltech) ... 267
Fig. 12.21 Resulting heading errors when traversing the test setup shown in Fig. 12.20 ([20], courtesy NASA/JPL-Caltech) ... 267
Fig. 12.22 Commanded path sequence ([10], courtesy Airbus) ... 268
Fig. 12.23 Planned, estimated and true rover position in the horizontal plane ([10], courtesy Airbus) ... 269
Fig. 12.24 Commands from the trajectory control module during the execution of the second half of the up-slope path (from [10], courtesy Airbus) ... 269
Fig. 12.25 Estimated slippage ratios for the commanded path segments (1 and 2: up-slope, 3: point turn; 4 and 5: down-slope, [10], courtesy Airbus) ... 270
Fig. 12.26 MDM rover at the end of the up-slope path ([10], courtesy Airbus) ... 270
Fig. 12.27 A 3D terrain model from breadboard rover testing. Note the knowledge gaps behind rocks, as these areas are not visible from the perspective of the navigation camera ([9], courtesy Airbus) ... 271
Fig. 12.28 JPL FIDO rover during field trials at Silver Lake (California, [21], courtesy NASA/JPL-Caltech) ... 274
Fig. 12.29 Web Interface for TeleScience (WITS) display, showing command sequence, panoramic view, 3D terrain model and overhead view of surroundings ([21], courtesy NASA/JPL-Caltech) ... 275
Fig. 12.30 FIDO rover at blind field tests at Black Rock Summit (Nevada). Left: rover after EGRESS from the platform, right: terrain data created from aerial footage ([21], courtesy NASA/JPL-Caltech) ... 276
Fig. 12.31 Data outputs from field trials at Black Rock Summit (Nevada). ([21], courtesy NASA/JPL-Caltech) ... 277
Fig. 12.32 ExoMars field trials (ExoFiT) in the Tabernas Desert in Spain. Front: rover 'Charlie'; Back: landing platform ([25], courtesy ESA/Airbus) ... 277
Fig. 13.1 E.Deorbit scenario: servicer satellite equipped with a robotic arm during the tracking of Envisat. Envisat tumbles about its intermediate axis with an angular velocity of 5 deg/s ... 284

Fig. 13.2 Robotic arm in stretched configuration with built-in gripper and stereo camera system at the end-effector 285
Fig. 13.3 Capability map of the manipulator arm for the e.Deorbit mission 285
Fig. 13.4 E.Deorbit scenario: example of end-effector camera image rendered by ASTOS 286
Fig. 13.5 The OOS-SIM facility: servicer robot (left) and client robot (right) 288
Fig. 13.6 Workspace available for simulation of space robotic operations with the OOS-SIM: sideview (left) and top-view (right) 288
Fig. 13.7 Overview of the system communication in the OOS-SIM. The dashed line divides simulated and mission-like elements 289
Fig. 13.8 Control architecture of the industrial robot for reproducing satellite dynamics 290
Fig. 13.9 The OOS-SIM facility in space light condition. A LED light mounted at the end-effector illuminates the target, and a spotlight simulates the sun illumination 291
Fig. 13.10 Tracking a target satellite: pose estimation for very close range approach phase with overlay of the model contours (in red) and image edges. The perfect overlapping of image edges and model contours indicates a precise orientation and position estimation 293
Fig. 13.11 Fuel consumption for different control strategies for a robotic manipulator in space 295
Fig. 13.12 Target scenario (left) and validation scenario of teleoperation with OOS-SIM (right) 297
Fig. 13.13 Hardware-in-the-loop simulators at DLR: EPOS facility for simulating rendezvous (left), and OOS-SIM facility for simulating the robotic capture of the target (right) 298
Fig. 13.14 Subsystems of an on-orbit-servicing mission simulation system 299
Fig. 13.15 Functional blocks of the robotics payload system 300
Fig. 13.16 OOS-SIM facility employed for simulating the approach and grasping of Envisat. The client robot is equipped with the LAR profile, and the servicer has an LBR robot with a gripper developed by PIAP. Top-left: simulated space scenario for the capture of Envisat 301
Fig. 14.1 Overview of domains, which are facing challenges in the development of sensor-based systems with regard to their verification, validation and optimization during development 306

Fig. 14.2 Use of X-in-the-loop (XiL) structured along the V-model of a typical development process showing hard boundaries between the different forms of XiL—where X can stand for Model (MiL), Software (SiL), Hardware (HiL) or Vehicle (ViL). Based on [11] 309
Fig. 14.3 Vehicle-in-the-Loop (ViL)—as special case of Final-System-in-the-Loop (FiL)—used within the automotive domain to verify and validate the developed driving functions as part of the real-world vehicle (i.e. overall system) driving within a safe and reproducible virtual environment 311
Fig. 14.4 Multidimensional scalability of HTBs to allow for seamless transitions within system development (development process plotted in horizontal direction)—compare use of XiL in V-model from Fig. 8.2 311
Fig. 14.5 Comparison of simulated camera images generated by rasterization (left) and ray tracing (right) for the very same scenario clearly show benefits of ray tracing in terms of physically correctness. Ray tracing results include reflections visible on the solar panels, indirect illumination by self-reflections of the ISS as well as ambient light from the atmosphere (leading to blue illumination visible at down facing surfaces of the ISS) 314
Fig. 14.6 Classification of different modeling methodologies (i.e. model types) regarding their level of detail and effort to be spent for modeling and simulation. Structured in idealized, phenomenological and physical models naming representative subjects of each model type within the appropriate boxes 314
Fig. 14.7 Building blocks for diffuse, glossy and specular reflection/transmission that are combined to mimic even complex scattering properties of arbitrary materials (left) defined within the surface reference coordinate system depicted on the right 315
Fig. 14.8 Idealized ray tracing with idealized reflections that only apply to mirror-like materials but can not mimic indirect illumination caused by diffuse or glossy reflections 315
Fig. 14.9 Physical ray tracing with complex scattering simulating the interaction at surfaces respecting full physical material properties including arbitrarily complex BSDFs 316
Fig. 14.10 Importance sampling of three differently wide opened BSDFs (visualized as lobes) with a finite number of rays at each hit point of the objects' surfaces 316

Fig. 14.11 Simulated lidar measurements visualized as point clouds for two differently scanning lidars. Independent of the specific scan pattern and internal processing steps used by each sensor, computations are always decomposed along the sensor-centric dimensions azimuth, elevation and time—being extended by the dimension of triangles when it comes to the simulation of sensor-environment interactions (see [8]). 3D model of ISS provided by NASA ... 317
Fig. 14.12 CUDA thread hierarchy consisting of grids and blocks, each of which can be declared by the programmer as one-, two- or three-dimensional. Based on [13] 318
Fig. 14.13 Computation times needed to generate 921,600 lidar measurements plotted over scenario time showing slight variations dependent on the complexity of objects that are within the field of view at the specific point in time. All GPU-based simulation tasks (*launch*, *intrinsics* and *processingSteps*) are solved in less computation time—thanks to high parallelism—while copying simulation results from GPU to CPU (*finalToDynStruct*) is noticeable high, even though still less than 40 ms 319
Fig. 14.14 Computation times of physical lidar simulation averaged over the simulated RvD scenario with the ISS as described for Fig. 14.13. This time additional computation times are depicted showing the time needed for using ten iterations for sensor-environment interaction (*launchLoop*) and the resulting overall computation time (*calcExecute*), which summarizes all averaged times needed for simulation and copy tasks in this scenario. Note that *launch* is included in *launchLoop* and not counting twice in this plot 319
Fig. 14.15 Spatial and temporal synchronization of real twin (left) and digital twin(right) including operating data of lidar sensor and Kuka robots. Here, two Kuka robots realize systematic and reproducible motions between lidar and ISS mockup with 1:100 scale 320
Fig. 14.16 Hybrid laboratory with real (left) and digital twin(right) both generating sensor data for the very same scenario. This allows for a direct comparison of both worlds based on sensor data that is perfectly synchronized in spatial and temporal domain 322
Fig. 14.17 Hybrid laboratory showing real sensor and ISS mockup with additional information of estimated pose and bounding box of the target fused within the image 323

Fig. 14.18 Automated Rendezvous- and Docking (RvD) with the virtual ISS. Simulated RVS3000(-3D) lidar sensor mounted on a spacecraft (left) scanning the ISS for distance measurement and pose estimation. Sensor data generated by simulated lidar visualized as color-coded point cloud. 3D model of ISS provided by NASA 324
Fig. 14.19 Hybrid closed-loop setup for testing real software and hardware components (right and top) within a simulated operational environment (left) 325
Fig. 14.20 Overview of mars sample return mission (left) and hybrid closed-loop setup to be used within sub-mission Earth Return Orbiter (right). Based on [18] 327
Fig. 15.1 Current robotic demonstrations of in-space assembly. From left to right: Dextre approaching the RRM module (Courtesy of NASA), Robonaut assembling truss structures on ground (Courtesy of NASA), Trusselator for SpiderFab (Courtesy of Tethers Unlimited), and initial ground demonstration of Phoenix with the FREND arms (Courtesy of DARPA) 334
Fig. 15.2 Different robotic concepts for in-space assembly. From left to right: SpiderFab (Courtesy of Tethers Unlimited), Archinaut (Courtesy of Made in Space), CIRAS (Courtesy of Orbital ATK), and Dragonfly (Courtesy of Space Systems Loral) 335
Fig. 15.3 OOS-SIM, an on-orbit servicing simulator developed at DLR 338
Fig. 15.4 Existing standard interfaces: top left: EE on SRMS (Courtesy of NASA), top right: LEE on SSRMS (Courtesy of NASA), bottom left: JEM EE (Courtesy of JAXA), bottom right: OTCM (Courtesy of NASA) 339
Fig. 15.5 Different standard interfaces: JAXAEE (Courtesy of JAXA), CTED (Courtesy of ESA) and DWIM (Courtesy of Tokyo Tech) 340
Fig. 15.6 European standard interfaces: From left to right: HOTDOCK (courtesy of Space Application Services), SIROM (courtesy of SENER) and iSSI (courtesy of iBOSS) 341
Fig. 15.7 European projects considering robotic assembly in space. From left to right: artistic concepts for PULSAR, MOSAR and PERIOD 341
Fig. 15.8 Workflow of the robotic assembly system. An assembly specification is automatically decomposed into a task sequence and then mapped to appropriate robotic skills. These skills are then executed in the robotic setup 344

Fig. 15.9 Two light-weight robot arms (KUKA LBR iiwa) collaborate in the assembly of aluminum structures 345
Fig. 15.10 Realization of various subtasks. Clockwise: slot nut insertion, placement of an angle bracket, placement of a profile, and fastening the angle bracket with screws using the screwdriver 346
Fig. 15.11 Components of the planning unit for the autonomous execution of assembly skills. The sequence planner, the grasp planner and the motion planner interact with a runtime world model to select and parameterize an appropriate sequence of skills 348
Fig. 16.1 Artistic concept of the PULSAR project for assembling a large space telescope 356
Fig. 16.2 Deployment of the James Webb Space Telescope. Left: folded configuration, for transportation in the cargo area. Right: fully deployed configuration in space (Courtesy of NASA) 357
Fig. 16.3 Primary mirror in the dPAMT demonstrator. Left: components of the assembly. Right: HOTDOCK standard interconnect, passive and active version 359
Fig. 16.4 Active segmented mirror tile of dPAMT 359
Fig. 16.5 Components of the dPAMT demonstrator 360
Fig. 16.6 Planning of the assembly sequence for dPAMT using a hybrid planning framework 361
Fig. 16.7 Visual reconstruction of the assembly using AprilTags. The image shows the alignment verification of SMT2 (bottom left) with respect to SMT1 (central tile) 361
Fig. 16.8 Single, double and triple mating connections during the dPAMT demonstration 362
Fig. 16.9 The outermost ring of the mirror structure lies outside the workspace of the robotic assembly system. To overcome this limit and facilitate the assembly, a unit consisting of five mirror tiles (green) is pre-assembled before being mounted to its final position on the outer ring 363
Fig. 16.10 dLSAFFE underwater spacecraft mock-up in the test facility. The robotic assembly system manipulates a subassembly consisting of three mirror tiles. Two tiles of the inner ring of the mirror structure are already assembled to the stationary central tile. Without mirror tiles, the mock-up is 3.5 m tall 363
Fig. 16.11 Simplified software architecture for dLSAFFE. Components in blue are shared among demonstrators, while yellow components are specific to dLSAFFE 365
Fig. 16.12 dLSAFFE demonstrating the manipulation of a subassembly with three mirror tiles 365

Fig. 16.13 dISAS final implementation of the attitude controllers 366
Fig. 16.14 Attitude deviation (deg) of the spacecraft and duration (s) without comlink during the full deployment with two sets of trajectories 367
Fig. 16.15 Behaviour tree (simplified representation) for the primary mirror assembly (left); current assembly state in Webots (right) 367
Fig. 16.16 Tile identification and localization using AprilTag fiducial marker detection 368
Fig. 16.17 Tile pose computation at the beginning of the operation (left), and close to convergence (right) 369
Fig. 16.18 Evolution of the visual features error during the servoing operation (top left), with the correct regularization to zero of the error function, and evolution of the end-effector velocities (top right). The corresponding commanded joint velocities are displayed at the bottom 370
Fig. 16.19 Large telescope assembly using the dISAS simulator, visualized at different steps of the assembly process 371
Fig. 17.1 Evolution of launch costs from 1985 to today [2]. The apparition of private companies has lowered down the prices by a factor of two in the last decade 377
Fig. 17.2 Number of launches in the last ten years [3]. The increasing contribution of commercial initiatives is observed, representing around 15% of the total launches per year in the beginning of the decade and reaching 25% by the end of it 378
Fig. 17.3 Schematic comparison of the development cycles of traditional spacecraft versus new space paradigm oriented to constellations 379
Fig. 17.4 Identified trends for the next decade across different actors of the space community 380
Fig. 17.5 The vicious circle about on-orbit servicing 382
Fig. 17.6 MOSAR technologies: walking manipulator, standard interface, spacecraft modules and dedicated software 386
Fig. 17.7 MOSAR demonstrator setup 387
Fig. 17.8 MOSAR exploitation roadmap 389
Fig. 17.9 Segmentation analysis of the main products of a large spacecraft manufacturer 390
Fig. 17.10 Schematic representation of the implementation of MOSAR modules in a product line of commercial spacecraft 391
Fig. 17.11 Overview of the life cycle of a commercial spacecraft implementing on-orbit servicing for payload reconfiguration purposes 392
Fig. 18.1 Cooperative scenarios description 398

Fig. 18.2 Mantis (*Credit* DFKI GmbH) ... 399
Fig. 18.3 IBIS mobile manipulation robot (*Credit* DFKI GmbH) ... 400
Fig. 18.4 Veles Mobile robot with a manipulator and gripper, grading blade and shovel ... 401
Fig. 18.5 Veles manipulator and the manipulator workspace volume for different weight payloads ... 401
Fig. 18.6 Mobile gantry deployment ... 402
Fig. 18.7 PRO-ACT system design ... 403
Fig. 18.8 Multi-Agent AutonomyComponent Diagram ... 405
Fig. 18.9 Task allocation with mission coordinator ... 406
Fig. 18.10 CREW multi-agent design, showing only two robot instances ... 407
Fig. 18.11 Multi-agent coordination ... 408
Fig. 18.12 ISRU mockup on lander detection and tracking ... 410
Fig. 18.13 Initial tests AruCo marker detect ... 410
Fig. 18.14 Different-sized AruCo markers on the demonstration ISRU mockups ... 411
Fig. 18.15 Visual representation of the estimated poses (blue dots) and map (point cloud) created by the CDFF's Visual SLAM DPFC ... 411
Fig. 18.16 Ground truth from the DLR dataset ... 412
Fig. 18.17 Initial DEM reconstructed from the DLR dataset ... 412
Fig. 18.18 CSLAM algorithm overview (**a**); Loop closure detection across different view points (**b**) ... 414
Fig. 18.19 I3DS architecture overview in PRO-ACT ... 414
Fig. 18.20 I3DS sensor architecture ... 415
Fig. 18.21 IP mesh Node—DTC SOLO8 Software Defined Radio (Concealment—left, Robust—middle), MIMO antenna right ... 415
Fig. 18.22 IP mesh communication architecture ... 416
Fig. 18.23 IPMesh node on mantis ... 417
Fig. 18.24 Monitoring and control system deployment ... 418
Fig. 18.25 Mapping Matrix of software interfaces ... 419
Fig. 18.26 Mantis in the space exploration hall ... 420
Fig. 18.27 Veles at PIAP SPACE lunar analogue ... 421
Fig. 18.28 Cooperative mapping performed remotely using Veles (left) and Mantis (right). Reconstructed maps from each robot shown below ... 422
Fig. 18.29 Veles arm connected to shovel after swapping the gripper via HOTDOCK ... 423
Fig. 18.30 Tool exchange procedure using HOTDOCKs ... 424
Fig. 18.31 Veles before flattening rough area in DEM ... 424
Fig. 18.32 Veles and its DEM after the grading task ... 424
Fig. 18.33 Two Panda robotics arms manipulating a wooden bar ... 426
Fig. 18.34 Grasping tests with veles ... 427

Fig. 18.35 Grasping approach to ISRU mockup 427
Fig. 18.36 Grasping of Mockup 428
Fig. 18.37 Mantis' perception and localization during the transportation scenario 428
Fig. 18.38 Gantry deployment sequence. Deployed (left) and stowed (right) configurations 429
Fig. 18.39 Gantry's end effector manipulating a Mockup 429
Fig. 19.1 Fertilizer usage in kilograms per hectare of arable land by farmers in World and UK 434
Fig. 19.2 AgriRover system Architecture 438
Fig. 19.3 Final AgriRover Platform system prototype and main control and sensing design feature components of the AgriRover 439
Fig. 19.4 Dynamic energy efficiency plot examples **a** at 30% of the maximum speed set running for 18.3 m, 63 s and with an average speed of 0.3 m/s; **b** at 50% of the maximum speed set running for 20 m, 43 s and with an average speed of 0.46 m/s 441
Fig. 19.5 Diagram of AgriRover Systems (implemented in ROS) 442
Fig. 19.6 Driller design: **a** Rendered views of the driller, **b** Detailed views of the driller, **c** Prototype of the driller consisting of a linear actuator to lift and lower the sample collector 444
Fig. 19.7 Image recognition class diagram 446
Fig. 19.8 An object recognition classification example 447
Fig. 19.9 **a** GUI design of the user interface; **b** Field Boundary Extraction 448
Fig. 19.10 Mapping and Loop Closure in Forest Grove (**a**) and field test navigating maps generated (**b**) 450
Fig. 20.1 Orbital test set up at ADS for PERASPERA OG5 project 456
Fig. 20.2 SIROM electro-mechanics upgrade 457
Fig. 20.3 Electronics subsystem block diagram 458
Fig. 20.4 Power PCA (left), motor driver PCA (right). (SENER©) 459
Fig. 20.5 Main steps of EROSS mission scenario 460
Fig. 20.6 EROSS servicer deployed configuration (Courtesy of Thales Alenia space) 460
Fig. 20.7 SIROM internal validation in SENER (December 2020) [18] (SENER ©) 461
Fig. 20.8 SIROM in EROSS demonstration at GMV premises in Madrid, Spain (April 2021) 462
Fig. 20.9 EROSS demonstration setup at Thales Alenia Space premises in Cannes, France (June 2021) (Courtesy of Thales Alenia Space) 463
Fig. 20.10 Stand-alone vibration test (left). Mated pair vibration test (right) 463

Fig. 20.11 Random vibration profile on Z axis on standalone configuration 464
Fig. 20.12 Triple docking functional test prior to SIROM delivery (SENER ©) 466
Fig. 20.13 PERIOD mission architecture 467
Fig. 20.14 RIDER fluid transfer connector 468
Fig. 20.15 SIROM-A for refuelling: latched position (left), connected position (right) 468
Fig. 20.16 SIROM interface close to latch 469
Fig. 20.17 Connectors plate distribution 471
Fig. 20.18 SIROM electrical and data diagram 472
Fig. 20.19 Capture switch 472
Fig. 20.20 Architecture diagram of the Electronic Unit inside each ORU 476
Fig. 20.21 Stackable connector from standard PC/104 476
Fig. 20.22 Visual servoing system 477
Fig. 20.23 Visual servoing preliminary test 478
Fig. 20.24 Active SIROM (left), Passive SIROM (right). Version not for resupply 479
Fig. 20.25 Resupply SIROM-A 480
Fig. 20.26 Resupply SIROM-P 481
Fig. 21.1 The SherpaTT rover in the morocco field trial 2018 490
Fig. 21.2 Rover SherpaTT equipped with avionics box during the Field Testing of the ADE project 2021 491
Fig. 21.3 Generic control architecture—the guidance and navigation layer, the motion generation and control layer, and the hardware abstraction layer are each interfaced through the lightweight RRC library with standardized interfaces to allow framework-independent access (orange). In this example, the approach allows the parallel use of ROS (green) and ROCK (blue) 495
Fig. 21.4 CREX walking on a 20° incline (left) and over a rubble field (right) 496
Fig. 21.5 Exploration view of coyote III showing its main elements: 1: Front Body with two directly driven hybrid legged-wheels and additional payload bay, 2: Centre Body with EMI for payload connection, 3: rear body with two directly driven hybrid legged-wheels, acting as passive boogie of the suspension system, 4: sub-system compartments, 5: sensor bench for navigation 497
Fig. 21.6 Coyote III navigating through boulders and showing its mobility capabilities 497
Fig. 21.7 RIMRES systems. rover deploys six-legged scout from lander mock-up 499

Fig. 21.8 TransTerrA systems, the exploration rover has a soil sampling module mounted to the manipulation arm. In the foreground, a stationary BaseCamp with another soil sampling module attached is visible 500
Fig. 21.9 Crex robot climbing a reproduction of a Moon crater (left) and Asguard v4 exploring a lava tube during the final field test of the project ENTERN (right) 503
Fig. 21.10 SLAM reconstruction of the explored 244 m long lava tube. between each color repetition the elevation corresponds to one meter 504
Fig. 21.11 The four consecutive phases of the CoRob-X Lunar Analogue Mission 505
Fig. 21.12 RH5 Manus grasping a tool on a table (left). It can also be teleoperated using an active exoskeleton (right) 509
Fig. 21.13 The four-finger gripper (left) of RH5 Manus developed in the project TransFIT has tactile and distance senors integrated in the finger segments (right), which can be exploited for manipulation tasks 510
Fig. 22.1 MOSAR mission concept artistic view and derived spacecraft design (client and servicer) 517
Fig. 22.2 MOSAR ground demonstrator derivate from MOSAR mission concept 519
Fig. 22.3 HOTDOCK passive (left) and active (right) standard interconnect 521
Fig. 22.4 Active and passive version of HOTDOCK with highlight of the four main interface functions 521
Fig. 22.5 Left: HOTDOCK mechanical interface elements; Right: Hall sensor configuration 522
Fig. 22.6 Left: Experimental result of HOTDOCK form-fit range of attraction; Centre: Test setup to validate HOTDOCK angular engagement; Right: Compatibility of HOTDOCK engagement with standard module shapes 524
Fig. 22.7 Left: HOTDOCK central power and data connector plate, equipped with POGO connectors; Right: HOTDOCK fluidic interface structure 524
Fig. 22.8 HOTDOCK avionics architecture and integrated PCB controller board 525
Fig. 22.9 Integration of active and passive HOTDOCK SI in MOSAR components 526
Fig. 22.10 MOSAR-WM relocatable manipulator (connected on the side of a MOSAR module) 528
Fig. 22.11 MOSAR-WM design and avionic integration 529
Fig. 22.12 MOSAR-WM actuator and avionics integration 530
Fig. 22.13 MOSAR-WM avionics architecture 530
Fig. 22.14 WM software architecture 532

Fig. 22.15 MOSAR manipulation sequence of spacecraft module from client to servicer spacecraft 534
Fig. 22.16 Walking on top of spacecraft modules 534
Fig. 22.17 Walking on side of spacecraft modules 535
Fig. 22.18 HOTDOCK triple orthogonal alignment and simultaneous connection 535
Fig. 22.19 Artisitc view and detailed design of the MIRROR Multi-Arm Robot concept 536

Chapter 1
Overview of Space Robotics for Sustainability

Xiu Tian Yan and Gianfranco Visentin

Writing a dedicated book on an exciting topic such as space robotics requires a lot of effort and thinking to capture the fundamental knowledge of the topic that has been published to dater and also understand and follow the latest developments of the topic and its rapid advancement in space robotics in this case. After several years of effort in addressing both of these requirements, editors believe we now have a book that can present the fundamentals of space robotics so as to lay a foundation from which those new to the topic can read and learn. In doing so it represents the latest technological advancement and the state of the art of space robotics as derived from major international initiatives that the editors have access to. As such, the book is structured to provide reading materials of these two groups of topics, which aims to disseminate knowledge of space robotics and advance future space robotics development by outlining future research and development directions and trends so that human beings can accelerate their desire to explore the universe and to discover new knowledge.

With the above in mind, the book is structured into four sections: namely, fundamentals of space robotics; essential unique tools for investigating space robotics; design and testing of space robotics where actual design, development and testing is covered; and finally, those future needs and research trends that we can report at this stage of a long journey for advanced and more fully autonomous space robotics to enable future space exploration missions. Figure 1.1 shows the key block representation of the above four sections and their detailed constituents for each section covered within the book.

X. T. Yan (✉) · G. Visentin
Robotics and Autonomous Systems Group, Department of Design Manufacturing and Engineering Management, University of Strathclyde, James Weir Building, 75 Montrose Street, Glasgow G1 1XJ, Scotland
e-mail: x.yan@strath.ac.uk

X. T. Yan and G. Visentin (eds.), *Space Robotics*,
https://doi.org/10.1007/978-3-031-39214-6_1

1.1 Fundamentals

This section covers the fundamental knowledge on space robotics and general robotics is presented as introduction to space robotics in Chap. 2 after this overview of the book. This consists of an introduction to space robotics by describing the endeavours by mankind on the general evolution of automation and robotics followed by an introduction to the unique characteristics of space robotics. The application fields of space robotics are also detailed to provide the context for the book and the exciting opportunities that these technologies can enable for both orbital servicing missions and planetary discovery missions. Based on the unique challenges and requirements for both orbital missions and planetary exploration, a number of the unique capabilities required for space robotics are outlined. A key to the success of space missions involving space robotics is the unique set of capabilities of robotic systems deployed.

Owing to the unique requirements for a high degree of autonomy for these missions, it is imperative that any space robotic system has a high level of perception capability both for the environment where the robot is operating, and the situation and the condition of the robotic system itself. A dedicated chapter on perception for autonomy as a fundamental space robotic technology is included as a key technological capability in the fundamental knowledge section. Sensing and sensor requirements define the hardware basis as a key enabler for perception autonomy, and these are described in detail and their working principles explained. This topic forms the first important topic of the fundamental knowledge section and is a broad introduction to the space robotics technologies as presented in Chap. 3. This chapter starts with a generic and holistic overview of the key functional components of space robotics and the key functions are autonomous perceptions enabled by a suite of advanced sensors based on their respective sensing working principles and methods. Building on the sensor data, perception processing provides the ability to understand the environment and to be aware of other robots if multiple robots are involved in a possible collaboration for completing a more complex task. Ultimately, the perception information enables a smart control of space robots in performing their planned tasks in a more autonomous manner. The perception capability hence provides intelligence enabling space robots to behave differently in different environments and they can therefore complete mission tasks in an optimal fashion with energy consumption minimised and mission time saved. Motion planning and control can be facilitated thorough technologies such as the localisation of robots in a typical unstructured environment where the robots need to understand the constraints of their operational environment so they can navigate in these complex environments and control their movements for a safe, energy efficient and productive space mission. With the understanding of these foundational capabilities, it is hoped that a reader will have an effective grounding in space robotics before moving to the next section of the book.

In comparison to other robotic investigation approaches, space robotics is characterised by its unique set of challenges as detailed in Sect. 2.2 of Chap. 2. This also means that space robots are expected to perform an array of tasks including manipulation in a harsh environment and mobility operations designed for a variety and possibly unknown terrains.

Chapter 4 provides a technical grounding on the design, control and prototyping of several robotic arms, including a small novel space robotic manipulator which was the first European robotic arm developed for a Mars mission and which was installed on the Beagle 2 lander. This 5 degrees of freedom robotic arm is a lightweight and human-like arm. Other robotic arms such as DEXARM manipulator by ESA, and MER robotic arm by NSAS's JPL are also introduced to show that a range of manipulators have been developed for space exploration missions.

Locomotion systems to enable a long-range mobility to explore part of a planet is another type of robotic actuation derived movement and is the focus of Chap. 5. There are a number of important concepts which need to be defined in order to provide a precise and deeper understanding of this group of space robots. These include trafficability defined as the ability to generate traction between the robotic wheels and the planet surface, terrainability which is about the ability of a mobile system to negotiate uneven terrains, and manoeuvrability to follow a planned path. In particular, a good level of understanding of the interaction between a wheel and soil is required through theoretical analyses, experiments and simulation to provide solid grounding for this unique requirement for space planetary robots. Energy modelling and usage monitoring is also important for planetary robots.

1.2 Essential Unique Tools

Having mastered these foundations and building on the hardware subsystems introduced in Sect. 1.1, it is necessary to discuss a range of essential, yet unique to a space mission and associated tools, which are critical to the success of space robotic design and testing.

The first essential tool is that a robotic control operating system for the operation of those of these space robots described in Sect. 1.1. Instead of describing a general robotic operating system, it was decided to describe a leading and current robotic system called European Space Robotics Control and Operating System (ESROCOS) in Chap. 6. An overview of the ESROCOS framework is provided and a number of tools to model heterogeneous real-time distributed systems and to model and analyse the Behaviour, Interaction, Priority (BIP) of such systems using BIP formal language, and other support tools are also detailed. It is hoped these will provide a reader with an appreciation of a real robotic operating system for space applications, as its applications in both planetary and orbital scenarios are later discussed in three case studies.

Further, a range of robotic concepts should be explored in relation to space missions, and Chap. 7 is dedicated to the efforts of designing and exploring a family of robotic concepts based on legged systems, which are included in this section to provide an interesting perspective of space robots. Both hardware of these legged systems and their control are reviewed and a number of field trials have been reported and discussed to demonstrate their potential for future space missions.

Unique to space robotics development on the Earth, it is necessary to create a simulated environment to test fully these robots before they are deployed in a mission. Robotic ground testing facilities play a significant role and these are covered in Chap. 8. Testing often leads to a significant improvement of success rate in a mission. Space robots experience a special effect under microgravity environments, either in an orbit or possibly on another planet where gravity will be significantly smaller than on Earth. As a result, the behaviour of these robots will be quite different and it is necessary to investigate fully these effects. A particular set of enabling technologies and associated testing kits have been designed and prototyped and these testing kits include test rigs made of air-bearing systems. These work on the principle of generating a thin air film between several aerostatic bearings, which are attached to and form a vehicle, and can move freely on a flat guide surface. These air films allow the vehicle to float frictionlessly as if the vehicle is moving in a microgravity environment. These frictionless movements can be extended to multiple degrees of freedom and more details can be found in Chap. 8.

Modular design and modularity is a new and alternative approach to space system design. Chapter 9 presents a set of definitions of terminologies in modular design of space robots. Building on these concepts, the chapter then explores the process of generating modules through modularisation of interfaces and functional components. Based on a set of modules, it is then possible to create a system through configuration of these modules by illustrating a brief reconfigurable system design approach to a multi-robot system for steep crater exploration at the lunar south pole. Further examples of heterogeneous modular reconfigurable robot systems are presented to show the manipulation capabilities for reconfiguration, and operation of these reconfigurable multi-robot systems.

One of the unique features and capabilities of space robotics in both planetary exploration and orbital missions is the ability to reconfigure the robot itself when a different mission task is planned. This ability becomes even imperative when the space robot faces with an unforeseen mission task which may have been defined after the robotic mission is launched. The ability of reconfiguration lies in two aspects, namely physical reconfiguration of robotic structure, and software reconfiguration. The key to physical reconfiguration is enabled by a universal and multiple functional interfaces which can connect the robot with different payloads to undertake different mission tasks. Chapter 10 provides an example of such a multiple functional interface example, called SIROM as a System Interconnect for robotic manipulation of payloads in future space missions. Functionalities include ability to transfer mechanical loading, multiple sensory data at high data bandwidth, electrical power to payloads, and thermal energy for system thermal management. All these are housed within one interface SIROM and provide many options, both active and passive, for reconfiguration of a payload. This System Interconnect has aimed to be developed as an international standard and has evolved from the first version mentioned in the chapter to a newer and more robust version detailed in Chap. 20, which is introduced to show the future research and development trends for space robotics.

Another common capability requirement for space missions, particularly for science missions, is the ability to mobilise a platform such as a rover and carry a manipulator for a number of mobile manipulation of objects of interests. For example, in the mission to fetch samples of scientific interest for a Mars mission, this capability is essential to collecting the samples prepared by another space rover. The planning of both the path trajectory of the mobile rover and manipulator movements are therefore needed to control these multiple functional robotic systems. In this case the planning should be coupled for both the path planning as well as the manipulator task planning as an integral system. Chapter 11 provides a detailed description of such planning for European Space Agency's ExoTeR system. The planning is undertaken in a platform called Vortex Studio so that it provides an effective simulation environment to validate the planning strategy pf the planning for both the rover path and manipulator's tasks. The fundamental knowledge and models such as terrains are applied in this essential tool to validate the planning strategy.

These unique space robotics tools covered in this Section, including legged systems for space exploration, robotic interfaces for modular and reconfiguration remotely, and testing facilities are depicted in blue section of Fig. 1.1. The technical details of these topics are detailed in Chaps. 6, 7, 8, 9, 10 and 11 respectively

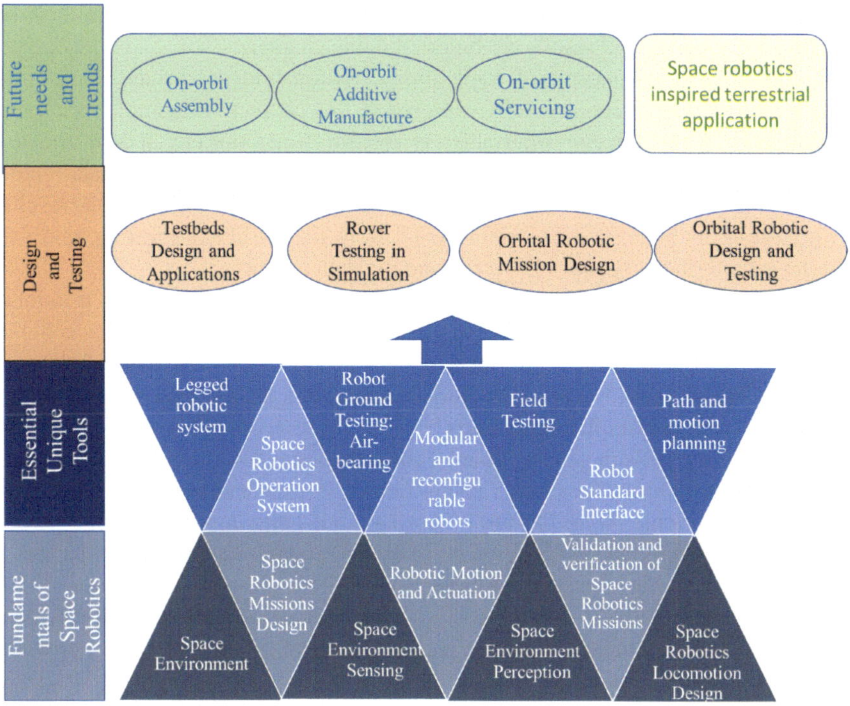

Fig. 1.1 A structure of space robotics themes, topics, future needs and trends and applications fields

1.3 Design and Testing

Building on the fundamentals and essential unique tools covered in Sects. 1.1 and 1.2, the book starts to describe some of leading design and testing work from recent years in Sect. 1.3. This section of 5 chapters aims to demonstrate how the fundamentals and essential tools have been applied in principle to the design of some space robotic centred missions and tasks. Chapter 12 then focuses on the design and testing of planetary rover. There are then three chapters which in turn describe on-orbit servicing (Chap. 13), assembly (Chap. 15 and a description of ultra large structure assembly robot in Chap. 16. In addition, the section also contains a detailed description of hybrid testbeds for both orbital robotic and planetary rover simulation and testing in Chap. 14.

Chapter 12 provides a description of several rover design solutions, including Mars Exploration Rovers (MER) Opportunity, Spirit and the Mars Science Laboratory (MSL) Curiosity, and European's ExoMars Rover. The focus is more on the higher level of autonomy and the driving modes of these rovers. Extensive testing has been performed in a simulator called the ROADMS simulator developed for MER and MSL rovers and detailed simulation results are presented to show the results of hundreds or thousands of scenarios through automated simulation. In addition, the actual breadboard rovers have also been tested in a Mars Yard at Airbus in the UK and testing scenarios and results are reported in the chapter.

The fundamental concepts of trafficability, terrainability and manoeuvrability introduced in Chap. 5 are also fully explored in this chapter with ample examples to illustration of traction control, heading errors, slippage ratios etc. in the Curiosity rover or the ExoMars rover. Fully automatous navigation testing of ExoMars' guidance, navigation and control (GNC) system has also been detailed to show the results of the testing including a 3 D terrain model from the breadboard testing, and the web-based interface to control the rover and the data outputs.

Chapter 13 focuses on the design and validation of orbital robotic missions and this provides an example of the fundamental knowledge on the space robotic mission design and validation introduced in Chap. 2 as a fundamental topic for space robotics. This chapter reviews the history of satellite launching and the associated problem of space debris after spacecrafts become non-functional but are still orbiting around the Earth. This increasing space debris challenge has provided the motivation for all main space agencies to define recovery plans and explore the concept of On-Orbit Servicing. Some On-Orbit Serving (OOS) missions such as Germany's orbital servicing mission DEOS have been explored and introduced. Similarly, some missions have been designed and explored to de-orbit a target non-functioning spacecraft. For example, the e.Deorbit project is designed to capture and deorbit the Envisat satellite and details of approaching and grasping the satellite are discussed to provide a practical perspective as to how these on-orbit debris removal could be realized in detail through simulation. The chapter then shows how the simulation is tested and demonstrated on ground using a servicer robot and a client robot to test the approaching and grasping operations. This demonstration is further simulated in a

OOS-SIM simulation environment where physical sensors provide real time data for developing various control strategies, including coordinated control of the base translation and rotation and the centre of mass of space robot in a semi-autonomous control, or teleoperation or shared control. The chapter also provides an example of validation of a mission operation in a close range involving the hardware in the loop of simulation using simulators at DLR. These simulations demonstrated and validated the rendezvous and capture of the target operation.

In addition to the robotic ground testing covered in Chap. 8 it is important to detail aspects that a physical ground testing facility may not be able to allow, that is continuous and early system testing in dynamic operational environments. Chapter 14 introduces new technologies such as digital twins and hybrid testbeds to enable more advanced and real time testing of space robotics, relying on physical sensors simulation, smooth synchronisation of the digital twin and real twin for seamless transition between simulation and reality. The overall architecture of the approach is also detailed, exemplified by two testing of Mars Sample Return mission Erath Return Orbiter (ERO), and an automated rendezvous- and docking with the virtual International Space Station (ISS).

Construction of a large space structure in an orbit in a new and cost-effective way would enable a new space age and remains a main motivating ambition for space communities to develop suitable technologies and approaches to facilitate such an ambition. Chapter 15 introduces a set of current technologies which have been investigated over last three decades towards a new solution for future large structure construction in orbit. These technologies have been designed and tested and their implementations are briefly introduced in this chapter. Additionally, a review of past efforts is provided in this chapter so a foundation on the design and testing of these missions and associated technologies can be established. These technologies include a special purpose dexterous manipulator for tasks such as opening and closing the covers of batteries and reconnecting their cables; a dual arm robotic system called Dextre; a humanoid robot called Robonaut 2 developed by NASA to undertake a number of operations that astronauts normally perform, such as assembling truss structures and cleaning; additive manufacturing, and a purposely developed robotic technology for assembling a satellite called "Chriysalis". Many missions are still at conceptual stage. A number of testing facilities have also been developed and are introduced in this chapter, including an on-orbit servicing simulator called OOS-SIM developed at DLR. Particular attention has been drawn to the connectivity and interfaces for space robotic enabled assemblies of large structures. A detailed description of assemblies using robots is given in this chapter to provide an in-depth understanding of the design, and planning and testing of these autonomous robotic solutions for in space assembly operations.

Building on these knowledge and missions, in-orbit construction of large structure is explored in Chap. 16. This chapter introduces a broad in space servicing, assembly and manufacturing in orbit and it provides details of a feasibility study for autonomous assembly of a large space telescope. Two sets of demonstrators are introduced to demonstrate the feasible solutions generated by the project team working on a European Commission funded project entitled PULSAR. This chapter presents three demonstrators. The first one is about the demonstration of precise assembly of mirror tiles to show the precision of standard connection interfaces enabled autonomous assembly, using an external measurement device. The second demonstrator is to demonstrate motion planning and control of mobility in assembling very large structures that exceed the workspace of the robotic manipulator. The final one is to test and simulate the assembly of a primary mirror with the highest accuracy possible.

Further exploration of constructing large structures both in orbit or on a planet are explored in Chaps. 18 and 22 later.

1.4 Future Needs and Research Trends

With all the fundamental knowledge introduced, essential unique tools for space robotics familiarised, and design and testing tools covered, the book moves to its final section on future needs and technological development trends in six chapters to show the diversity of future needs, conceptual solutions for both planetary and orbital missions, as well as terrestrial application of space robotics. The section starts with an analysis of the in-orbital servicing requirements considering the sustainable use of space in a new space age in Chap. 17. This is there followed by a presentation of a future project to build a lunar infrastructure in Chap. 18. Space technologies have much relevance to terrestrial applications and it is always desirable to explore these possibilities. Chapter 19 then shows an example of space robotics' application in agricultural sector Chap. 19. Chapter 20 presents a specific example of robotic interface design and application to provide an example of a future technological development trends as a discussion of an orbital robotic project in which a large telescope's assembly is investigated and solutions verified. The section, and the book, finishes with two on-going European projects one focusing on developing future robotic concepts for planetary exploration missions in Chap. 21, and the other striking to develop a future paradigm changing solution for on-orbit manufacturing and servicing in Chap. 22.

Specifically, Chap. 17 provides an overview of future space mission needs and requirements for a sustainable space age. A large number of space debris have already become a hazard to many critical space missions. With more and more small and in particular cube satellites launched, the increased number of spacecrafts in orbits, either in the form of functioning space systems or in the form of debris after their functional usage has expired, will only become even bigger and difficult-to-handle challenge. At a global level, it would be imperative and sensible for all space stakeholder to appreciate this growing challenge and explore new paradigm to space usage.

Chapter 17 also provides an approach to implementing this new paradigm change from constantly launching monolithic spacecraft system such as one use and none-serviceable satellite to extending the life-time of space assets through reconfiguration of modular sub-systems or main payloads to meet the new and sustainable needs.

This new paradigm is perceived to be an enabler to a sustainable space age and it will promote the serviceability of large and expensive space systems by providing new technological enablers to facilitate the life extension through servicing and repairing non-functioning modules of a space system, upgrading the functionality of system by introducing new and better performing modules. The chapter also provides a review of the number of the launches over last decade and forecasts new spacecraft operators will be increasing based on the market demands of services from space, and new and large constellations of small and less performing satellites as the future trends.

An interesting new space paradigm oriented to constellations is defined in contrast to traditional spacecraft programmes. The chapter then identifies the main challenge for on-orbit servicing, the cornerstone of the new paradigm, ranging from commercial operators' perspective and their common approach to new technologies, cost consideration, technological requirements on spacecraft design for serviceability, to preferred route for replacing or reconfiguring elements of the main payload on a spacecraft. The chapter also identifies the enabling technologies to support the paradigm shifting. The chapter then uses an example of a European Commission funded project to demonstrate how a walking manipulator could enable an autonomous assembly of spacecraft modules for life-extension as well as for replacement of failed modules in future.

Complex tasks in either orbital or planetary space robotic missions require cooperation of multiple robots to complete a common task. Chapter 18 describes such a collaboration among a team of robots to autonomously construct an infrastructure planned and intended on the Moon in future. These robots are heterogeneous from configuration, capability and other perspectives; hence they are all different but they need to work with each other to fulfil a common goal of building an infrastructure such as a living station or quarters for astronauts. Unique to the work is its plan to utilise the recourses on the Moon instead of transporting the materials, components and so forth from the Earth to the Moon. Another challenge to this ambition is the development of the sensor-based systems that are required to enable multi-robot cooperation as well as to perceive the environment in which these robots operate. This requires the sensor system to provide timely feedback on multi-robot manipulation based on the task allocation and planning execution. The chapter also demonstrates an excellent example of designing each heterogenous robotic system including a mobile robot with a manipulator, a six-legged mobile platform and a gantry mobile robot, and testing them with a detailed set of results presented. The design work deploys another robotic standard robotic interface called HOTDOCK. The perception capability is implemented using the InFuse common data fusion framework introduced in Sect. 3.2.5 to enable cooperation among heterogenous robots in co-operatively exploring and mapping a test region.

Space robotics has been developed and evolved such that they have matured a number of technologies to a high level technology maturity for harsh space environments. They can therefore operate autonomously in a new environment mostly if scenarios of a space mission have been fully considered, designed and tested. There are however situations where there may not be considered fully with sufficient scenarios studied due to the nature of exploring many unknowns. To warrant these levels of functionality and capability for unforeseen scenarios, it is only natural that space robotic system designers and engineers have to embed many redundancies to ensure high level of performances, reliability and the execution of fail-safe strategies. As a result of these considerations, space robotic systems tend to be much more expensive or simply over-designed, yet there are potential technologies which could be exploited for many terrestrial applications. Chapter 19 presents such an example as a way of showing the future space robotic development. AgriRover is inspired by planetary robotic technologies for planetary explorations and its key functionalities include soil nutrient analysis function, inspired by the capabilities of chemical analysis of soils or rocks for potential signs of life on other planets such as Mars. Other functions include autonomous navigation, mapping, animal or farming or agricultural object recognition. As a research prototype system, a number of field trials have been undertaken and details are reported in this chapter to demonstrate how space robotics technologies could be and have been exploited in terrestrial agriculture applications.

Generally speaking, space robotics is still at its early stage of the evolution for future space missions and there are many technologies or common building blocks that are still evolving. It is difficult to cover each of many technologies in one book. The editors have however decided to show one evolutionary example of such a key component of the future space robotics systems. Building on the initial introduction of the key functionalities and initial design of a spacer robotic connector called SIROM, Chap. 20 provides an evolutionary description of this space robotic connector. Emphasis in this chapter is on the maturing of the initial design and prototypes to be space qualified or space ready connector. This involves an extensive of testing from multiple perspectives including vibration testing, life testing, thermal testing and load testing. The applications of this space robotic connector are also introduced in several projects which are mostly targeted for orbital space operations, including assembly of tiled reflector and reconfiguration of a modular spacecraft. Another example application is the construction of a future robotised orbital factory for in-space manufacturing. Spacecraft refuelling will enable many spacecrafts out of fuel to be revitalised and given new life for a sustainable extended servicing life and a detailed example of refuelling is also provided in this chapter as a future research direction.

The next chapter of the section on the future needs and research trends sets out a range of robotic exploration concepts for future planetary explorations missions. These include a review of rover based, walking based and the hybrid legged-wheel robotic system, with each of these reviewed in terms of their mobility, controllability and navigation speed. The main challenge with the wheel based systems is its slow speed of exploration due to the limitations of the navigation and decision making systems of these rovers launched in many planetary exploration missions. By comparing the distance travelled so far in these missions, it is clear each rover can only travel a few tens of kilometres for each mission at the best. Some exploration of more advanced requirements such as robotic cooperation, collaboration and coordination have also been discussed for three types of planetary robotic systems. The hybrid legged-wheel system Coyote III is used as an example to demonstrate its design in overcoming the challenges faced by the wheel only based rovers, for instance the build-up of soil in front of wheels.

It is necessary for multiple robotic systems of heterogeneity to cooperate towards a common goal of complexity in a future space exploration mission. These robots could have different architecture, mechanical configuration and different numbers of degrees of freedom as well as different sensing capabilities. The chapter provides a feasibility study of such a cooperation among three teams of heterogenous robots of legged, wheeled or hybrid systems for a lunar polar crater mission and Mars explorations. This represents the evolution of the design concept shift from non-cooperative robotic system, including legacy systems, to cooperative and more coordinated task planning and execution approach to optimise the exploration distance between basecamp and sample collections.

The chapter then introduces a new and cooperative robotic exploration approach to meeting the desire of astrobiologists' and astrogeologists' to explore interesting and possibly challenging areas such as deep shady craters or caves for more likely discovery of water or organic matter. This approach involves the more complex modelling of environment introduced in Chap. 5 to extend the modelling of these caves or craters, and new type of task of establishing reliable communications among robotic exploration teams, and a controlled, coordinated and collaborative change of elevation of the exploration robot in the cave through planned ascending or descending processes. All these will be field tested in the near future in a lunar lava tube analogous site in Lanzarote. In-situ resources utilisation (ISRU), such as the use of regolith on the Moon for construction or extraction of minerals, remains an important area for future research and development and robotic manipulation will be an important function required for ISRU. The chapter also provides future direction of human robotic collaboration for object manipulation by a human controlling an exoskeleton robot or a humanoid robot with thirteen degrees of freedom with a two-figure right hand and a four-finger gripper as the left hand.

Chapter 22 of the future needs and research trends section of the book provides the details of the latest development of a new European robotic standard connector and their applications in an on-orbit assembly of a modular spacecraft. These include new standard interconnects to enable a walking manipulator which could potentially provide a game changing robotic solution for sustainable and meaningful space asset deployment, their life extension, function upgrading and ultimately significant improvement of affordability due to an overall cost reduction. It is believed that this ground demonstrated and tested walking manipulator could become a suitable solution for future space infrastructure construction in orbit and possibly on a planet. The chapter begins with an introduction to a European commission funded project called Modular Spacecraft Assembly and Reconfiguration (MOSAR) and its key mission concept, aiming to assemble a modular and reconfigurable satellite on-orbit by creating a walking manipulator. This whole concept is to be tested and demonstrated on the ground before further development. A key technological development is the design and prototyping of a standard robot interconnect called HOTDOCK and its development is discussed in terms of the motivation and technical overview from mechanical interface, fluid, power and data interfaces as well as its controller and

HOTDOCK's operations. The overview of the walking manipulator is then presented followed by description of the mechanical design, avionics, software and control of the walking manipulator. The MOSAR overall system including both the ground segment and the space segment has been successfully demonstrated and the testing and demonstration is given in the chapter to show what a future robotic on-orbit assembly system might look like and what types of operations a space robot could perform.

1.5 Unique Meta Table of Contents

Due to the nature of rapid and concurrent development of space robotics by multiple stakeholders and research organizations across the world, it is interesting to observe that multiple solutions and approaches have been followed by different leading research centres. It is therefore impossible to merge the same topic from different research initiatives funded and investigated by different research centres into a coherent chapter focusing on that topic. In order to facilitate readers to appreciate and compare different approaches reported by different author groups, and to stimulate further research exploration, the editors took a decision to create a meta table of contents that establishes hyperlinks among the sections covering the same topic to make identification of different contents of the same or similar topics being developed in parallel possible. This approach is built on the digital contents and technologies developed to support an easier navigation and search of information on a particular topic using a set of carefully selected key words structured for easy navigation. Readers will therefore see a new type of table of contents in the form of meta format in this book in the Index of space robotic technologies in the electronic version of the book. The editors hope this will help readers of electronics version to find the information and sections they are interested in more quickly and effectively. For those who have access only to a hard copy of the book, it is hoped this index will still provide an easier way of finding relevant topics quicker than normal indexing. We welcome feedback and comments from readers.

Finally, it is also worth highlighting authors of each chapter have also attempted to describe the state of the art and at the same time to define some future research and development trends in that specific topic for readers to have some pointers and further their research pursuit in future.

Chapter 2
Fundamentals of Space Robotics

Xiu Tian Yan and Gianfranco Visentin

2.1 Introduction

Since the birth of modern robotics concept as an academic discipline as defined by Isaac Asimov in the 1940s, robotics has made significant advancement. In contrast to mostly mechanical automated systems represented by the eighteenth century Mechanical Turk or Automaton Chess Player shown in Fig. 2.1 as an example, modern robotics is still evolving and can be defined as an increasingly multi-disciplinary academic subject where robots are designed and created and their use either alone or in interactions with human beings are also studied. The design and creation of robots involves a process in which a design methodology may be defined and applied to create new concepts of a potential robotic machine for a specific purpose. These concepts are then explored and evaluated based on the intended purposes of the robotic machine in order to identify the most appropriate conceptual solution. Selected concepts are further evaluated by using modern computational tools such as finite element analysis for mechanical structural analysis and optimization. Block diagrams can be used to investigate the dynamic behaviour of such a design. Both kinematic and dynamic behaviour can be also modelled in a computer system to study the predicted performance of the robotic concept.

During the conceptual design and evaluation, it is very likely designers will need to apply knowledge from a range of traditional academic disciplines such as mechatronics which encompasses mechanical engineering, control theories, electronics, computer sciences and informatics; mathematics; bioengineering and broad artificial intelligence.

X. T. Yan (✉) · G. Visentin
Robotics and Autonomous Systems Group, Department of Design Manufacturing and Engineering Management, University of Strathclyde, James Weir Building, 75 Montrose Street, Glasgow G1 1XJ, Scotland
e-mail: x.yan@strath.ac.uk

X. T. Yan and G. Visentin (eds.), *Space Robotics*,
https://doi.org/10.1007/978-3-031-39214-6_2

Fig. 2.1 Reproduction of the original Chess Player (The Turk) [1] (*Photo courtesy of* Marcin Wichary from San Francisco, U.S.A.)

Owing to their high efficiency, high repeatability, high precision and a relatively short return off investment cycle, industrial robots have been studied, designed and manufactured at a large scale. Such robots have been widely used within factory environments as for instance in automobile industry, where industrial robots automatically weld car chassis components with a high manufacturing quality in a production line, or assemble car engine components.

These industrial robots are typically secured on the factory floor in an articulated configuration to perform precision manufacturing operations. Mobile robotic systems have also been created in the form of so called automatic guided vehicles (AGV) and are typically used to transport components in a factory amount different manufacturing cells. These two types of industrial robots are specifically designed to be able to operate in a so-called structured environment, where information such as the dimensional information of a car chassis to be welded are known to the robots and embedded in their control programs before the welding takes place. No foreign and unforeseen objects are allowed in a typically caged production area so that there will be no possibilities of potential collision between the robot and any unknown objects intruding into the production space. This strict safety requirement ensures the safe and highly efficient operations of robots in a factory where there are human operators to manage the whole production process, check any potential issues or problems and resolve them if they arise in order to achieve successful manufacturing operations.

In recent years, robotics has advanced to be able to handle objects and operate in a unstructured environment, where no prior knowledge typically on the size of the objects or the presence of new objects is available. In the case of sending a robot to a new environment with unknown terrain properties, such as the surface hardness, friction coefficient of wheel rolling etc., it is necessary to expand the scope of study to include soil mechanics and terrain geology.

The new requirements for robots to work among themselves and interact safely with human beings raise the need of broadening the consideration of robotics in both design and operation, to include issues such as ethics, safety and legal requirements. Study of swarms operating in a distributed manner needs also considered where robots collaborate among themselves.

A robot can work in collaboration with the other robots. This collaboration can take place in a structured environment such as in a factory where robots may make cakes with skilled human operators and many environmental constrains such as the geometries of the manufacturing machinery along production lines are known. This information could be made available to the robots so they know these constraints. Alternatively, robots could also work in the unstructured environment such as those found in subsea oil and gas exploration or in space exploration.

With the introduction of compliance in robots and more advanced sensing technologies, it is more feasible to allow human operators to work alongside a robot which then assists operators to perform better in a number of manipulation-based operations. It is equally possible that robot can also assist a person with a reduced degree of mobility to improve their quality of living. These increasing robot human interactions generate a need for better understanding between robots and human beings in order to design better robots to support such interactions.

It is clear that robotics is still a rapidly evolving academic discipline and it is built on a number of academic disciplines such as sciences for various sensing, perception and task planning, engineering for creation of new and novel robotic solutions and reliable and efficient operations of robots designed,, technology development for more capable robots, and social sciences for safe and ethical introduction of robots for their intended roles and operations in a society. Social sciences have become an increasingly important consideration for socially acceptable robots and study perspectives include topics such as legal studies and ethics investigation to ensure the safe development of robots to meet societal, moral and ethical standards and ensure societal acceptance of new robots as developed. All these will contribute to the future advancement and evolution of the robotics discipline and help to create better robots to address more and wider societal challenges which require new and creative thinking and solutions. With the above understanding, robotics them becomes an engineering process of deploying scientific knowledge to create new robotic concepts, evaluate these concepts following a chosen design process, prototype the chosen concepts most likely in a computer virtual environment first, then into a physical format for testing and validation, then manufacture them in a batch size that is economically feasible. These new robots may then be deployed and operated in their intended environment, either controlled by human operators or working along with human operators to assist them for more efficient task execution.

A general trend is to make these robots more and more intelligent in undertaking the tasks they are designed and planned for. It is therefore necessary that knowledge from the following academic disciplines is integrated into the robots being designed: mechanical engineering, electrical engineering, information engineering, mechatronics, electronics, bioengineering, computer engineering, control engineering, software engineering, mathematics, etc. Exploration of the design space provided by all these multiple disciplines and their conjunctions will play an important role in the future research and development directions offer robotics.

With the introduction of robotic systems into many different applications, the requirements for robots to interact safely and reliable with the human operators within the same physical space become an important consideration. These recent re-emergence of more consideration of robotics from social science perspective reflect and consolidate the original thinking of Asimov's Laws of Robotics. Asimov defines these three Laws of Robotics as a set of rules in order to maintain its safety, the safety of human beings during interactions between the robots and human beings. The focus was on the safety rather than the technological advancement; hence these Laws still have much relevance in today's robotics research and development, despite Asimov built his stories around the ambivalences, inconsistencies and uncertainties in these 'laws;! These well-known Laws of Robotics are:

First Law of Robotics

- A robot may not injure a human being or, through inaction, allow a human being to come to harm.

Second Law of Robotics:

- A robot must obey the orders given it by human beings except where such orders would conflict with the First Law.

Third Law of Robotics:

- A robot must protect its own existence as long as such protection does not conflict with the First or Second Law.

The definitions of these Laws to some extent reflect the nature of mysteriousness of robotics until this day. It is a common myth on how complicated or technologically sophisticated robots will be and could be advanced to achieve or replace human operators in a number of industrial tasks. It is also a common concern, and even fear in some instances, that human employees are worried that as more industrial automation is introduced through industrial robots, their jobs will be to taken away. There are certainly logic and reasons for these concerns for robotic applications on Earth. There are however many specific challenges in space exploration missions and these shape the needs for the development of the specific branch of space robotics for research investigation and technological development.

2.2 What Makes Space Robotics Special

Space robotics is a special branch of general robotics with a number of unique application requirements and technological characteristics. These are determined by space robotic applications as detailed in Sect. 2.3. In a summary, space robotics needs to address a range of unique challenges only found in space applications. These include challenges imposed by environmental constraints, delays or latency in receiving information from a remote planet for example, unpredictability due to a lack of information relating to objects or environment where the robots need to operate.

2.2.1 Unpredictability Challenge

Due to the remoteness of a space mission's targets, in particular for deep space exploration, the information about the target planet is limited before a space exploration mission is launched. This potentially introduces many uncertainties into the mission operations. Potential unknown information of a target planet could include a lack of understanding of the terrain surface properties a space robot may have to traverse in order to explore the target planet.

A good and typical example of a lack of understanding for a robotic exploration is the European Space Agency's Rosetta space probe mission to explore Comet 67P/Churyumov–Gerasimenko. The comet is at a distance of about 7 billion kilometres or five times further between the distance the Earth and Sun and the mission was launched over ten years ago in 2004. Certain tools or technologies were not available then for this very uncertain mission full of unknowns such as the shape, surface properties and landing conditions of the comet, as well as the feasibility of robot mobile exploration. The comet surface could be softer than a blanket of new fallen snow or as hard as a sheet of ice.

When the spacecraft got closer to Comet 67P, the mission controllers gradually began to get an understanding of its geometry from more information sent by the spacecraft, particularly after its surveying flyover in an orbit about 30 km from the comet's surface to map and characterize the comet. The surface layout of the comet was unknown before Rosetta's arrival. Using its navigation camera as it gradually closes in on to its target Comet 67P, the understanding of the surface shape changes from a simple dot among many white planets in Fig. 2.2(1), then a single dot represented in several pixels in the image of Fig. 2.2(2), to a lump represented by several hundreds of pixels in Fig. 2.2(3), then to a unforeseen shape in Fig. 2.2(4), finally to a duck shape as shown in Fig. 2.2(5) with a very rugged surface shown in Fig. 2.2(6), which was never imaged before the launch of the mission. These can pose many challenges for the successfully design and implementation of a landing operation.

Based on this new surface and shape information and information from other cameras for mapping and characterisation of the comet, a suitable landing site Site_

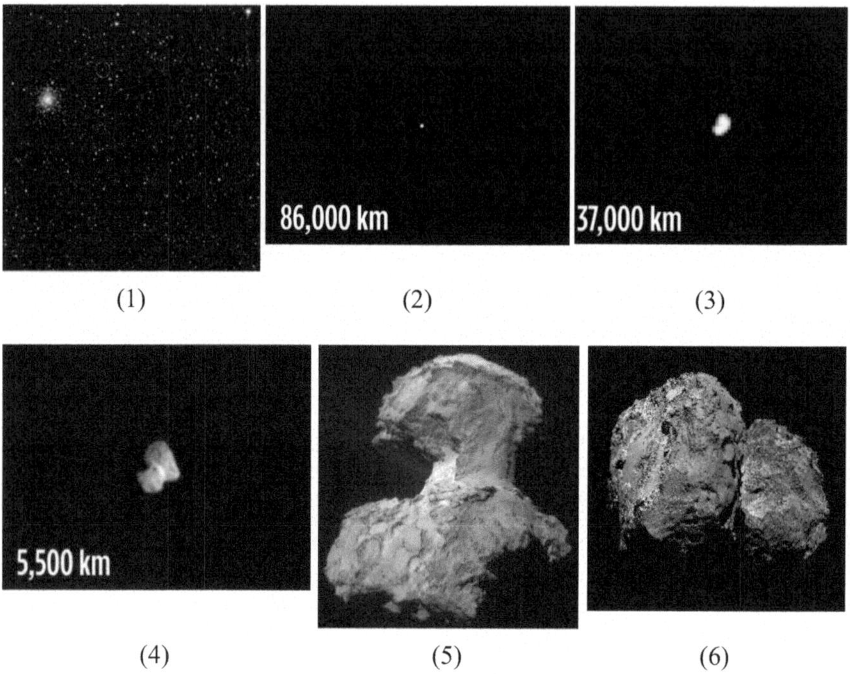

Fig. 2.2 Gradual approaching of Comet 67P/Churyumov–Gerasimenko (*Photo courtesy of* DLR)

J was selected based on the need to avoid some active areas to ensure long operating mission life. Some active areas are outgassing about 500 g of water per second, and in some areas, there are even sublimation process of solid matter undergoing a phase transition directly from a solid to gaseous form. These areas cannot be considered to be suitable sites for landing. The other consideration for the landing site selection is the geometry requirement of a relatively flat and large surface where the Philae lander can explore without the risk of falling into valleys, cliffs and crevices or even canyons.

Planning, replanning and testing of all sequences of landing operations repetitively were undertaken within a reference model built for the landing operation at the land control centre of German Space Centre DLR. Assumptions were made that the comet surface could be drilled by firing two harpoons to establish an initial anchoring of the Philae robot. This operation was to be followed by a drilling operation to further secure the attachment of the robotic probe Philae onto the landing site. Each of the three landing feet shown in Fig. 2.3 is embedded with a so-called ice screw and a thermal sensor called a Mupus device is also housed within the feet, The Mupus thermal sensor could be hammered into the surface.

When this landing operation plan was finalised, rehearsed and then executed, it was only to find out that the drilling operations on landing were not as easy as planned, possibly due to a much harder surface than anticipated for mechanical drilling. The

Fig. 2.3 Philae landing on Comet 67P/Churyumov–Gerasimenko (*Photo courtesy of* DLR)

drills could not establish initial penetration on the surface rapidly and sufficiently to land or attach the Philae robotic lander to the target landing area. This resulted in a Philae bouncing off the landing site, landed on and bouncing off onto another site with a reduced bounce height, before finally ending in an area of the comet where it was difficult for Philae to capture solar power and charge its onboard battery. This was described as triple jump landing by the mission team. The overall mission was however a great success even though the exploration of the comet was less successful due to this unpredictable and unknown landing surface when the landing mechanisms were designed. In the end, due to a lack of solar power to charge the battery, Philae had to be switched off.

2.2.2 *Environmental Challenge*

Unlike the environment we get used to as a human being on Earth, space robotic systems are required to perform a set of planned tasks and operational manipulations in an environment which is definitely different from those on Earth. Very often these tasks and manipulations were only designed and tested with limited knowledge of the target environment. Mission scientists and engineers can only review the unexpected changes in environment when the robot is either near the environment or actually in the environment, as illustrated by the Rosetta and Philae example described in Sect. 2.2.1. Categorically, the main differences include gravity, temperature, lighting and potential risks posed by radiation. For robotic applications in orbit,

Table 2.1 Distance of panets from the Sun and their average temepratures

Planet	Distance from the Sun (au)	Distance from the Sun (in kilometres)	The average temperature of the planet (°C)
Mercury	0.39	58,343,169.6	– 180° at night to 430 during the day
Venus	0.72	107,710,466.9	471
Earth	1	149,597,870.7	16
Mars	1.52	227,388,763.5	– 28
Jupiter	5.2	777,908,927.6	– 108
Saturn	9.54	1,427,163,686.4	– 138
Uranus	19.2	2,872,279,117.3	– 195
Neptune	30.06	4,496,911,992.9	– 233

space robotic systems will need to adapt to microgravity environment and associated all environmental challenges.

Spacecrafts and space robots also have to face and experience extreme temperatures. The temperature of a planet is proportional to the distance between the planet and the Sun. Generally speaking the further away a planet is from the Sun, the colder its temperature. For example, the black body temperature on Mercury is 430 °C during its day time, the temperature on the Saturn is about – 138 (°C). Further temperature information on each planet compiled from a NASA source [2] can be found in Table 2.1.

For planetary robotics, the morphology of different planets pose a significant challenge. These include both the shape and structure of the planet's surface conditions, which determine the suitability of a target region for a space robotic system to traverse. If a region of scientific interest is selected for an exploration and the region has terrains characterised by challenging shapes such as pits, valleys, cliffs, rocks, crater walls, lava tubes and so forth, special robotic system should be deployed to complete mission objectives. Chapters 7 and 21 cover more on these terrain issues typically facing space robotics.

2.2.3 *Latency Challenge*

As illustrated in the Rosetta and Philae mission example, the space robotic systems could be operating far away from a mission control centre on the planet Earth. In the case of Rosetta and Philae mission, the spacecraft was over 7 billion kilometres away from the Earth and the time it takes to transmit a signal to the Earth changes every day, depending on the relative position of Rosetta and the earth. The longest time was 52 min and 6 s on 1st December 2012. At its last touchdown on the comet shown in Fig. 2.3, the signal took over 28 min and 20 s to travel from the spacecraft to the Earth control centre.

It is clear any attempt to command the spacecraft in real time at such a distance will result in delays, or over 50 min in Rosetta's case, in receiving any command by the spacecraft or space robot and then implemented. These commands are most likely made by the mission controllers based on the information sent back by the spacecraft. In the case of Rosetta's mission during its landing, it would be at least 28 min before considering a decision. These make the whole latency in perceiving any change at the spacecraft end, sending the information back to the control Centre, making decisions and then issuing commands to the spacecraft and received over 57 min.

2.3 Space Environment Constraints

In space exploration, it is essential to observe and monitor the effects of the space environment on the astronauts and cosmonauts travelling in and through space and how these environmental conditions affect the human body. However, to ensure that a space mission is successful and is able to endure then careful consideration and design must be taken to create spacecraft technology that not only contains, where appropriate, technology to support life but technology that is robust and able to withstand operating and exploring in the hostile space environment. In the aerospace field, it is said to be joked about that once the paperwork weighs as much as the rocket you are designing then you are ready to launch.

The space environment boundaries that robots can operate can be classified into different categories; Low Earth Orbit (LEO), deep space (beyond the Van Allen belts), interstellar medium, planets and smaller celestial bodies. Figure 2.4 illustrates these boundaries.

The main threats in the space environment come from atomic oxygen, extreme varying temperatures, thermal cycling, vacuum, solar ultraviolet (UV) radiation, charged particle (ionsing) radiation, plasma, contamination and impact from orbital debris [3]. Table 2.2 summaries some space environment conditions for orbital operations, whereas Table 2.3 provides a summary of planetary environmental conditions.

The space environment can be very unpredictable and can cause varying damage and degradation to the components and materials of spacecraft and space robotics. The level of threat will depend on differences in material properties, thickness and stresses.

The specific type, duration and orbit of a mission will also determine the severity of the impact that the space environment will have on a spacecraft. These threats can cause space missions to fail or to have interruptions and delays as well loss of data and service outages. This section will explain in more detail the effect of some of these threats on spacecrafts in general [4].

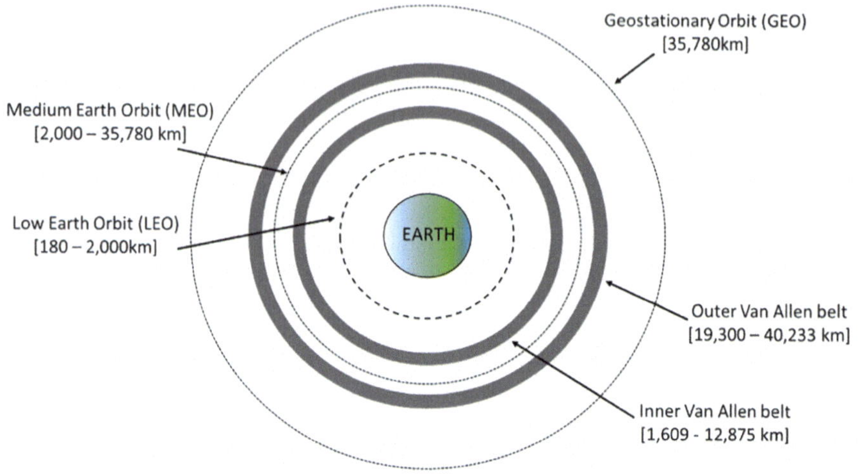

Fig. 2.4 Earth orbits and Van Allen belt boundaries

Table 2.2 Summary of space environment conditions: orbital environmental conditions

	From earth	From mars	
Gravity (g/ms^{-2})	~ 8.9 400 km from Earth (ISS)	~ 3 400 km from Mars	
Temperature range (T/K)	$-$ 120 °C < T < 120 °C 400 km from Earth (ISS)	$-$ 112 °C < T < $-$4 °C Phobos (Mars's moon)	
Atmospheric pressure range (p/Kpa)			1.33×10^{-7} to 1.33×10^{-10}

Table 2.3 Summary of space environment conditions: planetary environmental conditions

	Earth	Mars	Moon
Gravity (g/ms^{-2})	9.81	3.71	1.62
Temperature range (T/K)	184–331	140–300	113–400
Atmospheric pressure range (p/bar)	1014	6×10^{-3}	3×10^{-16}

2.3.1 *Radiation*

One of the most hazardous aspects of the space environment is the prevalent influence of charged particles. These charged particles come from a few sources, mainly solar wind and flares (coronal mass ejections (CME)), Galactic Cosmic Rays (GCR) and the Van Allen radiation belts [5].

To give context to these charged particles it is sensible to discuss that of which the universe it made, space plasma. Space plasma refers to the ionised particles in the Solar system, and further, which are found in the solar corona (outer portion of the suns atmosphere) and solar wind. The electrostatic and electromagnetic fields and forces that exist in space greatly influence the behaviour of this plasma and cause emission of radiation as well as considerable variations in the space environment weather. Charged particles, whether they come directly, from the solar wind, indirectly from the Van Allen belts, or from deep space they can lead to several problems for a spacecraft such as degradation or damage to surfaces and in particular damages to electronic components. The effects that radiation has on electronic devices and materials depends on the type of radiation, the type of material, the rate of interaction and the component involved.

A solar flare is an abrupt explosion of energy from the sun which causes a huge release of radiation. Often solar flares are accompanied by a Coronal Mass Ejections (CME) which are enormous releases of plasma and radiation. Galactic cosmic rays (GCRs) are particles similar to those found in the solar wind or in solar flares, but originate outside of the solar system and emit gamma rays. The Van Allen belts are more hazardous after the release of a CME or having been exposed to strong solar winds. The majority of the Sun's radiation is visible, infrared and ultraviolet (UV).

Prolonged exposure of a spacecraft to UV radiation will darken most materials and degrade spacecraft coatings, particularly if there is contamination. UV radiation will also damage exposed cermaics or polymers by either making it weak (chain scission) or by hardening (cross-linking). UV under high vacuum can also create atomic oxygen which is produced when ultraviolet radiation reacts with molecular oxygen in the upper atmosphere and causes materials to bleach which leads to significant colour changes. UV radiation is especially harmful to solar cells and can cause harm to electronic components leading to single event upsets, requiring them to be shielded, or hardened to be able to operate in the environment.

The particles in the Earth's Van Allen belts can cause radiation damage to the optical properties of inorganic insulators and semiconductors. Spacecraft operating in or outside the Van Allen belts are exposed to much greater radiation levels than those in a low Earth orbit. The surface of most exposed materials will be damaged by the radiation belts and by solar charged particle emissions.

These charged particles in space can typically affect a spacecraft in three main ways; single event effects (SEE), spacecraft electrical charging and sputtering.

Single event effects (SEE) occur when an ionizing particle interacts with electronic components. The effects of solar flares and GCR can cause a SEE to occur. These events can cause high current conditions, interruption to the functions of the electronics and corruption of data. In some situations, they can cause catastrophic failures. SEE's can be categorised into two forms, that of destructive and non-destructive.

A non-destructive SEE is classified as an event that changes, temporarily or permanently, the state of a device without affecting its functionality. These have three main types; Single Event Upsets (SEU), where a storage element changes state and causes corruption of the information stored, Single Event Functional Interrupts (SEFI), when an event causes temporary loss of the function of the device and Single Event Transients (SET) is where a transient causes incorrect data to be collected/observed.

Destructive SEE are described as an event which permanently damages a device and interrupts its function. These have four main types; Single Event Latch-up (SEL), produced by excessively high (or low currents) as the results of a charged particle travelling through sensitive electronic regions in devices. These events can be catastrophic. Single Event Burnout (SEB) is always destructive and is caused when a charged particle causes a localised high current and causes the device to burnout, Single Event Gate Rupture (SEGR) is also always destructive to the device. This is caused by high electrical fields that cause transistor gate dielectrics to break and Single Event Hard Errors (SHE), caused by very large energy depositions which cause single cells to function incorrectly and not be able to change state.

Spacecraft electrical charging (SEC), caused by low energy plasma, happens when an electrostatic charge builds up and causes the spacecraft to have a net charge on the surface. In GEO it is the predominant space weather effect. This can have critical effects on the operation of a spacecraft, it can cause damage to the electronic instruments that are on board or interfere with scientific measurements. As computer components become smaller they become more delicate which makes them more susceptible to charging effects. There are two categories of SEC which can be internal and surface charging.

If enough charge builds up then electrostatic discharges can occur which can have a variety of severe consequences such as surface/component degradation (SD), particularly of mirrors, solar panels, low-absorbance coatings and radiators. It can also cause issues with thermal control.

If these particles from solar flares accelerate towards the Earth's ionosphere into the auroral zone they will produce Auroral Kilometric Radiation (AKR) from being subject to magnetic compression. Scientists have been able to investigate this type of space radiation on a laboratory scale [ref myself]. If these particles continue into the magnetosphere they will produced the Aurora Borealis (northern hemisphere) or Aurora Australis (southern hemisphere).

As some of these charged particles travel at high speed there is a possibility of them causing damage when they come into contact with the spacecraft, this effect is known as sputtering. Sputtering can cause damage and unpredicted changes to the thermal coatings, the optical and electrical properties of the materials and the sensors. There is also a chance that the material sputtered off of the spacecraft will become re-deposited elsewhere on the surface.

To overcome and decrease the effects of radiation on spacecraft there are a few approaches that can be taken. These include shielding, the design of a mission, radiation hardening by either process, architecture or design. Engineers and scientists also research to find new materials that can withstand and are robust to the effects of the space environment.

2.3.2 Lighting

Lightning is a phenomenon that can both be an extraordinary spectacle and a destructive power. During launches lightning is a danger to spacecraft both on the launch pad and also throughout take-off. Therefore, protection from lightning must be a consideration for engineers.

When thunderstorms arise, thunderclouds can be formed. These thunderclouds are able to hold millions of volts which can discharge. This discharge is the reason for the concern for the safety and protection of the space vehicle. A thundercloud is able to discharge in a few different ways. These are namely between the charge centre of the cloud and either to an opposite charge in the cloud, the ground, between the air.

A space vehicle should be able to tolerate a lightning environment before and during a launch. For a space vehicle sitting on the launch pad or when it is in flight, discharge of a thunderstorm cloud to the ground is the major concern. Lightning can not only delay launches but can also cause considerable damage to spacecraft.

Once a space vehicle launches, it is possible for it to cause a lightning flash called triggered lightning. This phenomenon occurs as the vehicle acts like a conductor which changes the surrounding electric field causing a discharge from the ends of the vehicle. Apollo 12 experienced this phenomenon and triggered two lightning strikes despite there being no natural lightning.

When lightning strikes an object it can cause damage in two major ways; the lightning can strike an object directly in the path of discharge which can cause a variety of serious damage such as the electronics being short circuited and effects to the guidance and navigation systems. Damage such as this could cause the space vehicle to become out of control, or possibly crash. There is also a possibility that the electronics in the payloads such as scientific equipment could be badly affected. The second way that lightning can affect electronics is when lightning strikes nearby and the electromagnetic fields produced can cause damage to sensitive electronic systems.

The tower and vehicle must be designed with protection systems to ensure that the structure, equipment and all components are protected from the effects of lightning and the induced voltages and currents that it can cause. To protect from a lightning strike it is important for the tower and vehicle to be designed with a lightning bond (a current path) to carry the current on the outside of the vehicle. Therefore, apertures and joints between parts must be minimised to stop any electric field entering through the vehicle.

Launch pads can also be affected by lightning storms, given their height and structure. Launch pads contain several electrical circuits. The threat is different for the pad depending on whether there is a vehicle docked or not. A lightning strike to a pad without a vehicle causes damage to the electrical circuits. However, with a docked vehicle if lightning struck the pad then the electrical umbilical cables transport induced currents which cause can severe damage and interference.

The vehicle can be protected in a few different ways. One way is with mast lightning rods and shield wires which is called air termination. There is also Earth termination which uses lightning rods, this time, driven into the ground to dissipate current into the soil. It is also possible to connect the air and ground termination systems electronically which avoids secondary electrical flashover between any given lightning protection system and the vehicle system being protected. For example, at Kennedy Space centre have topped each launch pad tower with a fibreglass mast, a network of catenary wires and conductors to divert lightning away from the structures and isolate them from the large currents. It is also important that the towers are grounded.

Protection can similarly be achieved by ensuring that all components are suitable grounded, electrically bonded and shielded. Protection can also be achieved by covering exposed cables in conductive enclosures, any aperture to the vehicle should be covered with metal screens to stop any electromagnetic fields penetrating the structure. Should any metallic structure pierce the skin of the vehicle, they must be electrically bonded to the vehicle skin. It has been found that certain materials are more vulnerable to lighting damage, particularly composite materials, and so protection material choice is an important factor in all stages of design and should protection should be considered to compliment with structural requirements.

Lightning protection system concepts must be part of the development of the vehicle design and construction from the preliminary design stage and continue until completion.

2.3.3 Thermal

As previously seen, the space environment is a region of extreme conditions. This extreme behaviour is true for the thermal environment of space, where the temperatures can be extremely high or extremely low. The source of the thermal energy in space originates from solar radiation, long wave radiation (infrared radiation reflected off the Earth) and reflected solar energy named Aldedo.

Depending on a spacecraft's mission orbit and duration the specific temperature ranges and duration they are subjected will change. The temperature ranges a spacecraft can continuously change throughout its mission. For instance, when a satellite is in orbit it can experience warm temperatures as it orbits through sunlight and much cooler temperatures as it moves into shadow. This causes thermal cycling which can lead to mechanical problems and cracking.

For this reason, for engineers, the design of thermal control and thermal protection system can be exceedingly challenging as the spacecraft and components, including payloads, must be able to operate under their own safe operating temperature range and requirements at all stages of a mission. A system operating out with its design operating environment can become unreliable. The design of these thermal control systems (TCS) involves careful consideration and selection of the thermal hardware to achieve the required thermal requirements of all components and sections of the spacecraft. It also compromises of being able to predict the temperatures that the spacecraft may encounter on its mission [6–8].

Thermal control systems are normally designed for the environments that a spacecraft will encounter whilst in orbit, however all temperatures during transportation that may take place and also prelaunch must be anticipated to ensure that all temperature limits are not exceeded.

As space is a thermal vacuum the main methods of distributing thermal energy within a spacecraft are thermal radiation and heat conduction. However, some electronic equipment can be hermetically sealed to allow thermal energy distribution through convection. Thermal control systems must not only be designed to protect from and control excessive heat but also to insulate from exceptionally cold temperatures too. Thermal control methods can be passive or active.

Passive methods can provide reliable, low cost protection without any constraints to the spacecraft designs and testing. One of the simpler, methods to prevent from temperature variation effects is to cover the surface of the spacecraft with a paint or layer of a material that is appropriate to the environment conditions it is likely to encounter, called thermal control surfaces. Some of the criteria of these materials taken into account are its optical/reflective properties, if it will be susceptible to outgassing, its toxicity and the cost. Another effective passive method is the use of multi-layering insulation (MLI), also named a thermal blanket, is a high-performance insulator which hampers the flow of thermal energy through radiation-heat transfer barriers. These insulating blankets provide both protection from excessive heating caused by the environment as well as prevention of unnecessary heat loss from components. They can also protect against atomic oxygen and micrometeoroids.

Radiators are exposed to the environment and discard excess heat from a heat source to space through radiation and can be considered to be both passive and active devices depending on the method that the heat is transferred from heat source. Usually, radiators are made from aluminium and coated with a layer that provides a high infrared emittance and a low solar absorptance. Passive radiators make a connection between the heat source and the radiating surface by using heat pipes or through direct contact. Whereas, active radiators make a connection between the radiating surface and heat source through the use of fluid loops.

If passive thermal control methods are not suitable for the requirement, then active methods must be applied. These methods include heaters and heat pipes.

Of the most common thermal transfer methods used in spacecraft are heat pipes. They transport heat through condensation and evaporation and are able to operate at any orientation of the gravity field. They are typically used over short distances and they can be applied to heat loads to ~ 10 kW. They can vary in their shape and function which makes them versatile for the space constraints that can come with space system design. The structure of a heat pipe is a closed system that is lined with a porous material, which performs like a wick and a working fluid that is suitable to the application. The central volume is filled with the vapour of the fluid. They transfer heat due to a capillary force that is developed in the porous wick when heat is applied to one end.

If the heat load to be transferred is large then usually pumped fluid loops (PFL) are used. PFL are beneficial for spacecraft TCS as they are not difficult to integrate into the spacecraft mechanisms, they can transfer heat between multiple locations, they are scalable and they are relatively resistant to the effects of gravity. They operate by forced liquid convective cooling around pipes and can be single-phased or two-phase fluid loops. The thermal energy is transported from a component to a heat sink by being dissipated through the coolant which absorbs it.

As we have seen previously, components need to be functioning at their own specific operational temperatures to achieve their expected performance. As well as environments causing issues with excessive heat, space can also cause difficulties when the ambient environment is too cold. In these circumstances heaters are sometimes necessary to protect components from environment low temperatures or to provide additional heat to increase a component to its operating temperature. Typically, there are two types of heaters used on spacecraft. Namely, cartridge heater, which is a resistor surrounded by a metallic cover and patch heaters, which are flexible and can be used on surfaces that are both flat and curved.

2.3.4 Vacuum

The vacuum of space is described as a hard vacuum. The combined effect of thermal energy (whether on the surface or internal) and the high vacuum of space are such that they are able to cause missions to fail on their desired accomplishments.

This vacuum is one of the most critical constraints in the designing of robots, components and spacecraft for the space environment. As a consequence of this many of the design features, and material choice of spacecraft arise as a result of the known effects that hard vacuum can have on electrical, mechanical and thermal systems.

Prior to being launched, spacecraft are frequently meticulously tested in vast vacuum chambers. Here, design enrgineers are able to investigate for problems and issues that occur from spacecraft navigating in vacuum. These are outgassing, cold welding and heat transfer.

Outgassing is the release of gasses from spacecraft materials under high vacuum conditions. It occurs when small bubbles of gas get trapped in a material, commonly in composites, whilst in Earth atmospheric pressure and are then released when the pressure changes in the vacuum of space. This release of trapped gas is called outgassing. Outgassing does not always cause problems but when it does it can cause serious damage and interruption to a spacecraft's function, especially true of solar cells and optical instruments. The material that is outgassed can coat sensors and surfaces allowing contamination or degradation to full function. It can even form local clouds that will affect instrument readings.

Ways to prevent outgassing include making careful selection of materials used, NASA have space station external contamination control requirements and a data base of test results from their standard test methods for outgassing. Certain materials can encounter surface roughening or sublimation at varying temperature ranges in vacuum. Another way to reduce outgassing is to bake spacecraft components, at a temperature above that expected in orbit, in a thermal-vacuum chamber before flight.

Cold welding occurs when moving mechanical parts, with a small separation between them, operate in high vacuum for an extended period of time. In space there is no air between this tiny separation, as there would be on Earth, so the two moving mechanical parts are subject to effectively weld together or cause a significant increase in friction. In situations where there is an increase in friction, generally, a lubrication would be used. However, in the environment of space, the use of lubricats must be very carefully considered or removed as they may cause outgassing or evaporating. If two mechanical parts do cold weld together then they must be separated in order to allow them to regain their original function, which is a task that is much more challenging in space. In some cases, it is possibly to manoeuvre spacecraft to allow one side to face the sun and the other to be more shaded to allow them to separate as the temperature gradient will cause some parts to expand and some to contract. Finally, the vacuum of space produces problems in effective heat transfer. The previous section outlined the methods of transferring heat and cooling.

2.3.5 Mechanical System Design

As previously seen, the extreme environment of space can cause several destructive effects that must be considered in the designs of mechanical systems.

Several considerations must be made when designing mechanical systems for use in space. The lack of gravity and relative inertia and momentums must be accounted for. In particular, designs and concepts must be tested or simulated to analyse factors such as structural stresses, loads, thermal heat transfer, vibrations and material selection.

Challenge in mechanical design also comes from size and weight limitations for the spacecraft or robotic system. Due to payload restrictions and orbital paths mechanical components will often have to be designed to fit around existing technology or to strict design requirements which limits its tolerable shape, weight and volume.

Once a spacecraft or robotic system is in space it quickly becomes complex situation should it break down. Therefore, a key concern in design is to make sure that the final outcome is reliable, robust and possibly able to function for numerous years without failure. A stage after design and manufacture is testing. Systems and components will be rigorously tested before launch and integration into other sections of technology.

Any structural system designed for space must be able to stay intact without considerable deformation at all stages of launch and orbit. During launch a space vehicle or payload will experience violent vibrations and so the mechanical structures must be strong and thermo-elastically stable to endure these to eliminate fatigue damage, contact friction and impact damage.

Along with being able to withstand launch and orbit, a mechanical system must be able to optimally function in its destined environment. These adverse effects that the extreme environment of space can cause have been previously discussed.

Therefore, mechanical mechanisms must be coated with a protective layer to prevent from radiation damage that can cause serious weakening in mechanical properties. They must be designed as to have appropriate thermal control to stop parts becoming overheated or contrary to this too cold. Both conditions can have detrimental effects to mechanical systems. Sections that may move or slide need to adequately lubricated with a substance that does not evaporate quickly.

Some space and planetary environments can be exceptionally dusty. Dust can cause issues with the mechanical components of a spacecraft. Depending on the mission, design engineers will have to consider how certain mechanical properties will tolerate, avoid or remove the dust in these environments. This is particularly true for sensitive mechanisms such as bearings or gears.

When mechanical connections are necessary, then thought must be given to how these parts will be physically secure once connected. The most common mechanical connections (latches) are rotational, clamp, carabiner and hook. Depending on its requirement, each type of latch has its own advantage as well as disadvantage. Latches are commonly motor driven and can be locked through a mechanical or magnetic system. More details can be provided in Chap. 10.

2.4 The Applications of Space Robotics

The concept of using space robotic systems, e.g. manipulators to undertake complex various on-orbit operations has been piloted since the 1970's, and was first realised with the launch and implementation of the shuttle remote manipulator system (SRMS), also known as the 'Canadarm', in 1981 [9]. The arm was used in increasingly complex operations; including docking of the space shuttle, capture of satellites, and facilitating on-orbit repair. The concept was very successful, and a second, more complex version was launched in 2001, the mobile servicing system (MSS). This marked the maturation of large-scale robotic manipulator design for space applications, and also emphasised the increased importance of robotics in supporting extra vehicular activity (EVA). In 2008, the special purpose dextrous manipulator (SPDM) was launched to further enhance the capabilities of the MSS. Notably, the SPDM was intended to carry out some tasks that used to necessitate human EVA; directly replacing the astronaut's physical presence.

Another space robotic system was used in a robotic mission called Genesis, which was launched on August 8, 2001 by NASA to collect a sample of solar wind particles and bring them back to Earth for analysis. Since then, space robotics have also been developed rapidly for in-orbit operations in future in parallel to the rapid robotic development on many terrestrial applications represented by various Boston

Dynamic robotic systems, e.g. Big Dog, Spot and so on. Among many space mission goals for the next twenty years, robotic missions can be divided into two main categories: namely planetary exploration and on-orbit operations. Hence it makes sense to refer to robots for planetary explorations as planetary robots, and robots for in-orbit operations as orbital robots. Future robotics development for space may evolve further, but it is believed that space robotics applications will still fall into one of these two categories [10].

2.4.1 *Orbital Robots*

The evolution of robotic manipulator technology on the International Space Station (ISS) from a supporting tool like the SRMS, to a unit capable of human-like dexterity such as the SPDM, shows the potential for implementing a fully robotic on-orbit servicing system (OSS). Such a system would be most relevant to high-cost satellite missions that, although of critical commercial importance, have little chance of human intervention should a hardware failure, or depletion, occur. Critical to the success of an OSS is the design of the robotic manipulator that will carry out maintenance activities. Not only must the end-effector facilitate a variety of different satellite servicing operations, the structural linkages must also be designed with respect to the coupled dynamics of the complex system, comprising of the satellite and the active OSS.

Taking a closer look and consideration of launching a spacecraft such as satellite into an orbit, the following life-cycle stages are normally experienced from launch to graveyard of a spacecraft: launch of the spacecraft, transfer it to an operational orbit, deployment and activation of the spacecraft, its operational utilization, the nominal end of life, extension of the spacecraft's life, and finally transferring to the spacecrafts' graveyard.

Among these lifecycle stages, space robotics are likely to enable a range of operations to facilitate that the whole space mission in achieving the mission's goals. This list is not limited to these operations only and could be extended to other operations of future missions.

2.4.1.1 Rescue Mission/orbit Raising

Space missions can be designed to take place in multiple stages due the limitation in launch capacity of the current launch technologies. These staged launch process requires the payload is placed into the final operation orbit via several and different orbits. These mission requirements raise the need of changing the payload or spacecraft's orbit. This mission concept requires a general--purpose space tug to dock or robotically berth to a client satellite to pull it to a higher orbit. This approach can be an attractive mission concept for very big space infrastructure in which the payload to be delivered to orbit is bigger than the launcher capacity, hence it is difficult to send the space infrastructure into the correct orbit in one launch process. Staging of the payload is therefore necessary. By creating staged launch missions, the payload can be divided into smaller and more manageable ones which can then be sent to the right orbits through multiple launches. Large space infrastructure type of payloads can be gathered on a low orbit through initial launches and then raised to its final operational orbit by means of the tug and further elevation.

With the current International Space Station (ISS) still operational, it is envisaged that there will no immediate market needs for this particular operation. However, the next institutional infrastructure post ISS aimed at Earth Lunar L2 may benefit from such services.

2.4.1.2 Deployment and Assembly Aid

Launch vehicles are designed and optimised in their geometric profiles in order to minimise the flight drag and resistance and reach their mission orbit in a cost-effective manner. Due to these spatial constraints of launch vehicles, every spacecraft is tightly packed within available space within the launch vehicle at and during launch stage and this allows the spacecraft to be placed within the launcher fairings. The spacecraft therefore undergoes the transition between the tightly packed state at launch to the expanded deployed state in orbit, in which all appendages are extended to their operational state. The possibility of using spacecraft-mounted robot manipulators to deploy and or assemble the appendages has been studied and also tested successfully in orbit. There is also the possibility that great part of the spacecraft may be assembled in orbit by means of dedicated robots. This would allow improving the mechanical environment of critical components at launch making spacecraft lighter and more performing.

These operational concepts have certain advantages in long term, however the realisation of this concept requires further maturity of the technology involved and some technologies are covered in detail in Chap. 21. It is therefore likely to see some missions to consider and deploy this concept in the future missions, with further investigation of fundamental economic limitations.

2.4.1.3 Inspection and Support to Deployment

Spacecrafts such as satellites are often subject to unforeseen functional failures due to harsh launch conditions. Operational events due to microgravity environment could also cause part of the overall system's failures. Varies studies have been undertaken and shown that of these failures, 25% are caused by a flaw in design, 21% due to environmental influence, 30% are random most likely due to poor quality of components, 5% are from software and 19% are unknown [11]. A study was conducted on 1,584 satellite failures between 1990 and 2008 in order to find the percentage of failures caused on average by each subsystem over the course of the first 30 days of operation [12]. The results are shown in Fig. 2.5. Telemetry, tracking and command

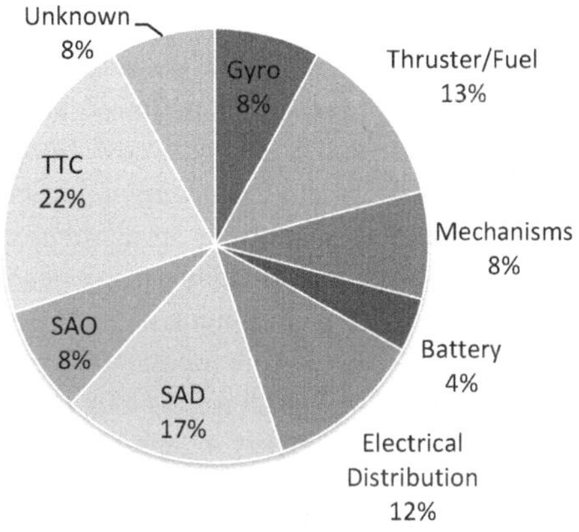

Fig. 2.5 A breakdown of subsystems causing satellite failures

(TTC) accounts for 22% of the total failures studied, followed by 17% of solar array deployment (SAD) related operations, and then by 13% of failures in the thruster and fuel sub-system. Electrical distribution subsystem, mechanisms including structural and thermal subsystems, Gyro including all sensor and reaction wheel (here after referred to simply as Gyro) and solar array operating (SAO) subsystems are the next group of subsystems with a high probability of failure at 12, 8, 8 and 8% respectively. Finally, battery failures and unknown failures account for 4 and 8% of the total failures. The results of proportion of each type failure vary from 30 day's observation to one year, 5 year and 10 years satellite operations. However, the overall failure proportion of the sensing subsystems represented by Gyro increases as time goes by, increasing from 8 to 12%, 11% and 18% from 1 month, to 1 year, 5 year and 10 years of satellite deployment. Telemetry, tracking and command subsystem remain more or less constant at 22, 23, 18 and 19% with a slightly reduced percentage over 10-year deployment. These results provide factual insights to these satellite systems and can be used for a general guideline as to what satellite functionality is required for inspection to identify details of the failure during its operation before a repair can be planned.

These inspection operations following deployment anomaly have been studied on a number of occasions. Unfortunately, this operation has proven to be economically unviable and is not likely to happen in the next fifteen-year period unless the cost of launch becomes sensibly lower.

2.4.1.4 Repair and Orbit Maintenance

The causes of spacecraft failures can be many and various. Thruster and fuel failures can be caused by build-up of combustion products, or by hydrazine freezing in the fuel lines (hydrazine has a melting point of 2 °C). It is necessary to study how to fix these failures once detected and the method of fixing an identified problem will very much depend on the subsystem and the accessibility of the subsystem.

As an example, access to the interior of the fuel system may be required to fix some of these faults, and this would necessitate a long, thin, flexible probe. In the case of freeing up frozen fuel lines, there would need to be a method of applying heat to the affected area. To remove build up or contamination of the fuel system, a fuel system cleaning procedure could be developed, which would be required to apply cleaning fluid to the appropriate area.

The simplest type of failure to fix is a mechanical failure, such as the problems found in solar array deployment (SAD) related operations. These failures can be corrected by either simple application of force to surfaces, or in the case of SAD failures, the use of a wire-cutting element to allow full deployment. As these types of correction require the prising apart of larger mechanisms, they dictate the minimum force and torque requirements for the manipulator system, while also requiring that the end effector be large enough to grasp the physical features on either part of the seized linkage.

It is worth highlighting that operations following failure or early depletion of a satellite fuel subsystem have been proven to be economically viable though marginally where the failed spacecraft belongs to a constellation and a general-purpose servicer can be stationed in advance in proximity of the constellation. However, these applications have not received the interest of their only potential customers: space insurers. In fact, commercial spacecraft are insured by operators against failures and early depletion, so costs are borne by the insurers, which increase premiums to the particular operator as well as to new insurance contracts, to recover the losses. There are trade-offs to be made by the operators on the options of accepting an increased insurance premium or supporting the development of new technologies such as space robotics in order service or refuelling these nearly or completely depleted satellites. There are also arguments for sustainable use of space assets and this will be further discussed in Chap. 17.

2.4.1.5 Planned Maintenance and Refuelling

With more and more spacecrafts launched into space, the sustainability issues become increasingly important for all stakeholders to consider. The amount of space debris has increased considerably over last thirty years and they now become hazards for many space missions. New thinking to reuse space assets through servicing and refuelling depleted fuel tanks become an attractive and potentially viable option for consideration. Servicing and refuelling operations require a client spacecraft to be re-engineered to facilitate these complex operations involving disassembling and re-assembling critical parts of a spacecraft in space.

Satellite fuel is usually in liquid form, and is either a low power monopropellant (such as liquid hydrazine) or a higher power bipropellant (utilizing such as liquid hydrogen as fuel and liquid oxygen as an oxidiser). A method of zero-gravity fluid transfer, accommodating various sizes of satellite fuel nozzles, must be integrated into the manipulator's operation in order to claim compatibility with the majority of satellites in operation. As the majority of satellites are not designed to be refuelled in-orbit, extra tooling and flexibility will be required in order to access the fuelling interface. The extra functionality required has been demonstrated by the Robotic Refuelling Mission (RRM) from NASA; which has been designed and conducted to test each individual function thought to be required in replenishment of a triple-sealed fuel valve of an orbiting satellite [13, 14]. The Robotic Refuelling Mission 3 (RRM3) has demonstrated an initial set of tool operations to robotically transfer liquid methane from one tank to another in space at the ISS in 2019. The RRM3 mission then completed its second set of robotic tool operations on the ISS in October 2020. The mission is set up to systematically demonstrate the capability of transferring cryogenic fluids used as coolants or propellants. It also aims to demonstrate that 42 L of cryogen can be safely stored in a tank for six months without any loss through a process called boil off. Cryogens have extremely low boiling points and this physical property makes it very difficult to store them in space [15]. During this demonstration, Dextre, a space robot, used its two robotic arms to employ two robotic refuelling tools simultaneously. The first tool is called the Cryogen Servicing Tool (CST) shown as the right arm in Fig. 2.6 and is used to manoeuvre a cryogen hose connected to a fuel tank. The second tool is actually a smaller robot called the Visual Inspection Poseable Invertebrate Robot 2 (VIPIR2) tool shown as the left arm in Fig. 2.6, which has an inspection camera to be manipulated to a suitable location for visual based inspection to understand the refuelling environment and verify the placement of the cryogen hose to the right location [16].

NASA is implementing a further mission called the Mobile Servicing System (MSS) Operations to service satellites with the use of four purpose-built end effectors and a tool. These adapt the general-purpose special purpose dexterous manipulator (SPDM) gripper to become a wire-cutter, a multifunction tool to remove fittings, a safety cap remover and a nozzle tool for fluid transfer. In addition, there is a Cryogen Servicing Tool to manoeuvre a cryogen hose. These separate functions should be considered for integration into the robotic manipulator design for the purposes of refuelling. A key idea of the RRM is to install a new 'quick disconnect' fuel access interface onto the customer satellite, enabling easier fuel transfer for future operations. The general concept of upgrading the satellite hardware to better

Fig. 2.6 NASA's robotic refuelling mission 3 [16] (*Photo courtesy of* NASA)

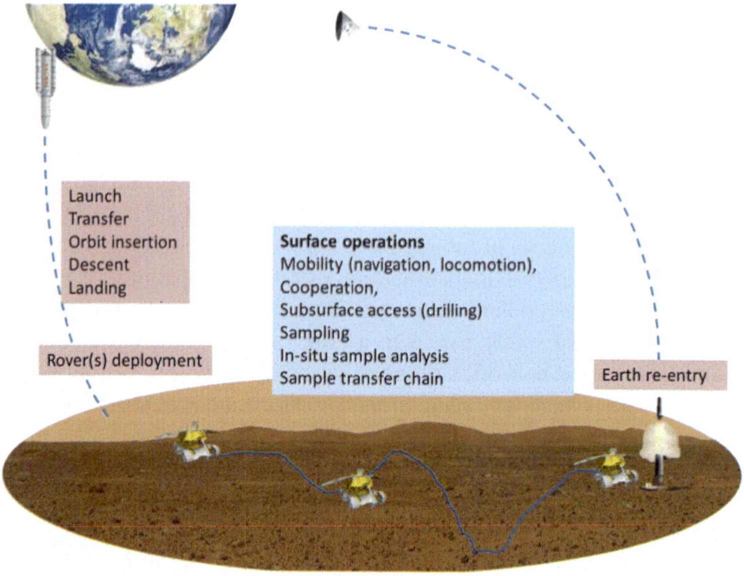

Fig. 2.7 A generalised scheme of a planetary sample return mission [10]

facilitate maintenance can be explored as an extension of the robotic manipulator system.

In order to be economically viable, these operations need to target a constellation of spacecrafts that all require to be operational at the same time, hence the urgency of reinstating their critical functionality. Of all these and other relevant operations, the one that appears more likely to happen in the next fifteen-year period is refuelling as it is the one that requires the least level of modification of a client spacecraft. In addition, most of satellites only have fuel to operate for between ten to fifteen years. Refuelling these satellites has the biggest potential to bring value to a space company as it is normally costlier to design, manufacture and launch a replacement spacecraft with similar functionality.

2.4.1.6 Performance Enhancement

As the new technologies emerge and new demands for services such as space data increases, traditional space industrial organisations also experience pressure from the market place to enhance the performance of their spacecraft more rapidly than the past. This also raises the need to upgrade the capability and functionality of current spacecraft with new technologies or additional capabilities which can be added by introducing additional payloads into an existing spacecraft. This type of mission evolution also has increasing impact on the demands for space robotic solutions to assist the upgrading of an existing spacecraft.

2.4.1.7 Lifetime Extension

In the same vein of spacecraft enhancement, lifetime extension of a spacecraft as a concept takes reuse of spacecrafts a step further for a more sustainable rejuvenation of in-service spacecraft. For example, lifetime extension performs an operation that allows a telecom satellite to continue its operation despite the fact it has depleted its fuel and therefore cannot remain autonomously stationary. It can be realised by different robotic means. One form studied in the past is the so-called captive carrying. An auxiliary servicing satellite is permanently attached to the client and performs station keeping. Another more speculative way foresees a servicing satellite that periodically attaches to the client satellite and recovers the orbital drift that has occurred since its last visit. This last form can only be advantageous if there is a constellation of closely placed satellites that can be visited with negligible delta-V. Lifetime extension may be financially viable only if the total cost of launching and operating a servicing satellite is well below the revenue that the additional lifetime of client satellite(s) generates. Studies show that even in the most benign analysis a Lifetime Extension mission would offer limited benefits but can come cross many risks, so the concept has failed to attract support of its possible customers such as satellite operators. It is not likely that in the near future this situation will change.

2.4.1.8 Re-Orbiting and/De-orbiting

With more and more spacecrafts having been launched into orbits and more planned to be launched in future, there has been an increased awareness of the risk posed by debris, such as nonfunctioning satellites, upper stages after launches and their fragments, to the operation of functional satellites. Self-disposal of satellites at the end of their operational life has become a formal practice and in some countries is an obligation for new satellites. For a satellite in a geosynchronous equatorial orbit (GEO), re-orbiting it to the Graveyard Orbit (GYO) is the required practice. In the past the use of a dedicated servicer to re-orbit constellations of telecom satellites (a sort of satellite undertaker) has been studied and found to be marginally financially viable. However, the concept has failed to attract support from the potential users as GYO re-orbiting was and still is not yet mandatory, so the percentage of operators that actually re-orbit their spent satellite is still low. For satellites in the low Earth orbits (LEO), the fact that there are infinite LEO orbits makes the concept of a common servicer impossible. New LEO spacecraft must therefore implement independent passive or active de-orbiting means. For all defunct satellites already in orbit, which constitute some form of threat to the usability of their orbit, the only available solution is to perform robotically Active Debris Removal (ADR) to interface a spacecraft and then drag it into the GYO. If the threat by a nonfunctioning spacecraft to highly critical orbits, such as the Sun Synchronous Orbit (SSO), is confirmed, by analysis and observation of the evolution of debris population, there will be a urgent need for the development of ADR in the future as the potential cost and impact could be very significant.

2.4.2 *Planetary Robots*

See Fig. 2.7.

2.5 Capabilities of Space Robotics

Whilst there are many robotic capabilities desirable for a space mission, it is necessary to define some essential capabilities based on the state art of the technologies available and also the consideration of the needs of all the space applications as detailed in Sect. 2.3. It is also recognised that the technologies associated with the robots are evolving rapidly and it is therefore not feasible to identify them all. The following space robotic capabilities are identified as a set of unique ones for space robotic applications.

2.5.1 *Navigation Autonomy*

Autonomy refers to an attribute of a space robot which is capable of making decision for itself based on its perceived environment conditions. There are many forms of autonomy and they can be classified based on their space mission tasks and technologies deployed into vision based and enabled perception autonomy, force and torque enabled actuation autonomy, perception enabled task planning autonomy, perception and actuation enabled navigation autonomy.

Vision based navigation autonomy involves a digital map building and merging process for autonomous map reading, direction finding and self-localization of a space robot in its environment. For an orbital robot, this means that the robot needs to be capable of walking around or moving safely on a space infrastructure platform without causing any potential collision or damage during its operations. In the context of planetary exploration this capability is of key importance due to communication delays as is in the case of MarsREX and ExoMars scenarios, where it takes roughly 15 min each way for a rover's downlink sessions when the rover sends information back to a control centre of the robotic mission on Earth. Autonomous navigation capability is therefore essential to avoid any potential risk to the rover as any human intervention will take about 30 min to reach the rover and it is very likely to be late for any such interventions to achieve desirable effects.

This autonomy is even more essential for missions such as Mars Sample Return (MSR) where a crucial aspect is the accuracy of the localisation. In the case of a vision-based navigation, the surface vehicle relies on a technique called data fusion, detailed in Chap. 3, from its on-board sensor set, which includes its cameras/stereo benches. The extraction of some visual features from the acquired images can be exploited to increase the accuracy in the vehicle relative position and orientation determination, with respect to its landing position or its operational goals, and/or its absolute position on the planet. The 3D environment reconstruction is an added output that the on-board visual sensor bench can provide, for benefit of the opportunistic science and on-board autonomy devoted to smartly and efficiently schedule the on-the surface vehicle exploration activities.

2.5.2 *Actuation and Locomotion Autonomy in Harsh Environments*

It is foreseen that future on-orbital mission may involve much activities for reuse and sustainable approach to life extension of existing spacecrafts in orbits. These require space robots to operate in a microgravity environment where robots need to operate safely without of risking being detached from carrier spacecraft to become a new item of space debris while they move about to perform construction or maintenance operations.

For the planetary explorations, future missions also foresee very complex and articulated on-ground activities, potentially executed in demanding scenarios and conditions, such as shadowed craters, highly sloped and irregular terrain, and low gravity bodies exploration. Moreover, selection of the most convenient area to perform a science focused detailed exploration, or module and plant settlement mission task, or sample collection can only be accomplished after reconnaissance over a vast area. Exploring quickly and efficiently large surface areas may also be required for vehicles better tuned for visiting large areas by deploying an unmanned aerial vehicle, rather than using 2D surface robots. Different surface vehicle types could be combined according to the exploration area scale and to the requirements of exploration velocity and area revisiting. Those scenarios ask for capabilities which can be categorized into:

- Advanced on-board autonomy: a decision-making skill on board the vehicle is fundamental to be reactive and able to cope with the environmental uncertainties for the sake of the vehicle and the mission, and to smartly schedule on-site its activities consistently with the best on board limited resource exploitation. Declarative and reactive reasoning mechanisms are foreseen on board, which also allow dealing with a more effective FDIR management timely. On board autonomy is also fundamental whenever a multi-vehicle scenario is supporting the mission for the on ground scientific and servicing operations, to robustly and timely manage the coordination of different vehicles in performing their distributed tasks harmoniously.
- Advanced Locomotion: the capability to move safely on the surface no matter of its configuration is needed to ensure robust evolution of the on-ground operations and to enhance the science feasibility; very harsh environments such as craters, caves, highly sloped regions, typically scientifically very important, can be reached.

Adaptable and flexible wheels, legs, tethered systems, rolling elements and bladed wheels are intended to be advanced locomotion solutions.

- Long range mobility including (1) aerial vehicles, such as balloons, airplanes, helicopters, for supporting surface robots and exploring large surface domains, and (2) surface robots, capable to support and adapt to long range transfers. Both aerial vehicles and advanced surface robots require long durability/high energetic efficiency, advanced materials for mechanisms and actuators, fusion of multiple sensors to accurately perceive the environment for navigation, control, decision making. Advanced surface robots must implement alternative locomotion concepts based on legs and/or poly-articulated wheels (such as configurable robot or robot teams with complementary capabilities) to access difficult environments such as steep slopes and terrains cluttered with obstacles.
- Advanced sensing: Harsh and unknown planetary environments include shadowed/dark areas (i.e. craters and caves), and highly dusty regions (i.e. Mars, Moon surfaces). There is a need for sensor kits capable of supporting robotic exploration in these demanding environments.
- Energy Efficiency: The more challenging requirement to the robotics for surface exploration relates to energy availability, both in terms of amount and duration. New and more efficient energy generators are needed together with multi-functional mechanisms capable to store and recover energy. These technological enhancements are fundamental to relaxing the strong constraints imposed by the on-board power availability on exploration of very demanding areas (i.e. cold traps, dark regions). Smart materials which can change their structure thanks to adaptable parameters, may represent a valid solution to reduce the power demand from the on-board mechanism, and to differently solve some of the actuation tasks on board.

2.5.3 Science Exploration Autonomy

The capability of a mobile vehicle to identify unforeseen or unpredictable opportunities for science investigation, detected while travelling on the planetary surface is key to maximising scientific return of a mission. Science exploration autonomy or autonomous science requires a robot to be able to detect interesting samples, phenomenon or regions in its surroundings, to judge and score their relevance according to the scientific mission objectives and goals. Based on these in-situ and real time information gathered whilst the robot is exploring, the robot needs to be capable of revising its own a-priori activity plan to include visiting and sampling of the interesting area, while knowing and respecting resource constraints.

2.6 Development of Space Robotics Missions

We have described what makes space robots special with respect to terrestrial robots. Another major difference is related to the fact that, so far, space robots are not serially produced for established uses, like most terrestrial robots.

A space robot is designed for a particular space robotic mission and needs to follow the establish practices of the space sector.

Due to the high risks space missions present, space agencies, industry and academia have studied all necessary known requirements from various stakeholder and standardised the development of space projects in a series of steps that aim at reducing risk, optimising resource use and eventually specifying what needs to be done and when. It is therefore necessary to be familiar with these stand design procedures before embarking on such a design and development of robotic system.

We have described what makes space robots special with respect to terrestrial robots. Another major difference is related to the fact that, so far, space robots are not serially produced for established uses, like most terrestrial robots. A space robot is designed for a particular space robotic mission and needs to follow the establish practices of the space sector.

The European Cooperation for Space Standardisation (ECSS) has developed a set of coherent standards for use in European space activities.

The standard "ECSS-M-ST-10C Space project management—Project planning and implementation" defines a standard life cycle for space projects, divided in 7 phases:

- Phase 0—Mission analysis/needs identification;
- Phase A—Feasibility;
- Phase B—Preliminary Definition;
- Phase C—Detailed Definition;
- Phase D—Qualification and Production;
- Phase E—Utilization;
- Phase F—Disposal.

Phases 0, A and B are where the functional and technical requirements and associated system concept are defined so that the mission goal can be achieved.

It is important to understand that the development of automation and robotics systems is somehow different from other space systems. In the following sections, these differences are briefly explained with respect to the design stage and the verification and validation. For what regards a Space Automation and Robotics system, phases 0 to C can be defined as design stages. Validation and verification activities are then performed across phases B to D.

2.6.1 The Design Stage

The Automation and Robotics (A&R) system will be part of the defined concepts and therefore will have functional and technical requirements associated. An A&R System is adopted in a space mission because of the flexibility it provides in coping with poorly defined process or poorly known environment or uncertainties defined in Sect. 2.2.1. The flexibility, which allows huge variation of activities at mission time, is enabled by the basic capabilities of the system and the intrinsic ability to replan operations.

Unplanned and untested behaviours can be implemented at mission time through the use of specific ground tools by qualified operators.

So a space A&R system is really an end-to-end control system where all the interplay between the robotic segment (i.e. the robot or rover) and the human segment (i.e. the Human Robot Interface and the operators) is undertaken and the responsibilities of each segment can vary from one mission to the next. Their basic abilities and roles must be defined to guarantee the needed flexibility as well as resilience to unpredictable events or situations.

Note that a Space A&R system normally has robotic segment on the flight segment of the mission. However the human segment can be in space (e.g. robots on the ISS are controlled from the inside of the ISS) and/or on ground.

ESA developed already in 1992, a Control Development Methodology (CDM) that aimed at a structured definition of the roles, responsibilities and capabilities of the end-to-end system.

The first step of the CDM aims at answering the following questions:

- "What is the process to be automated ?"
- "What are the functions (activities) to be performed by the A&R system?"

Answering these questions produces a "script" description of the process in the form of a hierarchical structure of activities with associated attributes to express parameters or constraints. The top level activities (the A&R missions) are broken down into lower level activities according to a well-defined layered hierarchy. The activities are the required A&R system functions. These functions are further analysed through the Activity Analysis Method, in order to arrive to a catalogue of commonly occurring tasks and actions, with associated attributes. The attributes represent the parameters (e.g. setpoints, performance) and constraints of each activity.

The result of this crucial step is the structured Activity Script with the description of the individual activities and attributes. This constitutes a great part of the A&R System Requirements Document.

The Activity Script will be used to design an architecture of the robot segment A&R devices, human segment HRI and roles and responsibilities of operators, to plan, execute and monitor the execution of the activities.

The design of the end-to-end system will also need to account for the communication constraints imposed by the distance between the robot segment and the human

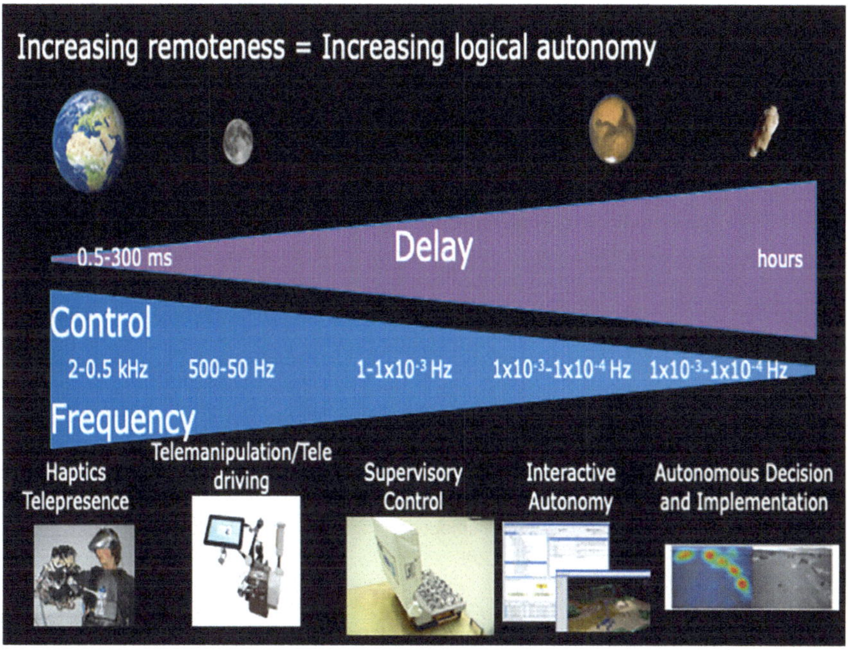

Fig. 2.8 A&R autonomy modes with regard to delays and control frequency

segment. The more limiting are the communication constraints the more logical autonomy needs to be implemented in the robot segment.

The levels of logical autonomy have a great impact on the A&R system design. There is no worldwide standard definition of the different levels of autonomy that can be implemented in space robots, also definitions that apply to some terrestrial sectors of A&R are hardly applicable to space.

At ESA the autonomy modes illustrated in Fig. 2.8 have been investigated in many R&D activities and are baselined for some ESA missions.

Table 2.4 provides the definition of the modes and the roles of the elements of the end-to-end system.

As it can be easily inferred, a low level of autonomy, which is made possible by the high rate control, can be attained with rather simple robot and overall faster, simple end-to-end system. Conversely the complexity of the whole end-to-end system increases if the more autonomy is demanded.

Due to the inherent complexity, the definition and design work of an A&R system, need to be supported at the earliest stage by a robust Validation and verification programme.

Table 2.4 ESA R&D autonomy modes

Mode name	Definition	Operator	HRI	Robot segment
Haptics Telepresence	The robot is 'slaved' to the operator The operator experiences the forces involved in the robot actions	The operator is immersed in the operating environment of the robot by complete sensorial feedback (visual, haptic,)	Must convey maximum situational awareness Requires immersive visual displays haptic devices (force reflecting joysticks or exoskeletons) fed with filtered robot segment telemetry	Provide a mapping of human commands to device commands Provides for low level safety features (e.g. collision avoidance)
Tele-manipulation/ driving	The robot is 'slaved' to the operator	The operator drives the robot without immersion Control rate does not allow for immersive feedback from robot as it would create discomfort	HRI produces situational awareness by synthetic feedback created by models predicting robot behaviour. Synthetic feedback may be provided also by force reflecting, vibration or soundification	Besides what provided in 1, the robot must close control loops (position, force) locally by means of adequate algorithms (e.g. object pose visual estimation, compliance control)
Supervisory Control	The robot executes commands initiated by operator. Operator can interrupt/ re-parametrise/ restart commands	Operator has complete responsibility over which action the robot will execute and on what subjects	HRI must provide means to select suitable action and predict the effect of the action HRI provides means to monitor the execution of the action by displaying model predicted telemetry and the (delayed) telemetry of the A&R system. This is done mostly by visual presentation	Besides what provided in 2, the robot must be able to independently plan and schedule in time and space motion trajectories The A&R system must filter, aggregate and increase level of abstraction of the sensory input

(continued)

Table 2.4 (continued)

Mode name	Definition	Operator	HRI	Robot segment
Interactive Autonomy	The robot executes high level tasks that are arranged in a plan. When the plan execution terminates the operator can review results of plan and upload a new plan for execution	The operator composes a plan made of individual tasks	The HRI has means to plan and schedule the individual tasks. HRI allows to simulate the execution of the plan and investigate its validity and robustness against environmental variation	The robot can execute a plan possibly also with the ability to replan tasks in case of failure of the plan
Autonomous Decision and Implementation	The operator provides a goal to the robot which tries to achieve it	The operator creates a goal with associated constraints The goal represents the sought result of the robot operation, but does not describe the plan of tasks	Th HRI provide means to create goals by creating variable plans of tasks and scheduling them	The robot can build a plan made of individual tasks and has also he ability to replan in case of failure of the plan, in case of failure of the plan

2.6.2 Validation and Verification of Space Robotics Missions

Verification and validation are important processes in many robotic and automation system design, hence they have been widely used and practiced in board engineering system design. It is however important to stress that in space system project development the precise meaning of these processes differs from other disciplines and the following definitions apply:

- verification: the activity that demonstrates, through a dedicated process, that the deliverable product meets the specified requirements.
- validation: the activity that demonstrates that the requirements set for the deliverable product are the adequate to attain the mission goals.

The activity of validation has a particular importance in space robotics missions. As underlined previously, an A&R System is adopted in a space mission because of the flexibility it provides in coping with poorly defined process or poorly known environment.

There is hence an inherent difficulty in defining requirements that are functional to the success of the mission (so being wide enough to allow the sought flexibility) and that can be verified (so having specificity).

Validation must therefore be performed in support of the definition of the conceptual architecture of the A&R system. Functional prototypes of the HRI and robot need to be built and tested in realistic operation scenarios.

The space automation and robotic systems' tests allow to:

- refine the conceptual architecture and interfaces between the end-to-elements;
- refine the conceptual operations (CONOPS);
- refine the requirements and associate to them specific figures.

Validation activities initially focus on the least understood functions of elements of the A&R system and take place in small scale test facilities.

Then the scope of the validation is enlarged to whole elements and finally to the end-to-end system. Tests on the whole system require often large scale facilities such as neutral-buoyancy pools (for orbital robots) and planetary analogues (for planetary robots).

Verification of space products relies on the methods: test, analysis, review of design and inspection.

In addition, the aspect of testing of space A&R systems has also some unusual aspects. Space robots have to function, with quite complex behaviours, in environments that cannot be reproduced fully in a test on ground. For example, a robot in orbital space will function while being exposed to vacuum, extreme temperatures, radiation and lack of gravity. All environmental factors that cannot be reproduced in full and in the same test. Similarly, a rover meant for Lunar use will need to work also in reduced gravity, vacuum, extreme temperatures, radiation and very fine dust.

The well-known guideline of "test as you fly and fly as you test" is actually never quite possible. All space system testing must be broken down in separate tests addressing one specific environment factor.

For space robot systems in particular the ability to perform functional tests in reduced or zero gravity imposes quite some complication.

For robots operating in planetary environments, where gravity is reduced but not zero, to account for testing, the robot may be designed to cope with the same loads it would experience in the target environment, but with Earth gravity. This leads to robots that are ~ three times (in the Mars case) or ~ six times (in the Moon case) stronger than need to be, just to simplify testing.

In other cases where making the robot sturdier would just not work, specific test set-ups need to be built. For example, to functionally test the European Robotic Arm (ERA), capable to move 10 metric ton payload in orbit, but absolutely incapable to even lift itself on ground, a test set-up involving weight compensation devices and air bearings hat to be built.

2.7 Conclusions

With the above introduction to space automation and robotic systems, a context has been established and scenes has been set. In the remaining twenty one chapters, editors have written partially, assembled and edited a range of topics to address the challenges discussed, technical development achieved, mission concepts demonstrated and future research and development trends identified.

Acknowledgements It is acknowledged that Sect. 2.3 on Space environment is contributed by Dr. Karen Donaldson.

References

1. Petkov, M. P. (2003). *The effects of space environments on electronic components*. The Turk: https://commons.wikimedia.org/wiki/File:The_Turk.jpg
2. NASA: solarsystem.nasa.gov/resources/681/solar-system-temperatures/
3. Petkov, M. P. (2003). *The effects of space environments on electronic components*.
4. Duzellier, S. (2005). Radiation effects on electronic devices in space. *Aerospace Science and Technology, 9*(1), 93–99.
5. Jursa, A. S. (2005). Handbook of geophysics and the space environment. Air Force Geophysics Laboratory Hanscom AFB MA.
6. Meseguer, J., Pérez-Grande, I., & Sanz-Andrés, A. (2012). *Spacecraft thermal control*. Woodhead Publishing.
7. Hengeveld, D. W., Mathison, M. M., Braun, J. E., Groll, E. A., & Williams, A. D. (2010). Review of modern spacecraft thermal control technologies. *HVAC&R Research, 16*(2), 189–220.
8. Gilmore, D. G. (2002). *Spacecraft thermal control handbook Volume I: Fundamental technologies* (2nd ed.). The Aerospace Corporation.
9. Sachdev, S. (1986). *Canadarm—A review of its flights* (Vol. 4, pp. 268–272).
10. PERASPERA Horizon 2020 Report: D3.4 Master Plan of SRC Activities, 2015.
11. Ellery, A. (2000). In-orbit servicing missions. In *An introduction to space robotics* (p. 61).
12. Castet, J.-F., & Saleh, J. H. (2009). Satellite and satellite subsystems reliability: Statistical data analysis and modelling. *Reliability Engineering System Safety, 94*(11), 1718–1728.
13. Brown, R. (2013). *Intern report*. University of Strathclyde.
14. ISS Utilization: RRM (Robotic Refueling Mission). https://www.eoportal.org/satellite-missions/iss-rrm#rrm-module-in-phase-1
15. NASA 2018. https://www.nasa.gov/feature/goddard/2020/nasa-s-refueling-mission-completes-second-set-of-robotic-tool-operations-in-space
16. NASA 2020. https://www.nasa.gov/feature/goddard/2018/nasa-to-launch-new-refueling-mission-helping-spacecraft-live-longer-and-journey-farther

Chapter 3
Space Robotics Technologies: Perception for Autonomy

Xiu Tian Yan, Jeremi Gancet, Shashank Govindaraj, Raúl Domínguez, Raphael Viards, Clément Bazerque, Thierry Germa, Simon Lacroix, and Michal Smisek

Abstract Space robotics technologies are a set of latest technologies designed specifically for space exploration missions which include both planetary, mostly science—driven ones, and on-orbit missions. This chapter provides a high-level introduction to these technologies with a focus on the advanced features of space robotics including perception of the space robotic systems, intelligent control for a potential full autonomous mission execution, and space robotic motion planning and real-time control. In the perception section, it is necessary to introduce the important

X. T. Yan (✉)
RAS Group, DMEM, University of Strathclyde, Glasgow, Scotland
e-mail: x.yan@strath.ac.uk

J. Gancet · S. Govindaraj
Space Applications Services, Leuvensesteenweg 325, 1932 Sint-Stevens-Woluwe, Zaventem, Belgium
e-mail: jeremi.gancet@spaceapplications.com

R. Domínguez
DFKI Robotics Innovation Center Bremen, Robert-Hooke-Str 1, 28359 Bremen, Germany
e-mail: raul.dominguez@dfki.de

R. Viards · C. Bazerque · T. Germa
Magellium, 1 Rue Ariane, 31520 Ramonville St Agne, France
e-mail: Raphael.Viards@magellium.fr

C. Bazerque
e-mail: ment.Bazerque@magellium.fr

T. Germa
e-mail: Thierry.Germa@magellium.fr

S. Lacroix
LAAS/CNRS, Robotics and Interactions, 7 Ave du Colonel Roche, BP 54200, 31031 Toulouse Cedex 4, France
e-mail: simon.lacroix@laas.fr

M. Smisek
German Aerospace Research Center (DLR), 82234 Wessling, Germany
e-mail: Michal.Smisek@dlr.de

X. T. Yan and G. Visentin (eds.), *Space Robotics*,
https://doi.org/10.1007/978-3-031-39214-6_3

requirements for perception for space missions. Based on these special requirements, the sensing methods are introduced focusing on the space-related environmental perception sensing method and associated sensing technologies and devices. This is followed by modelling in sensing and software engineering processes for perception. Similar to other sensing technologies, perception-focused sensing also requires calibration and this is briefly introduced. A key technique in perception is the data fusion of signals from several sensing devices and this is detailed in order to provide a good grounding of data fusion techniques. A particular research-based approach called InFuse is introduced here as an example data fusion system. Autonomous navigation and task execution is a high-level ambition for many space exploration missions. With an understanding of the data fusion techniques, a detailed understanding of the environment where a space robot operates can be achieved by applying perception processing. This is realised by generating a common data fusion framework, and data fusion techniques used within the framework for both planetary exploration and orbital missions are detailed in this chapter. Space mission planning with a view to autonomous mission execution is introduced and the multiple robotic planning and reconfiguration concept is introduced, which is further explained in another chapter. Finally motion planning and control of space robots are discussed.

3.1 An Introduction to Space Robotics Technologies

Space exploration can be broadly divided into planetary exploration and on-orbit servicing (OOS), each of which poses its own challenges mostly due to their operational environmental factors and operational missions. For example, in planetary exploration, the environment consideration focuses on factors such as terrains, surface mechanical properties of a space rover or robot traversing and so forth. In-orbit servicing or other related missions require consideration of other factors such as the impact of microgravity, lighting condition and the objects' properties that space missions are focusing on. Specifically, in the case of in-space assembly operations, for example, space robots need to have an adequate level of understanding of mechanical properties before these objects are manipulated for assembly or construction of a larger space system. Despite these differences, space robots need to operate safely and complete their intended tasks in these unforeseen environments. Hence, this chapter aims to take a generic and holistic view of space robotics technologies and introduce the generic aspects of these technologies. These include the crucial capability of perception -to use various sensors to capture initial data utilising different sensing methods, from which modelling approaches are deployed and calibrations are made in order to provide optimal results for understanding the environment.

3.2 Perception

Space robots, normally in the form of a rover, have been deployed for long-range exploration of planetary bodies, which poses stringent requirements on the platform's capabilities. With limited prior knowledge of the environment, the rover must have a high level of autonomy allowing it to continuously localize itself, detect potential hazards and obstacles, and plan a safe path through the rough terrain. Such tasks can only be achieved with the appropriate set of sensors which allow the rover to gather enough information about its environment and movements.

3.2.1 Sensors and Their Requirements for Space Robotics

Sensors are devices used in various mechatronic systems such as robots and rovers, that enable robots to see an environment, measure the geometric dimensions of objects within its perceived range, feel objects and their textures and, in some cases, even smell molecules in air. They can understand the surrounding environment by measuring the static information of position and orientation of objects as well as dynamic information such as distance, velocity and accelerations of objects of interest both in terms of translation and rotation. Sensors can also help to understand the behaviour of objects within the surrounding environment, by measuring many mechanical properties such as forces, torques, moment, temperature, weight, radiation exposed and so forth.

Sensors have been developed for different applications and the following are of particular relevance to space robotics. These are only some examples and new sensors are regularly emerging for future applications: namely, inertial measurement units (IMU), wheels encoders, wheels currents (or torque sensing), stereo-cameras (for navigation and/or visual odometry), 2D or 3D Lidar, and sun trackers.

Within space robotics applications, previous Mars exploration missions [1–3] have involved a range of sensors, used and combined in different situations. For long-distance drives in safe environments, relative localization of the rover is achieved with a set of wheel encoders for simple odometry, combined with attitude and heading information obtained from an inertial measurement unit (IMU). In more hazardous environments, visual odometry techniques using a set of stereoscopic cameras are applied at varying intervals to correct drift caused by wheel slippage. Local depth maps obtained with the stereo cameras are also used to compute traversability maps for path planning. Other sets of monocular or stereo cameras are usually also included for manual obstacle and hazard detection by operators on the ground. Additionally, an absolute rover heading measurement can be obtained through visual detection of the sun or stars.

Further research efforts [4, 5] have explored the integration of 2D or 3D LIDAR-type sensors on planetary rovers to overcome the limitations of stereo cameras. With such sensors, point-cloud data can be directly obtained and processed for terrain mapping and path planning. It is also possible to fuse LIDAR data with camera measurements for rover relative localization [6].

Interest for "alternative sensors" is growing in the robotics community: multi-spectral or hyperspectral cameras can provide precise information on the nature of the perceived scene, and hence augment the traversability maps that are currently only defined, based on geometric information. So far only the IR thermal spectral plane has been considered to augment the information provided by classic colour cameras [7, 8] but the recent advent of 16–25 spectral planes cameras opens interesting perspectives, not only to assess traversability, but also to identify potentially relevant scientific targets. Polarimetric cameras have been introduced in robotics, mainly to detect water hazards [9]: one could foresee their use in a planetary context to identify elements of a scene that exhibit (partial) specular reflections. Recent Lidars now deliver a multiple-echo return, which has been proved to be relevant information for terrain classification on earth [10, 11]: their interest in the context of a planetary rover remains to be assessed.

For OOS applications, active sensors such as LIDAR are effective for rendezvous, even under harsh lighting conditions (see for example the Neptec 3D Laser Carema System-LCS [12]), but their high weight and power consumption prevent their intensive use for future missions. Instead, passive, low-weight and low-power cameras are now widely used, and most often monochrome cameras are used because of the small bandwidth requirement, image processing speed and higher resolution. For example, Miravet et al. [13] demonstrated the use of a monocular, camera-based relative navigation sensor for rendezvous and docking of a non-cooperative client, by means of the Orbital Life Extension Vehicle (OLEV). The OLEV image processing module operates on-ground, on downloaded images acquired from the satellite at a 1 s interval. The major problems however are specular reflections due to multilayer insulation for close-range rendezvous. This can be addressed by mounting visual markers on the client satellite. For example, a camera-based tracking software (called Vis-STAR) which relies on image correlation algorithm was used in DARPA's Orbital Express [14] as a redundant rendezvous sensor. The client of the Orbital Express (NEXTsat) was designed for on-orbit servicing, which provides markers to aid visual tracking. Similarly, visual markers were used in ETS-VII [15] robotic satellite mission to aid 6 degrees of freedom (DOF) pose tracking for robotic visual servoing. Also, in the PRISMA mission [16], a vision-based sensor designed based on a star tracker microchip is used to estimate the relative attitude and position of the target.

3.2.2 Sensing Methods

Sensors work on different principles and different types of sensors are designed based on different laws of physics. Some exemplar sensors and their working principles are described in this section.

3.2.2.1 Inertial Measurement Units

Inertial measurement units (IMU) are an integrated set of three types of sensors including accelerometers, gyroscopes and magnetometers for up to 9 axes or DoF, which is composed of 3 axes from each of the three types of sensors included in an IMU. Six DoF IMUs made of an accelerometer and a gyroscope, each providing 3 DoF which can also be configured and become available. They are used to measure the angular velocity of a rotation by the gyroscope in IMU, motion properties such as the acceleration, speed and distance of a moving object by the accelerometer, and the magnetic field in the form of a vector by the magnetometer. Each of these adequately calibrated sensors hence produces data on the linear speed, acceleration, angular velocity and magnetic field vector in their own coordinate frames of the physical sensor. These data streams contain 3-dimensional information of three measured values, that is acceleration and angular velocity along the X, Y and Z axis.

Accelerometers

Accelerometers have three types based on their design and working principles:

- Capacitive also known as differential capacitance for a pair of electrode plates for each moving electrode plate attached to a moving proof-mass;
- Piezoelectric made of PZT material;
- Piezoresistive working on the principle of differential resistance in contracts to capacitance.

Capacitive accelerometers consist of a proof-mass which moves when there is an acceleration, electrode plates attached to the moving proof mass, and fixed electrode plates normally attached to the structure of the static electrode plates. In relation to the set of fixed electrode plates within the accelerometer, this set of moving plates with proof-mass causes the change in distance between matching plates, which results in a changing capacitance between these plates. The capacitance is governed by the following equation:

$$C = \frac{\varepsilon_r * \varepsilon_0 * A}{d} \tag{3.1}$$

where ε_r is a constant called the dielectric permittivity of free space between the capacitor, electrodes; ε_0—dielectric permeability of vacuum with a value of 8,8541 $\times 10^{-12}$ F/m; A is the area of overlap between two plates and it is normally a constant; and d is the plate separation distance. As the distance between plates changes, it causes changing capacitance due to a symmetric change of distance Δd in relation to two neighbouring capacitor matching electrode plates within the accelerator. The change of capacitance is expressed as

$$\Delta C = \frac{\varepsilon_r * \varepsilon_0 * A}{d + \Delta d} - \frac{\varepsilon_r * \varepsilon_0 * A}{d - \Delta d} = \varepsilon_r * \varepsilon_0 * A \frac{2\Delta d}{d^2 - \Delta d^2} \tag{3.2}$$

Rearrange Eq. (3.2) into the following quadratic polynomial equation in terms of change of distance between electrode plates Δd:

$$\Delta C \Delta d^2 + 2\varepsilon_r * \varepsilon_0 * A \Delta d - \Delta C d^2 = 0 \tag{3.3}$$

As the change of displacement Δd is very small and the term $\Delta C \Delta d^2$ can be considered as approaching 0. Equation (3.3) becomes:

$$\Delta d = \frac{d^2}{2\varepsilon_r * \varepsilon_0 * A} \Delta C = K \Delta C \tag{3.4}$$

from which the change rate of the distance provides the speed in a given direction. By measuring these changing capacitances, it is possible to calculate the acceleration as ΔC is proportional to the instantons speed expressed in Eq. (3.3). It is not normally easy to measure the capacitance change directly and further investigation is required to find an easier way to measure this capacitance change due to the movement of the proof-mass. Voltage is always easy to measure and process.

The capacitive accelerometer also works on the principle of Newton's second law and the proof-mass of the accelerometer is expressed as:

$$M \frac{d^2 x}{dt^2} + D \frac{dx}{dt} + Kx(t) = F_{ext}(t) = M * a_{ext}(t) \tag{3.5}$$

where M is the proof-mass of the accelerator moving part, D is the damping coefficient of the mass-spring-damper accelerometer system, K is the spring stiffness constant of the accelerometer system, and a_{ext} is the acceleration under a given external force F_{ext}. Acceleration, speed and distance are all time dependent during the use of the accelerometer.

Applying Laplace transform to Eq. (3.5), the following expression is generated

$$\frac{X(s)}{a_{ext}(s)} = \frac{1}{s^2 + \frac{D}{M}s + \frac{K}{M}} = \frac{1}{s^2 + 2\xi s\omega_n + \omega_n^2} \tag{3.6}$$

where $\xi = \frac{1}{2\pi}\sqrt{\frac{K}{M}}$ is the damping ratio and $\omega_n = \sqrt{\frac{K}{M}}$ is the natural frequency in rad/second.

For a capacitive the system capacitance can be normally simplified into two capacitors for a moving proof-mass with electrode plates on both sides of the mass as C_1 and C_2. With such a design, it is known that the voltage output V_{out} measured has the following relationship to the supply voltage V_{in}:

$$V_{out} = \frac{C_2 - C_1}{C_2 + C_1} V_{in} = \frac{\Delta C}{C_2 + C_1} V_{in} = \frac{x}{d} V_{in} \tag{3.7}$$

When the moving proof-mass moves Δd, $C_2 = \frac{\varepsilon_r * \varepsilon_0 * A}{d+\Delta d}$ *and* $C_1 = \frac{\varepsilon_r * \varepsilon_0 * A}{d-\Delta d}$. Replace C_1, C_2 and ΔC with Eq. (3.2), Eq. 3.7 becomes:

$$V_{out} = \frac{\Delta C}{C_2 + C_1} V_{in} = \frac{\varepsilon_r * \varepsilon_0 * A \frac{2\Delta d}{d^2 - \Delta d^2}}{\frac{\varepsilon_r * \varepsilon_0 * A}{d+\Delta d} + \frac{\varepsilon_r * \varepsilon_0 * A}{d-\Delta d}} V_{in} = \frac{\Delta d}{d} V_{in} \tag{3.8}$$

or

$$\Delta d = \frac{d}{V_{in}} V_{out} \tag{3.9}$$

Once the voltage V_{out} is measured, it is possible to calculate the displacement term Δd within a time interval to calculate the linear speed of the proof-mass displacement. An integration of the speed with regard to time can produce the displacement an object moved. Alternatively, a differentiation of the speed can obtain an acceleration.

Gyroscopes

Gyroscopes are designed to measure three rotational motions: namely pitch about X-axis, roll about Y-axis, and yaw about Z-axis. There are three types of gyroscopes based on their design working principles. Firstly, an angle can be measured by a gyroscope based on the principle of spinning which is a more traditional approach and uses mechanical means. As a result, they tend to be big so they are unsuitable for robotic applications. Secondly, an angle can also be measured using a vibrating mass and recent miniaturisation of gyroscopes using silicon Gyroscopes used in IMU systems are normally in the form of micro-electro-mechanical systems (MEMS) gyroscopes, which are miniaturised through various MEMS technologies and they are widely used due to its small size. One particular popular design of such a MEMS gyroscope is made in the form of a silicon-on-insulator and uses a mechanical vibrating element as its sensing mass to detect the angular velocity. Within these vibrating designs, there are no spinning elements hence there is no requirement for bearings, hence these designs provide a high-level advantage in miniaturising them by removing mechanical supporting components to facilitate rotation. MEMS gyroscopes also build on the capacitive element similar to the accelerometer hence they can be fabricated in the forms of wafers. Thirdly, optical fibre gyroscopes are functionally close to the mechanical gyroscopes as they can detect changes in position and direction using the Sagnac effect but operates with a coil of optical fibre using light. They are then known to be very precise.

To measure the performance of a gyroscope, the following performance parameters are normally used [17]:

Noise density in deg/second/$\sqrt{}$Hz is used to reduce the measured noise of a gyroscope's output, also known as down sampling as the noise signal is averaged at a lower vibration rate before outputting it. In some cases, it is insufficient to properly characterise the inherent noisiness of a gyroscope by simply using a single noise standard deviation. Noise density for a gyroscope hence is represented as deg/second/$\sqrt{}$Hz or ^{o}/s//$\sqrt{}$Hz in short to indicate its true noise in measuring the angle of a rotating object.

Temperature sensitivity of bias in a unit of deg/hour/C^{o} is an indicator of how a gyroscope's performance may be influenced by the change in ambient temperature over a period of time. Such a characterisation of the measuring performance can provide a clear technical indication of the performance of the gyroscope if it is used in an environment where the temperature may change significantly. Space exploration missions typically experience large temperature changes; hence it is important to consider this and select an adequate gyroscope to meet the mission requirements for a given set of temperature changes.

Angle Random Walk describes the accumulated errors due to the noise over time when the chosen gyroscope's output signal of an angular rate measure is integrated to determine an angle. This integration will drift over time hence produces this so-called random walk. Angle Random Walk is measured in units of degree/$\sqrt{}$s or ^{o}/$\sqrt{}$s or ^{o}/$\sqrt{}$hour.

Bias Offset Error measures the offset output reading of a gyroscope when there is no rotation or input. This measurement is normally undertaken in a standard environment at an ambient temperature of typically 25 °C or manufacturers' specification.

Bias Instability measures the performance of bias or offer-set error of a chosen gyroscope over a long period of time at a constant temperature and a constant environment. This performance measurement hence provides an indication of how much bias or off-set error will be generated on the overall angular velocity measurement, that is how unstable the gyroscope is and how much it will impact the angular velocity measurement over the period caused by the bias instability. For an application where a high level of accuracy is required over a long period of time, this becomes one of the most critical performance indicators for selecting a gyroscope.

A number of MEMS gyroscopes have been designed and investigated and a noticeable design by Trusov shows a design of silicon gyroscopes with interchangeable rate on a silicon-on-insulator [17]. There is an extensive review of other MEMS gyroscopes by Vittorio et al. [18] for further information.

For IMUs, it is also noted that the following considerations are of important to their performance. As each of the accelerator, gyroscope of an IMU has three orthogonal axes, it is important to alight these axes to create and maintain strict orthogonality between the sensing axes, that is X axis needs to be and remain orthogonal to Y axis and Z axis at 90 degrees. Any slight misalignment will result in an orthogonality error between two axes or errors in between several pairs of two axes. Such misalignment error can also exist between the IMU case and a space robot if the case is not mounted perfectly aligned to the robot. Similarly, when a camera is mounted to the rover, it is also possible to have this misalignment error if the camera is not perfectly aligned to the robotic coordinate frames.

Cross-axis sensitivity of any of the sensors within an IMU may also occur when the sensor for one axis is also sensitive to the inputs from another axis. This will result in an orthogonality error and it is therefore necessary to investigate if there is a cross axis sensitivity issue for a chosen IMU for a space exploration mission.

Magnetometers

The magnetometer within an IMU is an integral device that measures the direction and the intensity of a magnetic field of a planet, e.g. the earth or other planets in space exploration. It can detect the absolute orientation of the north, south, east and west plans of a planet. A compass is a common example of such a device. A magnetometer can typically act as and provide a heading reference in the direction it measures. There is a potential error between the measured heading and the actual reference value of the magnetic field if this knowledge of the reference value is known and a necessary calibration or correction of the estimated heading can be made. Based on the measured values, the Kalman filter deploys the linear regression iterations which are fundamental to its process. The Kalman filter and its variant methods such as the extended Kalman Filter (EKF) algorithm can be used to process these data in order to obtain more accurate, reliable and correct heading information effectively. The base Kalman filter algorithm has been developed into many variants, such as a new variant called the robust Kalman filter algorithm and its further variant the sequential robust Kalman filter (SRKF) [19]. Kalman filter based algorithms have shown their advantages of accuracy and effectiveness in measuring the magnetic fields and other parameters by an IMU and have been widely used to process these measured data of the heading. They however require very high sampling rates which can be less efficient and consume more energy in capturing the data and processing them. The state relationships of a space robot are described in three dimensions for its rotational kinematics and these relationships hence require large state vectors for implementing a Kalman filter based approach [20]. For applications where there is a need for slow motions a gradient descent algorithm is proposed and implemented to show it can also achieve an equivalent level of accuracy matching a Kalman based algorithm with a value of less than 0.8° static root mean squared (RMS) error, or less than 1.7° dynamic RMS error when measuring the orientation [21].

3.2.2.2 Stereo-Cameras

Stereo-cameras are the classical way of obtaining dense information about the environment for robotics applications. Indeed, this type of sensor is cheap, robust, passive, i.e. has low power consumption, does not contain any mobile parts, and relies on simple, well-established technologies. They can, at the same time, be used to provide semantic or 3D geometric information about the surroundings of a robot, and, the data they produce can also be interpreted directly by humans.

There exist a wide range of different camera sensors and optics, making them suitable and adaptable to various tasks. For example, planetary exploration rovers usually embed at least two different sets of stereo cameras: one with a wide field of view, located close to the ground, for hazard detection and/or visual odometry, and another mounted on a pan and tilt unit, aiming at providing accurate measurements in a more extended range of distances for navigation means.

However, on the downside, camera systems require light to work and are therefore not suitable for dark environments. They also suffer from a quick degradation in measurements precision the further the observed object is located from the sensor. In addition, they need to be calibrated precisely to compensate for lens distortions before any computer-vision task can be performed.

Intrinsically, a camera is only able to gather a 2D projection of the 3D world, meaning that 3D information must be reconstructed by triangulation from stereo images. This induces the need for pixel matching or stereo-correlation processes in order to obtain a disparity map. This process is unfortunately computationally intensive and usually corresponds to the bottleneck of a robotic perception in its applications. Once the disparity map is obtained, it can be converted easily to a depth map by applying the relation:

$$z = \frac{bf_x}{d} \tag{3.10}$$

where z is the depth, b the baseline or distance between the two cameras forming the stereo bench, f_x the focal length along the X axis and d the disparity value.

Basic trigonometry can then be used to project the depth map into a point cloud or a Digital Elevation Model (DEM), providing dense 3D information about the scene.

3.2.2.3 Time-of-Flight (TOF) Sensors

The Time-of-Flight (TOF) sensors are depth sensors able to directly provide a distance map between the sensor and the points in the environment. They consist in illuminating the scene with a controlled light source and then analysing the reflected light. The TOF sensors can be categorised in 2 types: Phase detection TOF sensors and Direct impulsion feedback measure TOF sensors (also called flash lidar). The TOF sensors can be described by the following optical chain:

LED or Laser Diode emitting in the near infra-red band (~850nm);

Emitted signal is forward propagated to the environment surface, reflected and backward propagated;

The Received signal is optically filtered at the same wavelength of the emitted signal to reduce noise from external light sources

The signal is then mapped to a matrix sensor

The phase detection TOF sensors work as follows. The amplitude of the signal emitted is modulated at a given frequency f_{mod} and the phase difference between the emitted wave and the received wave on the sensor is measured. The phase difference is directly linked to the time of flight of the signal and then allows to deduce the distance to the reflecting object. The precision of the estimation then depends on the frequency of the emitted signal, usually set between 10 and 100 MHz. One particularity of this type of sensor is that the phase cannot be directly measured via

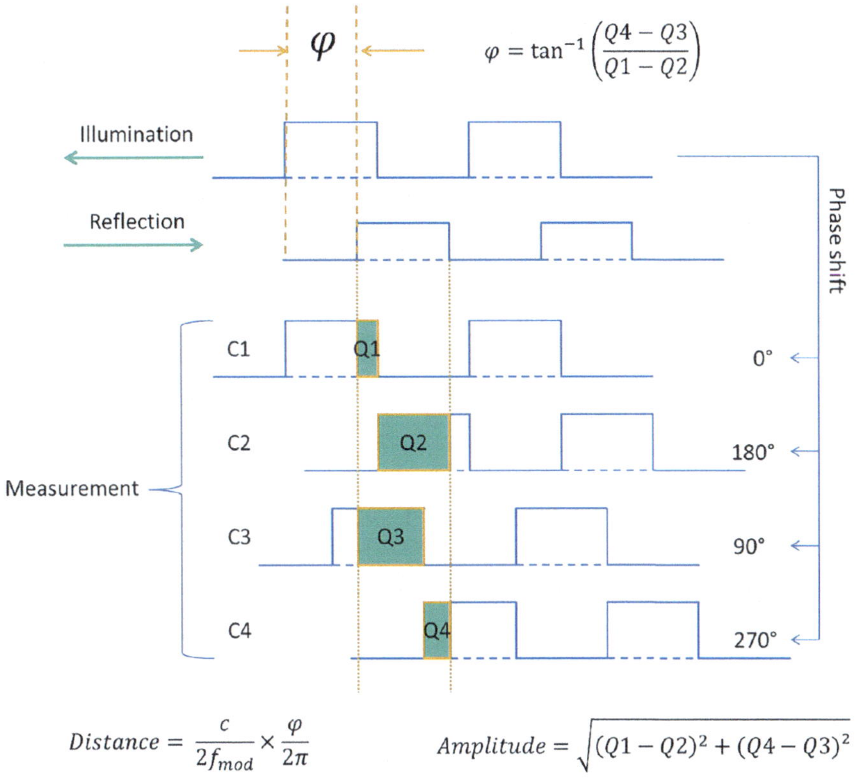

Fig. 3.1 Phase detection principle (*credits* Melexis)

a classical CMOS sensor. The method used to create depth images is then based on the use of 4 electronic shutters, each of them having a shifted phase with regard to the emitted one shown in Fig. 3.1.

The obtained 4 measurements (Q_1, Q_2, Q_3, Q_4) allow to compute the phase difference (Δ_φ), the amplitude of the received signal (a) and the distance (d) so that

$$\Delta_\varphi = tan^{-1}\left(\frac{Q_4 - Q_3}{Q_1 - Q_2}\right) \tag{3.11}$$

$$a = \sqrt{(Q_1 - Q_2)^2 + (Q_4 - Q_3)^2} \tag{3.12}$$

$$d = \frac{c}{2f_{mod}} \times \frac{\Delta_\varphi}{2\pi}. \tag{3.13}$$

The obtained phase measure is given up to 2π (phase wrapping). Then, the maximum distance that can be measured without any ambiguity is

$$d = \frac{c}{2f_{mod}}, \tag{3.14}$$

meaning that for a modulation frequency of 30 MHz, the maximum distance is 10 m.

Flash lidars are directly measuring the time of flight of the emitted signal. Two methods have been identified. The first one is based on the control of the emitting light by a door signal of period T. The sensing process is then closed to the one used by the Phase detection TOF sensors, but usually based on 2 shutters (instead of 4). Knowing Q_1 and Q_2, the 2 signals received for each emitted signal, the distance can be computed as follows:

$$d = \frac{1}{2}c\Delta_t\left(\frac{Q_2}{Q_1 + Q_2}\right) \tag{3.15}$$

with Δ_t corresponding to the time of emitting the light source.

The second method used by flash lidar is based on the emission of a laser pulse of T_{laser} duration at regular intervals. Then, the shutter of each pixel is synchronised with the rising edge of the pulse and 2 measures are performed on 2 consecutive pulses:

The first one is done with T_1 exposure time equal to T_{laser} giving V_{F1};

The second one with T_2 exposure time corresponding to the total length of the perceived signal V_{F2}.

The distance is then computed as follows:

$$d = \frac{c}{2}T_{laser}\left(1 - \frac{V_{F1}}{V_{F2}}\right) \tag{3.16}$$

3.2.2.4 LiDAR

The light detection and ranging (LiDAR) sensors are devices for remote sensing and are widely used for robotic digital mapping. They use the principle of Pulse Ranging Technology (PRT) which emits intense and focused beams of laser light pulses to a direction at a particular moment of time and then detects the reflection of the emitted laser light pulse if the light bounces back after hitting an object. PRT measures the time it takes between the emission of a light pulse and the receipt of the pulse from the reflection to compute the range or distances from the sensor to the object. As the speed of light is constant, this measured time is a proportional measurement of distance. The laser light emitter and the receiver of a LiDAR are normally mounted on a rotating device which can rotate 360° along a Z axis so that it allows a 360° scan of the surrounding environment. The laser light pulses are emitted in a rotating and controlled manner at a high frequency, for example the sampling rates can be greater than 20 kHz, or 20,000 pulses emitted per second. These calculated times and then distances allow the generation of distances information for all these points for a 360° scan, and a point cloud can be formed by representing all these points in the environment from the reflections from each angle of scanning. The smallest angular incremental angle is called the angular resolution of a LiDAR and the smaller the angle, the more points a LiDAR can generate, and denser the point clouds are for a higher resolution representation of the environment.

As a LiDAR emits light and it is therefore an active sensing system in comparison to the system which detects the information from an object based on the light sources from an external source such as the Sun. A LiDAR hence works better in an environment where there is less external light or less random light, such as night-time when there are fewer light noises in comparison to the daytime.

The point cloud generated by a LiDAR is stored in a file format called laser file format (LAS) and this is a widely accepted file format to store and exchange three-dimensional point cloud data between different point cloud processing and visualisation systems.

As most LiDARs are used on a mobile platform and if the platform moves at a reasonable speed during the LiDAR scanning process, the position where the emitter transmits the laser pulse will most likely be different from that when the receiver picks up the reflected light. This results in the receiver of the LiDAR receiving the reflected pulse at a new position. The faster the mobile platform moves, the bigger the change of position between emitting and receiving the signal. This results in a closer distance being calculated if the platform is moving toward the object in the environment. Alternatively, if the platform moves away from the object, the distance measured will be increased. This change of detected distance from the actual true distance to either reduced distance or increased distance when the laser light pulse is transmitted is called motion distortion. It is necessary to work out this distortion by applying the extended Kalman Filter (EKF) to fuse the information from the LiDAR, IMU and other sensing devices. Alternatively, auxiliary sensors are required to measure and compute the distorted distance moved between emitting and receiving the laser light pulse for each point. Sometimes it is necessary to interpolate linearly the distances for other points based on the known distortion distances. This removes any distortion for fast-moving platforms. If the speed of platform moving is not intense for an application where the accuracy of the point cloud is less critical, it is acceptable to ignore the motion distortion in the LiDAR data processing [22].

Compared with other distance measurement processes, runtime measurement is affected very little by disturbances in the measuring environment. As a result, this measurement process ensures a high level of accuracy even under tough everyday industrial conditions. The light emitter and light receiver are located in the rotating sensor head.

The traditional set of sensors will allow the application of mastered visual odometry as well as SLAM navigation techniques. The novel addition of LIDAR measurements can provide high-quality 3D environment information without the processing required by stereo visual measurements. Depth maps obtained from LIDAR can be converted into a local Digital Elevation Model (DEM). Fusion of multiple DEMs as well as wide-baseline stereo acquisition with the fisheye camera can allow experimentation with a multi-view reconstruction of the environment. The fusion of LIDAR and vision data for rover localization can also significantly improve the overall digital terrain maps and their representations.

A valuable addition will be the ability to identify partial objects (that could be partially shadowed) in 3-D by matching only some of the features detected with part of a model.

More prospective work could be investigated regarding the use of alternative sensing systems that are now or will be becoming available for robotics. These include multispectral cameras, polarimetric cameras, multi-echo Lidars and so on in future. Their use should be evaluated with respect to their added value for trafficability assessment and scientific target detection in future space exploration missions.

3.2.3 Sensing Modelling and Software Systems for Space Robotics

A space robot typically deploys a number of sensors in order to perform adequate perception tasks. Various data from different sensors require a strategy to handle all these real-time data in a timely fashion. This include the processing of each data stream from each sensor from a single perspective of the interest of the sensor and extracting the useful information of what the sensor has perceived. At the same time, it is critical to make use of all sensory information from many sensors to form a holistic perception of the environment a robot is operating in and the objects it interacts with. A technique for generating this perception is called data fusion and an example of such an approach is to define a Common Data Fusion Framework (CDFF) which aims to enable a holistic perception solution formulated dynamically and updated constantly during a robotic operation.

Perception is the main capability of a space robot and is the key entrance to decision-making followed by an action loop, upon which autonomous systems are built. In robotics, it encompasses all the processes that aim at acquiring and representing the information on the environment and the robot itself, which all come to integrate ("fuse") various sources of information available: data gathered by on-board exteroceptive and proprioceptive sensors, initial data regarding the environment, and initial knowledge pertaining to the environment and the robot itself (e.g. dynamic robot model, geological knowledge, mechanical laws of motion, etc.).

ECSS-E-40 "Space Engineering—Software" standard defines the principles and requirements applicable to space software engineering. All applicable processes to the software life cycle are defined in the standard. The standard will be tailored according to the software criticality level of the framework, simplified in order to be used by the partners easily and provided to the partners as a software development guideline. Applicable Reliability, Availability, Maintainability and Safety (RAMS) related requirements should also be identified for the CDFF in the guideline. Availability, Maintainability and Safety issues are mostly related to the application derived from the CDFF but several reliability requirements might be advised in the guideline such as; defensive programming techniques should be used in the CDFF, boundary values should be checked, all the variables should be initialized, etc.

The processes defined in the standard can be summarized as follows:

Systems Engineering: The software is considered as a part of the system that put requirements on it. The activity ensures a better transition of the requirements from system to software.

Software Requirements Engineering: The software requirements are analysed and high-level architecture is defined in this phase.

Software Design Engineering: Design of software items, coding, unit testing, integration and validation against the technical specification are carried out in this phase.

Software Validation and Acceptance: This phase consists of validation against the requirements baseline, delivery and installation and acceptance.

Software Operations Engineering: This phase comprises the preparation of software operations procedures, preparation of plans for operational testing, user support, etc.

Software Maintenance: This phase comprises software problem analysis, software problem correction, reacceptance, software migration, and software retirement.

The "Space Product Assurance" standard defines a set of software product assurance requirements to be used for the development and maintenance of software for space systems which are mainly classified as supporting and organizational processes. These requirements deal with quality management and framework, life cycle activities and process definition and quality characteristics of products.

Implementation of CDFF functionality to distributed and accelerated FPGA hardware are more easily possible by using High-Level Synthesis (HLS) techniques to convert C, C++, or higher-level program code such as MATLAB to a hardware description language such as VHDL, as has been done with several types of data processing systems including nonlinear controllers. Qualification of distributed devices such as those employing FPGAs must be done at both the component and system levels to ensure that the correct behaviour is reliably obtained.

3.2.4 Sensors Calibration

Despite advancements in sensing and data fusion, sensors still produce errors in perceiving or measuring a targeted phenomenon parameter, being distance, speed or temperature. This can be caused due to its inherent limitations such as designed sensing resolution, or constraints on its working environment, for example, if a sensor was not designed for a sensing task in space exploration, which was not foreseen at the time of the sensor design and selection. During data fusion and sensing information processing, it is also likely there are data processing errors due to the limitation of algorithms deployed.

These limitations raise the need to find ways of compensating for these errors either inherently from a sensor or computationally from data processing. Perception

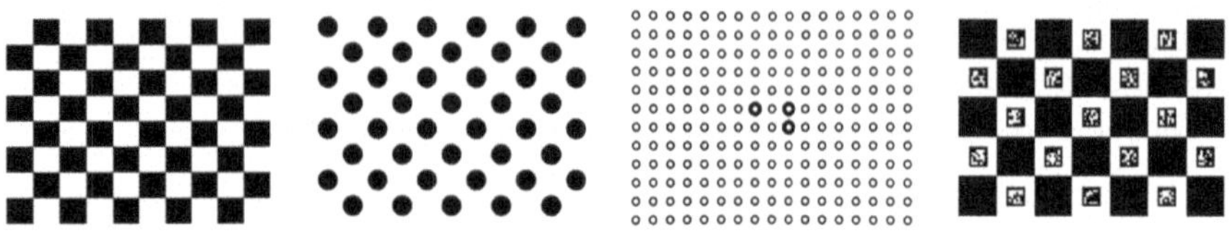

Fig. 3.2 Different pattern types for camera calibration

calibration is the process of compensating for these errors in order to provide a closer to truth view of both the environment and the objects that a space robot operates in. Both involves sensor hardware calibration to ensure that it can produce reliable and true data for what that sensor is designed to perform in varying operational conditions. Experimental tests are normally conducted to check and validate the data captured by the sensor or an array of sensors.

Perception calibration of intrinsic camera parameters is normally performed by observation of a known, planar geometric pattern in a sequence of still images. Typically, a calibration pattern, shown in Fig. 3.2, is used to calculate the intrinsic parameters of the camera. After capturing a series of frames of the calibration pattern, the camera matrix and distortion coefficients can be calculated by using a function designed to calibrate a camera. To obtain accurate results, the calibration process must cover the entire working distance and field of view of the cameras, i.e. a full set of frames shall be acquired with the calibration pattern located in different areas of the image, at different distances and with different tilts. After detection of the pattern, the extracted point positions are used in a simultaneous optimization of the camera pose corresponding to each image and its intrinsic parameters, including focal length, principal point, and distortion coefficients. The optimization process typically employs Perspective N-Point algorithms to determine an initial camera pose guess, then Levenberg–Marquardt for a final global minimization.

This allows finding the intrinsic matrix K that links a 3D point P_{cam} (expressed in the camera coordinates system) to its 2D projection on the image plane $p\prime$. In homogeneous coordinates:

$$p \sim KP_{cam} \tag{3.17}$$

$$\Rightarrow \begin{bmatrix} u \\ v \\ 1 \end{bmatrix} \sim \begin{bmatrix} f_x & s & c_x \\ 0 & f_y & c_y \\ 0 & 0 & 1 \end{bmatrix} \begin{bmatrix} X_{cam} \\ Y_{cam} \\ Z_{cam} \end{bmatrix} \tag{3.18}$$

With $[u, v]$ the pixel coordinates of the projected point, $[X_{cam}, Y_{cam}, Z_{cam}]$ the 3D coordinates of the point expressed in the camera coordinate system, f_x and f_y the focal length of the camera along the X and Y axes respectively, and $[c_x, c_y]$ the pixel coordinates of the principal point and s the skewness coefficient, usually negligible for modern cameras.

On top of the intrinsic matrix, it is possible to model the distortions occurring in the images. There are two main types of distortions, radial and tangential, caused respectively by differences in thickness along the profile of the lens and by misalignments between the lens and the sensor. These distortions are usually defined by using the Brown-Conrady model, which uses the coefficients $[k_1, k_2, p_1, p_2, k_3]$, with the k_i coefficients corresponding to the radial distortions and the p_i coefficients to the tangential distortions. The distortion coefficients are obtained at the same time as the intrinsic matrix and both are used to rectify the images before performing any computer vision task.

When working with multiple sensors, it is necessary to localise each of them in a common reference system (typically the robot base frame) to be able to fuse their data later on. This step is called extrinsic sensor calibration. There are different calibration methods depending on the sensors used and the configuration between sensors: hand-eye calibration to estimate the transformation between the end-effector of a robotic arm and another sensor, visual-inertial calibration to compute the transformation between an IMU and a camera, and multi-camera calibration. In most cases, the calibration can be made offline, supposing that each sensor was already intrinsically calibrated, and uses a calibration target (checkerboard or any other reference system/object shown in Fig. 3.2). The relationship between the sensors and the target frames is expressed as a mathematical model with several unknown variables and is solved by an optimisation algorithm in order to minimise the desired error function. Additionally, extrinsic calibration usually performs temporal calibration to synchronise the data (as sensors may have different rates).

In the case of a stereo bench, it is crucial that the two cameras are triggered at the same time to avoid this issue. Several acquisitions of the calibration target are necessary. The target must appear in both camera images and their positions must cover the whole field of view of the stereo bench to have good results. The distortion in the images is corrected using the intrinsic calibration, then the calibration target is detected in each pair of images and the 3D pose of the checkerboard points is computed for the left and right camera frames.

The following formula links together the projections p and $p\prime$ of a 3D point P on the image planes of the two cameras forming a stereo bench:

$$p^T F p' = 0 \tag{3.19}$$

where F is called the fundamental matrix such that:

$$F = K^{-T} E K'^{-1} \tag{3.20}$$

With K and $K\prime$ the intrinsic matrices of the two cameras, obtained by intrinsic calibration and E the essential matrix such that:

$$E = [t]_{\times} R \tag{3.21}$$

With R the rotation matrix corresponding to the rigid body transformation between the two cameras and $[t]_\times$ a matrix containing information about the translation part of the transformation:

$$[t]_\times = \begin{bmatrix} 0 & -t_z & t_y \\ t_z & 0 & -t_x \\ -t_y & t_x & 0 \end{bmatrix} \tag{3.22}$$

Given enough point-to-point correspondences, it is then possible to estimate the rotation and translation between the two cameras. This transformation can be used to rectify pairs of images by aligning their epipolar lines which is a necessary step before running stereo matching algorithms. The stereo matching process then allows finding correspondences between the images of both sensors to produce 3D information from pairs of 2D images.

3.2.5 Data Fusion for Perception

Data fusion is at the heart of many robotic systems, playing a key role in obtaining exploitable information on the environment which is commonly referred to as exteroception. It is the goal of a good exteroception that all necessary information on the terrain environment is fully constructed for a safe and efficient space exploration mission. This information includes the terrain geometrical information, and terrain mechanical properties relating to the traversing of a space rover. In this exteroception, a rover perceives all the above information from outside of the rover body and plans its actions and motion based on this exteroception. It is therefore important to build a comprehensive, easy to search and explore and risk-avoiding set of information normally represented in a format of maps.

On the other hand, it is also necessary to be aware of the robot's own behavioural states and this is normally referred to as proprioception in robotics. In space robotics, a wide range of data fusion techniques are required to achieve extremely challenging objectives. For planetary exploration purposes, planetary robotics require a full perception of robotic behaviour states so that this state information can be used to understand the health of the robot itself as well as predict the external environment which can be used to enhance the exteroception by adding terrain traversability information based on the friction forces captured from wheel sensors. Similarly, for on-orbit spacecraft applications of telecoms, Earth observation, science, etc., good proprioception of the robot will help to understand the health condition of the robot working in the harsh environment of the orbit where the robot experiences cyclic temperature changes which could cause mechanical failure. The robot is also exposed to in-orbit space conditions for other potential problems.

3.2.5.1 Data Fusion Techniques

While data fusion has been recognized as an independent research topic and reference area for some time, a general formal theoretical framework to describe data fusion systems is still missing due to the vast variety of data that could be fused [23]. The key challenge in sensor data fusion is the imperfection and variation of real-world data that must be consistently interpreted and fused in the context of the data captured. In addition, there are a number of flexible interpretation methods which have been proposed and developed and these large numbers of choices make it difficult to choose. These methods include fuzzy, probabilistic, possibilistic, set-based, and evidence-based systems which have been proposed to deal with the uncertainty inherent in sensor systems [23].

Data can be classified in terms of relationship: data sets can be complementary to existing ones; they can also be redundant as there are already high-quality datasets available; or they can be cooperative. Data can also be defined in terms of levels of abstraction of an environment at a low level for simple representation and computational efficiency, a medium level, a high level, or a multi-level for different purposes. Data can also be classified by extension, input and output abstraction (DAI-DAO, DAI-FEO, FEI-FEO, FEI-DEO, DEI-DEO). Another definition of multiple fusion levels has been provided in terms of signals/features, entities, situation, impact, and performance [24]. To accomplish both decentralized and centralized sensor fusion, it is necessary to have a common fusion methodology that can flexibly deal with many levels of abstraction in a modular fashion. Fusion at each of these levels at different unit scales and sampling rates, such as orbital, aerial, and ground data, is also possible through resampling or statistical methods [26]. Most satellite-based sensor fusion methods are filter-based and focused on precise position and attitude determination from multiple sources, such as radio and GNSS signals [23] or between inertial & external sensors to increase accuracy and robustness [24]. Pre-calibrated certainty grids can also be used with probabilistic confidence measures to identify features in multiple scales of observation [25]. The tier-scalable reconnaissance problem is an example of an open multi-sensor multi-scale fusion problem solvable by modular, flexible, and decentralized fusion.

A Decentralized Data Fusion (DDF) system is necessary to achieve portability and redundancy, generally comprised of a network of independent nodes that requires no central fusion centre or communications hub, using only peer-to peer data links [27]. This requires the use of distributed methods such as the Distributed Kalman Filter or Distributed Information Filter, and enables distributed SLAM and map-building such as in the case of DDFSLAM. Decentralized systems such as URUS have combined data at very distributed locations using a common framework [28].

Some of the earliest fusion methods involved pattern recognition algorithms, template methods, and clustering algorithms [29], but were limited by the number of data permutations they could be programmed with. Neural Networks that learn have also long been evaluated for sensor interpretation and fusion [30], most notably with Probabilistic Neural Networks [31] and Neural Network classifiers for determining Bayesian posteriors [32]. These methods have largely been superseded by fully probabilistic learning systems that require less computation to achieve better performance. Other common and successful sensor fusion methods include traditional averaging filters, Kalman filters, information filters, and particle filters [33, 34], for example in inertial measurements [35]. Centralized, Decentralized (DKF), and Federated (FKF) Kalman filter architectures provide multi-sensor data fusion (MSDF), as well as the addition of local fuzzy logic to the sensor inputs that allows uncertainty to be directly represented in sensor readings [36]. Fuzzy logic alone also has the capability to convert sensor readings to abstract quantities which can then be interpreted [37].

Bayesian networks have been used for many kinds of data fusion such as situation assessment [38]. The statistical information from sensor readings can also be directly fused using Bayesian methods and probability distributions in an opinion pool [39] or for applications such as navigation [40], classification [41] and detection [42], though complete Bayesian estimation may be computationally expensive. Bayesian fusion methods may also use Mixture-of-Gaussian (MoG) and Support Vector Machines (SVM) [43] as classifiers. Variational Bayesian importance sampling has been proposed to combine accurate Bayes approximations and fast Monte Carlo methods [44]. Particle filters have also been used frequently for sensor fusion [43], mainly to lower the processing requirements for accurate estimation. Variational Bayesian importance sampling (VBIS) can also be used to apply human-generated classifications to fused Bayesian data [45]. Once information has been obtained, it is often desired to autonomously act on it intelligently, which is more closely related to the more general superset of sensor fusion, known as "information integration" or High Level Information Fusion (HLIF) [46]. To draw logical conclusions from probabilistic operators, Bloch [47] proposed a classification of data fusion operators using Bayesian methods. Bayesian semantics and Multi-Entity Bayesian Networks (MEBN) also provide an uncertainty-based alternative to the conventional use of ontologies in High Level Information Fusion (HLIF) [48]. Dempster-Shafer evidence theory, which uses a more general concept of belief to achieve higher flexibility at the cost of accuracy [49], has also been frequently used for information fusion. Another proven method for information fusion is the use of the Dezert–Smarandache Theory of Plausible and Paradoxical Reasoning (DSmT).

Recently, optimization based SLAM solutions started to become more popular thanks to the new insights coming from probabilistic graphical models and the exploration of the sparsity intrinsic to SLAM problems. Strasdat et al. [50] clearly assessed the relative pros and cons of stochastic filtering and optimization-based approaches in the case of monocular-vision SLAM. While stochastic filters incorporate the measurements one by one into the state vector, optimization techniques try to find the estimate for the robot poses and map elements that minimize all the measurements acquired by the robot. One intuitive way to solve SLAM through optimization is via the graph-based formulation. It involves constructing a graph whose nodes represent robot poses or landmarks and whose edges encode sensor measurements that constrain the connected nodes. Although the graph-based formulation has originally been proposed by [51], it only became popular in recent years, especially due to new insights in the structure of the SLAM problem and advancements into the field of sparse linear algebra [52].

3.2.5.2 InFuse Data Fusion Approach for Perception

It is necessary to develop a comprehensive data fusion toolset for robot sensors (aka. Common Data Fusion Framework, or CDFF) that will serve in the context of many space robotics applications, on planetary surfaces as well as in orbit or other space environments. One such a CDFF is called InFuse, which has been developed relying on a team of experts with substantial experience. InFuse is developed by encapsulating a wide range of sensors data handling and processing techniques for perception and navigation so it is and will be able to be deployed in a wide range of robotic applications—both in space and terrestrial conditions. InFuse makes provision for a convenient and effective articulation with the other common building blocks—in particular: a European space robotics control and operating system called ESROCOS [53], a European Robotic goal-oriented autonomous controller for an autonomy framework, called ERGO [54], and a suite of inspection sensor supporting the sensing data capture and transmission called I3DS [55]. The suite of sensors I3DS is a generic and modular system addressing the needs of near-future space exploration missions in terms of exteroceptive and proprioceptive sensors with integrated pre-processing and data concentration functions.

These general requirements for the CDFF under consideration can be summarised as a framework within which one should be able to develop, integrate, blend, test and validate a whole suite of localisation and mapping solutions detailed in [56].

The CDFF must therefore embed a variety of processes, such as techniques deployed in data processing, including filtering of data sets and outliers detection and rejection, feature extraction from various data sets as an exemplified in rock feature detection shown in Fig. 3.3. All rocks are detected and marked by green squares within a visual image.

Fig. 3.3 Rock feature detection

Techniques relating to the data association, such as tracking features within the scenes of a visual sensing system, or matching a newly identified feature or object to a previously perceived or learned feature or object. Both stochastic or optimisation based estimation tools can also be used to generate useful object feature recognition and matching. It is also paramount to structure the data set into exploitable representations such as using spatial raster representations, graphs and so forth to support the efficient search of these data.

Besides, the CDFF should expose to, a robotic goal-oriented autonomous controller such as ERGO, the various spatial information it produces in a principled manner. This involves the management and incremental refinement of spatial information of the environment. For example, for the planetary exploration scenarios, the built terrain models need to be refined when drastically new localisation information is produced, after a SLAM loop-closure or an absolute position estimation after a long traverse. The CDFF must hence embed tools that make it akin to a Geographical Information System, which internally handles all the input information, and externally replies to various requests of information for planning and navigation purposes.

Finally, from the data fusion for perception viewpoint both mapping and localisation processes are mainly passive data-flow processes with the CDFF in terms of data flowing and processing. both mapping and localisation are however active processes taking place in real time in a mission, and these processes must be controlled by a robotic goal-oriented autonomous controller such as ERGO in order to make suitable decisions in actual mapping and localisation activities. Hence, besides producing map structures the robotic autonomous controller for localisation, data fusion for perception must expose models of the mapping and localisation processes for the robotic autonomous controller to reason on and make decisions in an active manner.

The solution proposed in InFuse to wrap and handle data fusion techniques and their produced data will make their adoption easy and effective by a wide range of users, both among the European stakeholders and in the wider space robotics community. In particular, InFuse not only provides access to an extensive set of robust data fusion capabilities, relevant both for On-Orbit and Planetary scenarios, but also include a data fusion orchestration and products management tool allowing to control the data fusion processes and to retrieve conveniently (on-demand) products such as maps, models of the environment or objects, science relevant data, etc. More details of the data fusion are covered in the next section in the context of robotic control with autonomy and intelligence.

3.3 Control, Autonomy and Intelligence

In robotics, localisation provides a full set of information about the position and orientation (pose) of a mobile robot in relation to a defined coordinate system at its highest level. At the robot control and actuation level, the localisation also considers many waypoints for path critical operations, environmental information such as terrain

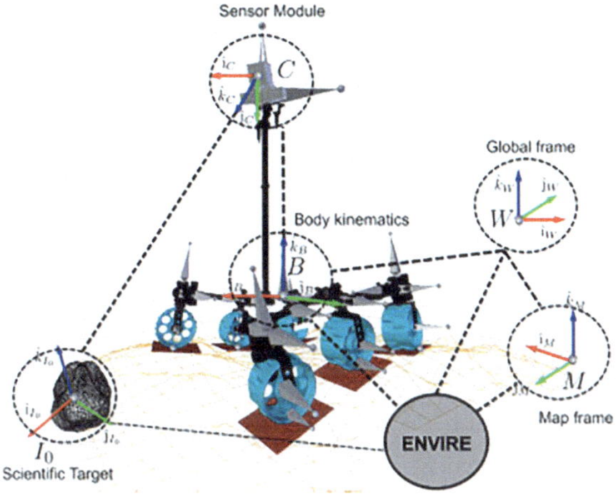

Fig. 3.4 A rover and its coordinate systems using Envire [57]—an open-source robotics library that allows to generate graphs connecting coordinate systems

maps and environmental models, and information at lowest level in actuating each of actuators in order reaching the robot goals from its current location. There are therefore a number of coordinate system frames typically defined for such a robot in order to understand the relationships among each frame and how the set of position and orientation information can be transformed from one frame to another frame. The whole coordinate frame system starts from a Global frame then to various individual coordinate frames for sensor devices as well as for actuators. The global coordinate frame needs also to relate to a target coordinate which is itself also a new coordinate frame in order to facilitate to perform more accurate handling of the target object. It is also necessary to relate the global coordinate frame to other objects in the surrounding environment. One example for a six wheeled rover is shown in Fig. 3.4: with a global coordinate frame marked as well other coordinate frames defined in order to control the rover to reach its targets.

For space robotics, localisation is required to:

ensure lowest level (locomotion and motion) controls;

ensure the proper execution of paths / trajectories;

ensure the spatial consistency of the built models (terrain maps, reconstructed 3D model of the objects to interact with);

be optimized to be compatible with the limited resources of the space qualified processors which typically have less memory space, minimal requirement for processing capability in terms of computing power, in order to produce light spacecrafts for cost-effective space missions;

ensure the achievement of the mission ("goto [goal]", “seize[object]”, etc.).

These functionalities impose different requirements to each of the criteria that define a localisation solution. A number of theoretical and technological approaches are required in order to maximise each of the criteria to advance the solution to beyond the state of the art. Hence a variety of localisation solutions with varying performance for each criteria need be developed and integrated.

3.3.1 Perception Processing

Based on the requirements of understanding the surrounding environment for either planetary exploration or on-orbit servicing or other on-orbit missions, it is essential for a space robot to have sufficient information in order to process them to achieve expected perception of the environment. For example, for an on-orbital operation, it is necessary to estimate through the generation of the position and orientation and both linear and angular speed from associated derivatives of a target satellite with respect to the chasing servicer. The above information will enable the robot have clear understanding of the objects in this case a satellite with the necessary accuracy and frequency to plan and execute a number of manipulations such as approaching the satellite for servicing. This requirement of perception continues during a space mission and can be required from a long distance range, to a middle range then to a close range in order to provide full and continuous perception.

The state estimation requires visual information, e.g. in the form of passive camera images or active range measurements. Due to the complexity of active range sensors for space, passive cameras are normally a valid option, that require little mass and power. For far and mid-range, a simple and single point range measurement might be an option. The correct choice of the sensor suite depends very much on the actual target and the specific mission.

In the case of camera-based operation, well illuminated images with good level of contrast are required as input in order to allow the detection of certain features or key-points on the target. Additionally, a guess of the target state from previous measurements, e.g. from ground based tracking for the far-range case, is important for the initialization of a timely perception processing. A good initialization can normally increase the success and accuracy of the target pose estimation.

At a close range, two cameras can be arranged to operate in a stereo vision configuration, which can make the perception more reliable. Alternatively, a single camera can be used to take multiple images for fast perception processing and as a result, this approach tends to result poor perception and be less reliable due to a lack of image data from which less useful information can be produced. These images however can still be used to build up a point cloud over time of a moving target which can then be compared with a model shape to determine the identification and orientation of a target.

A typical and generic flowchart of perception processing is illustrated in Fig. 3.5, through which the position and orientation of a target object within a surrounding environment can be estimated. The process consists of perception data capture module which consists of a 3-dimensional LIDAR, and three types of cameras including mono-camera, stereo-camera and a colour RGB camera with depth information generated at the same time.

A Kalman filter can also be used to improve the accuracy of relative object tracking by making use of ego-motion estimate from the camera as well as the movement of the tracked point cloud. Both Extended Kalman Filters (EKF) and Unscented Kalman Filters (UKF) can be used, though the UKF provides higher model flexibility and

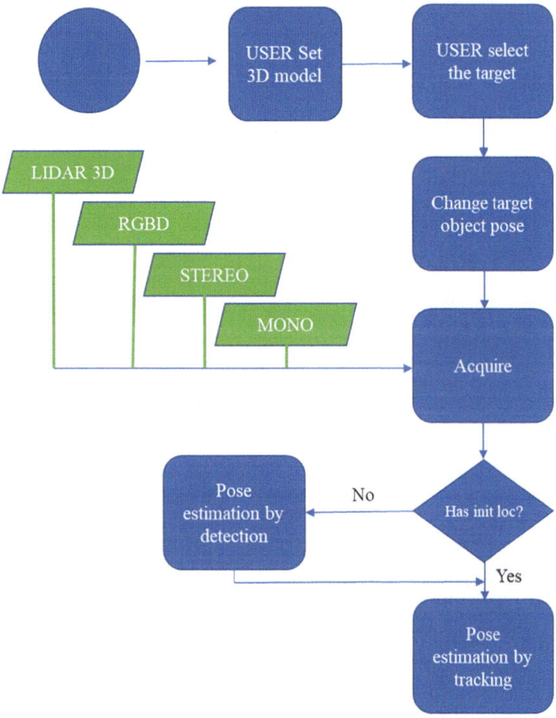

Fig. 3.5 A generic flowchart of the target state estimation for perception

better performance in exchange for higher computational load. The Cubature Kalman Filter (CKF) is a variant of the UKF with better performance in higher-order systems (Fig. 3.6).

The reconstruction and identification methods are highly processing intensive due to the amount of matching required to triangulate points and identify three dimensional point clouds (usually RANSAC is used in both cases) but can produce high fidelity point clouds of a target as long as enough visual features are available. As several algorithms are available at each stage of feature detection, triangulation, reconstruction, and identification, trade-offs can be made in terms of the computational time required versus the quality of results obtained. Table 3.1 provides a trade-off comparison of several of these options available.

In summary, the point cloud reconstruction methods are described below:

Performance: Centimetre level accuracy at close range with high resolution hardware. Accuracy decreases with triangulation distance and resolution, and is least accurate in the Z (away from camera) direction.

Maturity: This method has been tested using laboratory images of spacecraft movement with a simulated spacecraft movement model

Portability: Based on OpenCV functions and reproducible on other platforms as long as memory and processing power are sufficient for near real time operation.

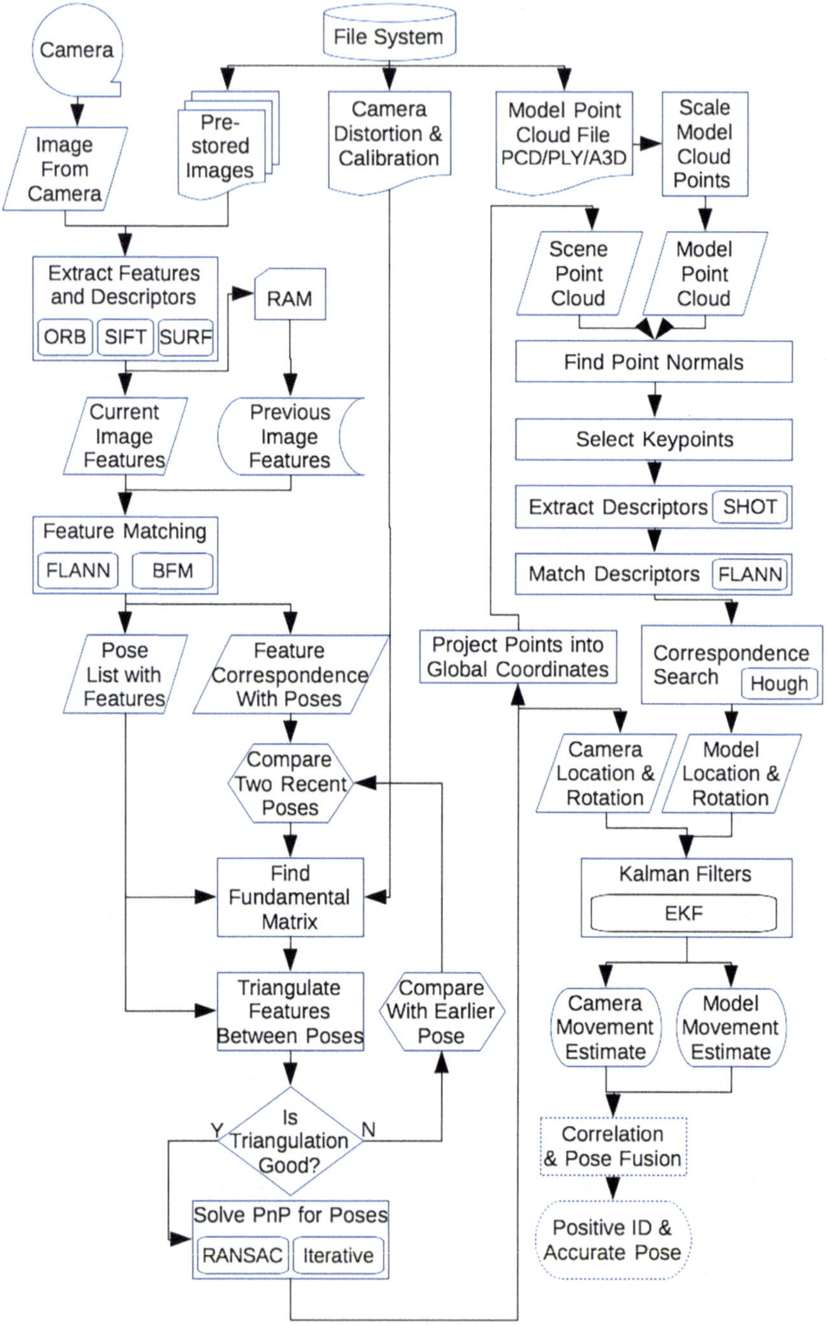

Fig. 3.6 Generic flow chart of 3D target reconstruction and tracking using 3D model-based point cloud recognition

Table 3.1 Trade-offs between different algorithms for the point cloud reconstruction process

Operation	Advantages	Disadvantages
Feature detection using ORB	Open-source, fast, invariant to rotation	Lower accuracy and scale invariance than SIFT/SURF
Feature detection using SIFT	Good quality and commonly used	Patent-encumbered, high computational load
Feature detection using SURF	Faster than SIFT and almost as good quality	Patent-encumbered, not as fast as ORB on fixed-point
Single-camera comparison of images	Less hardware needed	Triangulation harder and less reliable than multi-camera
Dual or Multiple-camera comparison of images	Better triangulations, more reliable reconstruction	More hardware with known baselines needed
Iterative (brute-force) matching of features, clouds, and PnP	Reliable and simple	Very high computational power needed
FLANN (Fast Approximate Nearest Neighbour) matching	Much faster for features and very accurate	Slightly less accurate but not noticeably
RANSAC (random sample) matching	Much more efficient than brute force, almost as good	More complex method, very occasionally incorrect
Extended Kalman Filter estimation	Well known, efficient, and robust estimator	More complex to design and not as flexible or high order
Unscented/Cubature Kalman Filter estimation	More flexible and performs for higher order systems	Higher computational load

3.3.2 Understanding of the Environment

In addition to define the perception process and a framework to enable such a perception, considerations are also need to be made on the representation of the environment where a space robot will operate, since the various functionalities that require them impose diverse requirements. This is particularly important for the case for planetary exploration where autonomous rover navigation require construction of a more realistic representation of the environment for safe and efficient space exploration missions. This is particularly important for the case of planetary exploration where autonomous rover navigation operations require the construction of a more realistic representation of the environment to improve the safety and efficiency of the missions. For instance, different layers of information about the terrain can provide critical information needed for dynamic motion planning, where different layers of information on the terrain would provide critical information for dynamic motion planning. Similarly, for orbital servicing, different layers of information on the objects to be tackled in the working environment are required for safe and reliable servicing of a spacecraft.

In principle, it is fundamentally more beneficial to create a new multiple sensing based common data framework in which data fusion of multiple sensing is enabled to make the understanding of environment possible. One such an approach has been proposed and implemented in the InFuse project, where such a common data fusion framework (CDFF) has been designed. InFuse is conceived to address Navigation (localization) and Perception (including map making, in particular) as standalone solution to space robotics [56].

InFuse was developed by a team of very knowledgeable and experienced researchers with the main existing "robotic operating systems" (ROCK, GenoM, ROS…). The consortium partners confidently worked with different solution options in order to select the most sensible solution, considering the approach proposed by the consortia working on the other related projects on robotic operation system and planning systems.

The main objective of the CDFF is to provide a framework for the development, evaluation and deployment of space robotics data fusion technologies in the context of the developing future European robotic solutions for space missions.

The InFuse common data fusion framework is illustrated in Fig. 3.7, and it is constructed in a modular fashion by bringing two key components namely the InFuse CDFF-Dev and CDFF-Core, and three interfaces into the data fusion framework. The CDFF-Core provides the "core" data fusion set of the state of the art algorithms and techniques currently in use and applicable to the planetary and orbital applications, as a collection of ready-to-use libraries or packages. These libraries represent the CDFF's processing methods necessary to fuse sensory data. The CDFF-Core is implemented in a modular fashion with a common interface to allow high flexibility in configuration as well as in operation as a distributed system on multiple platforms. CDFF-Core includes a number of modules for filtering initial raw data sets from various sensors, detecting outliers of features and rejection of false outliers and they are represented in purple colour in Fig. 3.7. In addition, data fusion modules for both planetary applications as well as for orbital applications should be included in order to make the data fusion common framework more generic and robust to deal with various application demands. The InFuse CDFF-Dev includes supporting tools for the orchestration, data product management, data product visualization, and generic tools dedicated to data filtering and outliers removal.

Within CDFF, there are two types of modules and they are data fusion nodes (DFNs) and Data Fusion Processing Compound (DFPC). A DFN is an atomic processing entity that fulfils a given basic function. It is the smallest unit of a complex task defined by its function, input and output. However, a DFN can be defined by a combination of elementary functions which may not expose their inputs and outputs. DFPC is a combination of several nodes designed to provide a certain data product such as pose estimation, map generation and so forth. Within a DFPC, suitable DFNs are identified and then assembled for a particular application. These chosen DFNs are then sequenced based on the mission tasks and can be triggered via pre-defined scripts, which are configured according to the parameters associated. The DFPCs are generally required to be defined for both the planetary and orbital reference implementations due to different environmental challenges and focus of interests. While both implementations mostly require different DFPCs, many DFNs can be shared between the two targeted contexts and this promotes the reuse of common DFNs for more efficient perception processing. The DFPC defines the connections

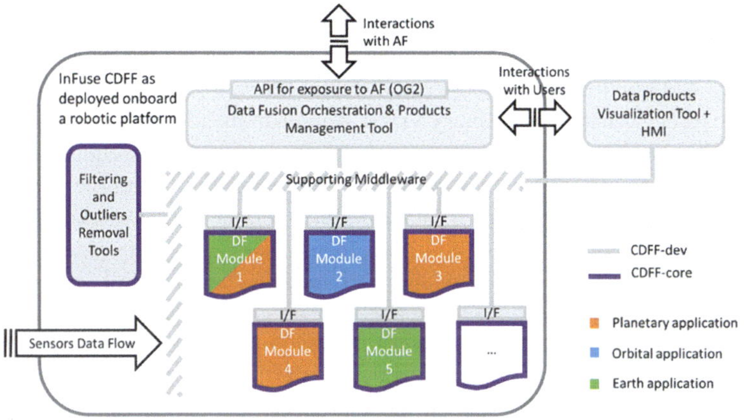

Fig. 3.7 Illustration of InFuse CDFF-Dev, CDFF-Core, and modular approach to data fusion processes

between input and output ports of different data fusion nodes (DFNs). A DFPC states the DFNs involved in the DFPC, their connections and the processing order. This defines the steps to produce a data product from a set of data product(s).

The orchestrator is an important functional component in the CDFF that has the main task of receiving queries from a robotic goal-oriented autonomous controller such as ERGO to activate certain Data Fusion Processing Compound (DFPC). The orchestrator provides the fused data products to a robotic autonomous controller in the desired format. It acts as the central coordinator in the target system to control the activation states of DFPCs.

One interface allows the sensors data captured to flow into the CDFF for processing. The interface on the top enables the CDFF to be exposed to by a robotic goal-oriented autonomous controller such as ERGO, and the final interface enables users to interaction with the various data products which can produce various information for users' better visualisation of the perceived environment or the objects within this environment.

Through the CDFF-Dev utilities, the developer will be able to: (1) connect data fusion nodes(DFNs) and evaluate them with logged data, (2) define DFPCs and evaluate them with logged data, (3) test the orchestrator with logged data, (4) test the DPM tool with logged data, (5) run performance analysis to evaluate at least data filtering and outlier removal nodes and (6) visualize data communicated through DFNs using the Envire visualizer. Envire is an example system which aims to be a generic library for environment representation and to provide a common interface for different data fusion algorithms [58].

The core of Envire essentially provides a directed graph with spatial–temporal relations between its nodes [59]. Each node is able to store environmental objects such as sensor data such as point clouds, visual features, etc., or local maps including elevation maps, octomaps, multi-level-surface maps, etc. All these sensor data sets and maps are either represented directly as Envire-type or by third-party libraries via lightweight wrapper classes. Envire graphs and entities can be serialized for logging/replaying and communication between different hosts, and they can be visualized using the Rock Visualization Library Vizkit. A main motivation is to ease the interaction of different algorithms to work on shared data structures, as is required for many high-level data fusion problems, especially if information from multiple sensors is to be merged. This is also important if robots perform complex mission scenarios where different sub-systems intervene and need to interact. These sub-systems include perception, planning, internal simulation, telemetry, ground control operation, and so on. Sharing of data structures happen with near-zero overhead, assuming that the interfaces of the involved algorithms are sufficiently compatible. Envire is currently well-integrated into the Rock robotic framework [60]. The integration of the MARS simulation tool [61] is currently under development.

Additionally, CDFF-Dev provides a Middleware Facilitator which eases the transition from the developer environment to the target system. To illustrate more specific targeted requirements for two types of space exploration missions, it is necessary to define specific requirements for these missions so that a clear set of purposely defined common sensing data fusion framework core functions can be defined below.

3.3.2.1 The Core CDFF Specifications for Planetary Exploration Scenarios

Main planetary track reference implementation scenario involves the deployment of simultaneous localization and mapping (SLAM) with regular sensors equipment, possibly with variations. Input data from IMU, wheels encoders, wheels currents (or torques sensing), nav-cams/stereovision (covering also visual odometry), 2D or 3D Lidar, sun tracker can all be used to build a map using SLAM so the rover can localise itself autonomously within the map.

For this localisation-related scenario, a back loop trajectory should be considered, that integrates visual odometry (or sliding-window SLAM) throughout the trajectory An absolute localisation technique from places to places, a loop-closure detection when returning to the start position, and, off-line, a reprocessing of the data using an optimization scheme can also be used so as to re-compute as precisely as possible the overall trajectory, and hence rebuild an as precise as possible terrain model.

Complementary scenario for planetary exploration mission P1 is to achieve an efficient and reliable detection, identification and tracking of a landmark feature such as a rock in an unstructured environment such as Moon or Mars;

Complementary scenario P2 considers the challenges of perception and navigation based on the use of alternative sensors. This scenario would focus on the perception part, taking localization for granted assuming that a GPS RTK ground truth can be truthfully maintained. A digital terrain map (DTM) is need to be built to show the various layers of terrain information and various kind of maps along a trajectory. These maps include DTM, traversability map, objects of potential scientific interest enhanced with the associated detailed and precise models.

3.3.2.2 The Core CDFF Specifications for Orbital Exploration Scenarios

For the orbital operations, the main track reference implementation scenario deals with on-orbit servicing, rendezvous with an uncooperative free-flyer whose 3D model is known and available a priori, approaching, catching and stabilization. Inputs data from: IMU, star tracker, camera/stereo, radar, Lidar, time of flight camera, manipulator force/torque sensors, manipulator encoders.

Complementary scenario for orbital mission O1assumes the target spacecraft is still operating. Hence a rigid connection has to be established for: Lifetime extension by refuelling the satellite, or repair/upgrade by hardware exchange of key components such as computers, communication components, solar panels, or adding other satellite modules to enhance its functions.

Complementary scenario O_2 is about a fly-by of a comet or asteroid, during which, it is aimed for a clear detection of such a comet or asteroid, followed by an identification of key areas of interest before tracking these areas. Based on the visual image data sets it is possible to construct a 3D computer model of the comet or asteroid.

For all space exploration missions, data fusion frameworks such as InFuse have the capability to reconstruct three-dimensional models of objects in view of one or more sensors from distinct features and position information. These objects can be tracked over time using both two and three dimensional recognition algorithms and pose can be identified with respect to the coordinate system of a model provided.

The lighting and shadows are the main areas to be further investigated in the orbital scenario, particularly the shiny smooth surfaces and specular reflection due to multilayer insulation (MLI). Furthermore, a poorly illuminated target in case of an eclipse or severe shadow condition, can be compensated by an additional source of light (target illumination system). On the other hand, we envision to avoid total saturation of visible camera images, supported by mission planning in such a way that the on-orbit serving could be performed during favourable lighting condition.

3.3.2.3 Data Fusion Techniques in Planetary Robotics

Digital Terrain Maps are the core maps of the rover environment, that can be used for a variety of purposes. The primary need they fulfil is for a rover planning tool to plan short distance paths, to detect obstacles while navigating, and to define navigation maps upon which long distance itineraries such as path defined with waypoints, speed of travel and stopovers for scientific mission tasks can be planned. Additional uses can be locomotion monitoring, or to build Localization Maps used in the "Getting Back to Base Location" scenario.

The core structure is a regular Cartesian grid to each cell of which elevations are assessed, built upon a point cloud. This representation can be complemented by information related to the soil (Soil Type Map), extracted from luminance information associated to the point cloud.

To produce the Soil Type Map, luminance information (in the case of stereo-vision) can be exploited. The literature abounds with classification processes that exploit texture/colorimetric attributes, the most classic ones (supervised classification schemes) requiring bounded processing time and memory, and hence being portable to space-grade CPUs. An example of a digital terrain map is shown in Fig. 3.8, which is built using stereo vision system to capture a sequence of three dimensional image data to construct the base digital map. The map is divided into many cells, for each of which, a luminance values is calculated to understand its soil type and such values are then added to the soil type map layer of the multiple layered map to represent the cell in colour to indicate potential challenges in traversing these colour areas in the digital terrain map and they should be avoided when planning. This fused map, consisting of the initial digital terrain map and soil type map, can form a critical navigation tool containing spatial and terrain surface mechanical property related information.

The choice of the attributes to compute and the selection of the classification approach to apply (naive Bayes, adaboost, SVM, random trees…) can be made according to a more detailed definition of the classes of terrain relevant for navigation. The detailed considerations include terrains such as sand, gravel, small rocks, rock outcrops and so forth. The specific techniques used for classification problem at hand should also consider the amount of data required and if high computational capability is expected, especially for the latest state of the art approaches. More relevant to this map construction is the exploitation of the geometric information of the point clouds to compute additional attributes that can yield more precise estimation of soil types, allowing in particular the use of prior knowledge that relate the soil geometry and type. Typical classic classification processes enable such an extraction of soil type information.

(a) (b)

Fig. 3.8 Illustration of Digital Terrain Maps built with a sequence of stereovision data. A luminance value has been associated to each cell (colours encode a notion of traversability), **a** an overview of the terrain of interest for long-distance planning purpose and **b** a local terrain map for detailed planning

The situation is different regarding terrain classification using Lidar remittance and reflectance values for the Lidar, as it is a problem that has not been thoroughly studied. This is mainly due to the lack of Lidar delivering calibrated remittance information in the roboticist community, such Lidars have only been available recently and open avenue for further investigation. While on a varied earth scene, this information has shown to allow the identification of *e.g.* vegetated vs. mineral soils, the benefits of the use of such information in a planetary context may be limited and remains to be assessed. This will be made by evaluating the application of classic classification processes on data acquired on representative sites of interests for space planetary exploration.

Both approaches using stereo cameras and Lidar to generate soil type maps are detailed in Fig. 3.9, where initial data from two types of sensors are captured but different image information is extracted based on luminance in case of stereo cameras, or reflectance information for Lidar. This soil type information is then classified using the soil type classification to generated classified image before the soli type is determined and stored in the digital terrain map also referred as Rover Map here.

3.3.2.4 Data Fusion in Orbital Robotics

The state estimation for orbital robotic operations requires visual information, e.g. in the form of passive camera images or active range measurements. Due to the complexity of active range sensors for space, passive cameras are a valid option, that require little mass and power. For far and mid-range, a simple, single point range measurement might be an option. The correct choice of the sensor suite depends very much on the actual target and the specific mission.

In the case of camera-based operation, well illuminated images with a good dynamic range of contrast are required as input in order to allow the detection of certain features or key points on the target. Additionally, a guess of the target state from previous measurements, e.g. from ground based tracking for the far-range case, is important for the initialization of space object perception. A good initialization can increase the success and accuracy of the target pose estimation.

At close range, either two cameras operating in a stereo vision configuration (more reliable) or a single camera taking multiple images (less reliable) can be used to build up a point cloud over time of a moving target which can then be compared with a model shape to determine the identification and orientation of a target.

A typical flow chart of data fusing process is shown in Fig. 3.10. In this case, external environmental data captured using a 3-dimensional Lidar which are complemented by three types of camera systems: namely a monocular camera, a pair of stereo cameras and a colour RGB-D camera with depth information. This image information is fed into a data fusion system such as CDFF proposed and developed in InFuse for data fusion. Objects can be recognised and reconstructed using 3D point cloud generated from these images. Based on the recognised object, their position and orientation within camera coordinate system can be estimated. During this process, sometimes it is necessary to allow or need human intervention in order to register

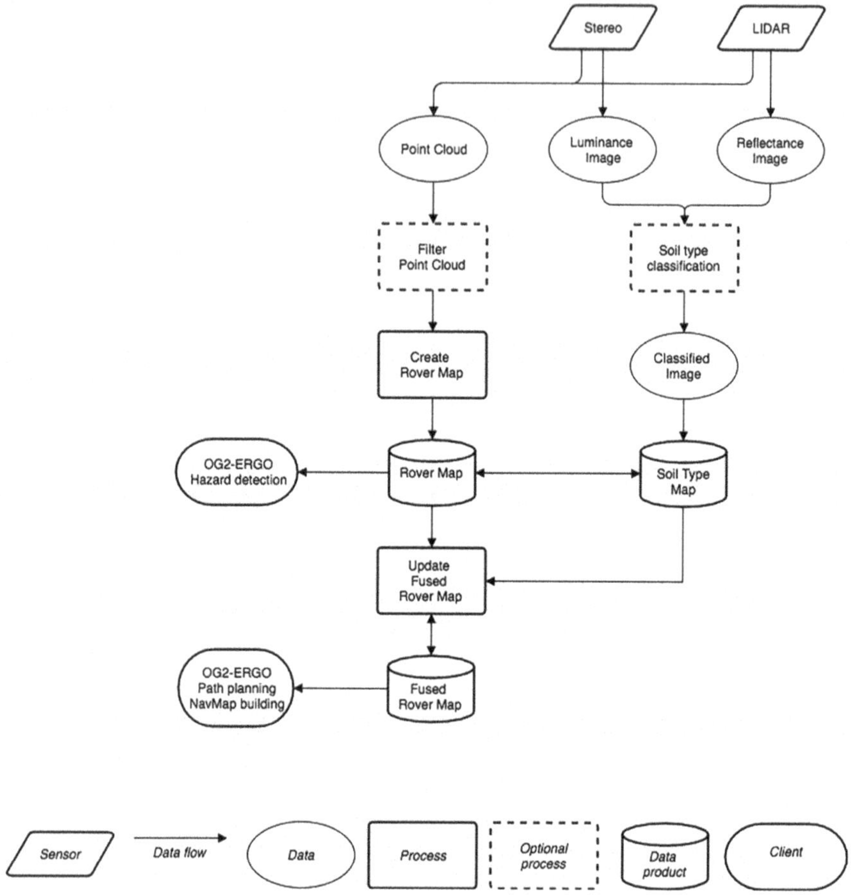

Fig. 3.9 A flow chart of the processes that generate the digital terrain map and the soil type map

features from images where some algorithms will struggle to recognise certain critical features due to poor image data captured, or in some cases automated algorithms may have missed some important features for a complete detection of objects or their anomalies within a target working environment.

There exist several methods of tracking which exploit 3D model and image features or pixel intensities. Based on a comparative analysis detailed in Table 3.2, and the stringent constraint of resources (power, speed, memory) for space applications, the edge-based tracking approach is considered to be the most suitable technique for on-orbit servicing. Hence, the reference implementation for the CDFF focuses mainly on the edge-based methods due to the fact that space objects are man-made and textureless, and spacecrafts hence can provide abundant image edge based features for their recognition. Feature-based tracking can also be kept as an option for more

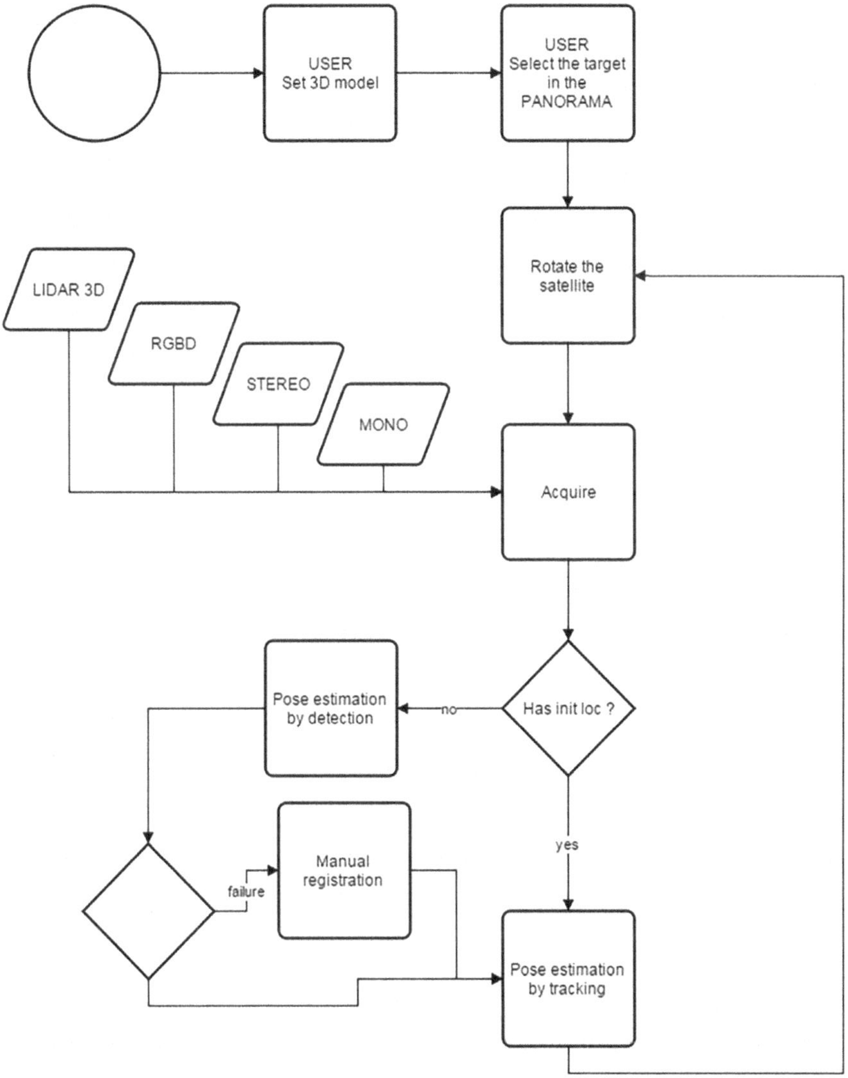

Fig. 3.10 Generic flow chart of the satellite state estimation

high performance and high memory processors to produce high resolution point clouds of objects for more accurate space object recognition and tracking.

For orbital robotic applications, the lighting and shadows are the main challenges for a suitable technical solution. On-going investigation in the orbital scenario is still being conducted by researchers, particularly for the objects with the shiny smooth surfaces and specular reflection due to multilayer insulation (MLI). Furthermore, a poorly illuminated target in case of an eclipse or severe shadow condition, can also pose challenge and may need to be compensated by an additional source of

Table 3.2 Comparison of image features to estimate the target satellite state

Features	Advantage	Disadvantage	Tracking method	Remark
Key points	Easy to detect Small computational cost	Few in number and unstable due to reflections Hard describe invariantly	Model and 3D sensor data from stereo registration	Can require significant processing time depending on size and precision
Edges	Available abundantly on spacecrafts easier to match model and image Faster	Ambiguous for symmetrical structures	Model-image edge matching along edge normal	Most spacecrafts consist of smooth structures and are poorly textured objects
Hybrid features	Exploit both edge and keypoint features Robust	Computationally expensive for tracking	Integrating colour statistics and intensity edges	Requires high quality camera

light using a system such as target illumination system. During the on-orbit mission planning stage, it is also possible to plan properly to avoid total saturation of visible camera images, and the servicing will be conducted and scheduled in such a way that the on-orbit servicing could be performed during a period of time when a favourable natural lighting condition is available.

3.3.2.5 Activity Planning for Autonomy

For planetary space mission, it is essential to plan exploratory mission activities ideally for autonomous and safe exploration. Based on the established digital terrain maps, a mission planner can then start to define exploration mission activities, prepare these mission activities and finally execute these activities.

At the mission definition stage, the objective of the mission for a space exploration rover is to reach autonomously a set of targets located at a certain distance away from the rover's starting point, defined by the absolute coordinates of these target locations. The mission is defined by an itinerary of a combination of exploration tasks and waypoints specified for the traverse. Two types of constraints are also defined in order to satisfy the mission requirements. These constraints define the technical requirements for the rover to achieve when it reaches the targets:

Localisation constraints: each part of the itinerary or more specifically each waypoint for the rover to reach is associated to a requirement on the precision of the localisation, that is how accurate is the final location of the rover as well as how accurate of its orientation once the rover reached a waypoint;

Time constraints: the itinerary is specified with bounds on time to satisfy.

- The itinerary path of the rover can be specified either by a corridor in which a bounding area is defined in order to ensure the rove will traverse within that bounding corridor, or a series of waypoints in which case there is no requirements for the rover to follow between waypoints.

During mission preparation, a human operator on the ground normally used or possibly assisted with planning tools to prepare the mission by studying the mission requirements and mission constraints. The mission planning operation needs to consider a range of information, such as terrain traversability, sun illumination, possibility to absolutely localize the robot. This information come from orbiter data and a priori knowledge from previous orbit survey missions or can also encompass information gathered by other prior space rover missions. All the information can be used to build models to support such a mission preparation process. From these models, it is possible to identify a series of constraints which need to be satisfied. With such clearly identified constraints and mission aim in terms of carrying out the specified tasks and within specified mission duration, the mission planner could use optimisation tools to optimize the mission through simulating the execution of the mission tasks in a simulation environment by investigating several what-if scenarios, whilst meeting the criteria identified.

Mission execution phase is when a well-considered and optimised mission plan is executed on the rover to embark on an exploration mission. The rover follows a nominal navigation cycle, which consists of two phases: namely a *short-term path planning* phase and a *trajectory following* phase, both making a sequence of several navigation cycles. Short term path planning phase requires the access to the "Fused Rover Map" or also known as the Digital Elevation Map structure. Based on the information stored in this map, it is possible to generate a path for a short distance navigation that focuses on reaching the target position or a defined length of a planned path.

The *trajectory following* phase is about to ensure an adequate execution of a navigation plan by following a predefined trajectory within a corridor. One key mechanism for this phase is to select a suitable frequency of updating the localisation information after the rover starts navigate along a trajectory path defined. The more frequent of this updating, the more accurate and closer to real-time the localisation information is updated for a better navigation where objects within the environment might move quickly and so is the rover. Associated with a higher frequency of updating localization information and maps, it also demands more computational power to process all the data captured. For applications such as space exploration in deep space, where human interaction becomes even more challenging due to the time latency in transmitting data, and computational power may be limited, it would be sensible to have a lower frequency of updating. The frequency is also measured in Hz and one example could be using 10 Hz for updating. Within the *trajectory following* phase, it is also necessary to execute a trajectory that the rover can avoid any potential hazards, such as rocks, pits and so on. This capability requires the production of the "Rover Map" which takes as input a point cloud registered in the rover coordinate reference frame, either produced by stereovision or Lidar, and sets the elevations for the map points. The elevations of the Rover Map can be associated with uncertainties if this information is more probable than deterministic for a given scenario.

3.3.3 Multiple Robotic Planning and Reconfiguration with Reconfigurable Multi-Robot System

This planning scenario involving multiple robots is the most complex in comparison with the single space rover exploration, and it is therefore also the most prospective. The definition of multiple robotic planning and reconfiguration could be and remains quite open as this is very much space mission dependent or depending on the scope of a research project. Chapter 11 deals with the reconfiguration of multiple robots. Instead of a complete and complex scenario, this section briefly depicts some cooperation schemes, aiming at defining the requirements for such a planning, involving multiple robots for planetary exploration. For orbital applications, multiple robots cooperation remains much in the research stage and this will be updated in future.

The consideration of such schemes is here mainly to assess that the actual definition and achievement of such scenarios can be supported in a normal space robotic application such as within InFuse framework. This approach will not require to update an existing system such as the InFuse architecture or to develop brand new functionalities. The structure of this section is therefore less precise that for the other scenarios.

3.3.3.1 Perception and Navigation Functionalities for Cooperation

Cooperative schemes can be defined to achieve the whole variety of missions by a group of space robots in cooperation with each other with a common goal defined for the mission. A single rover can be deployed and can only be tasked with relatively simple tasks and these tasks can only be executed singly and slowly. Complex exploration tasks such as exploring and mapping a large area, or in a long-range traverse exploration, or a declination of the getting back scenario, can be performed by a group of robots, among which one robot can copy and execute a trajectory formerly executed by another to improve the overall robot group's efficiency. The cooperating robots may be heterogeneous, e.g. a UAV can act as a scout for a rover, and the cooperation schemes are of course not limited to two robots. It is clear however the more robots are involved in such a cooperation, the more complex the coordination and planning for an efficient cooperation. Without going into further details, the following functionalities related to perception on the terrain modelling and navigation based on the localization that can be used to build cooperation scenarios can be considered:

Fusion of data gathered by the different robots into a Digital Elevation Map become possible and could be advantageous for such a swarm of robots in such an exploration scenario as much as more information can be captured by all robots involved. Here the fusion can be made along a tight scheme by incorporating point-clouds into the existing digital terrain map (DTM) and in some cases, the map may require updates as more accurate information of the terrain are acquired as each of the robots starts to explore and capture close-up information. The data fusion can also take place on a loose scheme through integrating DTMs independently produced by the robots with their own-acquired data. Note that the data/DTMs need to be registered beforehand as detailed below;

Inter-robot localization. This can be achieved by two different means:

- Direct localisation of one robot with respect to the other when both robots are in sight of each other;
- Indirect localisation, by registering point-cloud data of DTMs, or point-cloud data with a DTM. This can be achieved even when the robots are not in direct sight.

This inter-robot localization information can be used to refine each robot position and orientation estimates for better overall cooperative scheme generation and updating to ensure safe and effective cooperation among all robots.

As an example, two robots cooperate in their collaborative perception and discovery of a new terrain of interest as shown in Fig. 3.11 where two rovers cooperate with each other to explore a new terrain of scientific interests. The terrain information captured and processed by each robot can be stored and shared among all members of the cooperation team for an exploration mission. The shared information can be further fused and merged into a global digital terrain map with other terrain information for more informative exploration. For example, if a rover covered a certain area within the exploration region with some samples collected and later on there is a further interest of the samples collected from the region by scientists for further investigation, this prior exploratory information can be made available within an expanding map to facilitate a better re-investigation mission planning by another rover with better or different type of capabilities for further investigation. This could imply in collecting more samples for in-situ analysis at a higher resolution of sample sensing for more accurate measurement of mineral chemical composition of the sample. Alternatively, scientists might be interested by an associated chemical compound information of the sample so a new type of sample sensing capability may be required from a cooperating rover for exploration.

To extend the capability of perception further for wider area exploration, it is also pertinent to deploy unmanned aerial vehicle (UAV) as a member in a cooperative exploration swarm of robots. In this scenario, the UAV can fly at much higher speed to cover a much larger area of terrain within a short period of time. During its flight, the on-board sensors can capture terrain related information at a certain height with fast changing scenes normally. In comparison with the ground based rover, the information coming from these on-board sensors is of a nature of higher throughput and lower resolution. This information is therefore normally useful for high level lower resolution based perception.

It is necessary to consider the requirements of fusion of data sets captured from a UAV to those captured by a ground based rover due to significant difference in

Fig. 3.11 A multiple robot cooperation using different perception sensors mounted on each robot (*credits* CNRS-LAAS)

their coverage and perspectives and the speed of travel when data are captured. Figure 3.12 shows an example of deploying a UAV in cooperation with two rovers shown in Fig. 3.11 in a field trail of capturing terrain information in a Mars analogous terrain in Morocco.

Fig. 3.12 A multiple robot cooperation using perception sensors mounted on a unmanned aerial vehicle which is deployed to enhance the perception of the area terrain of interest, in particular covering a larger area in terrain perception (credit: ESA)

3.3.4 Space Qualification of Data Fusion Based Autonomous Control Software

Many traditional space robotics missions launched by several space faring nations or space agencies, past and in development, have relied in specifically developed software that absorbs a large percentage of the mission cost and allows very limited reuse. The data fusion architecture and algorithms are developed and verified for each individual mission, and no commonality is sought between different mission scenarios such as planetary rovers, in-orbit assembly and servicing, planetary landing and so forth.

The qualification of software implies verifying that the software meets all its requirements, is compliant with the specified interfaces, and has been developed, validated, documented and controlled following a process compatible with the intended quality. Qualification is therefore linked to the software requirements, as are the desirable properties of Reliability, Availability, Maintainability and Safety (RAMS). In consequence, it is not possible to state that a software framework such as the CDFF is qualified, only that it is "qualifiable", meaning it is intended for the qualification of the software products derived from it.

The "qualifiability" of the CDFF can be addressed from two complementary perspectives: the adequacy of the software development process, and the compatibility of the framework with the development platforms, tools and methods of on-board software development.

The software development process for European systems is governed by the ECSS standards [62, 63] and their accompanying handbooks. The standards are tailored according to the criticality level of the software to be developed. The data fusion functionality for perception and navigation normally relies on system-level capabilities to recover in case of failure, so the criticality level of the data fusion software is normally rated to be "C". A higher level might be allocated for data fusion in highly dynamic mission phases such as rendezvous, or planetary landing. Specific provisions are defined for the reuse of software in an operational environment [64].

As for the supported platforms and methods, the CDFF must be compatible with the performance of space processors (e.g. LEON, PowerPC RAD series), memories and radiation-tolerant FPGAs. Raw computing power is an important factor for data fusion algorithms. In addition to the particular hardware platforms, space systems have characteristic hardware interfaces (e.g. MIL-1553, CAN, SpaceWire), monitoring and control protocols (e.g. PUS [65]) and services (e.g. SOIS [66]). In the frame of developing space robotics technologies, the development methods and the runtime environment are mainly determined by the mission activities of a targeted mission, or a future mission. Nevertheless, it also depends on critically on a space robot control and operating system on which data fusion will be running.

The space robot control and operating system needs to be developed to be compatible with three increasing levels of quality: laboratory, high reliability and space quality. These levels differ in their RAMS features and also in their required software behaviour and interfaces. In order to allow its integration in systems with a high-quality level, the individual components or modules of the data fusion framework must be amenable to the types of verification that are required for onboard software, which include code analysis, unit testing, schedulability analysis, memory and communication budget characterization, and safety and dependability analysis. Verifiability imposes constraints, for instance, on the programming language, the memory allocation strategies, and the creation and interaction of software tasks impose all these constraints on verification of the modules of the data fusion framework and the framework itself.

Model-Driven Software Development (MDSD) can be used if possible to allow reasoning about the software properties and behaviour, and to ensure that the non-functional properties of the software are compatible with its requirements and that they are correctly translated from the software design to its implementation.

It is also worth considering that a number of navigation and perception functions may be properly running in decentralized, FPGA based architectures. There has been interesting application of these approaches. Perri et al. [67] describe a processor architecture for high-speed and reliable stereo matching based on adaptive window-size control of sum of absolute differences (SAD) computation. This SAD-based method for stereovision processing is introduced, that can process 512×512 pixel images with a disparity range of 255 at a frequency of 25.6 Hz. Jin et al. use a local census based matching method that can operate on VGA images of 640×480 pixels and a maximal disparity of 64 pixels with up to 230 frames per second (fps) [68]. In addition to the stereo matching the architecture also performs the rectification and post-processing. Banz et al. present an architecture which is able to calculate the disparity maps of VGA images with a 128 disparity range at 30 fps [69]. Gehrig report a similar architecture which is shown that uses all paths and the architecture can calculate disparity ranges of 64 pixels on image sizes of 340×200 pixels at 27 Hz [70]. For more information of the computation hardware requirements and their applications, a more complete review of stereo algorithms and implementations on CPUs, GPUs and FPGAs can be found in [71].

3.4 Motion Planning and Control

3.4.1 Navigation, Guidance and Control in Unstructured Environments

Localization of robots in unstructured environments poses challenges in robot navigation and has been the subject of extensive research. Since simple odometry derived from wheel movement is inevitably imprecise, robots have typically incorporated visual information to compute an accurate pose. To this effect, multiple types of algorithms, each with several varying implementations, have already been developed for ground robotics applications.

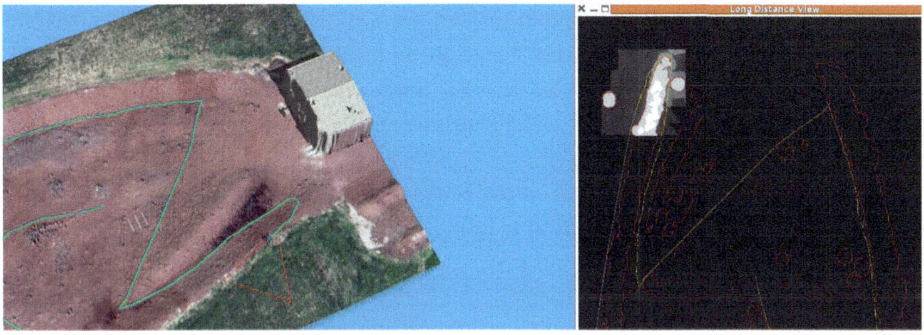

Fig. 3.13 An example SLAM used for navigation, guidance and control

Legacy Mars exploration missions have typically used visual odometry with stereo cameras [14] to estimate their movement from a given reference. However, the current trend in mobile robot localization, enabled by recent developments and access to more computing power, is the use of the Simultaneous Localization and Mapping (SLAM) navigation methods. Figure 3.13 shows an example of a SLAM used to construct a map for the planning of a robotic mission where a path plan is shown on the left in green line and the track map is recorded and generated on the right of the image.

SLAM allows the localization of a robot while incrementally building a map of its environment. The technique is applicable to various sensors, including visible-band cameras. Monocular SLAM [20] and stereo SLAM [21], can both be supplemented with other sensors, such as an IMU [22] or a GPS receiver, greatly reducing drift and improving the final accuracies of the trajectory track and reaching the target destinations.

SLAM can be implemented using various types of estimation processes. Due to its low complexity, the extended Kalman Filter (EKF) has typically been the estimator of choice for real-time SLAM applications [23]. Another filtering approach uses a Rao-Blackwellised Particle Filter [24]. Recent developments in bundle adjustment (BA) SLAM techniques, using methods derived from offline Structure from Motion (SfM), have enabled this type of estimator to nearly rival the performance of an EKF and be considered for real-time applications [25, 26]. Additionally, most modern SLAM implementations address specific mapping considerations such as loop-closure, where the system detects when a known map area is revisited, allowing an additional global correction of the map. Finally, hybrid methods [27], combining the performance and complexity of the EKF on local maps with the accuracy of BA on global maps have also been proposed.

Being an incremental localization method, any type of pure SLAM implementation remains subject to a certain amount of drift, especially over long-distance displacements. It is however possible to periodically perform an absolute localization of the vehicle using prior information about its environment, such as aerial or orbital imagery. Past Mars missions [28] have used orbital mosaics and DTMs to perform a manual correction of the rover's position, but great difference in viewpoints, resolution and uncertainties between orbital and ground images makes autonomous absolute localization from direct feature registration a challenging task. Attempts have been made to solve this problem by registration of higher-level, semantic features, such as objects or skylines [29].

The advent of commodity depth cameras has also brought about methods of Visual Dense SLAM such as [30], where a true incremental surface reconstruction of the environment can be obtained in real time.

3.4.2 Navigation, Guidance and Control in Structured Environments

The mission task of vision-based localization in a structured environment such as on-orbit servicing of a satellite, as opposed to an unstructured one, benefits from additional information that is either known prior the mission, for example a full 3D model of the target body, or can be relatively easily extracted from in-situ measurements, such as a reconstruction of simple geometric features.

Previous applications include multiple iterations, developed by NASA, of a relative navigation system for in-orbit rendezvous. Initially tested on a Hubble Space Telescope servicing mission [31], the system employed non-cooperative model-based tracking by edge extraction and matching in images obtained with a monocular camera. The system is capable of estimating the relative attitude of the telescope with regards to the space shuttle in real-time. Pose estimation with edge matching between a 3D model and images from a monocular camera have also been studied by ESA and others [35] for satellite servicing. Edge matching is done by Gauss–Newton minimization, and the extracted relative poses are directly used in the control loop for visual servoing. Further extensions of NASA's relative navigation system [32, 33] for application to capsule rendezvous with the International Space Station have also employed Iterative Closest Point (ICP) algorithms with point-clouds obtained with a LIDAR sensor for 6DOF relative pose measurement, fused with IR and visible-band camera measurements. Fusion of measurements and pose estimation is performed with Extended Kalman Filters, with a distinction between long-range and short-range operation.

Several in-flight tests campaigns [34] have additionally been carried out by NASA for various commercial applications, using either scanning or Time of Flight, LIDAR sensors with the purpose of qualification for proximity operations and rendezvous with the ISS. The preferred measurement scheme again appears to be model-based matching between an extracted point-cloud and a 3D model of the target.

Acknowledgements The authors would like to acknowledge and express their thanks to all the supporting staff and partners of the InFuse Project: Space Application Services, DLR, Magellium, CNRS/LAAS, Strathclyde University, DFKI Gmbh, and the financial support for InFuse as part of the Strategic Research Cluster on Space Robotics Technologies from the European Union's Horizon 2020 research and innovation programme under grant agreement No 730014.

References

1. Ali, K. S., Vanelli, C. A., Biesiadecki, J. J., Maimone, M. W., Cheng, Y., San Martin, A. M., & Alexander, J. W. (2005). Attitude and position estimation on the Mars exploration rovers. In *2005 IEEE International Conference on Systems, Man and Cybernetics* (Vol. 1). IEEE.
2. Maimone, M., Johnson, A., Cheng, Y., Willson, R. & Matthies, L. (2006). Autonomous navigation results from the Mars Exploration Rover (MER) mission. *Experimental Robotics IX, 3–13.*
3. Grotzinger, J. P., Crisp, J., Vasavada, A. R., Anderson, R. C., Baker, C. J., Barry, R., Blake, D. F., Conrad, P., Edgett, K. S., Ferdowski, B., & Gellert, R. (2012). Mars science laboratory mission and science investigation. *Space Science Reviews, 170*(1), 5–56.
4. Ishigami, G., Otsuki, M., & Kubota, T. (2012). Lidar-based terrain mapping and navigation for a planetary exploration rover. In *Proceedings of International Symposium on Artificial Intelligence, Robotics and Automation in Space (i-SAIRAS)* (Vol. 12).
5. Thrun, S., Thayer, S., Whittaker, W., Baker, C., Burgard, W., Ferguson, D., Hahnel, D., Montemerlo, D., Morris, A., Omohundro, Z., & Reverte, C. (2004). Autonomous exploration and mapping of abandoned mines. *Robotics & Automation Magazine, IEEE* 11(4), 79–91.
6. Moghadam, P., Wijesoma, W. S., & Feng, D. J. (2008). Improving path planning and mapping based on stereo vision and Lidar. In *2008 10th International Conference on Control, Automation, Robotics and Vision* (pp. 384–389). IEEE.
7. Kelly, A., et al. (2006). Toward reliable off road autonomous vehicles operating in challenging environments. *The International Journal of Robotics Research, 25*(5/6), 2006.
8. Dima, C. S, Vandapel, N., & Hebert, M. (2004). Classifier fusion for outdoor obstacle detection. In *Proceedings of the IEEE International Conference on Robotics and Automation.*
9. Matthies, L. H., Bellutta, P., & McHenry, M. (2003). Detecting water hazards for autonomous off-road navigation. In *Proceedings of SPIE Conference Unmanned Ground Vehicle Technology.*
10. Elseberg, J., Borrmann, D. & Nüchter, A. (2011). Full wave analysis in 3D laser scans for vegetation detection in urban environments. In *Information, Communication and Automation Tech. (ICAT)*, Sarajevo.
11. Reymann, C., & Lacroix, S. (2015). Improving LiDAR point cloud classification using intensities and multiple echoes. In *IEEE/RSJ International Conference on Robotics and Automation, Hamburg (Germany).*
12. Samson, C., English, C., Deslauriers, A., Christie, I., Blais, F., & Ferrie, F. (2004). Neptec 3d laser camera system: from space mission sts-105 to terrestrial applications. *Canadian Aeronautics and Space Journal, 50*(2), 115–123.
13. Miravet, C., Pascual, L., Krouch, E., & Delcura, J. (2008). An image-based sensor system for autonomous rendezvous with uncooperative satellites. In: *7th International ESA Conference on Guidance, Navigation and Control Systems.*
14. Shoemaker, J., & Wright, M. (2003). Orbital express space operations architecture program. *International Society for Optics and Photonics, 5088*(2003), 1–9.
15. Inaba, N., & Oda, M. (2000). Autonomous satellite capture by a space robot: World first on-orbit experiment on a Japanese robot satellite ETS-VII. *IEEE International Conference on Robotics and Automation, 2*(2000), 1169–1174.
16. Bodin, P., Noteborn, R., Larsson, R., Karlsson, T., D'Amico, S., Ardaens, J. S., Delpech, M., & Berges, J. C. (2012). Prisma formation flying demonstrator: Overview and conclusions from the nominal mission. *Advances in the Astronautical Sciences, 2012*, 441–460.

17. Trusov, A. A. (2011). Overview of MEMS gyroscopes: History, principles of operations, types of measurements. *Engineering*.
18. Passaro, V. M., Cuccovillo, A., Vaiani, L., De Carlo, M., & Campanella, C. E. (2017). Gyroscope technology and applications: A review in the industrial perspective. *Sensors, 17*, 2284. https://doi.org/10.3390/s17102284
19. Wang, S., Wang, J., Yu, H., & Li, H. (2018). A MIMU/GPS/magnetometer integrated alignment method based on sequential robust estimation. *The Journal of Global Positioning Systems*. https://doi.org/10.1186/s41445-018-0019-2
20. Sabatini, A. M. (2006). Quaternion-based extended kalman filter for determining orientation by inertial and magnetic sensing. *IEEE Transactions on Biomedical Engineering, 53*(7), 1346–1356.
21. Madgwick, S. O., Harrison, A. J., & Vaidyanathan, R. (2011). Estimation of IMU and MARG orientation using a gradient descent algorithm. In *2011 IEEE International Conference on Rehabilitation Robotics*, ETH Zurich Science City, Switzerland, June 29–July 1, 2011
22. Li, F., Liu, S., Zhao, X., & Zhang, L. (2021). Real-time 2-D Lidar odometry based on ICP. *Sensors, 2021*(21), 7162. https://doi.org/10.3390/s21217162
23. Allende-Alba, G., D'Amico, S., & Montenbruck, O. (2009). *Radio-frequency sensor fusion for relative navigation of formation flying satellites*.
24. Jia, H. (2004). Data fusion methodologies for multisensor aircraft navigation systems. Associated with document Ref. Ares (2016) 5785626. June 10, 2016.
25. Grocholsky, B., Keller, J., Kumar, V., & Pappas, G. (2006). Cooperative air and ground surveillance. *Robotics & Automation Magazine, IEEE, 13*(3), 16–25.
26. Fink, W., George, T., & Tarbell, M. A. (2007). *Tier-scalable reconnaissance: The challenge of sensor optimization, sensor deployment, sensor fusion, and sensor interoperability* (p. 2007). International Society for Optics and Photonics.
27. Sukkarieh, S., et al. (2003). The ANSER project: Data fusion across multiple uninhabited air vehicles. *The International Journal of Robotics Research, 22*(7–8), 505–539.
28. Sanfeliu, A., et al. (2010). Decentralized sensor fusion for ubiquitous networking robotics in urban areas. *Sensors, 10*(3), 2274–2314.
29. Hall, D. L., & Llinas, J. (1997). An introduction to multisensor data fusion. *Proceedings of the IEEE, 85*(1), 6–23.
30. Van Dam, J. W., Kröse, B. J., & Groen, F. C. (1996). Neural network applications in sensor fusion for an autonomous mobile robot. *Reasoning with uncertainty in robotics* (pp. 263–278). Springer.
31. Specht, D. F. (1990). Probabilistic neural networks. *Neural Networks, 3*(1), 109–118.
32. Richard, M. D., & Lippmann, R. P. (1991). Neural network classifiers estimate Bayesian a posteriori probabilities. *Neural Computation, 3*(4), 461–483.
33. Nakamura, E. F., Loureiro, A. A., & Frery, A. C. (2007). Information fusion for wireless sensor networks: Methods, models, and classifications. *ACM Computing Surveys (CSUR), 39*(3), 9.
34. Anjum, M. L., Park, J., Hwang, W., Kwon, H. I., Kim, J. H., Lee, C., ... & Cho, D. I. (2010). Sensor data fusion using unscented Kalman filter for accurate localization of mobile robots. In *2010 International Conference on Control Automation and Systems (ICCAS)* (pp. 947–952). IEEE.
35. Xu, T., Sutton, R., & Sharma, S. (2007). A multi-sensor data fusion navigation system for an unmanned surface vehicle. *Proceedings of the Institution of Mechanical Engineers, Part M: Journal of Engineering for the Maritime Environment, 221*(4), 167–182.
36. Piasecki, M. (1994). Mobile robot localization by fuzzy logic fusion of multisensor data. *Robotics and Autonomous Systems, 12*(3), 155–162.
37. Hanson, M. L., et al. (2004). Automated data fusion and situation assessment in space systems. *International Journal on Artificial Intelligence Tools, 13*(1), 255–271.
38. Punska, O. (1999). *Bayesian approaches to multi-sensor data fusion*. Cambridge University.
39. Pinheiro, P., & Lima, P. (2004). Bayesian sensor fusion for cooperative object localization and world modeling. In *Proceedings of 8th Conference on Intelligent Autonomous Systems*.

40. Prado, J., Cabrita, G., & Marques, L. (2013). Bayesian sensor fusion for land-mine detection using a dual-sensor hand-held device. In *Industrial Electronics Society, IECON 2013–39th Annual Conference of the IEEE* (pp. 3887–3892), IEEE.
41. Song, X., Cui, J., Zhao, H., & Zha, H. (2008). Bayesian fusion of laser and vision for multiple people detection and tracking. *SICE Annual Conference, IEEE, 2008*, 3014–3019.
42. Halatci, I., Brooks, C. A., & Iagnemma, K. (2007). Terrain classification and classifier fusion for planetary exploration rovers. In *Aerospace Conference* (pp. 1–11), IEEE.
43. Ahmed, N. R., Sample, E. M., & Campbell, M. (2013). Bayesian multicategorical soft data fusion for human–robot collaboration. *IEEE Transactions on Robotics, 29*(1), 189–206.
44. Ristic, B., & Clark, D. (2012). Particle filter for joint estimation of multi-object dynamic state and multi-sensor bias. In *2012 IEEE International Conference on Acoustics, Speech and Signal Processing (ICASSP)* (pp. 3877–3880).
45. Blasch, E. P., Lambert, D. A., Valin, P., Kokar, M. M., Llinas, J., Das, S., & Shahbazian, E. (2012). High level information fusion (HLIF): Survey of models, issues, and grand challenges. *Aerospace and Electronic Systems Magazine, IEEE, 27*(9), 4–20.
46. Pohl, C., & Van Genderen, J. L. (1998). Review article multisensor image fusion in remote sensing: Concepts, methods and applications. *International Journal of Remote Sensing, 19*(5), 823–854.
47. Costa, P. C., Laskey, K., Chang, K. C., Sun, W., Park, C., & Matsumoto, S. (2012). High-level information fusion with Bayesian semantics. In *UAI 9th Bayesian Modeling Applications Workshop*, Catalina Island, CA.
48. Bracio, B. R., Horn, W., & Moller, D. P. (1997). Sensor fusion in biomedical systems. Engineering in Medicine and Biology Society, 1997. In *Proceedings of the 19th Annual International Conference of the IEEE* (Vol. 3), IEEE.
49. Post, M. A., Li, J., & Quine, B. M. (2016). Planetary micro-rover operations on Mars using a Bayesian framework for inference and control. *Acta Astronautica, 120*(2016), 295–314.
50. Strasdat, H., Montiel, J. M. M., & Davison, A. J. (2010). Real-time monocular SLAM: Why filter? In *IEEE International Conference on Robotics and Automation (ICRA).*
51. Lu, A., & Milios, E. (1997). Globally consistent range scan alignment for environment mapping. *Autonomous Robots, 4*(4), 1997.
52. Grisetti, A., Kummerle, R., Stachniss, C., & Burgard, W. (2010). A tutorial on graph-based slam. *IEEE Intelligent Transportation Systems Magazine, 2*(4).
53. Arancón, M. M., Montano, G., Wirkus, M., Hoeflinger, K., Silveira, D., Tsiogkas, N., Hugues, J., Bruyninckx, H., Dragomir, I., & Muhammad, A. (2017). ESROCOS: A robotic operating system for space and terrestrial applications, ASTRA 2017.
54. https://www.h2020-ergo.eu/publications/
55. https://eross-h2020.eu/i3ds/
56. Dominguez, R., Post, M., Fabisch, A., Michalec, R., Bissonnette, V., & Govindaraj, S. (2020). Common data fusion framework: An open-source common data fusion framework for space robotics. *International Journal of Advanced Robotic Systems, 17*(2). https://doi.org/10.1177/1729881420911767
57. Environment Representation package to represent information of the environment for a robotic system: https://envire.github.io/
58. Schwendner, J., Carrió, J. H., Domínguez, R., Planthaber, S., Yoo, Y.-H., Asadi, B., Machowinski, J., Rauch, C., & Kirchner F. (2015). Entern—environment modelling and navigation for robotic space-exploration. In *Symposium on Advanced Space Technologies in Robotics and Automation* (ASTRA).
59. Hidalgo-Carrió, J., Arnold, S., Böckmann, A., Born, A., Domínguez, R., Hennes, D., Hertzberg, C., Machowinski, J., Schwendner, J., Yoo, Y. H., & Kirchner, F. (2016). EnviRe—Environment representation for long-term autonomy. In *AI for Long-term Autonomy Workshop on the International Conference on Robotics and Automation (ICRA), (ICRA-16)*, May 16–May 20 2016, Stockholm, o.A.
60. http://rock-robotics.org/
61. https://robotik.dfki-bremen.de/en/research/softwaretools/mars.html

62. European Cooperation for Space Standardization. Space Engineering—Software, ECSS-E-ST-40C. March 6, 2009.
63. European Cooperation for Space Standardization (ECSS). Space Product Assurance—Software Product Assurance, ECSS-Q-ST-80C. March 6, 2009.
64. European Cooperation for Space Standardization (ECSS). Space Product Assurance—Reuse of existing software (handbook), ECSS-Q-HB-80-01A. December 5, 2011.
65. European Cooperation for Space Standardization (ECSS). Space Engineering—Telemetry and Telecommand Packet Utilization, ECSS-E-70-41A, January 30, 2003.
66. Consultative Committee for Space Data Systems (CCSDS). Spacecraft Onboard Interface Services. Green Book. Issue 2. December 2013.
67. Perri, S., et al. (2006). SAD-based stereo matching circuit for FPGAs. In *13th IEEE International Conference on Electronics, Circuits and Systems.*
68. Jin, S., et al. (2010). FPGA design and implementation of a real-time stereo vision system. *IEEE Transactions on Circuits and Systems for Video Technology 20*(1), 15–26.
69. Banz, C., et al. (2010). Real-time stereo vision system using semi-global matching disparity estimation: Architecture and fpga-implementation. *IEEE Embedded Computer Systems (SAMOS).*
70. Gehrig, S. K., Eberli, F., & Meyer, T. (2009). A real-time low-power stereo vision engine using semi-global matching. In *Computer vision systems* (pp. 134–143). Springer.
71. Tippetts, B., et al. (2016). Review of stereo vision algorithms and their suitability for resource-limited systems. *Journal of Real-Time Image Processing 11*(1), 5–25.

Chapter 4
Space Robotics Technologies: Small Dexterous Manipulators

Andrea Rusconi and Piergiovanni Magnani

Abstract In this chapter a brief overview of the main features of small size manipulator arms is presented. These type of arms are in general of limited dimensions and mass and suited to be accommodated in bound spaces or installed as servicing arms onboard moving platforms or landers and rovers. Therefore these arms can find applications in very diverse extra-terrestrial environment from inside of modules, outer space or planetary surfaces.

4.1 Overview

Space robotic arms with a length in the range 0.5–1.5 m can be classified as small manipulators.

The kinematics and geometrical configuration is not only due to the operational capabilities, such as executing collision free trajectories in a narrow environment, but also to other aspects, as the need to be stowed in a very compact launch configuration. For these reasons, they usually require a high level of dexterity within the reachable workspace.

These manipulators must provide the capability of positioning objects in free space and work in contact with the environment. The requirements are usually expressed in terms of positioning accuracy, velocity, force, torque and payload handling capability.

Two different main classes of manipulators can be identified: very dexterous arms, used in manned missions as crew assistants, capable of performing fine manipulation and complex operations, and very lightweight arms employed for surface sample acquisition, sample retrieval or instrument deployment as part of a rover or a lander platform in a robotic exploration scenario.

A. Rusconi (✉) · P. Magnani
Leonardo S.p.A., Nerviano, Italy
e-mail: andrea.rusconi@leonardo.com

X. T. Yan and G. Visentin (eds.), *Space Robotics*,
https://doi.org/10.1007/978-3-031-39214-6_4

4.2 Highly Dexterous Human-Like Arms

These are robot arms comparable in size, force and dexterity to a human arm, to be used for space robotics applications in which the manipulation/intervention tasks were originally conceived for humans. These applications are typically external or internal servicing of manned orbiting platforms or robotic exploration scenarios, in preparation and support to human arrival and in support to long term human presence.

The kinematics is based on at least six degrees of freedom, allowing the positioning of the arm in the desired Cartesian pose within the reachable workspace. The need of high dexterity usually implies a seventh degree of freedom, which can be exploited to implement redundant kinematics schemes, to fulfill additional tasks such as collision avoidance.

The geometrical shape must be carefully studied in order to allow a wide joint angular range and to accommodate all internal components in compliance with required arm dimensions (length and lateral encumbrance).

These robot arms are required to provide significant performance in terms of positioning accuracy and force capability. Contact motion control and reasonable fast dynamics are also required. All of this has to be achieved with limited resources, such as mass, power and volume.

Since most of the resources are allocated to the joints, the design of high performance sensorized joints is a key aspect of these developments.

A good example of such developments is the lightweight robot joint technology developed by DLR. An experiment based on a two-joints robot, named ROKVISS (RObot Komponent Verification on ISS) [1] was conducted on the international space station to demonstrate these technologies in space.

On NASA side, the Robonaut system [2] is a humanoid robot that can use space tools and work in similar environments suited to astronauts. The latest Robonaut version, R2, has been flown on the ISS internal environment to conduct demonstration experiments.

On European Space Agency side, an important development is the Dextrous Robot Arm (DEXARM) [3]. It provides the technology needed by the Eurobot system [4], a robot concept similar to the Robonaut, with characteristics comparable in size and capability to astronauts, including anthropomorphic arms, that can use the same walking paths and assist astronauts in extra-vehicular activities. A ground prototype version of Eurobot was developed to prove the concept of a robotic assistant, designed to move around and work on a planet's surface either autonomously or in cooperation with astronauts. The DEXARM, integrated with the DEXHAND [5], was installed and operated as part of the Eurobot Ground Prototype in a demonstration session.

The DEXARM, shown in Fig. 4.1, is a manipulator with seven degrees of freedom, a length of 1.2 m, a mass of about 25 kg with a payload capability of 10 kg in 1 g over the full workspace (up to 500 kg at 0-g).

It can be controlled in position/speed (both Cartesian and joint spaces), impedance and torque. The arm positioning accuracy is in the order of 1 mm and the repeatability of 0.2 mm. It can produce a Cartesian velocity of 0.15 m/s and exert a force of 200

Fig. 4.1 DEXARM in launch and operational configuration (courtesy of ESA/Leonardo)

N and a torque of 20 Nm at the arm tip. The power consumption in under nominal workload conditions is 100 W.

The redundant kinematics and the wide joint angular range (±175° for roll joints, − 175° + 45° for pitch joints) result in a very high dexterity. Finally, it has been designed to be stowed in a folded configuration during launch, with three fixation points.

The arm is composed of one base, six limbs and seven joints.

The base provides mechanical and electrical connection with the user platform. It is designed to be attached/detached by the astronaut in extra-vehicular activity.

The full set of limbs has been developed in Aluminium. Design of the arm with Carbon Fibre Reinforced Plastic (CFRP) limbs has also been performed and two specific limbs have been developed also in CFRP to allow comparison between different technologies.

Inside the arm, high performance robotic joints are implemented in two sizes: the shoulder joint has a mass of 3.2 kg, with a torque capability of 200 Nm, while the elbow/wrist joint has a mass of 2.2 kg, with a torque capability of 100 Nm. Both joints can produce a maximum no load speed of 0.5 rad/s.

Each joint is composed of a brushless motor, an electromagnetic fail safe brake, a Harmonic drive gear, an electric encoder at the motor shaft, a double speed resolver and a torque sensor at output shaft. Furthermore, joint servo control electronics is integrated with the joints inside the arm. Only a power line, a digital data bus (CAN bus) and some service lines interface the arm to the central controller.

To comply with the requirements on arm length and dexterity, the arm geometry and the limbs have been designed to allow a tight integration with joints and electronics components, as shown in Fig. 4.2. The shape of the limbs is directly optimised to this purpose. The joints exploit a hollow shaft to allow cable routing inside the arm.

The joint servo electronics implements a cascade control scheme (see Fig. 4.3), with an inner motor velocity loop and an outer joint position loop. This scheme allows to achieve a balanced trade-off between the compensation of torque disturbances such as ripple and friction (which require a high bandwidth motor velocity control) and gearbox torsional flexibility (which require a large damping on the output position).

With the availability of the joint torque sensor, the proposed dual loop control scheme is further enhanced with the algorithm described in [6], a modified version of the Torque Disturbance Observer (TDO) control method, which allows a large bandwidth motor control loop without affecting the damping of the load. The classical TDO scheme is modified, by subtracting the measured joint torque (reported at

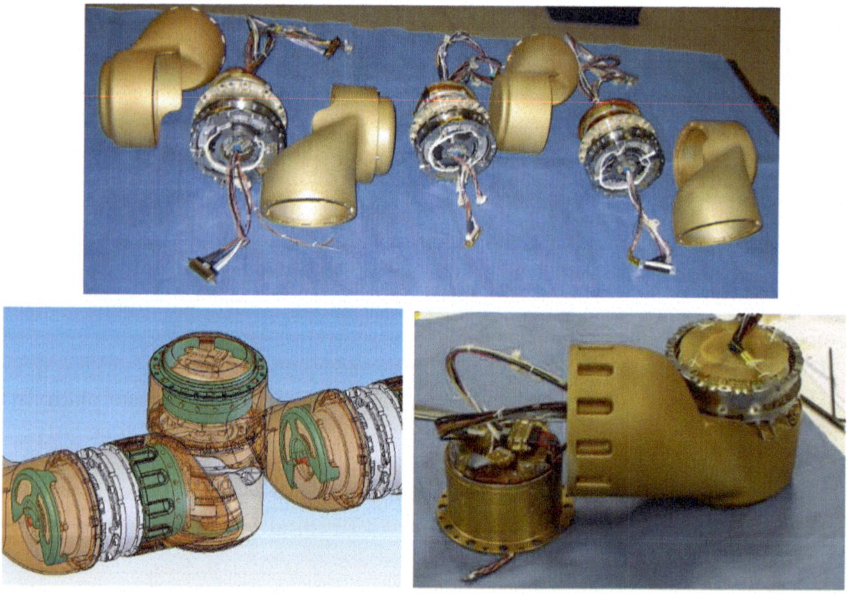

Fig. 4.2 DEXARM tight integration of joints, electronics and harness (courtesy of ESA/Leonardo)

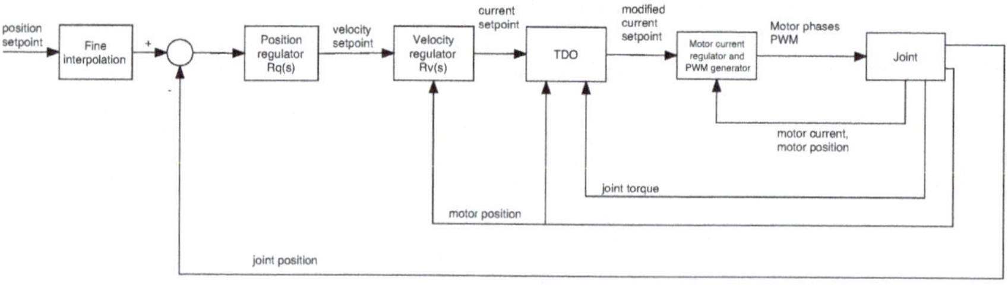

Fig. 4.3 Joint position control with inner motor velocity loop

motor shaft and converted into equivalent motor current) to the applied motor current setpoint. The effect is that the physical torque is no longer rejected as a disturbance. This simple modification, for elastic joints, outperforms classical TDO, in terms of well damped load behaviour.

For contact motion, two additional control modes are available:

- joint position loop with inner torque loop, which supports implicit impedance control at Cartesian level and mixed schemes controlling joint impedance (by simply setting the position loop parameters in accordance with the desired impedance);
- output torque loop, which supports explicit impedance and hybrid control at Cartesian level.

This gives maximum freedom for the implementation of the majority of control strategies present in the literature that can be used at Cartesian level.

4.3 Lightweight Arms for Exploration

Small robotic arms have been developed in the frame of Mars exploration studies and missions. The common aspect is always the constraint on resources, such as mass, power and volume. This leads to the development of lightweight joints and limbs, the careful selection of motors, gear combinations and sensors as strictly needed by the application.

On NASA/JPL side, it is worth mentioning the Spirit and Opportunity Mars Exploration Rovers (MER) arm [7], shown in Fig. 4.4. This arm has an overall mass of about 4 kg with a payload capability of 2 kg. The length is 0.75 m.

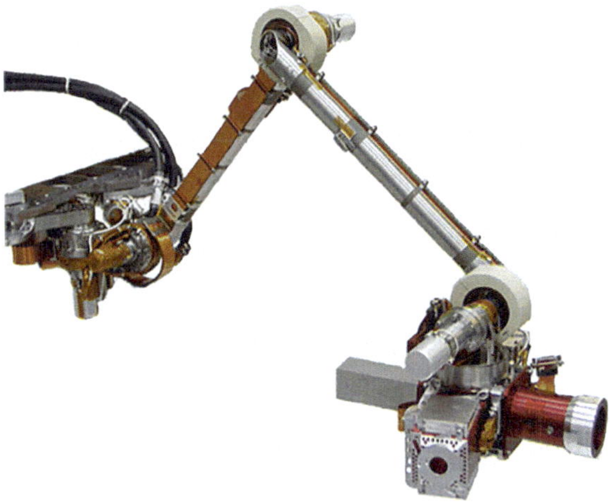

Fig. 4.4 Opportunity MER robotic arm (*credit* NASA/JPL/Caltech)

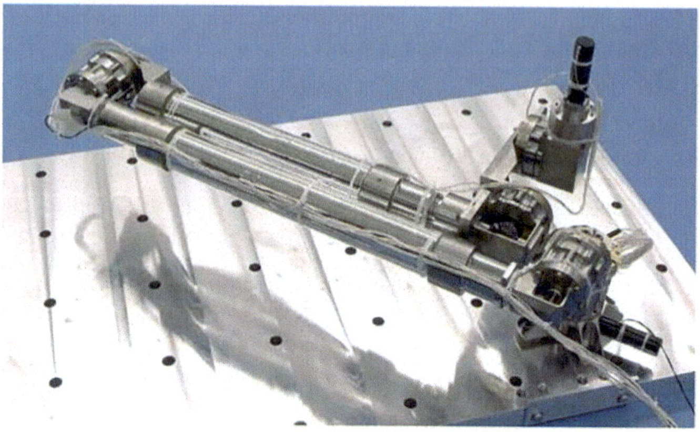

Fig. 4.5 Beagle 2 arm (*credit* Beagle2 team / ESA [8])

Each joint is equipped with a DC brushed motor connected to a planetary gearbox. On the first three degrees of freedom this assembly is further reduced by a Harmonic Drive gearbox. A modularity approach was taken by NASA, using the same motor and encoder in every unit. The Harmonic drive was introduced as secondary stage for joints requiring output torque beyond 9 Nm. In this configuration, the first stage planetary gear ratio was reduced in order to match the optimum overall reduction ratio. The joints are further equipped with a small detent brake, a motor encoder and a potentiometer on the output shaft.

The first European robotic arm developed for a Mars mission was a small robotic arm (shown in Fig. 4.5) embarked on the Beagle 2 lander. It has 5 DoF, a maximum reach of 0.7 m, and a mass of 2.2 kg. Each joint comprises a DC brushed motor, a planetary gear, a bevel gear and a Harmonic drive at the output stage, the position of the output shaft is detected by a potentiometer and the nominal joint speed is about 0.6 deg/s.

A recent ESA technology development project is the DExtrous LIghtweight Arm for exploratioN (DELIAN) [9]. The system level study allowed to define robotic arm concepts for each applicable mission scenarios, and then derive, through a scalability analysis, four families of joints which serve as building blocks for the arm.

The DELIAN joints contain a brushed DC motor (equipped with a motor position sensor), a planetary gear and a Harmonic drive. An accurate Anisotropic Magneto Resistive (AMR) angular position sensor with two indexes for motion limits has been integrated directly at the output shaft.

The six joints developed for the arm have a mass ranging from 0.36 to 0.73 kg, with a continuous torque ranging from 8 to 56 Nm and a peak torque from 10 to 78 Nm. The required speed is 0.7 deg/s.

Different motor variant was considered to provide sufficient friction to achieve the required holding torque, without needing the implementation of a detent brake. Joint 2 (the largest one) implements instead an active brake (the inherent motor friction

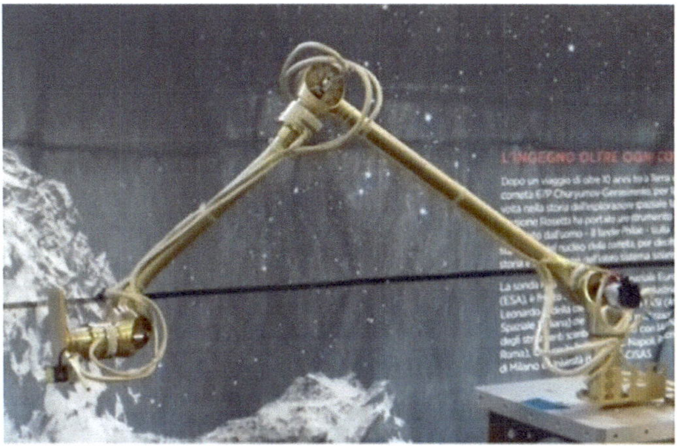

Fig. 4.6 DELIAN arm (courtesy of ESA/Leonardo)

was not enough to achieve the holding torque requirement). A detent brake may be considered in the flight version of this joint, to further optimise mass and power.

The arm has a length of 1.4 m, a mass of 4.85 kg with a payload capability of 2.5 kg, over full workspace in 1 g environment (without offload device).

The arm can be stowed in a folded configuration for launch (Fig. 4.6).

The arm is equivalent to a flight-like version in terms of joint and limb dynamics and flexibility. An end-to-end verification in laboratory conditions was performed, exploiting important control aspects such as positioning accuracy with deflection compensation.

4.4 Conclusions

Small dexterous manipulators can indeed provide an invaluable support to space and planetary exploration missions. Some have already been operatively utilized on Mars since two decades and notable examples are constituted by the robotic arms installed on MER-A, MER-B, Curiosity and Perseverance rovers. The main technological bricks of which such manipulators are made of, have been developed under the supervision of the main agencies and are currently in continuous improvement. Future complex Mars exploratory Missions encompass Sample Return scenarios where the handling and transfer of samples can be profitably performed by dexterous manipulators. More ahead in time, human outposts can certainly take advantage on the robotic technologies developed for the accomplishment of the various operations involved in building, upgrading and maintenance of the outpost itself.

References

1. Hirzinger, G., et al. (2005). ROKVISS—Robotics component verification on ISS. In *Proceedings of the 8th International Symposium on Artificial Intelligence, Robotics and Automation in Space (iSAIRAS)*, Munich, Germany.
2. Diftler, M. A., et al. (2011). Robonaut 2—The first humanoid robot in space. In *2011 IEEE International Conference on Robotics and Automation*, Shanghai, pp. 2178–2183.
3. Rusconi, A., Magnani, P., Della Pietà, A., Campo, P., Magnani, G., & Visentin, A. (2010). DEXARM integration and test results. In *Proceedings of the 10th International Symposium on Artificial Intelligence, Robotics and Automation in Space,* Sapporo, Japan.
4. Schoonejans, P. H. M., Stott, R., Didot, R. F., Allegra, A., Pensavalle, E., & Heemskerk, C. (2004). Eurobot: EVA-assistant robot for ISS, Moon and Mars. In *Proceedings of 8th ESA Workshop on Advanced Space Technologies for Robotics and Automation—ASTRA 2004*, ESA/ESTEC, Noordwijk (The Netherlands), November 2004.
5. Chalon, M., et al. (2011). Dexhand: A Space qualified multi-fingered robotic hand. In *2011 IEEE International Conference on Robotics and Automation*, Shanghai, pp. 2204–2210.
6. Magnani, G., Rocco, P., Bascetta, L., & Rusconi, A. (2013). On the use of torque disturbance observers in 2-mass systems with application to a robotic joint. In *IEEE International Conference on Mechatronics—ICM 2013*, Vicenza, Italy.
7. Fleischner, R. (2003). Concurrent actuator development for the mars exploration rover instrument deployment device. In *ESMATS 2003*, San Sebastián, Spain, 24–26 September 2003
8. Pullan, D., Sims, M. R., Wright, I. P., Pillinger, C. T., & Trautner, R. (2004). Beagle 2: The exobiological lander of mars express. *Mars Express: The Scientific Payload, 1240*, 165–204.
9. Rusconi, A., Molina, M., Fumagalli, A., & Foresti, L. (2018). DELIAN arm development and test for mars sample acquisition. In *69th International Astronautical Congress 2018*, Bremen, Germany.

Chapter 5
Locomotion Subsystem Design

Stéphane Michaud, Gerard Kruse, and Michel van Winnendael

Abstract A major challenge in planetary exploration robotics is to provide increased rover mobility, which is necessary to access to a wider range of scientific sites. The design of a locomotion subsystem depends on multiple constraints and in particular the mission profile and its interaction with the planetary surface. The aim of this chapter is to introduce the principle for designing adequate locomotion subsystems in particular for Moon and Mars missions with a main focus on motion performance.

5.1 Introduction

Vehicles for operation on planetary objects other than the Earth, also referred to as planetary rovers, have so far been successfully developed and operated on the Moon and on the planet Mars. Compared to in-orbit robots, the gravitational force allows rovers to interact with the surface for their motion. The locomotion subsystem consisting of structural, electro-mechanical elements and low-level controllers enable this mobility. In general, the following key design drivers have to be considered for this subsystem: (i) accommodate within the lander, (ii) deploy and egress and (iii) provide safe motion with limited power and computational resources.

In order to produce meaningful locomotion subsystem designs, the prerequisite is to have sufficiently detailed knowledge of the terrain in which the device has to operate [1]. In particular:

- The distribution of slopes and obstacles in the terrain drive the chassis kinematic layout in order to obtain a specified terrainability and manoeuvrability.

S. Michaud (✉)
Beyond Gravity, Zürich, Switzerland
e-mail: stephane.michaud@beyondgravity.com

G. Kruse
Deltares, Delft, The Netherlands

M. van Winnendael
European Space Agency, Noordwijk, The Netherlands

X. T. Yan and G. Visentin (eds.), *Space Robotics*,
https://doi.org/10.1007/978-3-031-39214-6_5

- The mechanical characteristics of the soil over which the vehicle has to traverse, for a given chassis kinematics, wheel or leg design and a given gravity field, determine trafficability.

5.2 Locomotion Performances and Operation

5.2.1 *Trafficability*

Trafficability can be defined as a robot´s ability to generate traction and overcome resistances [2]. This metric is directly related to leg- or wheel-soil interaction and depends mainly on (i) sinkage and thus resistance that affect required driving speed and power and (ii) pulling force available from the terrain.

A static wheel or leg subject to a load will sink into deformable soil and will exert a stress distribution on this soil along the contacting interface. This sinkage increases the resistance with increasing depth, until the load and the resisting stresses are balanced. Upon rotating the wheel, stresses change, and further sinkage can take place by further compaction and/or by removal of soil by the moving wheel. The normal and shear stress distributions and the resistive forces change by these actions. Various semi-empirical formulation have been established in [1, 3, 4], to approximate this complex interaction between wheel and soil. This require knowledge of terramechanical soil parameters available for certain region of Moon [5] and Mars [6]. Terramechanics is the study of soil properties, specifically the interaction of wheeled or tracked vehicles on various surfaces.

In early design phases, it is advantageous to have guidelines for preliminary wheel sizing. The image of the first step on the moon is well known and shows limited static sinkage. Such experience can be used to size a wheel. The aim of wheel design is to limit the average ground pressure (AGP) to the value exerted that time by the astronaut or by previous rover missions. The average ground pressure is defined as per Eq. 5.1 and test data performed on various Martian soil simulants indicate a wheel has a suitable behaviour for value below 7 kPa [4]. It has to be noticed that, in particular for flexible wheels on relatively hard soils, this value can be higher.

$$AGP = \frac{W}{w \cdot 2\sqrt{2rz - z^2}} \tag{5.1}$$

With W the load [N], z being the sinkage [m], w the wheel width and r the wheel radius in [m]. The AGP is expressed in N/m2 equivalent to Pascals [Pa].

This average ground pressure is defined in a similar way as the Effective Ground Pressure (EGP) formulated in some NASA publications. This represents the average pressure exerted by a wheel at defined sinkage in a substratum. For the EGP, "an assumption of wheel sinkage gives a contact patch area equal to the width of the wheel, w, multiplied by the wheel radius r" [7]. The EGP is thus equivalent to considering a defined sinkage of 13.4% of the wheel radius in the average ground pressure equation. Semi-empirical formulations for estimating more accurately the wheel sinkage are summarised in Sect. 5.3.2.

The drawbar pull (DP) is the key parameter for evaluating trafficability as it provides the net pulling force in the direction of motion. It is expressed as soil thrust (H) minus the sum of motion resistances (R).

$$DP = \sum_{j=1}^{j=m} \left(H_j - R_j\right) \tag{5.2}$$

With j = the wheel number, and m = total number of wheels.

This value represents the capability of a wheel or rover to counteract an increase of motion resistance as produced by the terrain (i.e. slopes, obstacles). The drawbar pull is mainly a function of the sinkage (z) and vary as a function of slip ratio. The soil thrust H multiplied by the wheel radius r allows to calculate the required input torque (T). This is an important parameter to size the drive actuator. Assuming a commanded wheel rotational speed (ω), the input power (P) can be determined for sizing the control electronic, harness and power sub-system. The parameters related to mechanics of wheel-soil interaction are shown in Fig. 5.1.

Apart from the load-sinkage of a leg or wheel, incurred as a result of the vertical load W, additional sinkage results from wheel slip when driving. This slip-sinkage is dominating the trafficability in loose soil as experience during Mars and Moon missions. The slip or slip ratio is defined as:

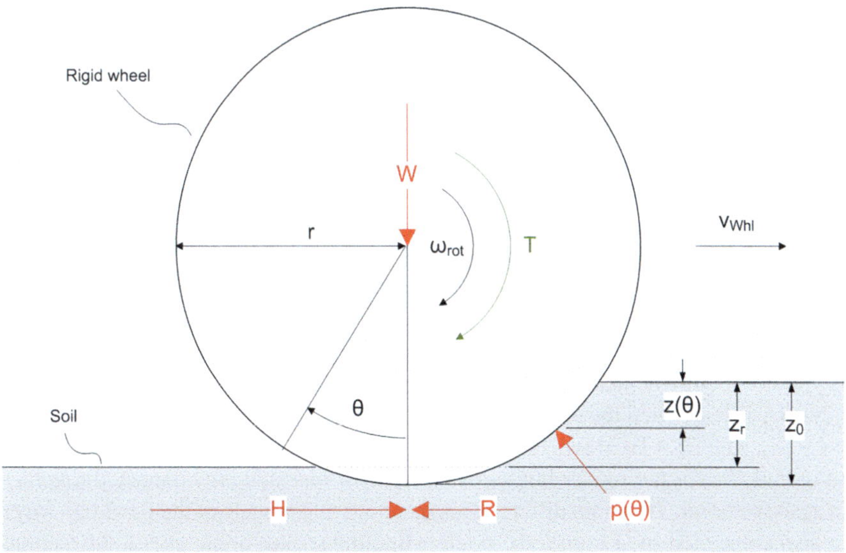

Fig. 5.1 Mechanics of wheel-soil interaction

$$\text{slip} = \frac{r \cdot \omega_{rot} - v_{Whl}}{r \cdot \omega_{rot}} \tag{5.3}$$

ω_{rot} = wheel rotational speed in [s^{-1}]

v_{Whl} = wheel linear velocity in the direction of motion [m/s].

r = wheel radius in [m] (for flexible wheel it is the undeformed wheel radius).

Due to this dependency, the trafficability is often represented in graphical form representing drawbar pull as a function of slip ratio. Because drawbar pull depends also on wheel load, comparison between different wheels can be achieved using a normalised drawbar pull (e.g. DP/W). Drawbar pull, slip ratio, and wheel load can be measured during ground testing with a dedicated test facility but are usually not available during a mission. Because processing rover odometry enables determining the rover slip ratio on a given terrain, wheel or rover slip is often plotted versus terrain inclination or slope gradeability in publications reporting rover observations. The relationship between the drawbar pull that can be delivered at a given slip ratio on levelled terrain and the slope gradeability can be approximated, for low terrain slopes, as per Eq. 5.4. For slope angles above ca. 12°, some deviations have been experimentally observed on loose soil due to difference in contact angle between the wheel and the soil as well as ingress of soil in front of the rotating wheel.

$$DP(slip) = W \sin(\theta) \tag{5.4}$$

With W being the wheel load [N] and θ the soil slope angle.

5.2.2 *Terrainability*

The ability to negotiate terrain irregularities is related to the chassis kinematics (also called suspension). The aim of the chassis is to allow proper wheel accommodation in uneven terrain and in particular to guarantee that the wheels are always in contact with the ground. Ideally, such kinematics should provide equal wheel load distribution and minimise wheel input torque when surmounting obstacles. For space exploration rovers, mass, volume and robustness are design drivers for selecting the suspension architecture. Those considerations so far have led to select passive suspension as they do not require actuators nor drive electronics for their function.

NASA JPL selected the so-called rocker-bogie chassis kinematics for all Martian rovers [8]. The ESA ExoMars rover is based on a three-bogie architecture as shown in Fig. 5.2. As for the rocker bogie, two wheels of one side are connected together via a passive joint. The main difference is that both rear wheels of the ExoMars rover are also connected by a transverse bogie while the rocker-bogie uses a differential mechanism connecting both sides.

Wheel load repartitions can be computed by various multibody dynamic tools like MSC-ADAMS, Simpack or Matlab with dedicated toolboxes. Newton–Euler

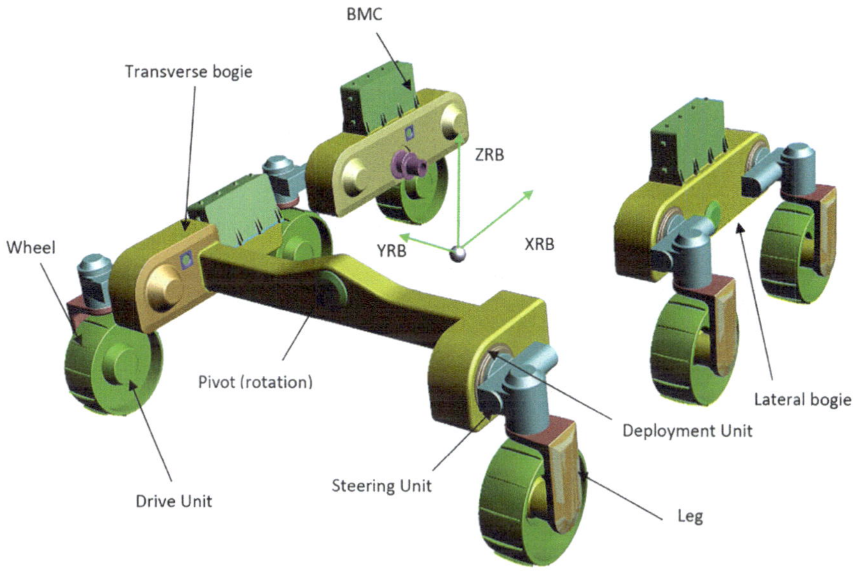

Fig. 5.2 Phase B Concept of ExoMars chassis kinematics

equations can also be established for all passive suspensions. The ExoMars rover pose calculation as a function of wheel height on the terrain can be found in [9] and is used in [10].

5.2.3 *Manoeuvrability*

The terrainability and trafficability determine the ability of the rover to safely traverse a given terrain. Based on this information, a guidance and navigation system or operator establishes a safe path and sends the corresponding motion commands to the rover. The capability of the locomotion subsystem to follow this path is essential for mission success. The manoeuvrability impacts the trajectory generation and therefore is to be considered when designing a locomotion subsystem.

The most basic way of changing direction, called "skid steering", is based on adapting individual wheel speed. For obtaining a point turn, the wheels of each side rotate at the same speed but in opposite directions. This technique has the disadvantage of requiring slippage at the wheel-soil interface. This consumes energy and becomes inaccurate on uneven terrain. As for cars, implementation of wheel steering capabilities is therefore the preferred approach for planetary exploration rovers exceeding a few kilograms.

Two steerable wheels in an Ackermann steering geometry provide mobility without the need for wheels to slip sideways (e.g. lateral slip). However, this configuration does not allow performing a point turn, which is considered essential on

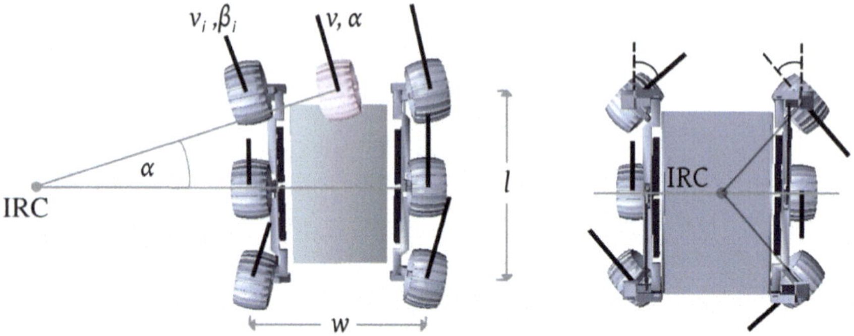

Fig. 5.3 Schematics of a six-wheeled rover with double Ackermann steering geometry. Credit: ETHZ

difficult terrain and for avoiding backward motion. To achieve a "clean" point turn maneuver, all wheels except for the wheels accommodated in the middle of the rover need to be steerable. For this purpose, dedicated actuators are accommodated within the chassis kinematics.

In general, a key aspect of manoeuvrability can be assessed by analysing the possible location of the instantaneous rotation centre (IRC) that describes the centre of the rotation of the rover movement. A four wheeled rover with Ackermann steering geometry has its IRC located on the line passing through the two rear wheels. For a six wheeled rover with double Ackermann steering geometry (i.e. four steerable wheels), the IRC is located on the line passing through the two middle wheels as per Fig. 5.3.

The ESA ExoMars rover further increase its manoeuvrability by having six steerable wheels. This allows locating the IRC anywhere on the XY plan. It also allows a so-called crab mode where the rover moves sideways without changing its orientation.

In conclusion, maneuverability is essential for the navigation system and can limit possible trajectories. Having high maneuverability provides better capability to avoid non-traversable regions but at the cost of additional electro-mechanical elements.

5.2.4 Energy Usage for a Given Motion

Space rovers have limited power, in particular when based on solar panels to perform motion and operation. Therefore, the daily travel range can be limited by the energy required by the locomotion subsystem. On a given terrain, this depends on (i) the effective travelling distance, (ii) the overall motion time and (iii) the motion efficiency. This section provides guidelines to estimate, in an early concept phase, the energy required by the locomotion subsystem for a given mission.

The effective distance is a function of the number of non-traversable regions that need to be avoided by the locomotion subsystem. For a given terrain, the performance of a given rover can be simplified to the mean free path (MFP) which is the straight-line distance which the vehicle can drive before encountering a non-traversable region. Those regions either features obstacles higher than the climbing capability / ground clearance, or slopes for which the inclination would lead to excessive slip ratio. The extra path for this avoidance can be estimated based on the size of this region and the rover width. Finally, the energy for changing the orientation (e.g. point turn) is to be added.

The overall daily travelling time is usually limited either by the available power, the maximum driving speed capability of the actuator or the navigation system. Based on those considerations, a maximum cruise speed can be determined that can be further impacted by the wheel slip ratio. The overall time is essential for determining the energy computation of subsystems for which the power is not necessarily function of the motion distance, such as OBC or heaters that consume energy even when the motors are not activated.

The key parameter to estimate motion efficiency are the requested wheel input torque, slip ration and MFP that can be based on test data or simulations as introduced in Sect. 5.3.2. The electrical power can also be considered. This can be derived from input torque, speed, and actuator efficiency (ca. 50% on current Martian rovers that are mass optimised and therefore feature high gear ratio).

5.2.5 Lander Accommodation and Deployment

Of particular concern for rover design is the ability to fit within the available volume inside the lander. This can be a severe constraint for designing the chassis kinematics and may require additional items (e.g. joint, actuators) in order to be stowed in the spacecraft. The deployment can be done either by the rover itself or using a deployment system accommodated in the lander as for the NASA MER [11]. On ExoMars, six dedicated deployment actuators accommodated in the bogies allow the rover to be stowed volume-efficiently in the lander (Fig. 5.4), and together with the wheel actuators provide self-deployment capability.

5.3 Wheel-Soil Interaction

Wheel-soil interaction has been briefly introduced as key element of rover trafficability. This section provides more in-depth knowledge related to current state of the art in simulation techniques and experiments.

Fig. 5.4 ESA ExoMars rover locomotion verification model in stowed configuration during deployment test at Beyond Gravity. Credit: ESA

5.3.1 Single Wheel Testing

There is a need to have accurate and extensive datasets of representative wheel-soil interaction to enhance theory or validate simulation tools. A soil channel is frequently used for this purpose. In such a facility, the wheel can be operated in towed or self-propelled state while running in a bin filled with a uniform soil simulant. Concurrently, forces, torques, vertical wheel displacement and rotation rate are measured to characterize the tractive behaviour in terms of motion resistance (R) and drawbar pull (DP) as function of wheel slip (s) and wheel load (W). Soil channels attempt to operate a wheel under controlled loading and controlled soil conditions to obtain reproducible measurements. This is in contrast with field trials where single wheels or rovers are operated on naturally occurring—and thus inhomogeneous—soils. The generated test data can be directly used for comparing different wheel designs or to improve theories or model parameters. For example, Fig. 5.5 shows measured sinkage versus slip data, an interpolation, as well as a new theoretical formulation of slip sinkage proposed in the ESA SWIFT project [12] and described in Sect. 5.3.2.

Compared to rigid wheels, for which vertical hub displacement is equivalent to sinkage, flexible wheels enable a deformation of the rim that allows increasing the wheel-soil contact area when loaded. This allows producing higher shear force and decreasing sinkage, and therefore can produce better performances than a rigid wheel of the same dimension. As shown in Fig. 5.6, the performance of the flexible ExoMars BreadBoard2 (BB2) wheel is better than the one of a rigid wheel with same dimension, number of grousers and grouser height. The delta in performance differs significantly as a function of the used soil simulant.

The drawbar pull versus slip curve has been interpolated based on single-wheel tests performed at different controlled slip ratios as explained in [12].

While flexible wheels provide better performances on loose soils, they lower the rover ground clearance and require slightly more input torque on hard soils than rigid wheels due to their deflection. Therefore, proper adjustment of the stiffness

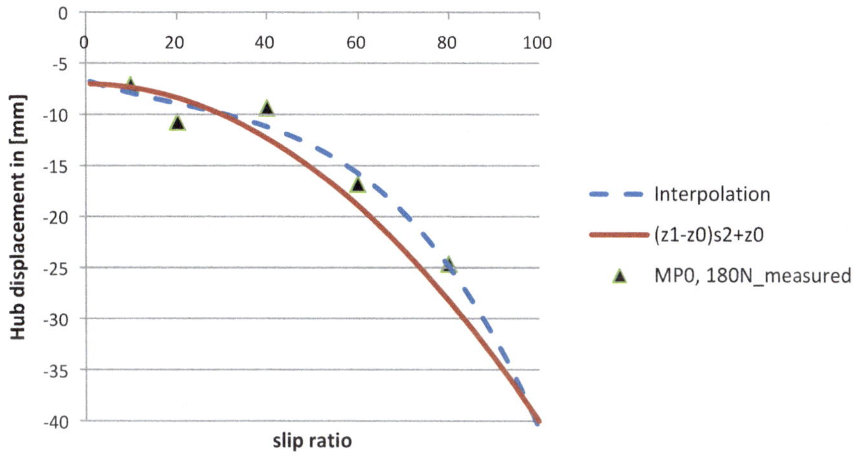

Fig. 5.5 Hub displacement versus slip for a rigid wheel of diameter 250 mm and width 50 mm at 180 N wheel load on martian soil simulant ES4-SS3590+MS6

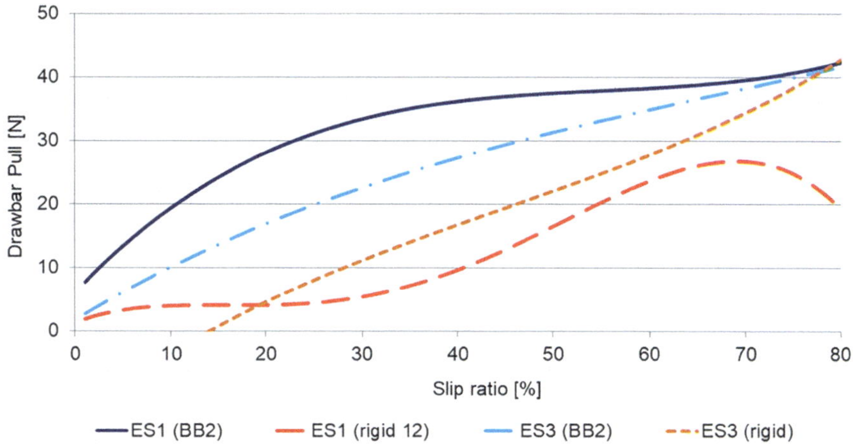

Fig. 5.6 Drawbar pull versus slip for flexible BB2 and similar rigid wheel at 180 N wheel load on soils ES1-D6DD and ES3-OMR

of a flexible wheel for a given mission is necessary. Due to a more complex inner structure, flexible wheels are heavier and thus mainly advantageous when limited storage volumes are available within the lander, or when shock transmission to the rover payload needs to be minimized during motion.

5.3.2 Simulation

Simulating wheel-soil interaction requires knowledge of soil mechanics and dedicated parameters that are not trivial to define before or even during the mission. Classical terramechanical formulations like in [1, 3] are often used, but have shown limitations when attempting to simulate small to medium-size flexible wheels as used on current rovers.

An interesting procedure for offline-identification based on enhanced Bekker-based soft soil contact model parameters is presented in [6]. Physical properties of terrains encountered by the Curiosity rover during the first 360 sols of operations have been inferred from analysis of data from the rover [13]. In all those approaches it is challenging to distinguish between the inaccuracy inherent to the approach itself and that related to the accuracy of the soil parameters used.

The main challenge is that mechanical behaviour of granular material such as soil is known to be complex, with nonlinear stress, deformation, and grain packing dependencies, which include among others the "phase transition" from a stiff elastic condition to a more or less easily deforming material. Upon passage of a wheel, soil incurs large stress variations bringing about packing density changes and related changes in strength and deformation. Even sophisticated soil mechanical approaches have to make use of approximations for this complex soil behaviour for practical simulations. It follows that a parametric tool can only function if a more generalized approach to the mechanics and mechanisms involved is used to incorporate soil behaviour in soil-wheel interaction.

Such an approach was followed in the development of the Wheel Parametric Module (WPM) [4], in the ESA SWIFT activity. The approach avoids the necessity for a specification of the stress distributions along the soil-wheel interface, which is required for capturing wheel performance with the Bekker approach [1], and the improvements by Reece and Wong [3]. This specified stress distribution is commonly used in branches of terramechanics, but the specification often deviates considerably from measured, and FEM modelled, stresses at moving wheels as shown in Fig. 5.7 from [14].

The different terramechanical formulations require to first estimate wheel sinkage. Sinkage can be separated in static (z_0) and slip sinkage (z_s). The static sinkage depends on the shape of the wheel and applied load:

- For a rigid cylinder on an elastic half space:

$$z_0 = \frac{4W}{\pi E^* b} \quad \text{with } W\text{: wheel load, and } b \text{ the wheel width} \tag{5.5}$$

For a sphere on an elastic half space (recommended for spherical wheels):

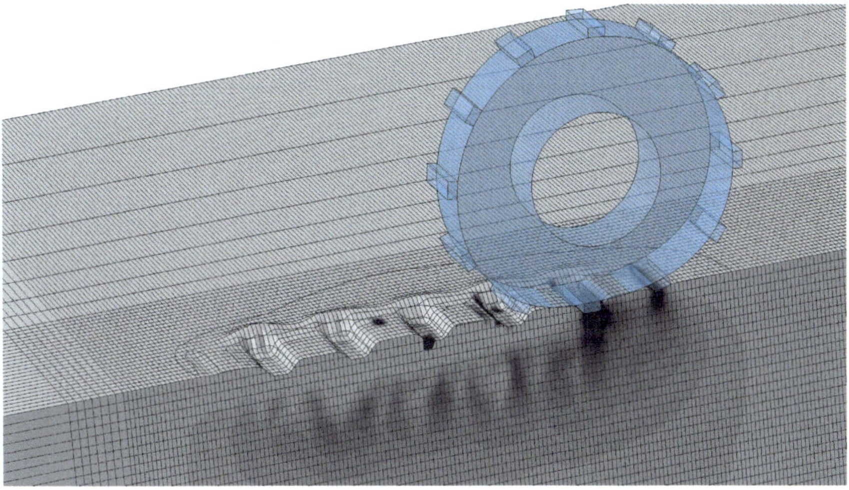

Fig. 5.7 Unpowered BB2 flexible wheel simulation on ES3-OMR soil and wheel load 180N. Contour shading denotes the pre-consolidation pressure. Credit: DELTARES

$$z_0 = \left(\frac{3W}{4E^*\sqrt{r}}\right)^{\frac{2}{3}} \quad \text{with } W\text{: wheel load, and } r \text{ the wheel radius} \tag{5.6}$$

With E*, the soil's Elastic Modulus, in [N/m^2]. This value can be optimised for a given range of wheel diameters. For Martian soil simulant, typical values between 110,000 and 800,000 have been proposed in [15].

For the slip sinkage estimation (z_s), the following options are proposed in the literature with s representing the slip:

Expression by Lyasko [16]:

$$z_s = \frac{1+s}{1-0.5\,s}\,z_0 \tag{5.7}$$

Expression by Reece [17]:

$$z_s = \frac{h\,s}{1-s} \quad \text{with } h \text{ the grouser height} \tag{5.8}$$

Expression by Richter [18]:

$$z_s = 2h\left(s - \frac{s^2}{2}\right) \quad \text{with } h \text{ the grouser height} \tag{5.9}$$

Expression determined in SWIFT [4] based on single wheel tests as in Fig. 5.5. The parameters in that formula can be derived for soils encountered in missions using for example dedicated stagnant wheel tests in situ and can be approximated from the tests on soil simulants.

$$z_s = (z_1 - z_0)s^2 + z_0 \text{ and } z_1 = p_z(r_{eq} + h) \tag{5.10}$$

with r_{eq} the equivalent wheel radius and p_z estimated to 0.7 in [4].

5.4 Simulating Motion on a Given Terrain

5.4.1 Terrain Generation

Terrain geometry can be represented in the form of a digital elevation model (DEM) where the value of each pixel of the image represents a height. Those models can be generated from in-orbit observations like HiRISE [19]. For example, the foreseen ExoMars landing site in Oxia Planum has been generated with a resolution of 25 cm per pixel. Small rocks can be added if considered relevant using probabilistic distributions as described for example by Golombek [20]. For the Moon, available DEMs can also be further enhanced with mathematical models of craters described in [5]. What is more difficult to establish is, as highlighted in Sect. 5.3.2, the nature of the soil in a given region to generate the soil map.

5.4.2 Rover Motion

Various simulation tools can be used to simulate a vehicle. When rover motion is slow (roughly < 5 cm/s), a quasi-static approximation can be made based on the Newton–Euler formulation as reported in [10]. This is helpful in early design phases when combined with wheel-soil interaction formulation (see Sect. 5.3.2) for trading-off different locomotion sub-systems and performing design iterations.

Dynamic rover motion simulations are usually conducted with multibody dynamics simulation software. The critical aspect to be considered for selecting a simulation environment for planetary exploration rover is the capability to properly capture the wheel or leg soil-interaction. Naturally, capturing the rover dynamics requires small time steps that enforce higher computation time demands, and thus tools are thus suitable for more advanced design phases. A more detail review can be found in [2].

Independently of the simulation tool used, visualization can be enhanced with 3D rendering engine. This can be of relevance for mission planning or for considering impact of the perception sensors necessary for the navigation. An open-source

Fig. 5.8 3D visualisation of the simulated ExoMars model after lander egress on Oxia Planum (Credit: Beyond Gravity)

rendering engine used in some ESA projects is Ogre3D. More recently, the ExoMars rover, lander and terrain have been imported in UNIGINE real-time 3D engine (Fig. 5.8) by Beyond Gravity engineers based on the CAD model used for manufacturing the parts. The conversion from CAD file to mesh used in most engine has been performed in Blender. This visualization can run real time and only requires positions information from the simulation that can be provided via TCP/IP.

5.5 Conclusions

Currently, there is only a relatively limited number of rovers that have been operated on Moon or Mars or that have reached a certain maturity level. Those missions demonstrate feasibility but also the challenge and limitations of moving on loose soils and naturally occurring obstacles. In particular from critical encountered situations emerged the need to get deeper understanding of soil mechanics. Such work is also essential for designing future locomotion subsystems that will require enhance motion capabilities, extension of the travel range and speed. Understanding wheel or leg interaction with the surface of planetary objects is therefore fundamental to assessing rover mobility performance and reliability, and thereby guaranteeing success of planetary rover exploration missions.

Progress in simulating wheel-soil integration as well as rover-level motion behaviour has been made in the last decades. This very brief review of terramechanics shows the range of skill and some guidelines needed to properly design the locomotion subsystem and generating key specification for the actuators, wheels, and chassis kinematic.

Acknowledgements The authors would like to acknowledge and express their thanks to all engineers and soil specialists at Beyond Gravity, Deltares, Airbus Stevenage, DLR Institutes, Swiss Federal Institute of Technology, BlueBotics, vH&S and Surrey Space Center that contribute to increase knowledge of design and operating locomotion subsystems. We also thank ESA's technical team for support in all robotic projects.

References

1. Bekker, M. G. (1969). *Introduction to terrain-vehicle systems* (p. 846). MI University of Michigan Press.
2. Thüer, T. (2009). *Mobility evaluation of wheeled all-terrain robots metrics and application.* Doctoral Thesis, DISS. ETH NO. 18160, Zürich.
3. Wong, J. Y. (2001). *Theory of ground vehicles* (3rd ed.). Wiley-Interscience. ISBN 0-471-35461-9.
4. ESA, RUAG, Deltares (2014). *Parametric tool design description. ESA project SWIFT, phase 2, TN12a.* Deltares report 1204498-003 version 2 for RUAG and ESA, 68 pp.
5. Handbook of Lunar Soils. (1982). SN4/NASA Johnson Space Center, Houston, TX 77058.
6. Gallina, A., et al. (2014). Parameter identification of a planetary rover wheel-soil contact model via a Bayesian approach. *Journal of Field Robotics, 31*(1), 161–175.
7. Heverly, M., Matthews, J., Lin, J., Fuller, D., Maimone, M., Biesiadecki, J., & Leichty, J. (2013). Traverse performance characterization for the mars science laboratory rover. *Journal of Field Robotics, 30*(6), 835–846.
8. Randel, A., et al. (2006). Mars exploration rover mobility development, mechanical mobility hardware design, development, and testing. *IEEE Robotics & Automation Magazine.*
9. Michaud, S., et al. (2011). Exomars locomotion subsystem analytical tool development and correlation. In *Proceeding of the ASTRA 2011*, ESA/ESTEC, Noordwijk, the Netherlands.
10. Oettershagen, P., & Michaud, S. (2012). Development and validation of a modular parametric analytical tool for planetary exploration rovers. In *Proceeding of the I-SAIRAS 2012*, Turin, Italy.
11. Iskenderian, T. (2004). Deployment process, mechanization, and testing for the mars exploration rovers. In *Proceedings of the 37th Aerospace Mechanisms Symposium*, Johnson Space Center, May 19–21, 2004.
12. Michaud, S., et al. (2012). Wheel level test data generation and utilization to predict locomotion performances of planetary rovers and validate simulation tools. In *Proceeding of the I-SAIRAS 2012*, Turin, Italy.
13. Arvidson, R. E., et al. (2014). Terrain physical properties derived from orbital data and the first 360 sols of Mars Science Laboratory Curiosity rover observations in Gale Crater. *Journal of Geophysical Research: Planets, 119*, 1322–1344. https://doi.org/10.1002/2013JE004605
14. Senatore, C., & Iagnemma, K. (2014). Analysis of stress distributions under lightweight wheeled vehicles. *Journal of Terramechanics, 51*, 1–17.
15. Wong, J. Y. (2012). Predicting the performances of rigid rover wheels on extraterrestrial surfaces based on test results obtained on earth, *Journal of Terramechanics* 49–61
16. Lyasko, M. (2010). Slip sinkage effect in soil-vehicle mechanics. *Journal of Terramechanics, 47*, 21–31.
17. Reece, A. R. (1964). Problems of soil-vehicle mechanics. US Army Land Locomotion Lab, ATAC, NO 8479, Mich: Warren.

18. Richter, L., Ellery, A., Gao, Y., Michaud, S., Schmitz, N., & Weiß, S. (2006). A predictive wheel soil interaction model for planetary rovers validated in testbeds and against MER Mars rover performance data. In *10th European Conference of the International Society for Terrain-Vehicle Systems, Budapest*.
19. HiRISE website: https://hirise.lpl.arizona.edu/
20. Golombek, M. P. (2002). *Rock statistics calculations for the MER landing sites, landing site workshop*, Pasadena, CA.
21. Bekker, M. G. (1985). The development of a moon rover. *JBIS, 38*(12), 537–543.

Chapter 6
ESROCOS: Development and Validation of a Space Robotics Framework

Miguel Muñoz Arancón, Malte Wirkus, Kilian Hoeflinger, Nikolaos Tsiogkas, Saddek Bensalem, Olli Rantanen, Daniel Silveira, Jérôme Hugues, Mark Shilton, and Herman Bruyninckx

Abstract The H2020 ESROCOS project has built an open-source framework for the development of space robotics software. The ESROCOS framework provides a model-based engineering approach and a collaborative development process. It integrates proven and new technologies, providing modelling and verification tools as well as reusable components that facilitate the production of high-quality robot control software. While other robotics frameworks exist, ESROCOS is designed

---

M. Muñoz Arancón (✉)
GMV Aerospace and Defence, Tres Cantos, Spain
e-mail: mmunoz@gmv.com

M. Wirkus
Robotics Innovation Center, Deutsches Forschungszentrum Für Künstliche Intelligenz GmbH (DFKI), Kaiserslautern, Germany

K. Hoeflinger
Deutsches Zentrum für Luft- und Raumfahrt e.V., Cologne, Germany

N. Tsiogkas · H. Bruyninckx
KU Leuven, Leuven, Belgium
e-mail: herman.bruyninckx@kuleuven.be

N. Tsiogkas
Intermodalics BVBA, Leuven, Belgium

S. Bensalem
Universite Grenoble Alpes, Grenoble, France

O. Rantanen
VTT Technical Research Centre of Finland Ltd, Espoo, Finland

D. Silveira
GMVIS Skysoft SA, Lisbon, Portugal
e-mail: daniel.silveira@gmv.com

J. Hugues
Software Engineering Institute, Carnegie Mellon University, Pittsburgh, USA

Institut Superieur de L'Aeronautique et de L'Espace, Toulouse, France

M. Shilton
Airbus Defence and Space Ltd, Stevenage, UK

X. T. Yan and G. Visentin (eds.), *Space Robotics*,
https://doi.org/10.1007/978-3-031-39214-6_6

from the ground up to fulfil the needs of the space robotics community, which requires sound software engineering processes. This paper presents the outcomes of the ESROCOS project, the evaluation activities performed in three case studies (on-orbit servicing, planetary exploration and nuclear industry), and the results of this evaluation.

6.1 Introduction

The development of software for robotic systems has become easier with the emergence of robotics frameworks, with ROS, the "Robot Operating System" [1], being the most popular among them. The term Robot Control Operating System (RCOS) has evolved, describing frameworks like ROS and others with a similar aim, such as OROCOS [2], GenoM [3], ROCK [4] and CLARAty [5]. It can be defined as a framework for the development of robotics applications that may provide a component model, a runtime environment and a set of tools and reusable components to build robotics software. An RCOS allows the user to easily combine generic and application-specific components to build applications, and provides tools to support development and testing.

The RCOS has become an essential tool in robotics research and in industry. The standardization of components and interfaces through a component model, together with the use of common tools and practices, fosters a software ecosystem that facilitates the development of applications. However, existing RCOS are not suitable for critical applications with stringent Reliability, Availability, Maintainability and Safety (RAMS) requirements. It is not feasible to modify an existing RCOS to comply with the requirements of critical software development. Therefore, to leverage the benefits of the RCOS concept in critical domains, specifically developed frameworks are needed. Yet, past efforts to develop a standard robot control software at ESA failed to be adopted by the community, even if they achieved their immediate technical objectives.

To address the issues that prevented a wider adoption, the PERASPERA project (Plan European Roadmap and Activities for Space Exploitation of Robotics and Autonomy) [6] identified the need for an open-source RCOS as one of the "building blocks" to be produced in the frame of the first call of activities of the H2020 Space Robotics Technologies Strategic Research Cluster (SRC). The ESROCOS project

H. Bruyninckx
TU, Eindhoven, The Netherlands

Flanders Make, Kortrijk, Belgium

K. Hoeflinger
thya aerospace, Munich, Germany

O. Rantanen
Metso, Tampere, Finland

[7] was selected to develop this building block, and it has aimed to answer the aforementioned needs by:

- Developing an RCOS specifically designed for space software processes and technologies.
- Integrating advanced modelling technologies to support the design and verification of the architectural and behavioural properties of the system from the early stages.
- Focusing on the space robotics community, addressing their needs and use cases.
- Coping with the complexity of robotics applications, both at development and at execution time.
- Avoiding proprietary solutions and vendor lock-in by using open-source licenses.
- Leveraging existing tools and reusable components, benefiting from their maturity and user base.
- Exploring the needs of other domains with stringent RAMS requirements, such as nuclear robotics.

The ESROCOS project was executed by a consortium of ten industrial companies and academic institutions led by GMV, and spanned over 27 months from Nov-2016 to Jan-2019.

The following sections present an overview of the ESROCOS framework, detail the supported development approach and the tools and components provided by the framework, present the evaluation of ESROCOS on three case studies, and summarize the status of the framework and the future work.

6.2 Overview of the ESROCOS Framework

ESROCOS is a framework for developing robot control software applications. It includes a set of tools that support different aspects of the development process, from architectural design to deployment and validation. In addition, it provides a set of core functions that are often used in robotics or space applications.

The ESROCOS framework is intended to support the development of software following the ECSS standards for space software. It does not by itself cover all the development phases and verification steps, but it facilitates certain activities and ensures that the software built can be made compatible with the Reliability, Availability, Maintainability and Safety (RAMS) requirements of critical systems.

Figure 6.1 summarizes the main elements of the ESROCOS framework. At the top of the figure there are tools for robot and software modelling. These are supported by the common data types for component interfacing, basic libraries for robotics functions, test tools, and monitoring and control services. The middleware layer allows for the management and communication of software components at runtime. Time and space partitioning capabilities are provided to isolate application components of

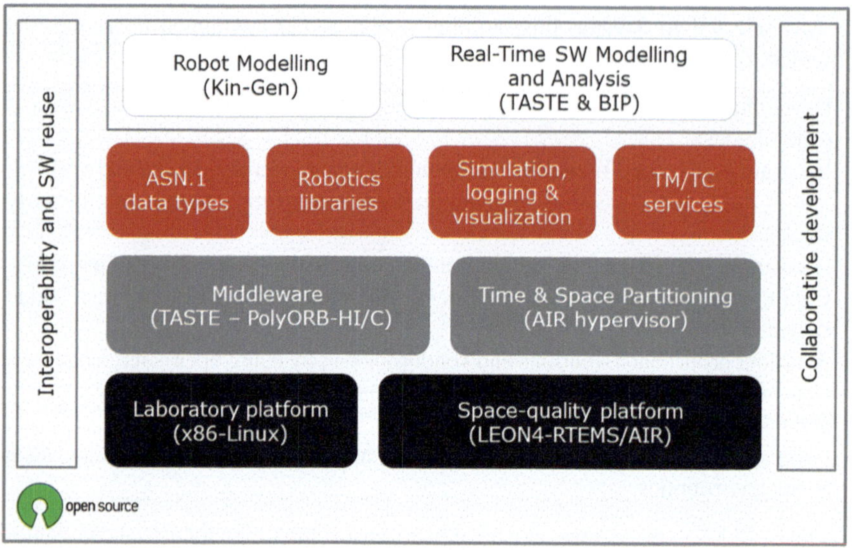

Fig. 6.1 The ESROCOS framework

at runtime and prevent failure propagation between components. Finally, the applications may run laboratory platforms, space-quality hardware or a combination of both.

The boxes on the sides of the figure represent orthogonal concerns. Firstly, ESROCOS integrates with third-party frameworks to leverage existing code and tools. Secondly, ESROCOS supports a collaborative development approach based on component reuse. Finally, the figure highlights that ESROCOS is open-source and relies on non-proprietary technologies in order to encourage usage and contributions from the community.

6.3 Development with ESROCOS

The requirements of space robotics systems may lead to complex software and hardware architectures that combine heterogeneous computer nodes and software functions. Capabilities such as fast control loops, sensor data processing, autonomy functions or monitoring and control, all of which may be needed in a typical application, have very different requirements. In addition, the design and validation of the application often requires specific tools for simulation, data recording and replaying, or visualization.

The ESROCOS framework provides a heterogeneous set of tools and reusable components that address different needs of the robotics applications. The framework relies on a model-based engineering approach, and in particular on the TASTE toolchain (see Sect. 6.3.1), to link together the different elements.

The model-based approach was a requirement of the ESROCOS project. It provides separation of concerns between the different design aspects and the deployment details, and enables early formal verification activities to reduce the amount and impact of design errors.

ESROCOS supports model-based development in two dimensions:

- Modelling of the robot system, in particular of the robot kinematics, using the Kin-Gen tool (see Sect. 6.3.3).
- And modelling and validating the software architecture and behaviour, using TASTE (see Sect. 6.3.1) and BIP (see Sect. 6.3.2).

In addition, ESROCOS provides an infrastructure to maintain software implementations divided in a number of individual packages, where each package can have its own development cycle. Combined with a set of common robotics data types for modelling the interfaces between components, this mechanism allows the user to easily combine existing and new software blocks to design an application. By enabling the reuse of components developed by the community, and facilitating that newly developed components are published and contributed back, ESROCOS makes it possible that a user community of commercial vendors and open-source projects emerges and supports a lively software ecosystem around the framework.

The following subsections describe in more detail the main tools and components provided by the ESROCOS framework.

6.3.1 The TASTE Toolchain

TASTE (The ASSERT Set of Tools for Engineering) [8] is a toolchain developed by the European Space Agency (ESA) to model heterogeneous real-time distributed systems. It is the backbone of ESROCOS. At development time, it allows the user to model the hardware and software architecture of the system.

The user models the system in TASTE using four views, each addressing a different aspect of the system:

- Data view: describes the data definitions of the system using the ASN.1 standard, as well as data encoding aspects.
- Interface view: models the functional aspects of the system, structured in components with defined real-time behaviour and interfaces; it is described using the AADL standard [9].
- Deployment view: describes the hardware architecture of the system as a set of nodes and data buses, and the allocation of software components and interfaces to these; it is also described with AADL.

- Concurrency view: generated from the previous two, this view represents the system in terms of real-time software artifacts (tasks, subprograms, etc.), enabling the schedulability analysis of the system by automated tools and the tuning of different parameters by the user.

The TASTE editor allows the user to model the system using these four views, and to validate the model. The interface and deployment views are edited graphically, while the data view is textual. From the model, TASTE generates code using the Ocarina tool. This code manages the system initialization, task execution and communication aspects, using the PolyORB-HI middleware. The user must then provide the implementation of the individual components. TASTE supports different languages such as Ada, C, C++, SDL state machines, VHDL (for supported FPGAs) or Simulink. TASTE supports several hardware platforms, including RTEMS on LEON processors and Linux on $\times 86$, and communication links such as TCP/IP and SpaceWire.

Individual TASTE components can be exported for reuse in different TASTE applications and deployed to different architectures and communication buses. The toolchain thereby takes care of the compilation, communication and runtime details. To allow the independent development of compatible software components, the ESROCOS framework provides a library of common data types and provides support in development with specific components, e.g., for monitoring and control and data visualization (see Sec. 6.4).

6.3.2 The BIP Tools

ESROCOS complements the TASTE modelling capabilities with BIP (Behaviour, Interaction, Priority) [10], a formal language for modelling and analysing heterogeneous real-time systems. The BIP language and tools go further than TASTE in capturing the behaviour of the software and the interaction between the components, allowing for new scalable formal verification techniques of the system model.

The following BIP tools can be used to support the modelling and analysis of robotics applications built with ESROCOS:

- BIP Compiler & Engines: the BIP compiler generates C++ code that can be linked to different engines that support model execution and simulation for analysis.
- iFinder/iChecker: it allows for the verification of requirements (safety, performance, etc.) on BIP models.
- Statistical Model Checking (SMC-BIP): it runs a representative number of simulations and statistically checks whether a requirement (safety, performance, etc.) is satisfied.
- FDIR tool: permits the generation of fault detection and recovery software components that can be embedded in the target system.

The integration of BIP in ESROCOS is done via an automated model transformation from TASTE to BIP models.

6.3.3 The Kin-Gen Kinematics Modelling Tool

The Kin-Gen tool [11] has been developed within the ESROCOS project to model the characteristics of the robotic platform. Kin-Gen allows the user to model the robot kinematics using an innovative approach based on model composability. The robot is modelled as a set of kinematic trees and an arbitrary frame. The model can be verified with semantic checks. Additionally, custom solvers can be defined as queries to the model, and code can be generated for these custom solvers, ensuring correctness.

The tool provides a complete toolchain for forward and inverse kinematics. Future evolutions of the tool will cover other aspects of the robot system and its environment.

6.3.4 The AIR Hypervisor

Robotics software must combine functions ranging from real-time control loops to non-deterministic planning algorithms, which differ in their criticality level and execution constraints. This may be an issue for the development of critical robotics software, which must satisfy the safe execution of certain critical functions.

The Time and Space Partitioning (TSP) paradigm consists of separating different applications running on the same on-board computer by enclosing them in isolated partitions with their own assigned memory and processor utilization. The aim is the execution of applications with different criticality in such a way that a failure in one cannot affect the others. ESROCOS includes the AIR hypervisor [12], an ARINC 653 [13] compliant hypervisor that uses paravirtualisation to enable the execution of partitioned applications and supports RTEMS and ARINC 653 APEX as guest operating systems.

6.4 Support Tools and Components

In addition to the elements described above, the ESROCOS framework includes other tools and components to support the development and testing of space robotics applications.

The most important is a collection of common robotics data types modelled in ASN.1. These types, derived from the ROCK and ROS frameworks, represent common robot and sensor data and allow modelling the interfaces between reusable components.

In addition, a set of libraries that support common robotics and space software functionalities are provided. These libraries are prepared for easy integration in TASTE applications.

- Transformer: a C++ library to model and compute at runtime geometric frame transformations.
- Stream Aligner: a C++ library to manage the synchronization of multiple data streams.
- Data Logger: a C++ library to allow for data logging and replay to support testing.
- Data Visualization: a C++ library for robot and sensor data visualization in 3D based on Vizkit3D.
- PUS Library: a missionizable C library that provides a subset of the ECSS Packet Utilization Standard (PUS) [14] for monitoring and control of space assets. The library implements the services at logical level (the packet encoding uses ASN.1 instead of the standard PUS frames). It includes an On-Board Control Procedure (OBCP) engine based on MicroPython [15], allowing for the simultaneous execution of several OBCPs that can interact with the on-board software system.

ESROCOS includes tools for interoperability with the ROS and ROCK frameworks so that prospective users can leverage the rich ecosystems existing around these frameworks as well as their existing software assets. The aim is to facilitate the transition from the laboratory to the level of quality and verification demanded by space systems. To this end, ESROCOS provides:

- Tools to import data type definitions from ROS and ROCK to the ASN.1 format used by TASTE.
- Tools to generate bridge components that enable seamless communication between TASTE functions and ROS or ROCK components running on a separate middleware environment.

Finally, ESROCOS integrates with external tools and libraries commonly used in robotics software development, such as the OpenCV image processing library, the Eigen linear algebra library, the Gazebo simulator or the RVIZ data visualization tool.

6.5 Installation of ESROCOS

ESROCOS is available as open-source software in public repositories (see Sect. 6.7 below). The framework relies on the Autoproj tool to manage both the components of the framework itself and those of the user application. Each component is hosted in its own repository and includes a manifest file declaring its dependencies. The Autoproj tool handles the building and installation of the framework and the applications using this information. The fact that the framework and the user application components are indistinguishable from each other makes it possible that individual components can be contributed back to the ESROCOS ecosystem, and that the users of the framework can seamlessly mix components provided by the framework, by other suppliers and by their own organisation.

The installation of ESROCOS creates a workspace hosting both framework components and user applications. The setup is organized in three package sets that define the components available in the workspace:

- Core: the components provided by the framework.
- External: existing tools and libraries that the core components depend on.
- Universe: contributed packages (drivers, algorithms, etc.) contributed by third parties.

As an ecosystem develops around the ESROCOS framework, the Universe package set will be extended with new capabilities and functions that can be reused in new applications.

6.6 Case Studies

The validation in the space reference scenarios took place in two test facilities provided by the H2020 FACILITATORS project [16] in the frame of the Space Robotics Technologies SRC. In addition, the Finnish Technology Research Centre VTT was part of the project consortium and provided a nuclear robotics test facility to perform a case study based on a terrestrial application.

The ESROCOS framework was evaluated in three case studies by developing representative applications targeting on-orbit satellite servicing, planetary exploration and nuclear robotics. The focus of the evaluation was the production and testing of a set of robotics applications that exercise the different elements of ESROCOS.

The reference applications addressed the functional layer of the robotics systems, specifically for a rover and two manipulator arms. The aim of the tests was not to build fully autonomous systems, but to demonstrate the software capabilities that will allow to do so in the future, considering that the future activities of the Space Robotics Technologies SRC will integrate other building blocks providing higher level functions, such as data fusion and autonomy.

6.6.1 Planetary Scenario

The planetary exploration scenario was demonstrated using the BRIDGET rover in Airbus DS Mars Yard facility at Stevenage (UK). The tests included the tele-operation of the rover, logging and replay capabilities and control of a CAN device. As in the orbital scenario, both laboratory (Linux PC) and space-representative (GR740) hardware were used (Fig. 6.2).

Fig. 6.2 Testing in the planetary exploration reference scenario: BRIDGET rover in the Mars Yard facility

The case study consisted of four test cases. For each of them, an application was modelled with TASTE and implemented using the different components provided by the framework. The test cases addressed:

- Rover teleoperation and control.
- Rover trajectory data logging and replay.
- CAN bus device management.
- Integration of data processing functions.

The test campaign demonstrated the functionality of the software modules Stream Aligner, Data Logger, Transformer, as well as the possibility to use BIP for modelling and executing safety routines for robotic systems.

6.6.2 Orbital Scenario

The on-orbit satellite servicing scenario was demonstrated at GMV's platform-art facility in Tres Cantos (Spain). The tests consisted in the teleoperation of a UR-5 manipulator with a camera as end effector to inspect a satellite mock-up. Different variants of the scenario were tested, including simulation aspects and targeting laboratory-type hardware (Linux PC) and space-representative avionics (GR740 board running RTEMS) (Fig. 6.3).

The case study consisted of four test cases with their respective ESROCOS applications:

- Control of the manipulator on simulation, using a laboratory platform (Linux on $\times$ 86).

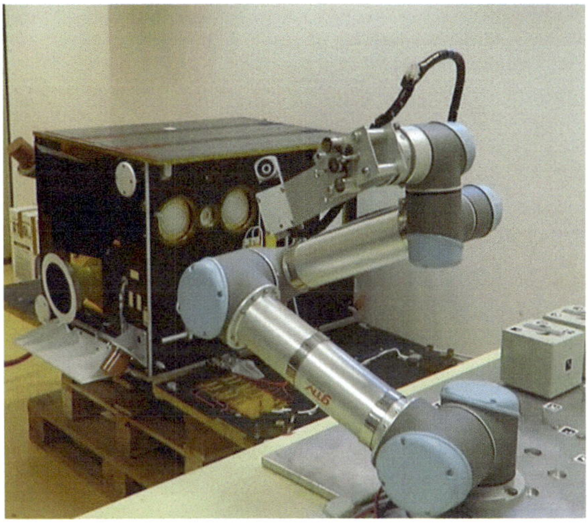

Fig. 6.3 Testing in the on-orbit servicing reference scenario: UR-5 manipulator and satellite mock-up in GMV's robotics laboratory

- Control of the actual manipulator and camera using a laboratory platform (Linux on $\times$ 86).
- Control of the actual manipulator using space-representative hardware (GR740 board running RTEMS) and SpaceWire data bus.
- EtherCAT driver testing (however, this test case could not be performed, see Sect. 6.6.4).

The test case applications were modelled using TASTE, BIP and Kin-Gen, and demonstrated the capabilities of the Data Visualization, integration of the Gazebo simulator, PUS Services and interoperability with ROS and ROCK provided by ESROCOS.

6.6.3 Nuclear Scenario

The nuclear robotics demonstration took place at the VTT Diverter Test Platform 2 at Tampere (Finland). A control application for the Cassette Multifunctional Mover (CMM) robot was developed with ESROCOS and deployed at the control centre. The CMM robot is a robotics demonstrator for the ITER experimental fusion reactor, intended for the replacement of the elements that compose the bottom part, so called divertor, of the toroidal enclosure of the reactor (Fig. 6.4).

The case study included two test cases with their respective ESROCOS applications:

- Manipulator trajectory planning and control using interpolation in Cartesian space.
- Manipulator trajectory planning and control using interpolation in joints space.

Fig. 6.4 Testing in the nuclear robotics scenario: CMM robot in the DTP2 facility at VTT

Both applications ran on laboratory hardware (Linux on ×86) and interacted with the CMM controller through a UDP bus. The applications were modelled in TASTE and included a kinematic solver generated with Kin-Gen. The tests also demonstrated the data visualization functionality and the integration with the Gazebo simulator.

6.6.4 *Test Results, Constraints and Limitations*

The results of the evaluation of ESROCOS have been in general satisfactory, proving that the framework is functional and useful to support the development of space robotics applications following a sound modelling, implementation and verification approach.

Nevertheless, not all the technical objectives of the project have been achieved, and the case studies have shown some limitations that should be taken into account by prospective users of the system:

- The installation of the ESROCOS framework is not fully managed by Autoproj, and some dependencies and components need to be manually installed.
- The TASTE toolchain has some limitations for large application models, exhibiting excessive memory usage and long build times.
- The driver configuration in TASTE is complex, with some elements configured at model level and some others at middleware source code level.
- The transformation of TASTE models to BIP is partially automated, and some concepts in the deployment view must be manually modelled.
- The integration of the AIR hypervisor in TASTE is not complete, and currently it is not possible to generate code for models that combine AIR and other types of hardware nodes.

- The SOEM EtherCAT driver could not be ported to the GR740 board selected as space-representative avionics board due to the use of a deprecated network stack in RTEMS for LEON4.
- The ESROCOS runtime components have not been verified to the level required by flight software in terms of unit testing, coding rules or code review.

Despite these shortcomings, the results of the validation of ESROCOS have been positive, and the framework is usable in its current state to support the development and verification of robotics applications.

6.7 Status and Future Work

The ESROCOS project was successfully completed in January 2019. The public results from the project, including the open-source software, can be found at the project's website [17] and GitHub page [18]. The evolution of ESROCOS continues in the second phase of the Space Robotics Technologies SRC that will use and extend the framework to support the development of several application demonstrators. The H2020 MOSAR project [19], in particular, is in charge of the maintenance of the ESROCOS building block during this phase of the SRC, coordinating the corrections and extensions of the framework in the different projects.

In addition to the usage and maintenance foreseen within the SRC, the fact that ESROCOS is based on technologies that have a trajectory outside the project will also contribute to the success of the platform. The ESROCOS framework may benefit from the further development of its constituent tools, and likewise improvements in the framework may reflect on the tools, creating a live ecosystem.

6.8 Conclusions

The ESROCOS project has developed a robotics framework that addresses the needs of the space robotics community. The framework was validated by developing robotic applications for three reference scenarios: on-orbit servicing, planetary exploration and nuclear robotics. The resulting RCOS will be used as a building block in future Space Robotics Technologies SRC activities, and is available for wider use by the community under open-source licenses.

References

1. Quigley, M., Gerkey, B., Conley, K., Faust, J., Foote, T., Leibs, J., Berger, E., Wheeler, R., & Ng, A. (2009). ROS: an open-source Robot Operating System. *ICRA Workshop on Open Source Software*, *3*(3.2), 5.

2. Bruyninckx, H. (2001). Open robot control software: the OROCOS project. In *IEEE International Conference on Robotics and Automation (ICRA)* (Vol. 3, pp. 2523–2528). IEEE.
3. Ceballos, A., De Silva, L., Herrb, M., Ingrand, F., Mallet, A., Medina, A., & Prieto, M. (2011). GenoM as a robotics framework for planetary rover surface operations. In *11th Symposium on Advanced Space Technologies in Robotics and Automation (ASTRA). ESA.*
4. The Robot Construction Kit (ROCK). Online at: https://rock-robotics.org
5. Nesnas, I. A. (2007). CLARAty: A collaborative software for advancing robotic technologies. In *Proceedings of NASA Science and Technology Conference* (Vol. 2). NASA.
6. PERASPERA (Plan European Roadmap and Activities for Space Exploitation of Robotics and Autonomy). Online at: https://www.h2020-peraspera.eu/
7. Muñoz Arancón, M., Montano, G., Wirkus, M., Hoeflinger, K., Silveira, D., Tsiogkas, N., Hugues, J., Bruyninckx, H., Dragomir, I., & Muhammad, A. (2017). ESROCOS: A robotic operating system for space and terrestrial applications. In *14th Symposium on Advanced Space Technologies in Robotics and Automation (ASTRA). ESA.*
8. Perrotin, M., Conquet, E., Dissaux, P., Tsiodras, T., & Hugues, J. (2010). The TASTE Toolset: turning human designed heterogeneous systems into computer built homogeneous software. *Embedded Real Time Software and Systems (ERTS2).*
9. Architecture Analysis & Design Language (AADL). (2017). SAE International. AS5506C.
10. Basu, A., Bozga, M., & Sifakis, J. (2016). Modeling heterogeneous real-time components in BIP. In *Fourth IEEE Conference on Software Engineering and Formal Methods (SEFM)* (pp. 3–12). IEEE.
11. Frigerio, M., Scioni, E., Pazderski, P., & Bruyninckx, H. (2019). Code generation from declarative models of robotics solvers. In *Third IEEE International Conference on Robotic Computing (IRC)* (pp. 369–372). IEEE.
12. Gomes, B., Silveira, D., Gouveia, L., & Mendes, L. (2019). Air hypervisor using RTEMS SMP. In *European workshop on on-board data processing (OBDP2019)*, ESTEC (ESA).
13. RTCA. (October, 2019). ARINC653, Arinc Specification 653-1. Avionics Application Software Standard Interface. RTCA.
14. Space Engineering: Telemetry and Telecommand Packet Utilization. ECSS-E-70-41C (April 15, 2016). European Cooperation for Space Standardization (ECSS).
15. MicroPython. Online at: https://micropython.org/
16. H2020 FACILITATORS project: Facilities for Testing Orbital and Surface Robotics Building Blocks. Online at: https://cordis.europa.eu/project/id/730068
17. The ESROCOS project: European Space Robotics Control and Operating System. Online at: https://www.h2020-esrocos.eu/
18. ESROCOS at GigHub. Online at: https://github.com/ESROCOS
19. H2020 MOSAR project: Modular Spacecraft Assembly and Reconfiguration. Online at: https://www.h2020-mosar.eu/.

Chapter 7
Legged Systems for Exploration

Hendrik Kolvenbach, Philip Arm, Giorgio Valsecchi, Nikita Rudin, and Marco Hutter

Abstract Walking robots have advanced rapidly over the last decade and have reached a maturity level where commercial applications have become viable. However, robotic exploration of celestial bodies has been performed solely with wheeled systems until today, making access to highly unstructured, compressible, and sloped areas very challenging. Thus, it is only a matter of time until walking robots will be used to explore so far unreachable regions of our solar system. In this chapter, we investigate legged systems in the context of space exploration and perform a brief overview of key aspects of the technology. In detail, we review state-of-the-art focusing on terrestrial machines and space-legged robot prototypes. We analyse the specifics of the hardware design of those machines, namely the actuation and leg design and control aspects, with a focus on model-based and learning-based control. Lastly, we analyse recent developments with regard to analogue mission deployments and locomotion tests.

7.1 Introduction

Future robotic exploration missions into our solar system are characterized by increasingly challenging terrain. Scientists are interested in investigating diverse geological areas, which are extremely difficult or even impossible to reach with

H. Kolvenbach (✉) · P. Arm · G. Valsecchi · N. Rudin · M. Hutter
Robotic Systems Laboratory, ETH Zurich, Leonhardstrasse 21, 8092 Zurich, Switzerland
e-mail: hendrikk@ethz.ch; Hendrik.Kolvenbach@mavt.ethz.ch

P. Arm
e-mail: Philip.Arm@mavt.ethz.ch

G. Valsecchi
e-mail: Giorgio.Valsecchi@mavt.ethz.ch

N. Rudin
e-mail: Nikita.Rudin@mavt.ethz.ch

M. Hutter
e-mail: Marco.Hutter@mavt.ethz.ch

X. T. Yan and G. Visentin (eds.), *Space Robotics*,
https://doi.org/10.1007/978-3-031-39214-6_7

traditional, wheel-based robots due to the highly scattered surface, steepness, or uncertainty of the terrain properties.

Scientists are interested in investigating diverse geological areas, which are extremely difficult or even impossible to reach with traditional, wheel-based robots due to the highly scattered surface, steepness, or uncertainty of the terrain properties. Scientists are interested extreme environments because outcrops and exposed bedrock, such as in craters or close to a cliff, give insights into the planet's geological history. Plainly speaking, the more diverse the site, the higher the potential scientific return. The visible layers within the Victoria crater (Fig. 7.1a) on Mars, for example, gave hints about the groundwater processes [1]. A closer observation or sampling of the cliff's material was not possible, as a full descend into the crater with the Opportunity rover was considered too dangerous. Cliffs or escarpments, in general, are also offering so-called micro-environments, which, because of the geometry, offer local variation in the climate that have the potential to harbour life (Fig. 7.1b, c, e) [2, 3].

The polar regions of Mars (Fig. 7.1d, f) are also attractive as they offer various forms of ice deposits that change depending on the season [4]. A potential mission scenario could involve a traverse towards the poles while passing by craters and gullies with forming ice deposits. Icy cliffs are an excellent example of open and actively eroding terrain, which offers insights into the climate of the planet [5].

However, due to the uncertainty of the encountered soil and the necessity to overcome hilly terrain, a wheeled system might be the limiting factor in such a mission.

The Moon, on the other hand, is mostly covered in Regolith, which means that geologically interesting locations at which exposed bedrock can be found are impact craters or caves. Impact craters feature exposed, relatively fresh material in the ejecta sheet and direct access to outcrops. The central peak within the Aristarchus crater (Fig. 7.1g), for example, features potentially some of the deepest crustal materials of the Moon [6]. In-situ investigations via sampling would offer insights into the composition, formation, and evolution of our closest neighbour's lunar volcanic deposits. Additionally, permanently shadowed craters like the Shackleton crater (Fig. 7.1h) offer potential water ice and unique geological environments, to which an exploration robot might traverse to investigate [7]. Water on the Moon's surface would be vital for a future human settlement since transport from Earth is very costly. Additionally, splitting water into hydrogen and oxygen via solar power plants would sustain a supply of breathable air and rocket fuel, which could be used to travel further into the solar system. The only way to reach those areas of interest, given that landing within the crater is too risky with current technology, is a challenging ascent down from the crater rim.

Lunar pits (Fig. 7.1i) are another unique environment since the circumference of the pit directly exposes visible sediment layers [8]. Lava tubes below the lunar surface might be a potential place where a human settlement could be placed due to the natural protection from radiation and impacts. Lava tunnels are also prone to be present on most planetary bodies with a history of volcanic activities. However,

Fig. 7.1 Several high-priority targets on the Moon and Mars are difficult to reach with traditional wheel-based locomotion. Exemplary shown are on Mars: **a** Edge of the Victoria crater at Meridiani Planum. **b** Cliffs at Meridiani Planum. **c** Aeolis Mons "Mount Sharp" in the Gale crater. **d** South pole ice cap. **e** Visible ice on cliff faces (interior structure is shown in false color). **f** Korolev crater at Planum Boreum (northern polar plain); and on the Moon: **g** Aristarchus crater. **h** Shackleton crater (The crater's interior structure is shown in false color based on data from NASA's LRO probe). **i** Lunar pit located in the Mare Tranquilatis (*Photo courtesy of* NASA/JPL-Caltech/UA /Cornell/ Ohio State/GSFC/ESA/DLR/FU)

those tunnels are likely composed of highly unstructured terrain and are requiring versatile and capable exploration robots.

Beyond the Moon or Mars, there are many more exciting worlds in our solar system waiting to be explored but are characterized by an extremely complex surface geometry. Examples include icy worlds like Enceladus and Europa or places that are partially covered by liquids such as Titan. All in all, those examples highlight the need for highly versatile surface exploration robots.

Until now, solely wheeled locomotion has been used for mobile robotic exploration of the Moon and Mars. While providing high stability and robustness on flat ground, those systems are limited in their area of operation.

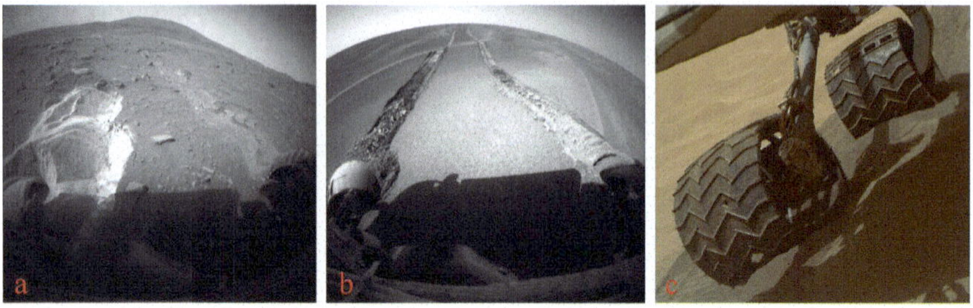

Fig. 7.2 Rovers on Mars experienced several high-slip and sinkage events. Depicted are: **a** Spirit. **b** Opportunity. **c** Curiosity (*Photo courtesy of* NASA/JPL-Caltech/MSSS)

Especially in unstructured environments with steep, sandy slopes and boulders patches, wheeled systems reach their limitations. In the past, several high slip and excessive sinkage events have been encountered with exploration rovers, which have severely disrupted the mission timeline [9] (Fig. 7.2). For example, it took five weeks to free the Opportunity rover from sand in 2006 [10], and rover trajectories are frequently adjusted to avoid challenging terrain [11]. The potentially worst situation occurred in 2009 when the Spirit rover got stuck in the sand and was unable to recover, ultimately ending the mission [12]. Similar observations have also been made on the Moon with Lunokhod [13] and the Apollo Lunar Roving Vehicle [14]. Future missions are also interested in providing faster mobility, traversing larger distances, or operating under time constraints, which constrain the challenge even more [15].

The highly demanding terrain in those regions, as mentioned above, creates an opportunity for versatile legged robots. Legged robots can traverse steep inclines, climb over obstacles, and open up previously unreachable sites for detailed scientific investigations.

In the following, we want to assess the current state of space legged robot prototypes and their dynamically walking counterparts. We further give an introduction about the particularities of those systems and ongoing field validation campaigns.

7.2 Legged Systems for Planetary Exploration

7.2.1 *Space Legged Robots*

The space legged robots developed so far are mostly multilegged walking robots [16], as seen in Fig. 7.3. Perhaps the first robots highlighting the potential of legged locomotion for exploration of steep, natural terrain were the eight-legged Dante robots developed at Carnegie Mellon University (CMU) in 1993 [17, 18]. The German Research Centre for Artificial Intelligence (DFKI) has started to develop a number of six-legged walking robots for space exploration around 2007 with Scorpion [19],

Crex [20] and SpaceClimber [21]. The German Aerospace Centre has introduced the six-legged walker DLR Crawler [22] in 2009, and the small quadrupedal robot Bert [23] in 2018.

The German Research Centre for Information Technology (FZI) has developed the six-legged Lauron robots, most notably Lauron V [24] in 2014, which can grasp objects with its limbs. Of these systems, SpaceClimber in specific is designed for a space exploration mission on the Moon and has space-qualified actuators and electronics in combination with low energy consumption and low computational cost. The robot has been tested under varying inclinations on a lunar slope test track covered by crushed, loose basalt.

The DFKI also developed the four-legged Mantis robot [25], which can use two of its six limbs for manipulation. Our group at ETH Zurich proposed using a quadrupedal

Fig. 7.3 Selection of space legged robot prototypes: **a** CMU Dante II. **b** DFKI Scorpion. **c** DFKI SpaceClimber. **d** NASA JPL LEMUR. **e** DLR Crawler. **f** DLR Bert. **g** FZI Lauron V. **h** ETH Zurich ALoF. **i** ETH Zurich SpaceBok (*Photo courtesy of* CMU/DFKI/NASA/JPL/DLR/FZI/ETHZ)

Fig. 7.4 Selection of leg-wheel hybrids for planetary exploration: **a** DFKI Asguard v4. **b** NASA JPL ATHLETE. **c** NASA JPL RoboSimian (*Photo courtesy of* DFKI/NASA/JPL)

robot ALoF for ESA's Lunar Robotics Challenge in 2010, during which the robot performed an inching motion to crawl on the demanding terrain [26]. Scaled plates on the bottom of the robot and the shanks helped hereby to generate traction but lead to significant side-slip on the slope. In 2019, we have developed SpaceBok, a dynamically walking robot with the goal of investigating dynamic locomotion for space exploration [27]. Another limbed system named Lemur has been developed by NASA JPL [28]. The system shows remarkable free-climbing ability by using microspine grippers to grasp onto rocks, allowing it to traverse cliffs.

All the aforementioned systems shown in Fig. 7.3 can walk on uneven and sloped areas by using solely static gaits (except for Bert and SpaceBok), whereby a minimum of three feet in ground contact ensures a stable stance at all times.

While the robots are impressive in terms of obstacle negotiation, statically stable walkers miss the advantage of legged systems when it comes to fast, energy efficient locomotion. Additionally, highly articulated systems introduce multiple joints, which increases the system's complexity and potential points of failure.

Another group of novel exploration robots is formed by leg-wheel hybrids (Fig. 7.4), where wheels are attached to an actuated linkage. In the context of space exploration, this concept has been demonstrated by NASA's ATHLETE [29] robot with 36 controlled DOFs shown in Fig. 7.4b or the latest iteration of RoboSimian [30] with 32 DOFs Fig. 7.4c. The multimodal systems show promising performance in flat terrain and high mobility in unstructured environments, but the high number of articulated joints drastically increases the system complexity and mass.

An alternative might be the use of wheeled rovers with equally spaced 'feet', which are rigidly connected to a rimless, spoked wheel (such as DFKI's Asguard shown in Fig. 7.4a [31]). However, these systems lack some of the critical advantages of fully controllable legs, such as switching between gait modes or precise foot placement.

7.2.2 Terrestrial Legged Robots

Meanwhile, starting from Raibert's simple 2D hopper in 1986 [32], dynamically balancing legged robots have gained increasing popularity. Since its founding in 1992, Boston Dynamics has been a significant driving force towards showcasing and developing this technology. Perhaps one of the breakthrough robots that highlighted the all-terrain negotiating skills of dynamic quadrupeds was the hydraulically actuated robot BigDog in 2008 [33], followed by a smaller all-electric, quadrupedal robot Spot Mini in 2016 [34].

Generally, this group of dynamically walking, dog-sized, general-purpose legged robots for terrestrial use has made remarkable progress in recent years. Prime examples of these systems include the latest (2018) version of the MIT Cheetah [35], and the smaller version of this robot, Mini Cheetah [36]. Our group at ETH Zurich has started the development of quadrupeds first with the statically walking quadruped ALoF [26], followed by the first dynamically walking robot StarlETH [37] and its successor ANYmal [38], which has been commercialized by ANYbotics in 2017. Hydraulically actuated quadrupeds have been investigated by the IIT, starting with HyQ [39] and the latest version HYQ2Max [40].

Other companies in the field are found in China, where most prominently, Unitree Robotics is recently a driving force to lower those robots' cost by introducing Laikago in 2017 and Aliengo in 2019 [41, 42]. An exemplary collection of those robots can be seen in Fig. 7.5. Compared to the multilegged space robots, this group of quadrupeds tends to have only three articulated DOF's per leg.

It has been shown that this group of dynamic legged robots can walk quickly over rough terrain [43], recover from falls [44], navigate unstructured environments [45], and operate for several hours on a single battery charge [34].

The increasing maturity of the technology has allowed the robots to leave the lab and start being used in real-world applications [46]. In this context, it is only a matter of time since this technology becomes suitable for exploring other celestial bodies.

7.3 Hardware of Legged Robots

This section reviews the basic structure and core components of legged robots. An overview of the robots' architecture is first laid out, together with a simplified functional breakdown. Then a more in-depth analysis of the most critical hardware aspects is carried out. From a very high level of abstraction, legged robots are machines with movable appendices, capable of exerting forces on the environment and self-propelling. They can be seen as a hybrid between vehicles and traditional industrial robots. On the one hand, their primary is locomotion, like vehicles; on the other hand, their physical capabilities closely resemble those of articulated arms. However, the combination of locomotion and multiple DOFs results in architecture and solutions that significantly differ from industrial robots and traditional vehicles. More specifically, the following challenges and opportunities drive the design of legged robots' hardware [38, 46]

- The system should be self-contained and autonomous from an energetic, computational, and sensory point of view.

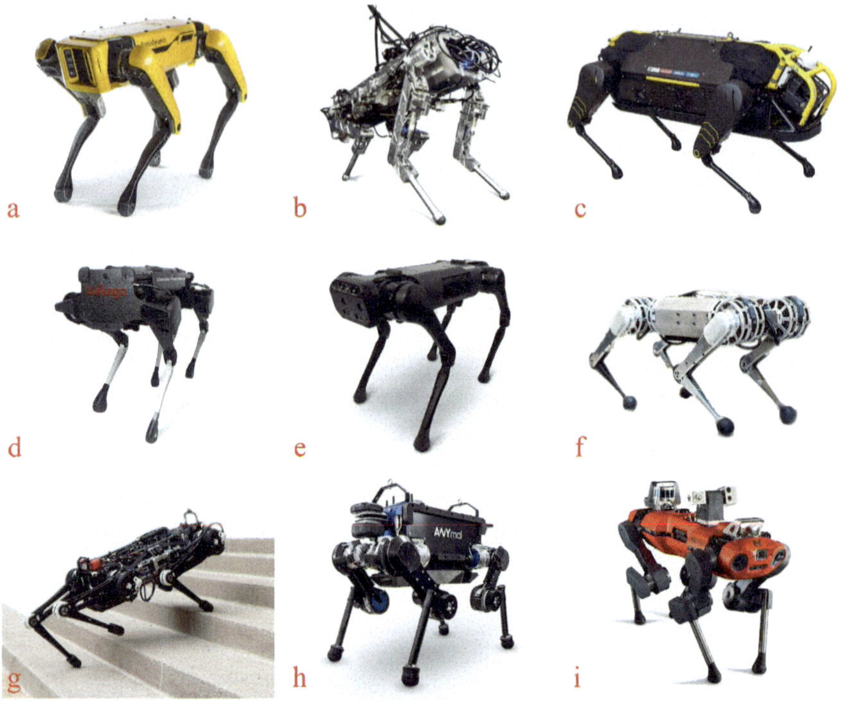

Fig. 7.5 Selection of dynamically walking legged robots: **a** Boston Dynamics Spot. **b** IIT HyQ. **c** IIT HyQReal. **d** Unitree Robotics Laikago. **e** Unitree Robotics Aliengo. **f** MIT Mini Cheetah. **g** MIT Cheetah 3. **h** ETH Zurich/ANYbotics ANYmal B. **i** ANYbotics ANYmal C (*Photo courtesy of* Boston Dynamics/IIT/Unitree/MIT/ETHZ)

- The legs should be able to control forces at the feet with good accuracy and high bandwidth.
- The weight and inertia of all components should be minimized for agility and energy efficiency.
- Impact tolerance is necessary because of the discontinuous and energetic interaction with the environment.
- The unstructured nature of the environment and the intrinsic instability of the systems create the need to protect the most sensitive components.
- High positional accuracy is less important than for industrial robots.

While being high-level considerations, these principles are well reflected in the general architecture of mobile robots.

Typically, there is a clear separation between the subsystems that should not move and those of the legs. A central structure (main body) connects all the components necessary for the autonomy goals and the legs. Batteries or fuel tanks (in combination with an electric generator) provide energy autonomy [33]. Integrated onboard computers or graphics processing units (GPUs) carry out the required computation.

Exteroceptive sensors such as cameras, depth cameras, and lidars, used to sense the environment for navigation and perceptive locomotion are also connected to the main body. Similarly, the inertial measurement unit necessary for proprioceptive information is rigidly attached to it [40, 47]. The main body is the attachment point also for payloads, which can range from specific, application-driven sensors [48] to robotic arms for manipulation [34, 49].

Legs can be divided into three subsystems: actuators, links, and feet. Unlike older units, modern actuators are designed for torque control, see sec. 3.1, and optimized for high torque density and control bandwidth. Additional sensors are integrated into the actuators to measure position and velocity. Impact mitigation devices and inertia reducing transmission [34–36, 41] might complement the architecture. Links are structural parts connecting the actuators and the feet, weight-optimized and produced with high strength-to-weight materials. Feet are designed for maximizing traction and damping impacts, but can they can also include sensors to acquire additional information [41, 50–52]. Protection elements, such as thermoformed shells and metal rollbars (Fig. 7.5), protect the most exposed parts of the robot.

The general architecture of dynamic legged robots is relatively consistent, and most robots are designed similarly from that perspective. Instead, the actuation principle and leg design present more variety and significantly impact the robots' performance, as will bed detailed in the following two paragraphs.

7.3.1 Actuation Concepts

The main function of actuators is converting a primary energy source into controlled motion. The nature of the primary energy source and the strategies used for achieving torque control results in different actuator architectures.

At scales relevant to this discussion, the primary energy source consists typically of electric energy stored in batteries or hydraulic oil, pressurized by a pump. Therefore, electric motors (brushless DC) and hydraulic cylinders transform the primary energy source into motion. Electric actuation is always preferred for planetary exploration systems; therefore it will be the focus of our analysis. As stated before, the possibility of controlling the torque of the actuators is necessary for successfully achieving dynamic locomotion. Stating specifications for the quality of the torque control is not easy because the controller's capabilities can influence that. Previous works however, showed that static torque errors up to 10% [35] are acceptable, and bandwidths above 24 Hz [47] are necessary.

The necessary control performance can be achieved either with feedforward control, if the characterization of the actuator is sufficiently good, or by closing the control loop with adequate force sensors.

The first approach is feasible with current control on electric motors in direct drive (DD) configuration, with the motor output directly connected to the joint [53]. Unfortunately, DD actuators operate in high velocity and low torques regimes that are not suitable for walking robot at this scale. The necessary gear reductions introduce friction that degrades the current to torque mapping. Low gear ratio, high efficiency transmissions, however, result in sufficiently good actuation performances [35, 36]. The resulting pseudo direct drive (PDD) architecture is typically composed of large gap radius motors and single or double stage planetary gearboxes [54], sometimes combined with chain or belt transmission. PDDs have relatively high power and velocity output, with rather significant winding losses and low torque accuracy. Given its simplicity and high power density, quasi direct drive is becoming the preferred choice for legged robots [36, 42, 55].

When the behaviour of the actuator cannot be modelled sufficiently well, additional sensing is necessary. Force sensors estimate forces and torques by measuring the deformation of an elastic element in series with the load. This can be done at high stiffness and small deformations measured with strain gauges or with low stiffness and large deformations measured by encoders (series elastic actuation, SEA) [56–58]. The two approaches differ primarily because of the low-pass effects of the low stiffness elements, which decrease the actuator bandwidth and decouple the transmission from impacts. Moreover, SEA offers the possibility of storing a considerable amount of elastic energy in the drive, a feature that control can exploit. In the case of electric actuators, differently from PDD, higher gear ratios can be used. The preferred architecture is therefore composed of smaller motors in combination with harmonic drives [34, 47]. Sensorized actuators have better torque accuracy and are not limited in the choice of components, resulting in more design freedom. However, they are affected by more complexity and lower mechanical efficiencies.

7.3.2 Leg Design

Legs are articulated structures that can transform actuators' torques into end-effector forces and motion. The legs' kinematic chain and design affect the robots' ability to operate in specific configurations and influence how the load is shared between actuators.

A wide range of topologies, geometries, and leg designs are possible. Instead of listing all the possible combinations, we will analyse the most characterizing features, bringing consequences and examples.

The first way to characterize a legged robot is to consider its number of legs and its degrees of freedom. As described in 1, more legs offer more stability and additional degrees of freedom, allow for larger workspace, but come at the cost of an increased number of actuators and risk of failures.

The legs can be attached to the main body with radial symmetry or according to planes of symmetry. The latter option opens the possibility of optimizing the robot for a preferential direction of motion, while the former makes the locomotion performance similar in all direction.

The orientation of the leg also has an impact by giving a particular direction to the manipulability ellipsoid. This results in different capabilities of applying forces and developing velocities for every point of the workspace and, therefore, suitability for specific tasks [59]. The two most common configurations for three degrees of freedom legs are mammalian (with the first axis parallel to the forward direction, or roll-pitch-pitch) and arachnid (first axis perpendicular to the ground, yaw-pitch-pitch). The former is more capable of producing forward velocities while the latter is better at lowering the base and maximizing tumble stability [60, 61]. More complex configurations have also been explored to improve the match between task and hardware capability [62, 63].

Kinematic chains can be divided into two categories: serial and closed chains. Serial chains are not branched, and all their joints are actuated, offering large workspaces and high-end-effector velocities. Instead, closed chains have multiple branches and unactuated joints; they can minimize inertia by having multiple joints fixed to the body. Moreover, the coupled actuation allows to combine the torque of actuators and achieves high end-effector forces [64]. Despite their advantages, closed kinematic chains are affected by antagonistic power, which decreases their efficiency [59].

As described in 3.1, actuators can be combined with transmissions, such as chains, belts, and linkages. This can help centralize masses and reduce leg inertia.

Finally, the design can include elastic elements, acting in parallel to the actuators to store energy and offset gravitational load [64]. This results in reduced energy consumption and loss of workspace because of the geometric constraints introduced by springs.

Despite the numerous possible designs, recently developed dynamic legged robots for mobility on difficult terrains converged to a similar architecture. All the robots in Fig. 7.5 have four legs, with three degrees of freedom, arranged in serial configuration. They also have a transmission to reduce the knee joint's inertia, except ANYmal, and none of them use parallel elasticity. They also all share the same mammalian configuration.

When a different or more specific application is targeted, the leg design can take a different form. For all the robots of Fig. 7.3, the locomotion task is specialized for planetary exploration context, resulting in more degrees of freedom and different leg orientations.

7.4 Locomotion Control for Legged Robots

Legged robots have the potential to cross much more diverse terrains, but they require sophisticated controllers to do so. The control of a mobile robot is typically split into multiple components. For example, a high-level controller generating a path. A path following controller steering the robot towards that path and finally the low-level locomotion controller actuating the motors to achieve the desired motion of the base.

A legged robot moves by taking consecutive steps during which it must decide where to place its feet, which ones to lift, and how the balance the weight of the base, with most combinations of these decisions resulting in a fall. The locomotion problem becomes even more complex in the case of space exploration. Low-gravity pushes the robot to adopt a dynamic gait with long flight phases, during which it has no contact with the ground, and as such no direct way to correct its balance [65]. Furthermore, the granular and unstructured terrain is an additional challenge for the locomotion controller.

7.4.1 Modelling

A legged robot is modelled as an articulation of rigid bodies, where consecutive link are attached through joints and each of these links has its own mass and inertia. We can then use this model to perform a non-linear optimization to compute the torques needed to move the different links in the desired way [67]. However, this model breaks down once a leg touches the ground. We need a different set of dynamical equations for each possible contact case and we need to know when to switch between them. Thus we need to perform contact detection between the feet of the robot (restricting contacts on other bodies) and the terrain. Similarly, we need to detect self-collisions between the links of the robot. We thus end-up with a complex model requiring a sophisticated physics simulator. Unfortunately, this model would be non-linear, non-convex, and non-differentiable making it ill-suited for efficient optimization.

Another possibility is to simplify the model as much as possible. For example, we may say that the weight of the legs is negligible and only consider the inertial properties of the base (Fig. 7.6) [68]. Furthermore, we can decide that we enforce a particular contact pattern and thus know which version of the dynamics to use at any given time. This allows us to have simpler, potentially convex and differentiable models, which can be used for online optimization.

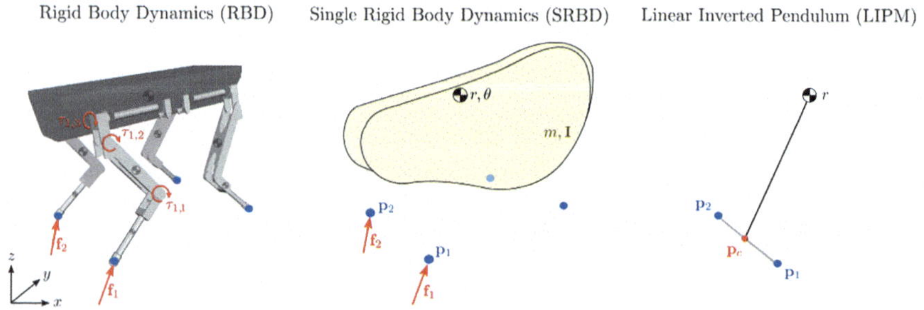

Fig. 7.6 Models of a legged robot with three different levels of simplification [66]

7.4.2 Model-Based Control

Each of the two modelling approaches described above has a corresponding control technique. The field of off-line trajectory optimization [69, 70] tackles the full problem using heavy computation. The complete model is used to optimize a motion of multiple seconds. It has the advantage of requiring a restricted number of assumptions and simplifications. Unfortunately, it means that the robot precomputes an open-loop trajectory before moving and it cannot be adapted the plan to react to unforeseen disturbances or model mismatches, making the motion extremely brittle.

On the other hand, Model predictive control (MPC) [71, 72], recomputes a short horizon plan at every step. Since the plan is always computed based on the current state of the robot it can react to unforeseen circumstances. However, the computations happen online and many times per second. As such, these methods are restricted to simplified models. For example, the gait pattern, contact schedule and trajectory of the feet in the air are fixed, self-collision is disregarded and it is assumed that the feet do not slip. This approach has two main disadvantages. First, by fixing the parts of the motion, we enforce a potentially sub-optimal behaviour. Additionally, in the real world, some of the assumptions cannot be respected exactly. Contacts with the ground may be hard to predict due to an offset in the terrain measurements or soft ground, for example. Some amount of slip is inevitable when walking on different materials or once again on soft ground. A violation of these assumptions leads to instabilities which are very to recover from.

7.4.3 Learning-Based Control

Recently, deep reinforcement learning has emerged as a powerful tool for robotic control. Just like a neural network can be trained to detect dogs and cats in images, it can learn to control the motors of a robot [44]. Every time the neural network, called a policy in this setting, selects an action it receives a reward based on how well this action full filled the desired task. Next, the parameters of the policy are updated to maximise actions leading to the highest rewards and avoid those that lead to bad outcomes. This process repeats many times until the policy converges to the desired behaviour. Unfortunately, current algorithms require millions of trial-and error steps to converge making it impractical to train on the real robot. Not only would it take years, but the robot would fall and damage itself many times in the process. Instead, training is performed in a simulated environment, where the robot can fall without any damage (Fig. 7.7). Additionally, current simulators can run thousands of time faster than real-time [74] allowing us to simulate years of experience in just a few minutes of computation [73]. Once the training in simulation has converged we can freeze the parameters of the policy and deploy it on the real robot.

Deep reinforcement learning algorithms typically do not make assumptions about the linearity, convexity or differentiable of the underlying model. They can work with

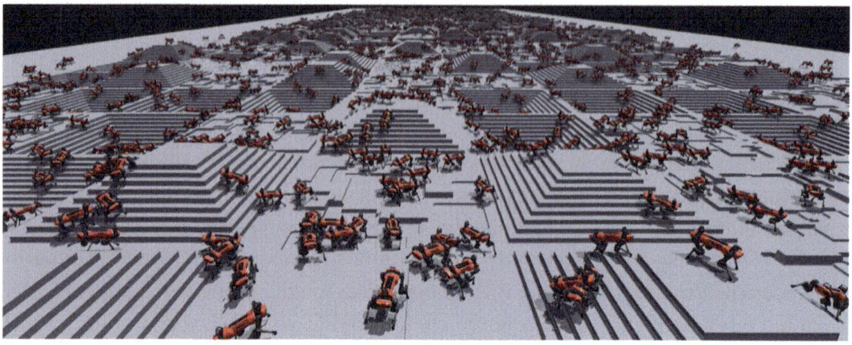

Fig. 7.7 Thousands of robots simulated in parallel [73]

complex simulators that closely mimic the real world. However, policies are prone to over-fitting to the environment they are trained in and do no not immediately transfer to the real system, which might have slightly different physical properties. Multiple approaches have been developed to facilitate this sim-to-real transfer. One of the simplest and most widely used techniques is domain randomization. During training, some physical quantities are randomized and the policy must learn to deal with all of them. For example, since the friction of the ground might vary depending on where the robot is deployed, we randomize it within a reasonable range in simulation. The policy then learns to adopt a more robust gait and recover if the feet start slipping. Compared to model-based approaches described above, reinforcement learning does not require any simplification of the model and can be trained to handle diverse disturbances, noise and model-mismatches. Furthermore, it allows the policy to find behaviours which would not be possible with the restrictions we impose on model-based approaches. Finally, even though the training is computationally expensive. The deployment on the real robot requires only the inference of a relatively small neural network, which does not require high computational power and can be done at high frequency on the on-board computer.

However, there are also drawbacks to the approach. As most deep learning models, a neural network policy can only be treated as a black box. It lacks the interpretability of model-based approaches. Similarly, there are no theoretical guarantees on the performance, robustness or quality of transfer from simulation to reality. We can only evaluate the whole system empirically. Nevertheless, as described in the next section, experimental data suggest that learned policies outperform model based controllers in terms of efficiency, robustness and traversability capabilities.

7.5 Field Results

To provide value in planetary missions, for example in exploration or prospection missions, legged robots need to be able to solve a broad variety of tasks: They need to map the terrain for future mission planning, record footage of the extra-terrestrial environments and prospect or collect potential resources for in-situ resource utilization. To succeed in these tasks, the robot first and foremost needs to be able to reach the site of interest, that lies potentially in a hard to access terrain as described earlier in this chapter. Furthermore, the robot needs to carry and deploy a wide range of payloads in this terrain, for example ranging sensors, cameras, and spectrometers to name a few. To raise the technology readiness level of these robots and spot potential deployment issues early in the development process, the underlying technologies have to be validated in analogue environments and not just in the laboratory. Recently, research groups around the world have started to develop the specialized technology required to enable legged robots the applicability in space [75]. In this section, we show how some of the presented legged robots have been deployed in field trials in recent years, demonstrating the state of the art in terrain negotiation capabilities and mission utility of legged robots.

7.5.1 DARPA Subterranean Challenge

A core task of future planetary robots will be autonomous exploration of unknown, harsh environments. A prominent recent example of terrestrial exploration missions is the DARPA Subterranean Challenge [76]. The challenge was initiated in 2017 with the goal to advance the state of the art in robotic underground exploration, for example in natural cave environments or in human-made urban systems such as tunnels or buildings. More precisely, to score points in the challenge, the robots had to detect artifacts in the underground environment and precisely report their position. Some examples of artifacts to be found were backpacks, phones or mannequins simulating a human. Only one human supervisor was allowed to interact with the robots during the mission. The robots had to operate under adverse underground conditions, such as poor lighting conditions, dusty environments and rough, hardto- traverse terrain. While at the beginning of the challenge, only one team, team CERBERUS, used legged robots, the top 6 teams each had at least one legged robot in their fleet in the final run. The rough terrain negotiation capabilities of legged robots were a crucial element in the success of team CERBERUS [77].

7.5.2 Lemur Martian Analog Deployment

Although the DARPA subterranean challenge was a great platform to demonstrate robot capabilities in harsh environments, it is not really representative of extraterrestrial environments in terms of environment structure and terrain composition. To validate the locomotion performance of legged robots, they have to be deployed in analogue environments with terrain compositions as they occur on the Moon or on Mars. However, only few researchers have deployed legged robots in analogue missions, despite the recent success of legged robots in terrestrial environments. One example of a successful analogue deployment was demonstrated with Lemur 3 [78]. A team from NASA JPL outfitted Lemur 3 with an astrobiology payload suite consisting of a micro-X-ray fluorescence instrument, a deep-ultraviolet fluorescence instrument and a near-infrared acousto-optic tunable filter-based point spectrometer. The system was deployed in more than ten field tests in lava tube caves and canyon rock walls, emulating potential target sites for the search of life on Mars. The robot was able to access targets on vertical walls and position the instrument suite within 10 mm of a target location.

7.5.3 Locomotion Validation on Martian Soil Simulant

Another field test on relevant terrain was conducted at the robotic test facility of RUAG Space, Zurich (Fig. 7.8). This site was specifically designed to validate the locomotion subsystem of ESA's ExoMars rover Rosalind Franklin and has later been used to test the locomotion subsystem of the Sample Fetch Rover. The testbed consists of a 6 m by 6 m container that can be filled with several types of Martian soil simulant. It can be tilted from 0 to 25 deg to test on a variety of slopes. SpaceBok was the first legged robot to walk on this test site [79]. Although slippage occurred on high slopes, the robot managed to traverse a slope of up to 25 deg, the maximum of the testbed, both in ascent and descent. A model-based control approach was used for locomotion control. Additionally, transverse trajectories and locomotion on rocks submerged in the soil simulant was successfully tested.

More recently, ANYmal was deployed successfully on slopes up to 25 deg in the same facility. Unlike SpaceBok, ANYmal used a state-of-the art reinforcement learning control policy [80]. Although foot-soil interaction cannot properly be … in simulation for the training of the policy, the domain randomization on the friction and link lengths was sufficient to provide robust locomotion on the soil simulant.

Fig. 7.8 SpaceBok (**a**) and ANYmal (**b**) at the robot testing facility of RUAG Space

7.5.4 ESA/ESRIC Space Resources Challenge

A more recent example of a successful analogue deployment is the ongoing ESA and ESRIC Space Resources Challenge [81]. The goal of the challenge is to assess the state-of-the-art in robotic lunar prospection technologies and accelerate the development of key technologies. The first field trial of this challenge took place in November 2021. The trial simulated a lunar environment with a challenging traverse region, including artificial boulders and steep slopes, and a sand-covered region of interest that resembled a lunar crater. The robots had to traverse the traverse region to reach the region of interest and then analyse rock samples in this region.

Two teams successfully used legged robots in their approach to solve this challenge. Team GLIMPSE, consisting of researchers from ETH Zurich, University of Zurich and University of Basel built upon an ANYmal robot [38]. They integrated a payload suite consisting of a pan-tilt context imager, a Raman spectrometer and a custom microscopic imager. Additionally, a team from the German Forschungszentrum Informatik (FZI) deployed legged robots in their fleet, namely one ANYmal C and one Spot besides their Husky robot. They used the robot's onboard camera systems to inspect the rock samples. Both these teams completed the first field trial of the Space Resources Challenge successfully and were among the five winning teams advancing to the second stage of the challenge.

7.6 Conclusion

This chapter highlighted the current developments and potential of versatile legged robots for future planetary exploration missions. We aimed to give a general overview of the specifics of such systems from a hardware and control perspective. The first promising developments of transitioning legged robots towards space application have been launched by performing analogue tests in a terrestrial setting. Based on our findings, we conclude that dynamic, all-terrain (quadrupedal) walking robots have high potential in cluttered, sloped, and sandy environments. Such systems will be able to venture into regions that were so far out of reach and will ultimately allow us to increase our knowledge of the worlds that surround us.

While the robots are becoming more mature and the technology better understood, the following next steps are required for future development.

- To generate more accurate locomotion policies that can properly account for slippage and sinkage effects, modelling soil behaviour in a high-fidelity simulation would drastically improve the performance prediction. Predicting the robot's performance is useful to accurately size actuators and validate the estimated locomotion performance in high-sloped and unstructured terrain. Furthermore, once obtained, the simulation results need to be validated through locomotion validation campaigns.
- As legged robots venture into unstructured environments, an elevated level of autonomy and situational awareness is required to take frequent actions without direct oversight from an earth-based operator. While such capabilities are an ongoing research effort, the applicability and realization for planetary robots have yet to be developed.
- The majority of the systems presented in this contribution are not space-graded. To account for space's hostility, hardware adaptations must be performed. More specifically, the highly articulated robots must be ingress protected and energy-independent, for example, by placing solar arrays without impeding the locomotion capability. Furthermore, thermal aspects of the design have to be considered.

We believe that considering legged robots for a defined mission scenario would drastically help the scalability of a current legged system for planetary exploration, as more targeted developments could be performed.

References

1. Squyres, S. W., Knoll, A. H., Arvidson, R. E., Ashley, J. W., Bell III, J. F., Calvin, W. M., Christensen, P. R., Clark, B. C., Cohen, B. A., de Souza Jr, P. A., Edgar, L., Farrand W. H., Fleischer, I., Gellert, R., Golombek, M. P., Grant, J., Grotzinger, J., Hayes, A., Herkenhoff, K. E., Johnson, J. R., Jolliff, B., Klingelhofer, G., Knudson, A., Li, R., McCoy, T. J., McLennan, S. M., Ming, D. W., Mittlefehldt, D. W., Morris, R. V., Rice, J. W., Schroder, C., Sullivan, R. J., Yen, A., & Yingst, R. A. (2009). Exploration of Victoria Crater by the mars rover opportunity. *Science, 324*(5930), 1058–1061.
2. Lederberg, J., & Sagan, C. (1962). (1962), Microenvironments for life on mars. *Proceedings of the National Academy of Sciences of the United States of America, 48*(9), 1473.
3. McMahon, S., Bosak, T., Grotzinger, J. P., Milliken, R. E., Summons, R. E., Daye, M., Newman, S. A., Fraeman, A., Williford, K. H., & Briggs, D. E. G. (2018). A field guide to finding fossils on mars. *Journal of Geophysical Research: Planets, 123*(5), 1012–1040.
4. Head, J. W., Marchant, D. R., & Kreslavsky, M. A. (2008). (2008), Formation of gullies on mars: Link to recent climate history and insolation microenvironments implicate surface water flow origin. *Proceedings of the National Academy of Sciences, 105*(36), 13258–13263.

5. Dundas, C. M., Bramson, A. M., Ojha, L., Wray, J. J., Mellon, M. T., Byrne, S., McEwen, A. S., Putzig, N. E., Viola, D., Sutton, S., Clark, E., & Holt, J. W. (2018). Exposed subsurface ice sheets in the Martian mid-latitudes. *Science, 359*(6372), 199–201.
6. Mustard, J. F., Pieters, C. M., Isaacson, P. J., Head, J. W., Besse, S., Clark, R. N., Klima, R. L., Petro, N. E., Staid, M. I., Sunshine, J. M., Runyon, C. J., & Tompkins, S. (2011). Compositional diversity and geologic insights of the Aristarchus crater from moon mineralogy mapper data. *Journal of Geophysical Research: Planets, 116*(E6), 2011.
7. Hayne, P. O., Hendrix, A., Sefton-Nash, E., Siegler, M. A., Lucey, P. G., Retherford, K. D., Williams, J. P., Greenhagen, B. T., & Paige, D. A. (20115). Evidence for exposed water ice in the moon's south polar regions from lunar reconnaissance orbiter ultraviolet albedo and temperature measurements. *Icarus, 255*, 58–69.
8. Haruyama, J., Morota, T., Kobayashi, S., Sawai, S., Lucey, P. G., Shirao, M., & Nishino, M. N. (2012). Lunar holes and lava tubes as resources for lunar science and exploration. Springer, pp.139–163
9. Gonzalez, R., & Iagnemma, K. (2018). Slippage estimation and compensation for planetary exploration rovers. State of the art and future challenges. *Journal of Field Robotics, 35*(4), 564–577.
10. Young, K. (2006). Mars rover escapes from "the bay of lamentation." *New Scientist, 6*, 2006.
11. Arvidson, R. E., Iagnemma, K. D., Maimone, M., Fraeman, A. A., Zhou, F., Heverly, M. C., Bellutta, P., Rubin, D., Stein, N. T., Grotzinger, J. P., & Vasavada, A. R. (2016). Mars science laboratory curiosity rover Megaripple crossings up to Sol 710 in Gale Crater. *Journal of Field Robotics, 34*(3), 495–518.
12. Guy, W., & McGregor, V. (2009). *NASA's mars rover has uncertain future as sixth anniversary nears.*
13. Florenskii, C. P., Basilevskii, A. T., Bobina, N. N., Burba, G. A., Grebennik, N. N., Kuzmin, R. O., Polosukhin, B. P., Popovich, V. D., Pronin, A. A., Ronca, L. B. (1978). The floor of crater Le Monier: A study of Lunokhod 2 data. In *Lunar and Planetary Science Conference Proceedings*, vol. 2, pp. 1449–1458.
14. Costes, N. C., Farmer, J. E., & George, E. B. (1972). United States. National aeronautics, space administration, and George C. Marshall Space Flight Center. In *Mobility performance of the lunar roving vehicle: Terrestrial studies, Apollo 15 Results.* NASA TR R. NASA.
15. Rodríguez-Martínez, D., Van Winnendael, M., & Yoshida, K. (2019). High-speed mobility on planetary surfaces: A technical review. *Journal of Field Robotics, 36*(8), 1436–1455.
16. Kolvenbach, H., Breitenstein, M., Gehring, C., & Hutter, M. (2017). Scalability analysis of legged robots for space exploration. In *International Astronautical Congress (IAC).*
17. Wettergreen, D., Thorpe, C., & Red Whittaker, R. (1993). Exploring Mount Erebus by walking robot. *Robotics and Autonomous Systems, 11*(3), 171–185.
18. Bares, J., & Wettergreen, D. (1999). Dante ii: Technical description, results and lessons learned. *International Journal of Robotics Research, 18*(7), 621–649.
19. Spenneberg, D., & Kirchner, F. (2007). The bio-inspired scorpion robot: Design, control & lessons learned. In *Climbing and walking robots: Towards new applications*. InTech.
20. DFKI. (2017). Systemblatt CREX.
21. Bartsch, S., Birnschein, T., Römmermann, M., Hilljegerdes, J., Kühn, D., & Kirchner, F. (2012). Development of the six-legged walking and climbing robot SpaceClimber. *Journal of Field Robotics, 29*(3), 506–532.
22. Gorner, M., Wimbock, T., Baumann, A., Fuchs, M., Bahls, T., Grebenstein, M., Borst, C., Butterfass, J., & Hirzinger, G. (2008). The dlr-crawler: A testbed for actively compliant hexapod walking based on the fingers of dlr-hand ii. In 2008 *IEEE/RSJ International Conference on Intelligent Robots and Systems* (pp. 1525–1531).
23. Lakatos, D., Ploeger, K., Loeffl, F., Seidel, D., Schmidt, F., Gumpert, T., John, F., Bertram, T., & Albu-Schäffer, A. (2018). Dynamic locomotion gaits of a compliantly actuated quadruped with SLIP-like articulated legs embodied in the mechanical design. *IEEE Robotics and Automation Letters, 3*(4), 3908–3915.

24. Roennau, A., Heppner, G., Nowicki, M., Zoellner, J. M., & Dillmann. R. (2014). Reactive posture behaviors for stable legged locomotion over steep inclines and large obstacles. In *IEEE/RSJ International Conference on Intelligent Robots and Systems (IROS)*, pp. 4888–4894.
25. DFKI. (2017). *Systemblatt Mantis*.
26. Remy, C. D., Baur, O., Latta, M., Lauber, A., Hutter, M., Hoepflinger, M. A., Pradalier, C., & Siegwart, R. (2011). Walking and crawling with alof: A robot for autonomous locomotion on four legs. *Industrial Robot: An International Journal*.
27. Arm, P., Zenkl, R., Barton, P., Beglinger, L., Dietsche, A., Ferrazzini, L., Hampp, E., Hinder, J., Huber, C., Schaufelberger, D. Schmitt, F., Sun, B., Stolz, B., Kolvenbach, H., & Hutter, M. (2019). SpaceBok: A dynamic legged robot for space exploration. In *IEEE International Conference on Robotics and Automation (ICRA)*.
28. NASA/JPL. (2019). Lemur. https://www.jpl.nasa.gov/news/news.php?feature=7449
29. Wilcox, B. H., Litwin, T., Biesiadecki, J., Matthews, J., Heverly, M., Morrison, J., Townsend, J., Ahmad, N., Sirota, A., & Cooper, B. (2007). ATHLETE: A cargo handling and manipulation robot for the moon. *Journal of Field Robotics*, *24*(5), 421–434.
30. Reid, W., Meirion-Griffith, G., Karumanchi, S., Emanuel, B., Chamberlain-Simon, B., Bowkett, J., & Garrett, M. (2021). Actively articulated wheel-on-limb mobility for traversing Europa analogue terrain. In *Field and service robotics*. Springer International Publishing.
31. DFKI. (2016). *Systemblatt Asguard IV*.
32. Marc, H., & Raibert, M. H. (1986). *Legged robots that balance*. MIT Press.
33. Raibert, M., Blankespoor, K., Nelson, G., & Playter, R. (2008). Bigdog, the rough-terrain quadruped robot. *IFAC Proceedings*, *41*(2), 10822–10825.
34. Boston Dynamics. (2020). Spot Mini.
35. Bledt, G., Powell, M. J., Katz, B., Di Carlo, J., Wensing, P. M., & Kim, S. (2018). MIT Cheetah 3: Design and control of a robust, dynamic quadruped robot. In *IEEE/RSJ International Conference on Intelligent Robots and Systems (IROS)* (pp. 2245–2252).
36. Katz, B., Carlo, J. D., & S. Kim. (2019). Mini Cheetah: A platform for pushing the limits of dynamic quadruped control. In *IEEE International Conference on Robotics and Automation (ICRA)* (pp. 6295–6301).
37. Hutter, M., Remy, C. D., Hoepflinger, M. A., & Siegwart, R. (2013). Efficient and versatile locomotion with highly compliant legs. *IEEE/ASME Transactions on Mechatronics, 18*(2), 449–458.
38. Hutter, M., Gehring, C., Lauber, A., Gunther, F., Bellicoso, C. D., Tsounis, V., Fankhauser, P., Diethelm, R., Bachmann, S., Blösch, M., Kolvenbach, H., Bjelonic, M., Isler, L., & Meyer. K. (2017). ANYmal—toward legged robots for harsh environments. *Advanced Robotics, 31*(17), 918–931.
39. Semini, C., Tsagarakis, N. G., Guglielmino, E., Focchi, M., Cannella, F., & Caldwell, D. G. (2011). Design of HyQ–A hydraulically and electrically actuated quadruped robot. *Proceedings of the Institution of Mechanical Engineers, Part I: Journal of Systems and Control Engineering, 225*(6), 831–849.
40. Semini, C., Barasuol, V., Goldsmith, J., Frigerio, M., Focchi, M., Gao, Y., & Caldwell, D. G. (2016). Design of the hydraulically actuated, torque-controlled quadruped robot HyQ2Max. *IEEE/ASME Transactions on Mechatronics, 22*(2), 635–646.
41. Unitree Robotics. (2019). Laikago.
42. Unitree Robotics. (2019). Aliengo.
43. Lee, J., Hwangbo, J., Wellhausen, L., Koltun, V., & Hutter, M. (2020). Learning quadrupedal locomotion over challenging terrain. *Science Robotics, 5*(47), 2020.
44. Hwangbo, J., Lee, J., Dosovitskiy, A., Bellicoso, D., Tsounis, V., Koltun, V., & Hutter, M. (2019). Learning agile and dynamic motor skills for legged robots. *Science Robotics, 4*(26), 2019.
45. Fankhauser, P., Bjelonic, M., Dario Bellicoso, C., Miki, T., & Hutter, M. (2018). Robust rough-terrain locomotion with a quadrupedal robot. *IEEE International Conference on Robotics and Automation (ICRA), 5761–5768*(5), 2018.

46. Bellicoso, C. D., Bjelonic, M., Wellhausen, L., Holtmann, K., Günther, F., Tranzatto, M., Fankhauser, P., & Hutter, M. (2018). Advances in real-world applications for legged robots. *Journal of Field Robotics, 35*(8), 1311–1326.
47. Hutter, M., Gehring, C., Jud, D., Lauber, A., Bellicoso, C. D., Tsounis, V., Hwangbo, J., Bodie, K., Fankhauser, P., Bloesch, M. Diethelm, R., Bachmann, S., Melzer, A., & Hoepflinger, M. (2016). Anymal—A highly mobile and dynamic quadrupedal robot. In *IEEE/RSJ International Conference on Intelligent Robots and Systems (IROS)* (pp. 38–44). IEEE.
48. Tranzatto, M., Mascarich, F., Bernreiter, L., Godinho, C., Camurri, M., Khattak, S., Dang, T., Reijgwart, V., Loeje, J., Wisth, D., & Zimmermann, S. (2022). Cerberus: Autonomous legged and aerial robotic exploration in the tunnel and urban circuits of the DARPA subterranean challenge. arXiv preprint arXiv:2201.07067
49. Bellicoso, C. D., Krämer, K., Stäuble, M., Sako, D., Jenelten, F., Bjelonic, M., & Hutter, M. (2019). Alma-articulated locomotion and manipulation for a torque controllable robot. In *2019 International Conference on Robotics and Automation (ICRA)* (pp. 8477–8483). IEEE.
50. Käslin, R., Kolvenbach, H., Paez, L., Lika, K., & Hutter, M. (2018). Towards a passive adaptive planar foot with ground orientation and contact force sensing for legged robots. *IEEE/RSJ International Conference on Intelligent Robots and Systems (IROS).*
51. Valsecchi, G., Grandia, R., & Hutter, M. (2020). Quadrupedal locomotion on uneven terrain with sensorized feet. *IEEE Robotics and Automation Letters, 5*(2), 1548–1555.
52. Kolvenbach, H., Wisth, D., Buchanan, R., Valsecchi, G., Grandia, R., Fallon, M., & Hutter, M. (2020). Towards autonomous inspection of concrete deterioration in sewers with legged robots. *Journal of Field Robotics, 37*(8), 1314–1327.
53. Kenneally, G., De, A., & Koditschek, D. E. (2016). Design principles for a family of direct-drive legged robots. *IEEE Robotics and Automation Letters, 1*(2), 900–907.
54. Wensing, P. M., Wang, A., Seok, S., Otten, D., Lang, J., & Kim, S. (2017). Proprioceptive actuator design in the MIT cheetah: Impact mitigation and high-bandwidth physical interaction for dynamic legged robots. *IEEE Transactions on Robotics, 33*(3), 509–522.
55. Deep Robotics. (2022). Products.
56. Paine, N., Oh, S., & Sentis, L. (2013). Design and control considerations for highperformance series elastic actuators. *IEEE/ASME Transactions on Mechatronics, 19*(3), 1080–1091.
57. Pratt, J., Krupp, B., & Morse, C. (2002). Series elastic actuators for high fidelity force control. *Industrial Robot: An International Journal, 29*(3), 234–241.
58. Hutter, M., Remy, C. D., Hoepflinger, M. A., & Siegwart, R. (2012). High compliant series elastic actuation for the robotic leg scarleth. In *Field robotics* (pp. 507–514). World Scientific.
59. Abate, A., Hurst, J. W., & Hatton, R. L. (2016). Mechanical antagonism in legged robots. *Robotics: Science and Systems (RSS).*
60. Kashiri, N., Ajoudani, A., Caldwell, D. G., & Tsagarakis, N. G. (2016). Evaluation of hip kinematics influence on the performance of a quadrupedal robot leg. *ICINCO, 1*, 205–212.
61. Uno, K., Valsecchi, G., Hutter, M., & Yoshida, K. (2021). Simulation-based climbing capability analysis for quadrupedal robots. In *Climbing and Walking Robots Conference* (pp. 179–191). Springer.
62. Roennau, A., Heppner, G., Pfotzer, L., & Dillmann, R. (2013). Lauron v: Optimized leg configuration for the design of a bio-inspired walking robot. *Nature-Inspired Mobile Robotics, 563–570*, 2013.
63. Nicholson, J., Jasper, J., Kourchians, A., McCutcheon, G., Austin, M., Gonzalez, M., Pusey, J., Karumanchi, S., Hubicki, C., & Clark, J. (2020). Llama: Design and control of an omnidirectional human mission scale quadrupedal robot. In 2020 *IEEE/RSJ International Conference on Intelligent Robots and Systems (IROS)* (pp. 3951–3958). IEEE.
64. Kolvenbach, H., Hampp, E., Barton, P., Zenkl, R., & Hutter, M. (2019). Towards jumping locomotion for quadruped robots on the moon. In *IEEE/RSJ International Conference on Intelligent Robots and Systems (IROS)*, 5459–5466, 2019.
65. Kolvenbach, H., Bellicoso, D., Jenelten, F., Wellhausen, L., & Hutter, M. (2018). Efficient gait selection for quadrupedal robots on the moon and mars. In *International Symposium on Artificial Intelligence, Robotics and Automation in Space (I-SAIRAS).*

66. Winkler, A. (2018). *Optimization-based motion planning for legged robots*. PhD thesis.
67. Featherstone, R. (2014). *Rigid body dynamics algorithms*. Springer.
68. Orin, D. E., Goswami, A., & Lee, S. H. (2013). Centroidal dynamics of a humanoid robot. *Autonomous Robots, 35*(2), 161–176.
69. Apgar, T., Clary, P., Green, K., Fern, A., & Hurst, J. W. (2018). Fast online trajectory optimization for the bipedal robot Cassie. *Robotics: Science and Systems, 101*, 14.
70. Winkler, A. W., Bellicoso, C. D., Hutter, M., & Buchli, J. (2018). Gait and trajectory optimization for legged systems through phase-based end-effector parameterization. *IEEE Robotics and Automation Letters, 3*(3), 1560–1567.
71. Kim, D., Di Carlo, J., Katz, B., Bledt, G., & Kim, S. (2019). Highly dynamic quadruped locomotion via whole-body impulse control and model predictive control. ArXiv. abs/1909.06586.
72. Neunert, M., Stäuble, M., Giftthaler, M., Bellicoso, C. D., Carius, J., Gehring, C., Hutter, M., & Buchli, J. (2018). Whole-body nonlinear model predictive control through contacts for quadrupeds. *IEEE Robotics and Automation Letters, 3*(3), 1458–1465.
73. Rudin, N., Hoeller, D., Reist, P., & Hutter, M. (2022). Learning to walk in minutes using massively parallel deep reinforcement learning. In A. Faust, D. Hsu, & G. Neumann (eds.), *Proceedings of the 5th Conference on Robot Learning, volume 164 of Proceedings of Machine Learning Research* (pp. 91–100). PMLR, November 08–11, 2022.
74. Makoviychuk, V., Wawrzyniak, L., Guo, Y., Lu, M., Storey, K., Macklin, M., Hoeller, D., Rudin, N., Allshire, A., Handa, A., & State, G. (2021). *Isaac gym: High performance GPU-based physics simulation for robot learning*.
75. Dettmann, A., Planthaber, S., Bargsten, V., Dominguez, R., Cerilli, G., Marchitto, M., Fink, G., Focchi, M., Barasuol, V., Semini, C., & Marc, R. (2022). Towards a generic navigation and locomotion control system for legged space exploration. In *16th Symposium on Advanced Space Technologies in Robotics and Automation (ASTRA)*, ESA.
76. DARPA. (2021). *DARPA subterranean challenge*.
77. EEE Spectrum. (2022). Inside DARPA's subterranean challenge.
78. Uckert, K., Parness, A., Chanover, N., Eshelman, E. J., Abcouwer, N., Nash, J., Detry, R., Fuller, C., Voelz, D., & Hull, R., et al. (2020). Investigating habitability with an integrated rock-climbing robot and astrobiology instrument suite. *Astrobiology, 20*(12), 1427–1449.
79. Kolvenbach, H., Arm, P., Hampp, E., Dietsche, A., Bickel, V., Sun, B., Meyer, C., & Hutter, M. (2021). Traversing steep and granular Martian Analog slopes with a dynamic quadrupedal robot. In *Field robotics*.
80. Miki, T., Lee, J., Hwangbo, J., Wellhausen, L., Koltun, V., & Hutter, M. (2022). Learning robust perceptive locomotion for quadrupedal robots in the wild. *Science Robotics, 7*(62), eabk2822.
81. ESA. (2022). Space resources challenge.

Chapter 8
Air-Bearing Microgravity Simulators for Space Robotics

Tomasz Rybus, Hendrik Kolvenbach, and Xiu Tian Yan

Abstract Ground testing is required to ensure that robotic systems will operate properly in the orbital environment or on the surface of other planets, asteroids or comets with microgravity. Air-bearing microgravity simulators allow to emulate the dynamic behaviour of the satellite or the satellite-manipulator system operating in zero-g conditions. This chapter presents key functionalities and design parameters of planar and spherical air-bearing microgravity simulators. Applications of such simulators include, i.a., testing of manipulator hardware for on-orbit operations, trajectory planning and control of satellite-manipulator systems, demonstration of capture with a manipulator, close proximity operations and formation flight experiments. Various possible applications are reviewed and an example design of a low-cost, spherical air-bearing satellite simulator is presented. The chapter also outlines future directions of possible research and open issues.

8.1 Introduction

Extensive ground testing of all space systems is required to ensure that they will operate properly in the orbital environment or their intended planetary environments [1]. One of the major challenges is that testing is performed under Earth's gravity conditions (1-g), while the actual system will be working in zero-g or close to zero-g in so called microgravity environment. Microgravity is defined as a gravitational field in which the gravity is significantly less than that of the Earth and is very small. A common unit of measure of microgravity in physics is one-millionth (10^{-6}) of the

---

T. Rybus (✉)
Centrum Badań Kosmicznych Polskiej Akademii Nauk (CBK PAN), Warsaw, Poland
e-mail: trybus@cbk.waw.pl

H. Kolvenbach
Robotic Systems Laboratory, ETH Zurich, Leonhardstrasse 21, 8092 Zurich, Switzerland
e-mail: Hendrik.Kolvenbach@mavt.ethz.ch

X. T. Yan
RAS Group, DMEM, University of Strathclyde, Glasgow, Scotland
e-mail: x.yan@strath.ac.uk

X. T. Yan and G. Visentin (eds.), *Space Robotics*,
https://doi.org/10.1007/978-3-031-39214-6_8

force of gravity at Earth's surface or 9.8 m/s^2. In the context of broad space robotics for both orbital operations and planetary explorations, the definition of microgravity is broadened to include and cover some of future deep space exploration mission scenarios. The microgravity conditions are particularly important for validation of robotic technologies [2] and satellite attitude control systems [3]. Various approaches can be used to simulate the microgravity environment: parabolic flights as shown in [4], suspension systems [5], drop towers [6], hardware-in-the-loop tests based on industrial manipulators [7], underwater tests with neutral buoyancy vehicles [8] and air-bearing microgravity simulators [9]. Review of methods for ground verification of robotic systems can be found in: [10, 11] and [12].

In this chapter we focus on air-bearing simulators. There are two basic types of such simulator: planar air-bearing microgravity simulators and spherical air-bearing simulators. The former use air bearings to achieve frictionless motion in one plane (providing two translational and one rotational degrees of freedom), while the latter use a spherical bearing to provide torque-free environment and allow free rotations of the satellite mock-up (three rotational degrees of freedom). Planar air-bearing microgravity simulators are used i.a., for validation of robotic technologies and satellite proximity operations (including formation flight), while the spherical air-bearing simulators are used for validation of the satellite attitude control systems. Some test-beds combine planar and spherical air-bearings. The general theory of air-bearings is presented in [13], while review of air-bearing microgravity simulators can be found in: [14–17].

This chapter is organized as follows. In Sect. 8.2 we present key functionalities and design parameters of air-bearing microgravity simulators. Applications of air-bearings in space robotic research and development are reviewed in Sect. 8.3. Section 8.4 presents a proposed design of a low-cost spherical air-bearing satellite simulator. Future directions of research and open issues are discussed in Sect. 8.5. The chapter concludes with a summary given in Sect. 8.6.

8.2 Key Functionalities and Design Parameters of Air-Bearing Simulators

8.2.1 Key Functionalities of an Air-Bearing Simulator

Air-bearing simulators are classified as dynamics simulators [11]. The main purpose of an air-bearing microgravity simulator is to emulate the dynamic behaviour of the satellite or the satellite-manipulator system operating in the orbital microgravity environment. Testing facility with the air-bearing simulator provides an environment in which certain systems, subsystems and components can be tested and validated in microgravity conditions. The key functionalities for testing include capability to resist gravity effect of a space system through suspension using an air-bearing, control and trajectory planning capability of simulating various space system manipulation, docking and capturing to name a few; various sensing capability for accurate control and monitoring; and sufficient space to allow a space system to be manoeuvred. Tests in relevant conditions are required to raise the Technology Readiness Level (TRL) [18]. Air-bearing simulators can be used at various stages of technology development: from tests of basic concepts and demonstrations of practical applicability of control algorithms, to validation of qualification models of hardware components designed for specific missions. Tests performed on air-bearing simulators can be focused on software (e.g., trajectory planning and control algorithms), on hardware (e.g., actuators) or on both.

8.2.2 Air-Bearings and Air Supply

Air-bearing utilizes a thin fluid film of pressurized gas to provide a frictionless interface between two surfaces of typically a fixed base and a moving structure [19]. Air-bearing microgravity simulators are based on aerostatic bearings that relay on the external source of pressurized gas to maintain the gas film. The gas supplied to the air-bearing flows through the bearing face and into the gap between the bearing face and the guide surface. As the fluid is pushed through the gap it creates a pressure profile across the area of the air-bearing [20]. The compressed fluid transfers forces and allows free motion of the air-bearing on the guide surface so that a coefficient of friction around 10^{-5} can be obtained [21]. Gas can be delivered to the bearing gap by several different methods. Two commonly used types of planar (flat) air-bearings are orifice air-bearings and porous air-bearings [22]. In the former, gas escapes through small openings in the bearing face where the gas flowing from the orifice expands and its pressure drops as it flows across the bearing face. The latter are based on the porous media technology, where the gas is flowing through the entire face of the bearing and more uniform pressure distribution can be obtained. The principle of air-bearing operation and photo of an exemplary planar air-bearing are shown in Fig. 8.1.

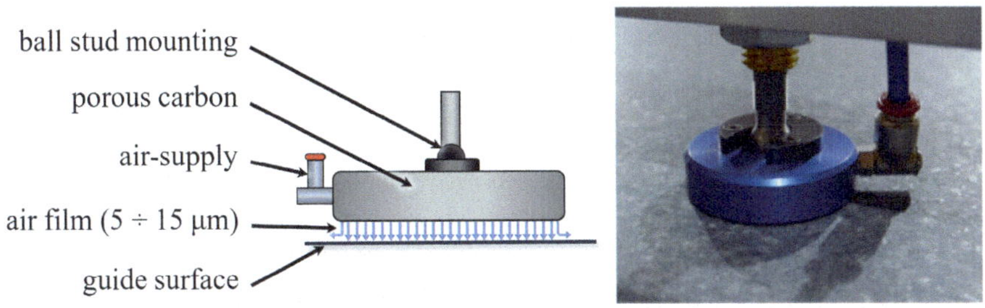

Fig. 8.1 Scheme of planar air-bearing (left) and photo of a 50 mm diameter planar round air-bearing (right)

The design of the air-bearing microgravity simulator depends on the considered application. Key design parameters include the number of bearings, assumed loads and operating pressure, the type of gas used to create the gas film, as well as gas storage and delivery system. Vehicles (mock-ups of satellites) used on planar air-bearing microgravity simulators are usually supported on three bearings that form a three-point stance [23, 24]. A larger number of bearings is rarely used to support a vehicle (four air-bearings are used in some designs, e.g., [25]). The vehicles used in experiments are usually equipped with actuators such as momentum wheels and cold-gas thrusters [26–29]. These actuators allow control of the vehicle orientation and position. Mock-ups of satellites that move on a flat surface in a controlled way are used in experiments related to formation flight and close proximity operations.

Some of the proposed concepts of the active debris removal [30] and in-orbit servicing missions [31] relay on the use of a robotic manipulator for grasping of the target object. Planar air-bearing microgravity simulators are used for various experiments in the field of space robotics. In such experiments, the satellite mock-up is equipped with a robotic manipulator. If a relatively light-weight and short manipulator is used, there is no need to support it with additional air-bearings. The manipulator is attached to the satellite mock-up and hangs above the surface of the test-bed [32–34]. However, when long and heavy manipulator is used, it must be supported by additional air-bearings [35]. In such case precise levelling of air-bearings is necessary, because all air-bearings that support the satellite-manipulator system must be located in one plane. Use of resilient suspension plates in the manipulator joints was proposed to compensate for possible vertical misalignments between components of the system [36]. In such a case precise levelling of air-bearings is not required. Experiments with manipulator base fixed at one of the edges of the test-bed surface are also performed [37]. In these experiments links or joints of the manipulator are supported by planar air-bearings and no satellite mock-up is used.

Aerostatic air-bearings require a continuous supply of compressed gas to produce the gas film that allows frictionless motion. In experiments in which the base of the manipulator is fixed, the compressed gas can be supplied to all air-bearings from an external source (large gas tank or compressor). The flexible hoses or rotational pneumatic connectors used on the manipulator joints allow the transport of gas to the bearings without causing disturbances. Experiments with satellite mock-ups that are free to move on the test-bed surface require the gas canister to be located on the vehicle. The external tank is only used to supply gas to a free-floating satellite in very few test set-ups (e.g., [38]), because the pneumatic hoses leading to the external gas tank could have a significant impact on the dynamics of the system. Use of on-board gas canister limits the possible duration of a single experiment. The maximum duration of the experiment depends on the volume and pressure of the canister, the number of air-bearings and the required gas-flow. The same canister can be used for supplying gas to thrusters, which further reduces the maximum duration of the experiment. The durations that can be achieved span from single minutes [35] to over three hours [39]. The use of on-board canister requires a system for filling this canister between experiments or such a design that allows convenient exchange of the empty canister with a full one.

Each air bearing model has a specific range of operating pressures. The required pressure depends on the carried load and properties of the test-bed surface. Air-bearings in planar microgravity simulators typically operate at a pressure of several bars (operating pressures range from 1 bar [40] to 12 bars [41]). Achieving high capacity and the small size of the on-board canister requires the gas to be stored at high pressure (even up to 300 bars [41]). A pressure reducing regulator is used to lower the pressure of the gas stored in the canister to the operating pressure of air-bearings. Compressed air is the gas used in most planar air-bearing microgravity simulators. However, other gases, such as carbon dioxide [42] and nitrogen [43], are also used is some test-beds. The choice of carbon dioxide may be dictated by safety reasons and lower storage pressure [32].

8.2.3 Test-Bed Surface Material and Construction

The surface of the planar air-bearing microgravity simulator can be made of one of several different materials. The surfaces made of granite [9, 41], glass [44, 45] and epoxy [46, 47] are the most common. Other materials that can be used include metal [48, 49] and marble [50, 51]. Each of these materials has certain advantages and disadvantages. The choice of material influences other parameters such as the possible size of the test-bed, levelling system and surface properties.

The test-beds with a granite surface have a construction similar to optical tables. Thus, the standard optical tables available on the market can be used. The test-bed is based on a monolithic granite slab, which is placed on three or four adjustable legs. To ensure the required stiffness of the slab, its thickness must be sufficiently large, which results in a high weight of the slab. The maximum feasible dimensions of planar air-bearing microgravity simulators with a granite surface are limited by the dimensions of granite slabs that can be mined and transported. Typical test-beds of this design are rather small (e.g., 2.2 m $\times$ 1.8 m [32]), but there are also test-beds of larger dimensions (e.g., 4 m $\times$ 2.5 m [41], 4 m $\times$ 4 m [39]). The glass surface is formed by a plate of glass which is usually about 10 mm thick. This plate is mounted on a frame that ensures the required stiffness. Adjustable legs allow for levelling the frame. The production technology does not allow to obtain large high-quality glass plates. Thus, the test-beds with a glass surface are small in size (e.g., 1.8 m $\times$ 1.8 m [44]). However, multiple glass panels can be stacked together to create larger surface area [52]. In such a case, it is difficult to obtain the appropriate surface parameters along the connection line. The epoxy surface is constructed by mixing the epoxy resins with a hardener (catalyst). The compound hardens and, under the influence of the gravitational force, forms a flat surface. The epoxy surface can have a very large area and its construction is not expensive. The test-beds with epoxy surface have sizes that exceed 20 m^2 (e.g., 4.9 m $\times$ 4.3 m [53], 5.0 m $\times$ 9.0 m [54]. Examples of test-beds with granite and epoxy surfaces are presented in Fig. 8.2.

Requirements on the surface finish depend on the bearing type and operating parameters (i.a., pressure). A surface roughness below 16 root mean square average

Fig. 8.2 Test-bed with a 2 m × 3 m granite surface at the Space Research Centre of the Polish Academy of Sciences (left) and 5 m × 9 m epoxy floor at the Orbital Robot test Bench for Integrated Technology (ORBIT) in ESTEC, ESA [54] (right)

(rms) of the profile height deviations from the mean height is usually recommended by bearing manufacturers, while the local surface flatness should be less than 50% of the air gap [21, 55]. The required global flatness and surface perpendicularity to the gravity vector is determined by the considered application. The highest global flatness is needed in experiments with free-floating satellite mock-ups as imperfect flatness disturbs the experiments and its influence must be compensated by the control algorithms [56]. Experiments with fixed-base manipulators require lower flatness of the test-bed surface. Granite and glass surfaces have the lowest roughness and highest flatness, while epoxy surfaces are characterised by moderate roughness and flatness. Imperfections in the construction of the epoxy surface may result in local undulations and gradients that lead to variations in the surface height and flatness. The height map showing flatness of the test-bed surface made from granite is presented in [9], such height map of the glass surface is shown in [57], while height maps of the surfaces made from epoxy are shown in [54] and [58]. The lack of a levelling system is another disadvantage of epoxy surfaces. The adjustable legs in case of granite and glass surfaces allow tilting of the surface for experiments in low-g conditions [59]. An approach that involves dynamical changing of the inclination of the test-bed surface for a better physical emulation of the relative orbital dynamics was proposed in [60].

8.2.4 Spherical Air-Bearings

Planar air-bearing microgravity simulators allow rotations of a satellite around one axis perpendicular to the surface of the test-bed. Spherical air-bearings are required to simulate rotational motion of a satellite in all three axes. The spherical air-bearing consists of two parts, which are portions of concentric spheres [61]. Compressed gas provided from an external gas tank is pushed between the two parts and forms the gas film. One section of the bearing rotates freely on the gas film bounded by the other section. Spherical bearings create a torque-free environment, provided the centre of mass of the system coincides with the bearing's centre of rotation. Simple manual system can be used to align the centre of mass with the centre of rotation [62] or automatic mass balancing method can be employed [63]. Commercially available or in-house designed and constructed spherical air-bearings are used in test-beds [64]. The range of motion of spherical air-bearing microgravity simulators is usually limited by the equipment affixed to the bearing [65]. However, in some of the proposed solutions full rotations are possible [66]. Reaction wheels and cold-gas thrusters are used as actuators and allow closed-loop control of the satellite attitude [67].

In some test-beds planar air-bearing microgravity simulators are combined with spherical air-bearing simulators to obtain more Degrees of Freedom (DOF). In a 5 DOF set-up a spherical air-bearing is mounted on a vehicle supported by planar air-bearings. The vehicle can move on the test-bed surface (two translational DOF), while the spherical air-bearing provides rotations in all axes. Such test-beds are operated, e.g., at School of Aerospace Engineering at Georgia Institute of Technology [68], University of Isfahan [69] and Aerospace Science and Technology Department of Politecnico di Milano [70]. Various concepts of new 5 DOF test-beds are proposed, and one example can be found in [71]. An additional sixth DOF can be added by an active system that allows vertical motion. In the concept proposed in [72], a cylindrical air-bearing is mounted on the vehicle and counterweight system allows vertical translation using suspension mechanism with pulleys and wires. In another concept, presented in [73], ball screw actuators are used in the vertical stage to provide translational motion in the direction perpendicular to the test-bed surface. A 6 DOF test-bed based on a comparable concept was developed at DLR Institute of Space Systems [41] and at Caltech's Aerospace Robotics and Control Lab [74]. Although such systems do not emulate free-floating dynamics in the vertical direction, the inclucion of vertical motion in experiments is important for formation flight research. A full 6 DOF test-bed was constructed at Rensselaer Politechnic Institute [75]. In this test-bed, a translation stage is supported by three planar air-bearings, whereas an attitude stage is based on a spherical air-bearing, while a system of air pulleys on the translational stage connects the central column and the attitude stage to a counterbalance deck.

8.2.5 Combination of Air-Bearings with Other Systems

Finally, it is worth noting that some test-beds combine the use of air bearings with other systems that are used for moving objects and interacting with vehicles supported on air-bearings. A fixed-base manipulator located next to the test-bed surface can be used in contact experiments [50, 76]. In an ESA facility, a robotic manipulator mounted on a rail can simulate a chaser, while a target is supported on air-bearings [77]. In a test set-up presented in [78], three air bearings are mounted on a mobile robot. These bearings are pointing upwards and the payload is mounted on a flat plate that floats on air-bearings. As the payload moves the mobile robot follows respectively using the relative-position information from a contactless sensor (the payload is decoupled from the platform). A 2 DOF motion table capable of positioning a payload (small industrial manipulator or pan-tilt mechanisms) above the air-bearing test-bed surface is used in the Orbital Robotic Interaction, On-orbit servicing, and Navigation (ORION) laboratory at the Florida Institute of Technology [79].

8.2.6 Experiments with Scaled-Down Models

Testing full-scale models of space hardware was one of the first applications of planar air-bearing microgravity simulators [80]. If the size of the tested hardware is small, then tests can be carried out on small-sized test-beds (e.g., experiments with CubeSat class nanosatellites [81]). Hardware models that are large in size require test-beds with a large surface area (e.g., the 11.3 m European Robotic Arm was tested on a 18 m × 14 m test-bed [82]). An alternative solution is to perform tests using scaled-down models of space hardware. The behaviour of two physical systems will be equivalent if certain conditions derived from dimensional analysis are fulfilled [83]. Experiments with scaled-down models have been conducted from the early days of the space age [84]. A general overview of various approaches that are used to design reduced-scale models can be found in [85].

Application of dimensional analysis for designing scaled-down model of a manipulator for experiments on a planar air-bearing microgravity simulator is presented in [86]. In this general approach, based on the Buckingham π theorem, a complete and dimensionally independent subset of dynamic parameters is constructed. In the considered case the subset of parameters of the space and ground manipulators has three elements. Selection of these elements is not unique. Dimensional analysis allows to formalize the dynamic equivalence conditions and control scaling laws between a real-size space manipulator and ground manipulator to be designed for operations on the microgravity simulator. Application of this approach for simulation of the orbital relative motion on the air-bearing microgravity simulator is described in [87]. While dimensional analysis gives the freedom to design the appropriate scaling, the following law is commonly used (e.g., [88]):

$$P_s = k^w P \tag{8.1}$$

where P_s and P denotes the values after scaling and before scaling, respectively, k is the selected scaling coefficient, while w is the scaling exponent. The list of scaling exponents for selected physical properties is presented in Table 8.1. As evident, various parameters should be scaled by various factors [89]. Scaling all physical dimensions by a factor of 1/2 results in the necessity to scale the mass by a factor of 1/8 and the inertia by a factor of 1/32. The time is also scaled. Scaling is especially problematic in experiments that involve contact between the gripper and the target

Table 8.1 Scaling exponents for selected physical properties [91]

Physical property	Distance	Time	Mass	Inertia	Velocity	Acceleration
Scaling exponent	1	1	3	5	0	-1

object, because the contact parameters also need to be scaled [90]. In order to obtain the same dynamic behaviour of the scaled system, the material properties of the contacting bodies should be changed according to the scaling law.

8.3 Applications of Air-Bearings in Space Robotic Rresearch and Development

8.3.1 Testing of Manipulators Hardware for On-Orbit Operations

Robotic manipulators intended for on-orbit operations are designed to work in microgravity environment. Thus, they are not able to sustain their own weight [92]. Tests performed on planar air-bearing microgravity simulators are one of the available solutions for validating such manipulators in laboratory conditions. However, such tests are limited to one plane, while manipulators are designed to operate in three-dimensional space. This requires separate testing of the individual planes of motion. Moreover, real-size models are typically used in such tests and base of the manipulator is fixed. Taking into account the size of the manipulators, tests can only be performed on the largest test-beds. During the tests the links or joints of the manipulator are supported by air-bearings with the use of a dedicated support equipment. The influence of this support equipment on the dynamics of the manipulator is usually neglected. However, the method proposed in [93] and validated in experiments allows to quantitatively assess the influence of an air-bearing based support equipment on the results of performed tests.

There are several notable examples of the use of planar air-bearing microgravity simulators for tests of manipulators intended for on-orbit operations. The Shuttle Remote Manipulator System (SRMS) was a 15.2 m long manipulator designed for use on the Space Shuttle [94]. Multi joint operations of the SRMS were tested on a flat floor constructed in a 100,000 class clean room. The SRMS was fixed at its base and supported on air-bearings [95]. Tests performed on the air-bearing microgravity simulator allowed verification of the manipulator controls, multi-joint straight-line controllability, tip positioning accuracy and tip force. These were the most comprehensive full system tests performed before the SRMS was checked on-orbit during the STS-2 mission [94].

The European Robotic Arm (ERA) [96] was delivered to the International Space Station (ISS) in 2021. The Engineering Qualification Model and the Flight Model of this arm were tested on a planar air-bearing microgravity simulator [82]. During tests, the hand of the ERA was supported by a vehicle moving on air-bearings, while a suspension system was employed to reduce the weight of the manipulator in its middle section. The ERA was mounted to the support vehicle by a spherical bearing, which was moving with minimal friction on the second set of air-bearings over a glass plate located at the top of the support vehicle. This set-up allowed double decoupling of the ERA motion.

The prototype of the Chinese Intelligent Space Robotic System was also tested on a planar air-bearing microgravity simulator [97]. Two-step testing method was used to measure the positioning accuracy of the manipulator. In this method the three-dimensional motion of the manipulator was divided into two planar motions that were conducted independently.

8.3.2 Control and Modelling of Flexible Systems

Manipulators designed for on-orbit operations are light-weight and highly flexible. Their flexibility mainly comes from transmission flexibility of harmonic gear reducers in joints and structural flexibility of links [98]. The dynamic behaviour resulting from high flexibility is specific to zero-g conditions of orbital environment. High flexibility makes it difficult to obtain high positioning accuracy. Tests performed on planar air-bearing microgravity simulators with flexible fixed-base manipulators allow verification of control algorithms and dynamic models. Open-loop control based on the motion input pre-shaping technique was validated in experiments performed on a planar air-bearing microgravity simulator at the Università di Pavia [99]. This control was applied to a 2 DOF manipulator in a configuration resembling a human hand. Links of the manipulator were highly flexible in the operating plane, but rigid in the plane perpendicular to the surface of the test-bed. In such tests the air-bearing support equipment may influence the dynamic behaviour of flexible manipulator. This problem is analysed in [100], where procedures are presented to remove the influence of air-bearings and identify the true equivalent joint stiffness and damping from the results of experiments.

Large payloads transported by the manipulator can also have high flexibility. In the experiments performed at the Tokyo Institute of Technology a 6 DOF planar manipulator was used to transport massive payload on a flexible structure [101]. The manipulator was supported by air-bearings and fixed to a flexible base. Input Shaping method was applied to suppress the residual vibration of the flexible base and the flexible payload. Other experiments focus on estimation of the shape, motion and dynamic model parameters of vibrating space structures [102]. This data is needed for effective trajectory planning and control of the manipulator. An algorithm that removes residual vibration from flexible structure undergoing large motion and reduces fuel consumption of the vehicle was succesfully demonstrated in [42]. A method that allows control of the end-effector position and vibration suspension of a flexible appendage was presented in [103]. In this approach the state of the appendage is estimated from the measurements obtained from the force/torque sensor.

8.3.3 Trajectory Planning and Control of Free-Floating Satellite-Manipulator Systems

Tests conducted with satellite-manipulator systems constitute the largest group of experiments performed on planar air-bearing microgravity simulators. Control of such systems is challenging due to the dynamic coupling between the manipulator and the satellite [104]. The free-floating nature of the satellite-manipulator system must be taken into account during trajectory planning and control [105]. Moreover, precise knowledge of geometrical and mass parameters of the system is required. Thus, various algorithms are proposed to perform identification of these parameters during the phase of orbital operations. Demonstration of such algorithms in experiments performed on planar air-bearing microgravity simulators is presented in [106] and [107]. Figure 8.3 shows the Robotic Space Emulator "Cepheus" used in the identification experiments performed at the National Technical University of Athens.

One of the first experiments aimed at validation of control strategies for free-floating satellite equipped with a 2 DOF manipulator was performed on a planar air-bearing microgravity simulator at Stanford University [111]. Comparison between the PD-control, the resolved rate control and the resolved acceleration control for trajectory tracking of a free-floating satellite-manipulator system is presented in [112]. A model-based control algorithm that does not require non-holonomic path planning was demonstrated in experiments performed with a 3 DOF manipulator mounted on a satellite at the University of Padova [34]. The proposed approach

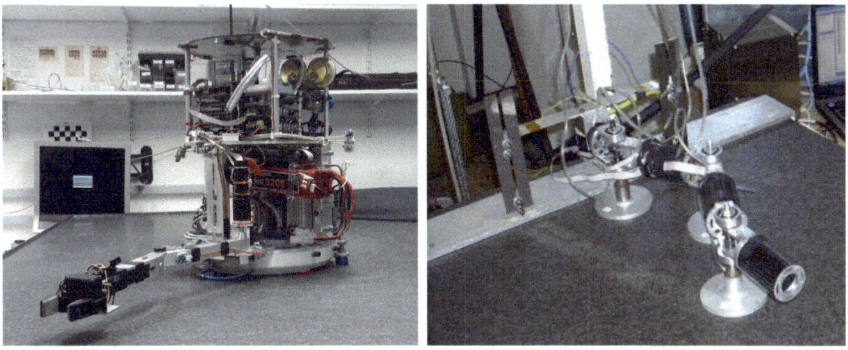

Fig. 8.3 Robotic Space Emulator "Cepheus" operated at the National Technical University of Athens [106] (left) and 3 DOF fixed-base manipulator used for demonstration of redundancy resolution schemes aimed at locally minimizing the reaction torque transferred to the satellite [37] (right)

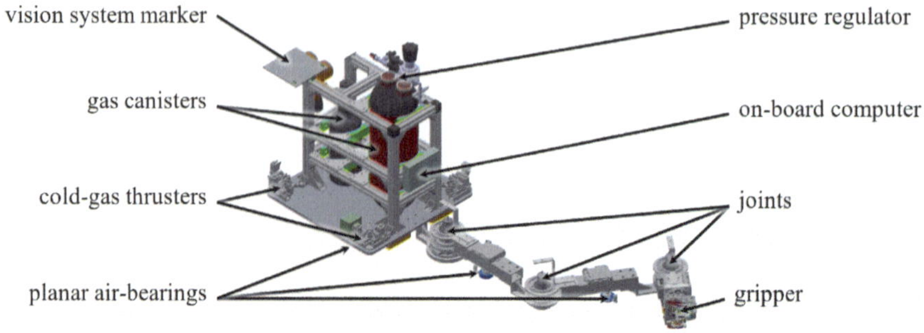

Fig. 8.4 Scheme of the satellite-manipulator system used in experiments performed on the planar air-bearing microgravity simulator at the Space Research Centre of the Polish Academy of Sciences

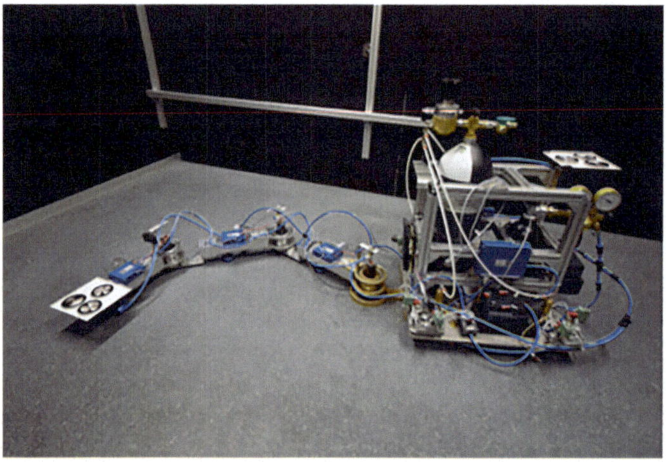

Fig. 8.5 Photo of the satellite-manipulator system operated at the Space Research Centre of the Polish Academy of Sciences

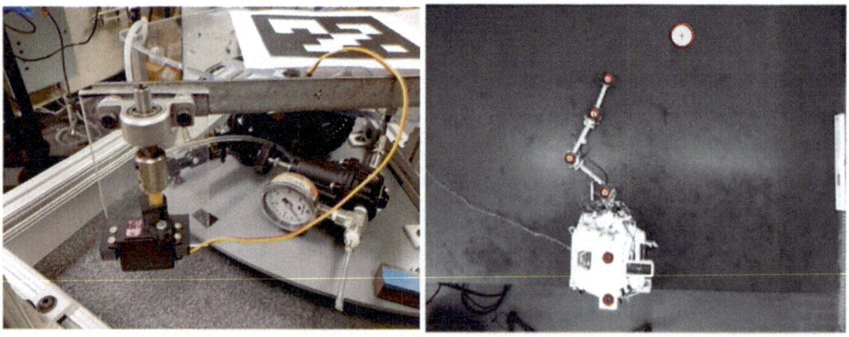

Fig. 8.6 Test set-up used in experiments with flexible-joint manipulator [108] (left) and view (acquired by the external vision system mounted above the test-bed surface) of the PINOCCHIO platform operated at the Sapienza Università di Roma [109] (right)

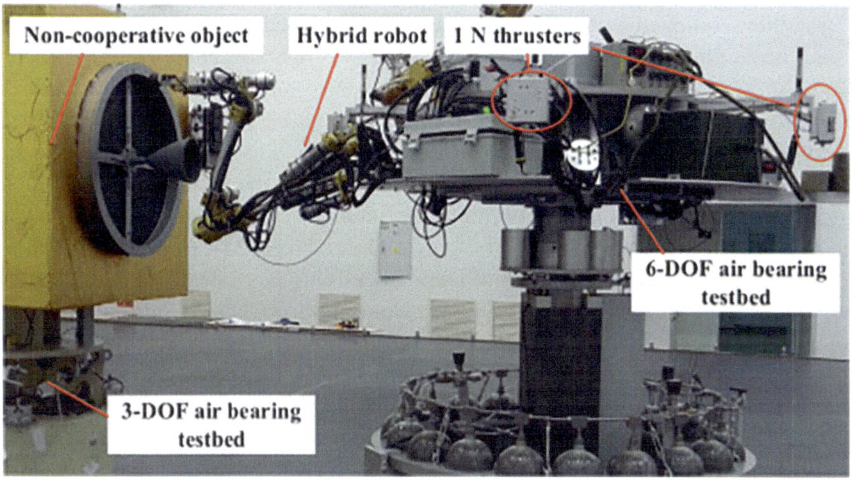

Fig. 8.7 A 6 DOF air-bearing microgravity simulator used for demonstration of a capture operation performed with a novel hybrid manipulator [110]

allows minimization of the satellite rotation during the motion of the manipulator. Two redundancy resolution schemes aimed at locally minimizing the reaction torque transferred to the satellite during the manipulator motion are demonstrated in [37]. The test set-up used in these experiments is shown in Fig. 8.3.

A satellite equipped with a relatively long and heavy 3 DOF manipulator is operated on a planar air-bearing microgravity simulator at the Space Research Centre of the Polish Academy of Sciences [113]. Scheme of this satellite-manipulator system is presented in Fig. 8.4, while its photo is shown in Fig. 8.5. The links of the manipulator are supported by additional air-bearings. This test-bed was used to validate a control strategy for a free-floating robot performing rendezvous manoeuvres [114]. Transformation of an end-effector trajectory given in the velocity space to velocities of manipulator joints requires inverting of the body Dynamic Jacobian [35]. This Jacobian cannot be inverted in singular configurations of the manipulator. Thus, a method using Bézier curves for planning trajectories that avoid dynamic singularities of the manipulator was proposed and demonstrated on this test-bed [115]. The planar air-bearing microgravity simulator at the Space Research Centre of the Polish Academy of Sciences was also used to conduct experiments aimed at demonstrating manipulator trajectory optimization algorithms. The algorithm presented in [116] allows minimization of manipulator control torques, while the approach proposed in [117] allows to generate a collision-free trajectory that results in the desired position and orientation of the end-effector and the desired orientation of the satellite. Recently, the same test-bed was used for validation of a collision-free trajectory planning method based on the Bi-directional Rapidly-exploring Random Trees algorithm [118].

Satellites equipped with multiple manipulators are used on several test-beds, e.g., a satellite equipped with two 4 DOF manipulators is operated at the National Technical University of Athens [32]. In experiments performed at the Osaka University assembly of a truss structure was performed by a space robot with two manipulators [119]. In the presented approach if the robot fails to accomplish the task in a planned way, it can still accomplish it by re-planning the task through the reinforcement learning technique. Different experiments were focused on controlling a team of robots that were used to transport a large object [33]. Experimental verification of algorithms for multi-robot team assembling large flexible structure was performed at the Space Robotics Laboratory at the Massachusetts Intitute of Technology [46]. A model-based trajectory generation methodology for flexible-joint manipulator was proposed in [108]. This methodology incorporates constraints on joint velocity, motor torque and base actuation. The proposed approach was verified on a planar air-bearing microgravity simulator with 1 DOF manipulator. Test set-up used in these experiments is shown in Fig. 8.6.

The simultaneous control of the satellite and the manipulator is another important topic. A neural network aided by a systematic complexity control scheme was trained using a supervised learning technique to emulate a complex thruster mapping function of a satellite [120]. This mapping was implemented on a free-floating satellite-manipulator system and combined with a simple neural network control component trained using backpropagation through time to provide accurate tracking of the manipulator. The coordinated control of the satellite and the manipulator was demonstrated in [109]. In the proposed approach all the required actions (forces and torques applied on the satellite and torques in the manipulator joints) are evaluated simultaneously using a single expression. The Platform Integrating Navigation and Orbital Control Hosting Intelligence On-board (PINOCCHIO) used in these experiments is presented in Fig. 8.6.

8.3.4 Capture with a Manipulator

Large group of experiments is focused on demonstration of the capture operation performed with the use of a gripper mounted on the manipulator. These experiments are usually performed with two objects on air-bearing test-bed: chaser satellite and target satellite (exemplary design of an air-bearing supported target satellite for demonstration of capture operation is presented in [121]).

The manipulator control based on the Generalized Jacobian Matrix (GJM) was tested in experiments performed on a planar air-bearing microgravity simulator at the Tokyo Institute of Technology [122]. The satellite was equipped with a 2 DOF manipulator. It was demonstrated that the manipulator is able to capture a stationary or a moving target. Experiments related to the contact dynamics during the capture operation were conducted on the same test-bed using satellite equipped with two manipulators [35]. A target object supported on air-bearings is used on the planar air-bearing microgravity simulator at the National Technical University of Athens and enables simulations of complex on-orbit servicing tasks [123]. Recent upgrades introduced to this test-bed allowed successful demonstration of a capture operation. A set of laboratory experiments was conducted on the planar air-bearing microgravity simulator at the Naval Postgraduate School to demonstrate the autonomous capture of a target object by a satellite equipped with a 4 DOF manipulator [124]. Three different approaches to control the satellite during the capture operation (floating, flying and rotation-flying control) were compared in these experiments. In [125] a planar air-bearing microgravity simulator was used to demonstrate capture performed in a tele-operation mode. During these experiments the orientation of the satellite was controlled. Successful capture of a target-object by a manipulator mounted on a satellite is also presented in [43] and [126].

A 6 DOF air-bearing microgravity simulator that combines planar air-bearings and a spherical air-bearing was used for demonstration of a capture operation performed with a novel hybrid manipulator [110]. This test set-up is shown in Fig. 8.7. A non-cooperative free-floating target object also supported on planar air-bearings was successfully grasped in these experiments. In experiments performed at the Sapienza Università di Roma a fixed-base planar 2 DOF manipulator was used to capture a free-floating vehicle supported by air-bearings [40]. The PINOCCHIO platform was used in these experiments as the target object. Capture of a free-floating target object by two 7 DOF manipulator was presented in [127]. Impedance-based contact control of a robot using a compliant wrist for capturing non-cooperative target was presented in [128]. In these experiments a target satellite was mounted on air-bearings, while a fixed-base 7 DOF manipulator was used to simulate a manipulator mounted on a servicing satellite. The proposed contact dynamics modelling of a compliant wrist was successfully verified [129].

8.3.5 Close Proximity Operations and Docking

Planar air-bearing microgravity simulators are used for validation of control algorithms for close proximity operations and docking procedures, validation of numerical models of contact dynamics and for tests of docking mechanisms. Relative navigation plays an important role in such experiments. A method for determining the relative position and velocity with respect to the target object was experimentally validated on a planar air-bearing microgravity simulator [130]. Control of the satellite during close proximity operations with the combined use of thrusters and reaction wheels is considered in [131]. The performed experiments demonstrated that such approach allows minimization of the fuel consumption. Application of the Rapidly-exploring Random Trees (RRT) algorithm for collision-free trajectory planning of the satellite with attached manipulator during close proximity operations is presented in [113]. Demonstration of close proximity operations for a 3U CubeSat Engineering Model is described in [132]. The satellite used simple cold-gas propulsion system to navigate in close proximity of other satellites. A gyroscopic obstacle avoidance algorithm was combined with a standard stabilization control laws to handle unexpected obstacles during trajectory execution.

In an experiment performed in the frame of the Synchronized Position Hold Engage Reorient Experimental Satellites (SPHERES) programme an H-infinity control algorithm robust to disturbances was successfully applied for docking operation performed on the air-bearing simulator [133]. In [134] the artificial potential field method was employed for control of a satellite during approach and capture of space debris. The chaser satellite and the mock-up of space debris were supported on air-bearings. Vehicles used in these experiments are shown in Fig. 8.8. Another application of the artificial potential field method for guidance during approach and docking was demonstrated in [135]. The adaptive artificial potential function allowed reduction of fuel consumption and the system was able to react to a dynamic environment. Other methods applicable for the final approach during the docking operation include the Model Predictive Control (MPC) and control based on the inverse dynamics in the virtual domain [136]. Nonlinear model predictive tracking control scheme for control of satellite position was also evaluated in experiments performed at Luleå University of Technology [137]. Docking of a chaser satellite to a rotating target object was demonstrated in experiments presented in [138]. The proposed controller takes into account the collision avoidance constraint.

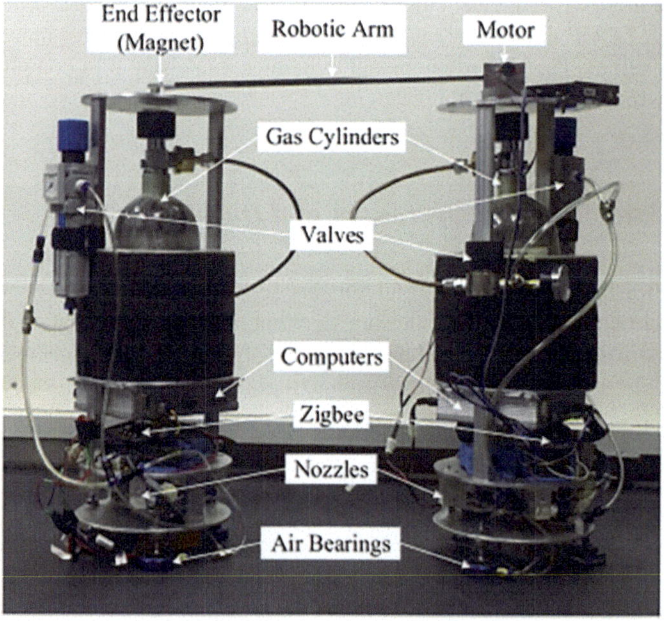

Fig. 8.8 Two vehicles used in approach and capture experiments performed at the Nanjing University of Aeronautics and Astronautics [134]

A planar air-bearing microgravity simulator was used for tests of a docking interface developed for the Orbital Express mission [139]. Experimental demonstration of docking using an apogee motor nozzle as a capture interface was presented in [140]. In the test campaign a full-scale docking mechanism was used, while mock-ups of chaser vehicles and target object were scaled down. Each vehicle was placed on a separate granite table. This test set-up was used for experimental validation of a generic contact-dynamics toolkit in a software simulator of docking manoeuvre based on Matlab Simulink [141]. Experimental validation of a different contact dynamics simulation software was performed at the University of Victoria [142].

8.3.6 Formation Flight

Large planar air-bearing microgravity simulators are used for formation flight experiments. Such experiments are usually performed with dedicated vehicles equipped with thrusters and reaction wheels. The planar air-bearing microgravity simulator at Stanford University was used for experiments related to the proposed Orion mission [143]. Low-level, multi-vehicle, station keeping algorithms and a specific control architecture were used to keep the vehicles aligned in a formation. Re-targeting and re-sizing of the formation was performed with a high-level planner that ensured equal consumption of resources for each vehicle. Results obtained in numerical simulations were verified by experiments performed on the microgravity simulator. A series of experiments related to the formation flight was carried out as part of the SPHERES program [144]. A navigation system that utilizes a single camera to estimate relative state of two vehicles in a formation was tested [145] and different approaches to human-spacecraft interaction with varying levels of autonomy were investigated [146].

A nonlinear controller based on State-Dependent Riccati Equation was used to handle relative position tracking and attitude synchronization problem in rendezvous between two satellites [147]. A PID control law and pulse width modulator were used to command cold-gas thrusters of the chaser vehicle to follow assigned trajectory for rendezvous with a semi-active target vehicle [148]. The guidance algorithms calculated the desired velocity and position profile starting from a pre-defined trajectory and compared them with the navigation output. Data fusion between multiple sensors was performed with a Kalman filter. Experiments performed at the Naval Postgraduate School were related to the orbital assembly performed with four vehicles [149]. These vehicles are presented in Fig. 8.9. Guidance and control strategies based on LQR were designed to take into account the evolving mass and shape of the assembling spacecraft. Four vehicles were also involved in closed path proximity manoeuvres [53]. The robustness of proposed algorithms against disturbances and sensor noise was demonstrated. In another group of experiments performed on the same test-bed a Lyapunov-based thrusters selection method and a commanding technique providing control of rotational and translational dynamics of a vehicle with only on–off actuators was validated [150]. The planar air-bearing microgravity simulator

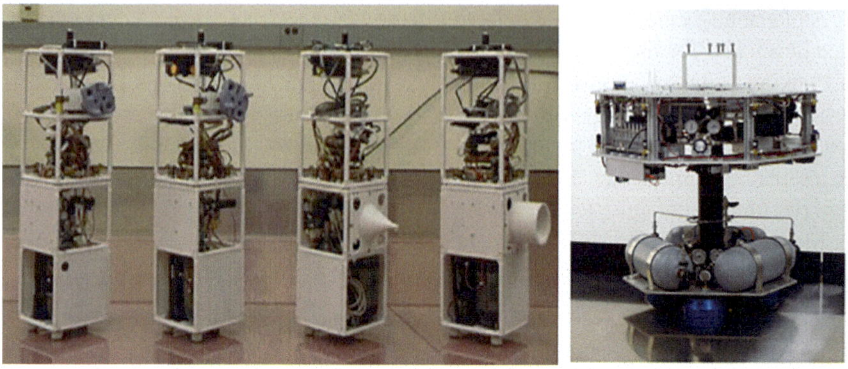

Fig. 8.9 Vehicles used in the formation flight experiments at the Spacecraft Robotics Laboratory of the Naval Postgraduate School [149] (left) and vehicle operated at the 5 DOF test-bed at Astrodynamics and Control Laboratory, Yonsei University [48] (right)

in Technology and Engineering Center for Space Utilization, Chinese Academy of Sciences was used for demonstration of tight formation control [151]. Four satellites were used and high positioning accuracy was demonstrated.

Test-beds that provide 5 DOF are also used for formation flight experiments. Figure 8.9 shows vehicle used for such tests described in [48]. Adaptive Saturated Neural Network Tracking Control of satellite was demonstrated on a 5 DOF test-bed, where the satellite tracked the desired planar trajectory, while synchronizing its attitude with the desired angle [152].

8.3.7 *Attitude Control*

Experiments performed with spherical air-bearing microgravity simulators support a variety of research areas, including satellite attitude control, precision pointing, satellite rotational dynamics, vibration suppression and integrated controller design [153]. Several existing test-beds are dedicated to nanosatellites, e.g., CubeSat class [62]. The experiments are usually performed with functional mock-ups of selected satellite subsystems. Some experiments are focused on actuators and sensors. Identification of parameters of the spherical air-bearing simulator usually precedes the actual test campaign. An approach for experimental identification of parameters was demonstrated in [154].

Experimental comparison of attitude control algorithm based on feedback linearization with an algorithm based on the State-Dependent Riccati Equation technique is presented in [155]. Two modes of the attitude control system (detumbling and pointing) were tested in experiments described in [156]. A compensation method used in Chinese Fengyun-4 satellite to counteract the line-of-sight influence by attitude motion during imaging was demonstrated on an air-bearing microgravity simulator [157].

A new inertia actuator that can generate toques in all three principal axes of a rigid satellite using a spinning wheel and a double tilt mechanism was proposed in [158]. Use of this actuator for satellite attitude control was demonstrated in experiments. Spherical air-bearing microgravity simulators are also used to emulate tumbling motion of the uncooperative space debris. A dedicated control algorithm that allows to obtain such motion was proposed in [159].

8.3.8 Other Applications

There are many other applications of planar and spherical air-bearing microgravity simulators [16]. Planar air-bearing simulators are used to perform tests of various devices designed for satellites, such as antenna pointing mechanism [160] and antenna deployments [161]. Other devices, tested in such experiments, include magnetic locking flux-pinned revolute joint for use on CubeSat class satellites [162]. Large group of experiments is related to tethers, e.g., tethered formation flight [163] and tether deployment [164]. Air-bearing microgravity simulators are used for testing sampling devices developed for low-g bodies [165, 166], landing gears [59, 167] and methods for locomotion in low-g environment (e.g., on the surface of an asteroid) [168].

8.4 An Example Design of a Spherical Air-Bearing Satellite Simulator

8.4.1 System Overview

Exemplary, we present the design of a 2.7 kg spherical air-bearing platform that allows for controlled, frictionless rotations (Fig. 8.10). Equipped with reaction wheels, thrusters, and a camera, the air-bearing is meant to serve as an educational platform [169]. The development was driven by low-cost and modular design, making it easy to recreate and customize the robot. The robot is simple to handle and only requires pressurized air to operate. A PID controller allows the control of the yaw and pitch orientation using an off-the-shelf gamepad.

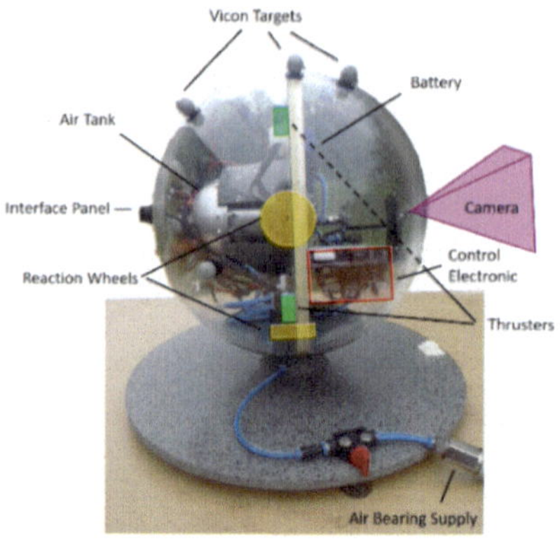

Fig. 8.10 An overview of the spherical air-bearing platform (left) and charging and control interface (right)

8.4.2 Hardware Overview

All components are integrated into an acrylic sphere of 300 mm in diameter. The sphere has a dismountable interface plate, through which the robot's interior can be assembled. Plates mounted rigidly inside the sphere serve as a structure to mount the components. The sphere is placed on top of a concave bowl, which is supplied with a constant stream of pressurized air. The air creates a small gap between the sphere and the bowl, thus allowing for frictionless rotation. We added small balancing weights to align the centre of mass along the Yaw axis of the sphere.

Thrusters are placed at the side of the sphere and allow for control of the yaw and roll motion. The thrusters are aluminium blocks with two small openings each to release compressed air. The compressed air is stored in a 0.75 l aluminium (Al2024) propellant tank inside the sphere. The tank is filled with compressed air up to 10 bar. We attached a pressure regulator (SMC ARJ1020F-M5-04) with a small pressure gauge (Festo MA-15–10-QS-4) to the tank and set it to a static pressure of 4 bar. Four miniature electromagnetic valves (Kuhnke MPP91 PV-T4-L36-04) are used to control the thrusters.

Two reaction wheels control the yaw and pitch axis of the robot. We chose commercially available brushless outrunner motors (Hacker A20-12 XL EVO) to spin the reaction wheels. Each reaction wheel is made of brass, weighs 67 g and provides inertia of 204 gcm^2 (excluding motor mass/inertia).

The motors are driven by consumer market electronic speed controllers (ESC) (Blue Robotics Basic 30A ESC v3). The speed controller must allow for both forward- and backward motion since most hobby speed controllers for the car or drone market have one preferred direction of travel. We implemented a microcomputer (Beaglebone Black) to control the platform. The computer has the convenience of combining low- and high-level interfaces. It is powerful enough for running the Robotic Operation System (ROS) while maintaining a low power consumption and small size.

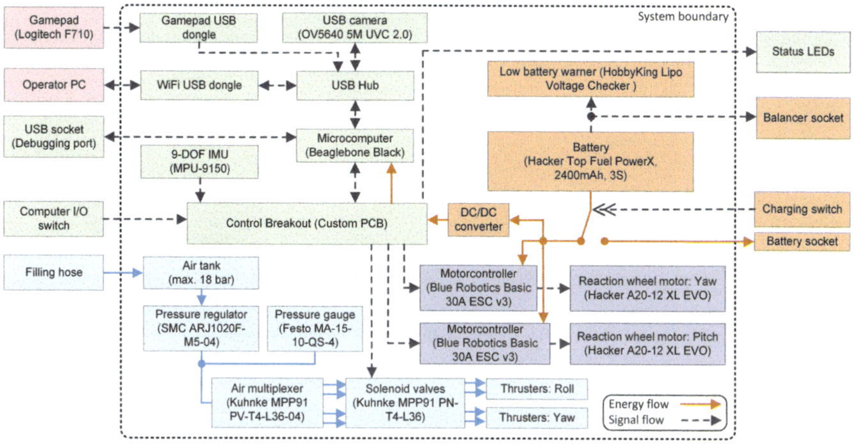

Fig. 8.11 Electrical and hardware system diagram

A custom-designed PCB is stacked on top of the microcomputer to interface the electromagnetic valves, the speed controllers, the interface panel and a 9-DOF IMU (Invensense MPU-9150). A USB hub is connected to the Beaglebone to connect a WiFi dongle and a dongle for the gamepad (Logitech F710). A small camera (OV5640 5 M USB Camera UVC 2.0) with a lens f(XH- 1820, f = 1,8) is also connected and pointed outwards. The system is powered by a LIPO battery with 2400 mAh (TopFuel ECO-X 20C, 2400 mAh), equipped with a battery monitor and low voltage alarm. Electrical and hardware system diagram is shown in Fig. 8.11.

8.4.3 Control

We chose to use the robotic middleware ROS on the microcomputer because of the broad support of open-source software packages. The ROS framework is component-based and follows modular design principles, which makes it easy to integrate new packages and keep the software up-to-date. The software structure is depicted in Fig. 8.12 (left).

We implemented a PID-controller to control the yaw axis (e_z^R) and a PD-controller to control the pitch axis (e_y^R) of the sphere of the form:

$$\begin{aligned}\Omega_1^* &= -k_{p,pitch} \cdot (\theta_{des} - \theta) - k_{d,pitch} \cdot \dot{\theta} \\ \Omega_2^* &= -k_{p,yaw} \cdot (\psi_{des} - \psi) - k_{d,yaw} \cdot \dot{\psi} - k_{i,yaw} \cdot \int_0^t (\psi_{des} - \psi) dt\end{aligned} \tag{8.2}$$

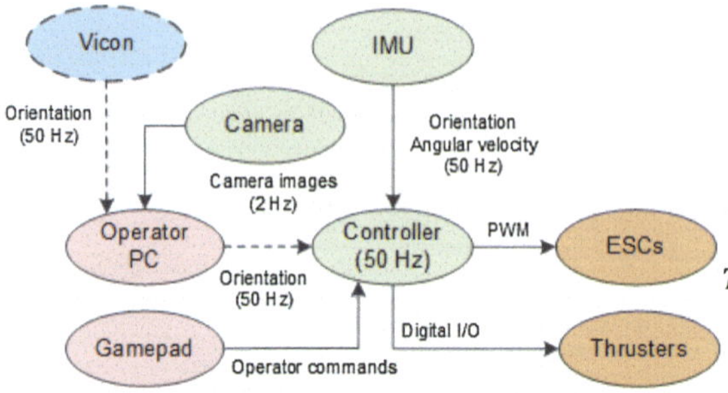

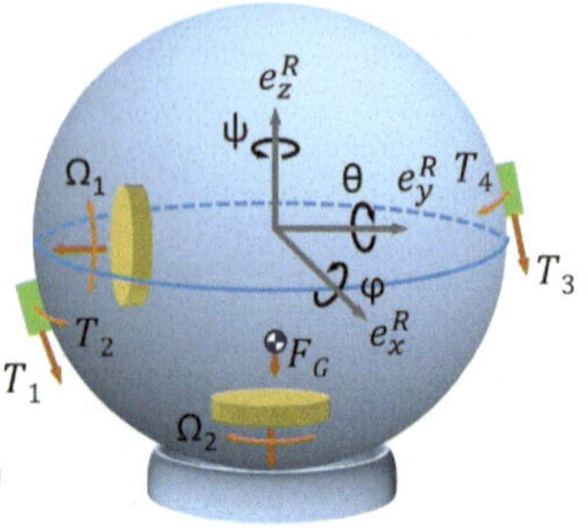

Fig. 8.12 Control system overview (left) and schematic overview (right)

The yaw-controller directs the robot to any orientation around the z-axis. The pitch controller stabilizes oscillations around the y-axis. The robot naturally oscillates around the x- and y-axis since the COM is located along the z-axis, as shown in Fig. 8.12 (right). Gravity thus creates a moment when the robot rotates around either x- or y-axis. A reaction wheel can create torque for a limited amount of time before it reaches maximum speed. However, the moment introduced by gravity is constant. Thus, position control around the pitch axis is not possible. The pitch axis is controlled around its equilibrium point, which is at the angle where the moment introduced by gravity is zero. The yaw orientation of the robot (ψ_{des}) is intuitively commanded using an analogue joystick of a gamepad. The orientation estimates are supplied either by the IMU or by an absolute positioning system (i.e. Vicon or similar).

The resulting reaction wheel velocities are transferred to the ESCs via PWM signals. The thrusters are not taken into account in the control loop, as their force is negligible in comparison to the torque of the reaction wheels. However, they can be switched on and off via the gamepad.

8.4.4 Remarks

We describe the design of a spherical air-bearing satellite simulator. The spherical robot can float frictionless in a bowl and is controlled via reaction wheels. The implemented controller can smoothly orient the robot in the desired direction and minimize the oscillation amplitude around the pitch axis. While the robot generally works well, further improvements are required to increase the fidelity of the platform. The pressure supplied to the thrusters is relatively low and can only be fired for a short time (<20 s). A possible solution would be to use a small, high-pressure paintball air tank with an integrated pressure regulator.

8.5 Future Directions and Open Issues

The active debris removal and in-orbit servicing missions that are currently planned will rely heavily on the use of robotic technologies. Complex robotic systems will replace astronauts in many tasks. With recent advances in space robotics, the need for testing of robotic systems is increasing. Tests in simulated microgravity environment play crucial role in development of robotic technologies intended for orbital missions. The demand for air-bearing microgravity simulators has increased in recent years as such simulators offer cheap and effective way to perform tests at various stage of technology development.

Furthermore, the demands of sustainable and affordable use of space assets points to a need to reuse the space assets such as satellites by upgrading their functionalities instead of launching a new satellite. Recent modular satellite design approach explores a paradigm shift from designing a monolithic satellite to modular one. This provides significant opportunities for space asset reuse and sustainable space exploration. At the same time, this trend presents much more complex testing requirements for space robotic systems in air-bearing microgravity simulation environments.

Many new facilities are constructed (e.g., [170]). Some of the new facilities are very large, e.g., 5.0 m $\times$ 9.0 m epoxy floor at the ORBIT test-bed in ESTEC [54]). Large surface areas allow tests with 1:1 models or real hardware elements. The fact that scaling is not required is a significant advantage in many cases. Moreover, complex formation flight experiments with multiple vehicles can be performed on such test-beds. However, obtaining low roughness and high global flatness is still a challenge in case of large test-beds. Construction of new test-beds based on epoxy floors with improved surface parameters would be an important accomplishment in the field. It is also worth noting that the newly built test-beds based on granite slabs are also larger than the existing ones (e.g., [39]). New systems are introduced in test-beds designs, e.g., advanced levelling system that allows dynamical changing of the surface inclination [60]. Low-cost solutions based on commercial off-the-shelf (COTS) components are proposed for some of the newly-designed test-beds [171]. This is an important trend that increases the availability of planar air-bearing microgravity simulators. The availability of such test-beds is important to both academia and industry.

Existing test-beds are being upgraded and their capabilities are being extended. New generations of vehicles are introduced (e.g., [149]), while the existing vehicles are modified (e.g., [113]). The introduced modifications and improvements open up new areas of research and allow for more advanced experiments. Coordinated experiments performed between two test-beds in remote laboratories are one of the most interesting examples of extending the capabilities of existing facilities [172].

The motion on the planar air-bearing microgravity simulators is limited to one plane only. This is a major drawback of such test-beds because the dynamics in the spatial case is more complex than in the planar case. The spatial case is more challenging for the trajectory planning and control algorithms. Spherical air-bearing adds 3 rotational DOF and the resulting 5 DOF test-beds are used for various formation flight experiments, e.g., [48]. Some test-beds allow motion in the direction perpendicular to the test-bed surface. However, this motion is performed by an active system. Construction of test-beds that allow full emulation of 6 DOF dynamics is very challenging. One example of such test-bed is presented in [75]. However, no results of experiments performed on such test-beds could be found. Due to the design challenges, it is not known whether new 6 DOF test-beds will be constructed in near future.

In order to obtain torque-free environment with the use of spherical air-bearing, the centre of mass of the system must coincide with the bearing's centre of rotation. Thus, 5 DOF test-beds are used for experiments with satellites but not satellite-manipulator systems. A major obstacle to be overcome in this case is balancing a satellite model with attached manipulator on a spherical joint. Overcoming this challenge will significantly expand the testing possibilities in the field of space robotics. One example of experiments with a manipulator on a 6 DOF air-bearing microgravity simulator is presented in [110].

8.6 Conclusions

Experiments in simulated microgravity conditions play an essential part in the development and testing of robotic systems intended for in-orbit operations. Despite some limitations, air-bearing microgravity simulators offer cheap and effective way to emulate the dynamic behaviour of the satellite or the satellite-manipulator system. Such test-beds provide an environment in which various subsystems, components and algorithms can be tested and validated in microgravity conditions. In this Chapter, we have discussed main design parameters of air-bearing microgravity simulators with special focus on test-bed construction, surface material and air supply for air-bearings. The range of applications of air-bearing microgravity simulators is very wide and includes, i.a., testing of manipulator hardware, trajectory planning and control of satellite-manipulator systems, capture operations, close proximity manoeuvres and formation flight. Experiments conducted in various areas were reviewed. The proposed active debris removal and in-orbit servicing missions will require complex robotic systems. Thus, the need for testing of robotic technologies in simulated microgravity conditions is increasing and new air-bearing microgravity simulators are constructed. Some of the new facilities are based on COTS components. We have presented an example design of a low-cost, spherical air-bearing satellite simulator. Finally, we have outlined future directions of possible research and open issues.

References

1. Peterson, M. R., & Rodberg, E. H. (2005). Spacecraft Integration and Test. In V. L. Pisacane (Ed.), *Fundamentals of space systems* (pp. 725–753). Oxford University Press.
2. Papadopoulos, E., Aghili, F., Ma, O., & Lampariello, R. (2021). Robotic manipulation and capture in space: A survey. *Frontiers in Robotics and AI, 8*, 686723.
3. Ley, W., Wittmann, K. & Hallmann, W. (Eds.) (2009). Handbook of space technology. John Wiley & Sons.
4. Menon, C., Aboudan, A., Cocuzza, S., Bulgarelli, A., & Angrilli, F. (2005). Free-flying robot tested on parabolic flights: Kinematic control. *Journal of Guidance, Control, and Dynamics, 28*(4), 623–630.
5. Fujii, H., Uchiyama, K., Yoneoka, H., & Maruyama, T. (1996). Ground-based simulation of space manipulators using test bed with suspension system. *Journal of Guidance, Control, and Dynamics, 19*(5), 985–991.
6. Watanabe, Y., Araki, K., & Nakamura, Y. (1998). Microgravity experiments for a visual feedback control of a space robot capturing a target. In *Proceedings of IEEE/RSJ International Conference on Intelligent Robots and Systems*, Victoria, Canada.
7. Ma, O., Flores-Abad, A., & Boge, T. (2012). Use of industrial robots for hardware-in-the-loop simulation of satellite rendezvous and docking. *Acta Astronautica, 81*(1), 335–347.
8. Sun, C., Chen, S., Yuan, J., & Zhu, Z. (2017). A six-DOF buoyancy tank microgravity test bed with active drag compensation. *Microgravity Science and Technology, 29*(5), 391–402.
9. Rybus, T., Seweryn, K., Oleś, J., Basmadji, F. L., Tarenko, K., Moczydłowski, R., Barciński, T., Kindracki, J., Mężyk, Ł., Paszkiewicz, P., & Wolański, P. (2019). Application of a planar air-bearing microgravity simulator for demonstration of operations required for an orbital capture with a manipulator. *Acta Astronautica, 155*, 211–229.
10. Xu, W., Liang, B., & Xu, Y. (2011). Survey of modeling, planning, and ground verification of space robotic systems. *Acta Astronautica, 68*(11–12), 1629–1649.
11. Wilde, M., Clark, C., & Romano, M. (2019). Historical survey of kinematic and dynamic spacecraft simulators for laboratory experimentation of on-orbit proximity maneuvers. *Progress in Aerospace Sciences, 110*, 100552.
12. Ding, X., Wang, Y., Wang, Y., & Xu, K. (2021). A review of structures, verification, and calibration technologies of space robotic systems for on-orbit servicing. *Science China Technological Sciences, 64*(3), 462–480.
13. Al-Bender, F. (2021). *Air Bearings: Theory, Design and Applications*. John Wiley & Sons.
14. Pronk, Z., & van Woerkom, P. T. L. M. (1996). Flat-floor facilities in support of configurable space structures development. *Acta astronautica, 38*(4–8), 277–288.
15. Schwartz, J. L., Peck, M. A., & Hall, C. D. (2003). Historical review of air-bearing spacecraft simulators. *Journal of Guidance, Control, and Dynamics, 26*(4), 513–522.
16. Rybus, T., & Seweryn, K. (2016). Planar air-bearing microgravity simulators: Review of applications, existing solutions and design parameters. *Acta Astronautica, 120*, 239–259.
17. Yeşilay, R., Ankarali, A., & Afşar, M. A. (2021). A review paper: The dynamics, kinematics, design and control of satellite simulators with spherical air bearing. In *Proceedings of GLOBCER'21, Global Conference on Engineering Research,* on-line.
18. Schwendner, J., & Kirchner, F. (2014). Space robotics: An overview of challenges, applications and technologies. *KI-Künstliche Intelligenz, 28*, 71–76.
19. Majumdar, B. C. (1980). Externally pressurized gas bearings: A review. *Wear, 62*(2), 299–314.
20. Gao, Q., Chen, W., Lu, L., Huo, D., & Cheng, K. (2019). Aerostatic bearings design and analysis with the application to precision engineering: State-of-the-art and future perspectives. *Tribology International, 135*, 1–17.
21. Air Bearing Application and Design Guide (2020). *IBS Precision Engineering*.
22. Devitt, A. J. (1999). Porous vs. orifice air bearing technology. *New Way Air Bearings*.
23. Zappulla, R., Virgili-Llop, J., Park, H., Zagaris, C., & Romano, M. (2016). Floating spacecraft simulator test bed for the experimental testing of autonomous guidance, navigation, &

control of spacecraft proximity maneuvers and operations. In *Proceedings of AIAA/AAS Astrodynamics Specialist Conference,* Long Beach, CA, USA.
24. Kwok-Choon, S., Buchala, K., Blackwell, B., Lopresti, S., Wilde, M., & Go, T. (2018). Design, fabrication, and preliminary testing of air-bearing test vehicles for the study of autonomous satellite maneuvers. In *Proceedings of 31st Florida Conference on Recent Advances in Robotics*, Orlando, FL, USA.
25. Zebenay, M., Boge, T., Krenn, R., & Choukroun, D. (2015). Analytical and experimental stability investigation of a hardware-in-the-loop satellite docking simulator. *Proceedings of the Institution of Mechanical Engineers, Part G: Journal of Aerospace Engineering, 229*(4), 666–681.
26. Kindracki, J., Tur, K., Paszkiewicz, P., Mężyk, Ł., Boruc, Ł., & Wolański, P. (2017). Experimental research on low-cost cold gas propulsion for a space robot platform. *Aerospace science and technology, 62*, 148–157.
27. Papadopoulos, E., Kaliakatsos, I. K., & Psarros, D. (2007). Minimum fuel techniques for a space robot simulator with a reaction wheel and PWM thrusters. In *Proceedings of IEEE European Control Conference*, Kos, Greece.
28. Ioniţă, A., Guţă, D. D., Van Nguyen, T., Lungoci, A., Stoian, Ş, & Nichifor, S. E. (2019). Air cushion robots ground testing bed experiments and control algorithm for autonomous rendezvous and docking. *International Journal of Modeling and Optimization, 9*(5), 277–284.
29. Huang, Z., Zhang, W., Chen, T., Wen, H., & Jin, D. (2022). Characterizing an air-bearing testbed for simulating spacecraft dynamics and control. *Aerospace, 9*, 246.
30. Shan, M., Guo, J., & Gill, E. (2016). Review and comparison of active space debris capturing and removal methods. *Progress in Aerospace Sciences, 80*, 18–32.
31. Li, W. J., Cheng, D. Y., Liu, X. G., Wang, Y. B., Shi, W. H., Tang, Z. X., Gao, F., Zeng, F. M., Chai, H. Y., Luo, W. B., & Cong, Q. (2019). On-orbit service (OOS) of spacecraft: A review of engineering developments. *Progress in Aerospace Sciences, 108*, 32–120.
32. Papadopoulos, E., Paraskevas, I. S., Flessa, T., Nanos, K., Rekleitis, G., & Kontolatis, I. (2008). The NTUA space robot simulator: Design and results. In *Proceedings of ASTRA'2008, 10th ESA Workshop on Advanced Space Technologies for Robotics and Automation*, ESTEC, Noordwijk, The Netherlands.
33. Dickson, W. C., & Cannon, Jr R. H., (1995). Experimental results of two free-flying robots capturing and manipulating a free-flying object. In *Proceedings of IEEE/RSJ International Conference on Intelligent Robots and Systems*, Pittsburgh, PA, USA.
34. Marchesi, M., Angrilli, F., & Bettanini C. (2001). On ground experiments of free-flyer space robot simulator in intervention missions. In *Proceedings of i-SAIRAS'2001, 6th International Symposium on Artificial Intelligence, Robotics and Automation in Space*, St-Hubert, Quebec, Canada.
35. Yoshida, K. (1994). Space robotics research activity with experimental free-floating robot satellite (EFFORTS) simulators. In T. Yoshikawa & F. Miyazaki (Eds.), *Experimental Robotics III* (pp. 561–578). Springer.
36. Rybus, T., Barciński, T., Lisowski, J., Nicolau-Kukliński, J., Seweryn, K., Ciesielska, M., Grassmann, K., Grygorczuk, J., Karczewski, M., Kowalski, M., Krzewski, M., Kuciński, T., Przybyła, R., Skup, K., Szewczyk, T., & Wawrzaszek, R. (2013). New planar air-bearing microgravity simulator for verification of space robotics numerical simulations and control algorithms. In *Proceedings of ASTRA'2013, 12th Symposium on Advanced Space Technologies in Robotics and Automation*, ESTEC, Noordwijk, The Netherlands.
37. Cocuzza, S., Pretto, I., & Debei, S. (2011). Novel reaction control techniques for redundant space manipulators: Theory and simulated microgravity tests. *Acta Astronautica, 68*(11–12), 1712–1721.
38. Du, B., Liu, J., Yang, Z., Wang, Y., & Deng, T. (2018). A dimension reduced ground test method of dual-arm space robot for on-orbit assembly. In *Proceedings of ICMA'2018, IEEE International Conference on Mechatronics and Automation*, Changchun, China.
39. Zappulla, R., Virgili-Llop, J., Zagaris, C., Park, H., & Romano, M. (2017). Dynamic air-bearing hardware-in-the-loop testbed to experimentally evaluate autonomous spacecraft proximity maneuvers. *Journal of Spacecraft and Rockets, 54*(4), 825–839.

40. Sabatini, M., Farnocchia, M., & Palmerini, G. B. (2012). Design and tests of a frictionless 2D platform for studying space navigation and control subsystems. In *Proceedings of IEEE Aerospace Conference*, Big Sky, MT, USA.
41. Schlotterer, M., & Theil, S. (2010). Testbed for on-orbit servicing and formation flying dynamics emulation. In *Proceedings of AIAA-GNC'2010, AIAA Guidance, Navigation and Control Conference,* Toronto, Canada.
42. Ono, M., Boning, P., Nohara, T., & Dubowsky, S. (2008). Experimental validation of a fuel-efficient robotic maneuver control algorithm for very large flexible space structures. In *Proceedings of ICRA'2008, IEEE International Conference on Robotics and Automation*, Pasadena, CA, USA.
43. Toda, Y., Iwata, T., Machida, K., Otsuka, A., Toriu, H., & Shinomiya, Y. (1991). Development of flying telerobot model for ground experiments. *The Journal of Space Technology and Science, 7*(2), 15–22.
44. Yoshida, K., & Umetani, Y. (1990). Control of space free-flying robot. In *Proceedings of 29th Conference on Decision and Control*, Honolulu, HI, USA.
45. Ledebuhr, A. G., Ng, L. C., Jones, M. S., Wilson, B. A., Gaughan, R. J., Breitfeller, E. F., Taylor, W. G., Robinson, J. A., Antelman, D. R., & Nielsen, D. P. (2001). Micro-satellite ground test vehicle for proximity and docking operations development. In *Proceedings of IEEE Aerospace Conference*, Piscataway, NJ, USA.
46. Boning, P., Ono, M., Nohara, T., & Dubowsky, S. (2008). An experimental study of the control of space robot teams assembling large flexible space structures. In *Proceedings of i-SAIRAS'2008, 9th International Symposium on Artificial Intelligence, Robotics and Automation in Space*, Los Angeles, CA, USA.
47. Saulnier, K., Perez, D., Tilton, G., Gallardo, D., Shake, C., Huang, R., & Bevilacqua, R. (2013). Operational capabilities of a six degrees of freedom spacecraft simulator. In *Proceedings of AIAA-GNC'2013, AIAA Guidance, Navigation, and Control Conference*, Boston, MA, USA.
48. Eun, Y., Park, S. Y., & Kim, G. N. (2018). Development of a hardware-in-the-loop testbed to demonstrate multiple spacecraft operations in proximity. *Acta Astronautica, 147*, 48–58.
49. Scharf, D. P., Keim, J. A., & Hadaegh, F. Y. (2008). Ground demonstration of synchronized formation rotations for precision, multi-spacecraft interferometers. In *Proceedings of 3rd International Symposium on Formation Flying, Missions and Technologies*, ESTEC, Noordwijk, The Netherlands.
50. Uyama, N., Lund, H., Asakimori, K., Ikeda, Y., Hirano, D., Nakanishi, H., & Yoshida, K. (2010). Integrated experimental environment for orbital robotic systems, using ground-based and free-floating manipulators. In *Proceedings of IEEE/SICE International Symposium on System Integration*, Sendai, Japan.
51. Gao, Z., Gu, Y., Lv., Y., Xu, Z., & Wu, Q. (2018). A monocular vision measurement system of three-degree-of-freedom air-bearing test-bed based on FCCSP. *Photonic Sensors, 8*(2), 119-126.
52. Regehr, M. W., Acikmese, A. B., Ahmed, A., Aung, M., Clark, K. C., MacNeal, P., Rasmussen, R. E., Shields, J., & Singh, G. (2004). *The formation control testbed*. In *Proceedings of IEEE Aerospace Conference*, Big Sky, MT, USA.
53. Hall, J. S., & Romano, M. (2010). Laboratory experimentation of guidance and control of spacecraft during on-orbit proximity maneuvers. In A. Milella, D. Di Paola, & G. Cicirelli (Eds.), *Mechatronic Systems Simulation Modeling and Control* (pp. 187–225). InTech.
54. Kolvenbach, H., & Wormnes, K. (2016). Recent developments on ORBIT, a 3-dof free floating contact dynamics testbed. In *Proceedings of i-SAIRAS'2016, 13th International Symposium on Artificial Intelligence, Robotics and Automation in Space,* Beijing, China.
55. Air bearing application and design guide (2006). *New Way Air Bearings*, rev. E.
56. Chen, X., Yao, W., Huang, Y., & Zhao, Y. (2009). A common ground experiment testbed for synthetic mission demonstration of small satellites. In *Proceedings of 23rd Annual AIAA/USU Conference on Small Satellites,* Logan, UT, USA.
57. Kwok-Choon, S. (2019). Hardware-in-the-loop testbed for the dynamics of spacecraft rendezvous and capture, PhD Thesis, Florida Institute of Technology.

58. Tsiotras, P. (2014). ASTROS: A 5DOF experimental facility for research in space proximity operations. In *Proceedings of 37th Annual AAS Rocky Mountain Section Guidance and Control Conference*, Breckenridge, CO, USA.
59. Chitu, C. C., Stefanescu, R., Bajanaru, P., Galipienzo, J., Ortega, C., Barciński, T., Rybus, T., Seweryn, K., & Visentin, G. (2015). Design and development of an active landing gear system for robotically enhanced surface touchdown. In *Proceedings of ASTRA'2015, 13th Symposium on Advanced Space Technologies in Robotics and Automation*, ESTEC, Noordwijk, The Netherlands.
60. Ogundele, A. D., Fernandez, B. R., Virgili-Llop, J., & Romano, M. (2019). A tip-tilt hardware-in-the-loop air-bearing test bed with physical emulation of the relative orbital dynamics. In *Proceedings of 29th AAS/AIAA Space Flight Mechanics Meeting*, Ka'anapali, HI, USA.
61. PIglide HB Hemispherical Air Bearings User Manual (2019), PI (Physik Instrumente) L.P., ver. 1.3.
62. Ustrzycki, T., Lee, R., & Chesser, H. (2011). Spherical air bearing attitude control simulator for nanosatellites. In *Proceedings of AIAA Modeling and Simulation Technologies Conference*, Portland, OR, USA.
63. Kim, J. J., & Agrawal, B. N. (2009). Automatic mass balancing of air-bearing-based three-axis rotational spacecraft simulator. *Journal of Guidance, Control, and Dynamics, 32*(3), 1005–1017.
64. Prado, J., Bisiacchi, G., Reyes, L., Vicente, E., Contreras, F., Mesinas, M., & Juárez, A. (2005). Three-axis air-bearing based platform for small satellite attitude determination and control simulation. *Journal of Applied Research and Technology, 3*(3), 222–237.
65. Peck, M. A., Miller, L., Cavender, A. R., Gonzalez, M., & Hintz, T. (2003). An airbearing-based testbed for momentum control systems and spacecraft line of sight. *Advances in the Astronautical Sciences, 114*, 427–446.
66. Gavrilovich, I., Krut, S., Gouttefarde, M., Pierrot, F., & Dusseau, L. (2016). Innovative approach to use air bearings in cubesat ground tests. In *Proceeding of 4S Symposium (Small Satellite Systems and Services)*, Malta.
67. Agrawal, B. N., & Rasmussen, R. E. (2001). Air-bearing-based satellite attitude dynamics simulator for control software research and development. In *Proceeedings of SPIE 4366, Technologies for Synthetic Environments: Hardware-in-the-Loop Testing VI*, pp 204–214.
68. Cho, D-M., Jung, D., & Tsiotras, P. (2009). A 5-dof experimental platform for spacecraft rendezvous and docking. In *Proceedings of AIAA Infotech @ Aerospace Conference*, Seattle, WA, USA.
69. Rasouli, K., Shahbazi, H., Ariaei, A., & Malekzadeh, M. (2016). Mechatronic design and construction of a five axes satellite simulator. In *Proceedings of ICROM'2016, 4th IEEE International Conference on Robotics and Mechatronics*, Tehran, Iran.
70. Visconti, P., Silvestrini, S., & Lavagna, M. (2018). Dance: A frictionless 5 DOF facility for GNC proximity maneuvering experimental testing and validation. In *Proceedings of IAC'2018, 69th International Astronautical Congress*, Bremen, Germany.
71. Rybus, T., Oleś, J., Osica, P., Seweryn, K., Tarenko, K., Moczydłowski, R., Kindracki, J., Mężyk, Ł., Paszkiewicz, P., & Wolański, P. (2017). Planar air-bearing microgravity simulator for testing satellite GNC subsystem. In *Proceedings of EuroGNC'2017, 4th CEAS Specialist Conference on Guidance, Navigation & Control*, Warsaw, Poland.
72. Okamoto, O., Nakaya, T., & Pokines, B. (1994). Concept verification of three-dimensional free motion simulator for space robot. In *Proceedings of i-SAIRAS'1994, 3rd International Symposium on Artificial Intelligence, Robotics and Automation in Space*, Pasadena, CA, USA.
73. Viswanathan, S. P., Sanyal, A., & Holguin, L. (2012). Dynamics and control of a six degrees of freedom ground simulator for autonomous rendezvous and proximity operation of spacecraft. In *Proceedings of AIAA-GNC'2012, AIAA Guidance, Navigation and Control Conference*, Minneapolis, MN, USA.
74. Nakka, Y. K., Foust, R. C., Lupu, E. S., Elliott, D. B., Crowell, I. S., Chung, S. J., & Hadaegh, F. Y. (2018). A six degree-of-freedom spacecraft dynamics simulator for formation control research. In *Proceedings of AAS/AIAA Astrodynamics Specialist Conference*, Snowbird, UT, USA.

75. Saulnier, K., Pérez, D., Huang, R. C., Gallardo, D., Tilton, G., & Bevilacqua, R. (2014). A six-degree-of-freedom hardware-in-the-loop simulator for small spacecraft. *Acta Astronautica, 105*(2), 444–462.
76. Caon, A., Feltrin, F., Bran, F., Sansone, F., & Francesconi, A. (2020). Ground facility for validation of proximity operations: A hardware–in–the–loop experiment. In *Proceedings of 7th IEEE International Workshop on Metrology for AeroSpace*, Pisa, Italy.
77. Zwick, M., Huertas, I., Gerdes, L., & Ortega, G. (2018). ORGL–ESA's test facility for approach and contact operations in orbital and planetary environments. In *Proceedings of i-SAIRAS'2018, 14th International Symposium on Artificial Intelligence, Robotics and Automation in Space*, Madrid, Spain.
78. Tomasek, J., Kolvenbach, H., Pagnamenta, M., & Wormnes, K. (2016). A robotic testbed for low-gravity simulation. In *Proceedings of i-SAIRAS'2016, 13th International Symposium on Artificial Intelligence, Robotics and Automation in Space*, Beijing, China.
79. Wilde, M., Kaplinger, B., Go, T., Gutierrez, H., & Kirk, D. (2016). ORION: A simulation environment for spacecraft formation flight, capture, and orbital robotics. In *Proceedings of IEEE Aerospace Conference*, Big Sky, MT, USA.
80. Hewes, D. E., & Glover, K. E. (1975). Premission and postmission simulation studies of the foot-controlled maneuvering unit for Skylab experiment T-020. *NASA Technical Note TN D-7963.*
81. Wilson, W. R., Jones, L. L., & Peck, M. A. (2013). A multimodule planar air bearing testbed for cubesat-scale spacecraft. *Journal of Dynamic Systems, Measurement, and Control, 135*(4), 045001.
82. Didot, F., Oort, M., Kouwen, J., & Verzijden, P. (2001). The ERA system: Control architecture and performance results. In *Proceedings of i-SAIRAS'2001, 6th International Symposium on Artificial Intelligence, Robotics and Automation in Space*, Montreal, Canada.
83. Kline, S. J. (1986). *Similitude and approximation theory*. Springer-Verlag.
84. Hilderman, R. A., Mueller, W. H., & Mantus, M. (1966). Landing dynamics of the Lunar Excursion Module. *Journal of Spacecraft and Rockets, 3*(10), 1484–1489.
85. Coutinho, C. P., Baptista, A. J., & Rodrigues, J. D. (2016). Reduced scale models based on similitude theory: A review up to 2015. *Engineering Structures, 119*, 81–94.
86. Zong, L., & Emami, M. R. (2020). Control verifications of space manipulators using ground platforms. *IEEE Transactions on Aerospace and Electronic Systems, 57*(1), 341–354.
87. Ciarcia M., Cristi R., & Romano M. (2017). Experimental emulation of the scaled clohessy-wiltshire dynamics on a flat air-bearing testbed. In *Proceedings of AIAA-GNC'2017, AIAA Guidance, Navigation, and Control Conference,* Grapevine, TX, USA.
88. Oleś, J., Kindracki, J., Rybus, T., Paszkiewicz, P., Moczydłowski, R., Barciński, T., Seweryn, K., & Wolański, P. (2017). A 2D microgravity test-bed for the validation of space robot control algorithms. *Journal of Automation, Mobile Robotics and Intelligent Systems, 11*(2), 95–104.
89. Blanchard, U. J. (1968). Evaluation of a full-scale lunar-gravity simulator by comparison of landing-impact tests of a full-scale and a 1/6-scale model. *NASA Technical Note TN D-4474.*
90. Oleś J., Rybus T., Seweryn, K., Surowiec, M., Wojtyra, M., Pietras M., & Scheper, M. (2017). Testing and simulation of contact during on-orbit operations. In *Proceedings of ASTRA'2017, 14th Symposium on Advanced Space Technologies in Robotics and Automation*, Leiden, The Netherlands.
91. Dexler, K. (1998). *Nanosystems: Molecular Machinery, Manufacturing, and Computation.* Wiley.
92. Turnbull, J. (1983). Acceleration environment of payloads while being handled by the Shuttle Remote Manipulator System. In *Proceedings of Shuttle Environment and Operations Meeting,* Washington, DC, USA.
93. Yao, H., Ren, W., Ma, O., Chen, T., & Zhao, Z. (2018). Understanding the true dynamics of space manipulators from air-bearing based ground testing. *Journal of Guidance, Control, and Dynamics, 41*(11), 2425–2434.
94. Jorgensen, G., & Bains, E. (2011). SRMS history, evolution and lessons learned. In *Proceedings of AIAA SPACE 2011 Conference & Exposition,* Long Beach, CA, USA.

95. Wagner-Bartak, C. G., Middleton, J. A., & Hunter, J. A. (1980). Shuttle remote manipulator system hardware test facility. In *Proceedings of 11th Space Simulation Conference,* Houston, TX, USA.
96. Boumans, R., & Heemskerk, C. (1998). The European Robotic Arm for the International Space Station. *Robotics and Autonomous systems, 23*(1–2), 17–27.
97. Gao, X. H., Jin, M. H., Xie, Z. W., Jiang, L., Ni, F. L., Wei, R., Zhu, Y. Y., Cai, H. G., Liu, H., Beyer, A., Kraemer, E., Landzettel, K., Schott, J., & Hirzinger, G. (2006). Development of the Chinese intelligent space robotic system. In *Proceedings of IEEE/RSJ International Conference on Intelligent Robots and Systems,* Beijing, China.
98. Meng, D., She, Y., Xu, W., Lu, W., & Liang, B. (2018). Dynamic modeling and vibration characteristics analysis of flexible-link and flexible-joint space manipulator. *Multibody System Dynamics, 43*(4), 321–347.
99. Mimmi, G., Rottenbacher, C., Frosini, L., & Negri, A. (2004). An experimental test-bed for microgravity simulation in robotic arm dynamics. In *Proceedings of 6th International Conference on Dynamics and Control of Systems and Structures in Space*, Riomaggiore, Italy.
100. Li, H., Wei, Q., Liang, J., Ren, W., Tang, Z., & Li, D. (2021). Estimation of vibration characteristics of a space manipulator from air bearing supported test data. *Frontiers in Robotics and AI, 8*, 641165.
101. Matsunaga, S., Sawada, H., Konoue, K., & Asami, M. (2005). Vibration suppression experiments of flexible payload transportation by space robotic arm with redundant DOF. In *Proceedings of i-SAIRAS'2005, 8th International Symposium on Artificial Intelligence, Robotics and Automation in Space*, Munich, Germany.
102. Lichter, M. D., Ueno, H., & Dubowsky, S. (2006). Vibration estimation of flexible space structures using range imaging sensors. *The International Journal of Robotics Research, 25*(10), 1001–1012.
103. Hirano, D., Fujii, Y., Abiko, S., Lampariello, R., Nagaoka, K., & Yoshida, K. (2014). Simultaneous control for end-point motion and vibration suppression of a space robot based on simple dynamic model. In *Proceedings of ICRA'2014, IEEE International Conference on Robotics and Automation*, Hong Kong, China.
104. Xu, Y. (1993). The measure of dynamic coupling of space robot systems. In *Proceedings of ICRA'1993, IEEE International Conference on Robotics and Automation,* Atlanta, GA, USA.
105. Papadopoulos, E., & Dubowsky, S. (1991). On the nature of control algorithms for free-floating space manipulators. *IEEE Transactions on Robotics and Automation, 7*(6), 750–758.
106. Christidi-Loumpasefski, O.-O., Ntinos, Ch., & Papadopoulos, E. (2017). Analytical and experimental parameter estimation for free-floating space manipulator systems. In *Proceedings of ASTRA'2017, 14th Symposium on Advanced Space Technologies in Robotics and Automation,* Leiden, The Netherlands.
107. Virgili-Llop J., Drew J., & Romano, M. (2016). Design and parameter identification by laboratory experiments of a prototype modular robotic arm for orbiting spacecraft applications. In *Proceedings of ICATT'2016, 6th International Conference on Astrodynamics Tools and Techniques*, Darmstadt, Germany.
108. Carabis, D. S., & Wen, J. T. (2022). Trajectory generation for flexible-joint space manipulators. *Frontiers in Robotics and AI, 9*, 687595.
109. Sabatini, M., Gasbarri, P., & Palmerini, G. B. (2017). Coordinated control of a space manipulator tested by means of an air bearing free floating platform. *Acta Astronautica, 139*, 296–305.
110. He, J., Zheng, H., Gao, F., & Zhang, H. (2019). Dynamics and control of a 7-DOF hybrid manipulator for capturing a non-cooperative target in space. *Mechanism and Machine Theory, 140*, 83–103.
111. Ullman, M. A., Cannon, R. H., & Rock, S. M. (1993). A modular system architecture for multi-manipulator, free-flying space robots. In *Proceedings of 8th IEEE International Symposium on Intelligent Control, Chicago,* IL, USA.

112. Murotsu, Y., Senda, K., Mitsuya, A., & Fujii, K. (1993). Experimental studies for control of manipulators mounted on a free-flying space robot. In *Proceedings of IEEE/RSJ International Conference on Intelligent Robots and Systems,* Yokohama, Japan.
113. Basmadji, F. L., Chmaj, G., Rybus, T., & Seweryn, K. (2019). Microgravity testbed for the development of space robot control systems and the demonstration of orbital maneuvers. In *Proceedings of SPIE 11176, Photonics Applications in Astronomy, Communications, Industry, and High-Energy Physics Experiments*; 111763V.
114. Seweryn, K., Barciński, T., Ciesielska, M., Grygorczuk, J., Rybus, T., Skup, K., & Wawrzaszek, R. (2013). Validation of the control strategy of the free flying robot – application to the rendezvous manoeuvres. In *Proceedings of ASTRA'2013, 12th Symposium on Advanced Space Technologies in Robotics and Automation*, ESTEC, Noordwijk, The Netherlands.
115. Rybus, T., Barciński, T., Lisowski, J., Seweryn, K., Nicolau-Kukliński, J., Grygorczuk, J., Krzewski, M., Skup, K., Szewczyk, T., & Wawrzaszek, R. (2013). Experimental demonstration of singularity avoidance with trajectories based on the Bézier curves for free-floating manipulator. In *Proceedings of RoMoCo'2013, 9th International Workshop on Robot Motion and Control,* Wąsowo, Poland.
116. Rybus, T., & Seweryn, K. (2018). Manipulator trajectories during orbital servicing mission: Numerical simulations and experiments on microgravity simulator. *Progress in Flight Dynamics, Guidance, Navigation, and Control, 10*, 239–264.
117. Rybus, T., Wojtunik, M., & Basmadji, F. L. (2022). Optimal collision-free path planning of a free-floating space robot using spline-based trajectories. *Acta Astronautica, 190*, 395–408.
118. Rybus, T., Prokopczuk, J., Wojtunik, M., Aleksiejuk, K., & Musiał, J. (2022). Application of bidirectional rapidly exploring random trees (BiRRT) algorithm for collision-free trajectory planning of free-floating space manipulator. *Robotica, 40*(12), 4326–4357.
119. Senda, K., & Matsumoto, T. (2000). Truss assembly by space robot and task error recovery via reinforcement learning. In *Proceedings of IEEE/RSJ International Conference on Intelligent Robots and Systems,* Takamatsu, Japan.
120. Wilson, E., Rock, S. M. (1993). Experiments in control of a free-flying space robot using fully-connected neural networks. In *Proceedings of World Congress on Neural Networks*, Hillsdale, MI, USA.
121. Zhang, X., & Feng, Y. K. (2012). Design and research of a microgravity floating target satellite system. *Applied Mechanics and Materials, 170–173*, 3009–3016.
122. Umetani, Y., & Yoshida, K. (1989). Experimental study on two dimensional free-flying robot satellite model. In *Proceedings of NASA Conference on Space Telerobotics*, vol. 5, Pasadena, CA, USA.
123. Machairas, K., Andreou, S., Paraskevas, I., & Papadopoulos, E. (2013). Extending the NTUA space robot emulator for validating complex on-orbit servicing tasks. In *Proceedings of ASTRA'2013, 12th Symposium on Advanced Space Technologies in Robotics and Automation*, ESTEC, Noordwijk, The Netherlands.
124. Virgili-Llop, J., Drew, J. V., Zappulla II, R., & Romano, M. (2017). Laboratory experiments of resident space object capture by a spacecraft–manipulator system. *Aerospace Science and Technology, 71*, 530–545.
125. Wei, X., Fuchun, S., & Huaping, L. (2011). Design and development of a ground experiment system with free-flying space robot. In *Proceedings of 6th IEEE Conference on Industrial Electronics and Applications,* Beijing, China.
126. Bettanini, C., Aboudan, A., Francesconi, A., Marchesi, M., Menon, C., & Angrilli, F. (2003). Improving the free-floater Space robot simulator for intervention missions. In *Proceedings of i-SAIRAS'2003, 7th International Symposium on Artificial Intelligence, Robotics and Automation in Space*, Nara, Japan.
127. Matunaga, S., Yoshihara, K., Takahashi, T., Tsurumi, S., & Ui, K. (2000). Ground experiment system for dual-manipulator-based capture of damaged satellites. In *Proceedings of IEEE/RSJ International Conference on Intelligent Robots and Systems,* Takamatsu, Japan.
128. Uyama, N., Nakanishi, H., Nagaoka, K., & Yoshida, K. (2012). Impedance-based contact control of a free-flying space robot with a compliant wrist for non-cooperative satellite capture.

In *Proceedings of IEEE/RSJ International Conference on Intelligent Robots and Systems,* Vilamoura, Portugal.
129. Uyama, N., Fujii, Y., Nagaoka, K., & Yoshida, K. (2012). Experimental evaluation of contact-impact dynamics between a space robot with a compliant wrist and a free-flying object. In *Proceedings of i-SAIRAS'2012, 11th International Symposium on Artificial Intelligence, Robotics and Automation in Space*, Turin, Italy.
130. Sabatini, M., Palmerini, G. B., Monti, R., & Gasbarri, P. (2014). Image based control of the "PINOCCHIO" experimental free flying platform. *Acta Astronautica, 94*(1), 480–492.
131. Flessa, T., Paraskevas, I. S., Rekleitis, Y., & Papadopoulos, E. (2012). Localization and fuel management techniques for the NTUA space servicer emulator system. In *Proceedings of i-SAIRAS'2012, 11th International Symposium on Artificial Intelligence, Robotics and Automation in Space*, Turin, Italy.
132. Mishra, S., Kobilarov, M. (2014). An autonomous proximity navigation testbed for nanosatellites. In *Proceedings of 28th Annual AIAA/USU Conference on Small Satellites*, Logan, UT, USA.
133. Andrade, C., Ramirez-Mendoza, R., Giacoman-Zarzar, M., Morales, R., Fejzic, A., Saenz-Otero, A., & Miller, D. W. (2009). Robust control applied towards rendezvous and docking. In *Proceedings of IEEE European Control Conference,* Budapest, Hungary.
134. Huang, Z., Lu, Y., Wen, H., & Jin, D. (2018). Ground-based experiment of capturing space debris based on artificial potential field. *Acta Astronautica., 152*, 235–241.
135. Zappulla, R., Park, H., Virgili-Llop, J., & Romano, M. (2018). Real-time autonomous spacecraft proximity maneuvers and docking using an adaptive artificial potential field approach. *IEEE Transactions on Control Systems Technology, 27*(6), 2598–2605.
136. Virgili-Llop, J., Zagaris, C., Park, H., Zappulla, R., & Romano, M. (2018). Experimental evaluation of model predictive control and inverse dynamics control for spacecraft proximity and docking maneuvers. *CEAS Space Journal, 10*(1), 37–49.
137. Banerjee, A., Satpute, S. G., Kanellakis, C., Tevetzidis, I., Haluska, J., Bodin, P., & Nikolakopoulos, G. (2022). On the design, modeling and experimental verification of a floating satellite platform. *IEEE Robotics and Automation Letters, 7*(2), 1364–1371.
138. Wei, Z., Wen, H., Hu, H., & Jin, D. (2020). Ground experiment on rendezvous and docking with a spinning target using multistage control strategy. *Aerospace Science and Technology, 104*, 105967.
139. Romano, M., Friedman, D. A., & Shay, T. J. (2007). Laboratory experimentation of autonomous spacecraft approach and docking to a collaborative target. *Journal of Spacecraft and Rockets, 44*(1), 164–173.
140. Ma, O., Yang, G., & Diao, X. (2005). Experimental validation of CDT-based satellite docking simulations using SOSS testbed. In *Proceedings of i-SAIRAS'2005, 8th International Symposium on Artificial Intelligence, Robotics and Automation in Space*, Munich, Germany.
141. Yang, G., & Ma, O. (2013). Validation of satellite docking simulations using an air-bearing-based experimental testbed. *International Journal of Space Science and Engineering, 1*(3), 253–267.
142. Van Vliet, J., Sharf, I., & Ma, O. (2000). Experimental validation of contact dynamics simulation of constrained robotic tasks. *The International Journal of Robotics Research, 19*(12), 1203–1217.
143. Robertson, A., Inalhan, G., & How, J. (1999). Spacecraft formation flying control design for the Orion mission. In *Proceedings of AIAA-GNC'1999, AIAA Guidance, Navigation and Control Conference,* Portland, OR, USA.
144. Saenz-Otero, A., & Miller, D. W. (2005). SPHERES: A platform for formation-flight research. In *Proceedings of UV/Optical/IR Space Telescopes: Innovative Technologies and Concepts II Conference*, San Diego, CA, USA.
145. Tweddle, B. E. (2011). Relative computer vision-based navigation for small inspection spacecraft. In *Proceedings of AIAA-GNC'2011, AIAA Guidance, Navigation and Control Conference,* Portland, OR, USA.

146. Stoll, E., Jaekel, S., Katz, J., Saenz-Otero, A., & Varatharajoo, R. (2012). SPHERES interact—Human–machine interaction aboard the International Space Station. *Journal of Field Robotics, 29*(4), 554–575.
147. Di Mauro, G., Schlotterer, M., Theil, S., & Lavagna, M. (2013). Experimental implementation of SDRE method for autonomous rendezvous and docking maneuvering. In *Proceedings of 5th International Conference on Spacecraft Formation Flying Missions and Technologies*, Munich, Germany.
148. Guglieri, G., Maroglio, F., Pellegrino, P., & Torre, L. (2014). Design and development of guidance navigation and control algorithms for spacecraft rendezvous and docking experimentation. *Acta Astronautica, 94*(1), 395–408.
149. Bevilacqua, R., Romano, M., Curti, F., Caprari, A. P., & Pellegrini, V. (2011). Guidance navigation and control for autonomous multiple spacecraft assembly: Analysis and experimentation. *International Journal of Aerospace Engineering*; 308245.
150. Curti, F., Romano, M., & Bevilacqua, R. (2010). Lyapunov-based thrusters' selection for spacecraft control: Analysis and experimentation. *Journal of Guidance, Control, and Dynamics, 33*(4), 1143–1160.
151. Yang, C., Zhang, H., & Zhong, W. (2020). Air-bearing testbed of spacecraft tight formation: System description and experiments. *The Romanian Journal of Technical Sciences: Applied Mechanics, 65*(3), 235–252.
152. Xia, K., Lee, T., & Park, S. Y. (2019). Adaptive saturated neural network tracking control of spacecraft: Theory and experimentation. *International Journal of Aerospace Engineering*; 7687459.
153. Colebank, J., Jones, R., Nagy, G., Pollak, R., Mannebach, D., Kramer, S., & Agnes, G. (1999). SIMSAT-A satellite simulator and experimental test bed for Air Force research. In *Proceedings of AIAA Space Technology Conference and Exposition*, Albuquerque, NM, USA.
154. Ousaloo, H. S., Sharifi, G., & Akbarinia, B. (2021). Extended validation of a ground-based three-axis spacecraft simulator model. *Proceedings of the Institution of Mechanical Engineers, Part G: Journal of Aerospace Engineering, 235*(2), 151–170.
155. Costa, R. F., Saotome, O., & Rafikova, E. (2019). Simulation and validation of satellite attitude control algorithms in a spherical air bearing. *Journal of Control, Automation and Electrical Systems, 30*(5), 716–727.
156. Pasand, M., Teshnizi, M. M., & Aghaei, R. (2017). Low-cost hardware-in-the-loop test of a spacecraft attitude control system. In *Proceedings of KBEI'2017, 4th IEEE International Conference on Knowledge-Based Engineering and Innovation*, Tehran, Iran.
157. Lyu, W., Dai, S., Dong, Y., Shen, Y., Song, X., & Wang, T. (2017). Attitude motion compensation for imager on Fengyun-4 geostationary meteorological satellite. *Acta Astronautica, 138*, 290–294.
158. Inumoh, L. O., Forshaw, J. L., & Horri, N. M. (2015). Tilted wheel satellite attitude control with air-bearing table experimental results. *Acta Astronautica, 117*, 414–429.
159. Zhang, Q., Zhang, K., Lu, Y., Liu, X., & Yin, Y. (2019). Pneumatic attitude control of the air bearing tesbed to simulate the three axis free tumbling motion of an uncooperative target. In *Proceedings of ICMA'2019, IEEE International Conference on Mechatronics and Automation*, Tianjin, China.
160. Yang, G., Wang, H., Ling, L., & Gao, F. (2018). An air bearing facility for antenna pointing mechanism. In *Proceedings of CSECE'2018, International Conference on Computer Science, Electronics and Communication Engineering*, Wuhan, China.
161. Jia, B., Bian, W., & Peng, N. (2019). Experiments on simulation microgravity of composite deployable antenna. *IOP Conference Series: Materials Science and Engineering, 631*, 042019.
162. Wilson, W. R., Shoer, J., & Peck, M. A. (2009). Demonstration of a magnetic locking flux-pinned revolute joint for use on cubesat-standard spacecraft. In *Proceedings of AIAA-GNC'2009, AIAA Guidance, Navigation, and Control Conference*, Chicago, IL, USA.
163. Nakaya, K., Iai, M., Mori, O., & Matunaga, S. (2004). On formation deployment for spinning tethered formation flying and experimental demonstration. In *Proceedings of 18th International Symposium on Space Flight Dynamics*, Munich, Germany.

164. Gates, S. S., Koss, S. M., & Zedd, M. F. (2001). Advanced tether experiment deployment failure. *Journal of Spacecraft and Rockets, 38*(1), 60–68.
165. Moreland, S. J., Backes, P. G., Badescu, M., Riccobono, D., Mongelli, M., Vieira, P., Brinkman, A., Rehnmark, F., Wei, R., Adams, G., Toda, R., Cervantes, W., & Zacny, K. (2018). Full-scale dynamic touch-and-go validation of the BiBlade comet surface sample chain. In *Proceedings of IEEE Aerospace Conference,* Big Sky, MT, USA.
166. Seweryn, K., Paśko, P., & Visentin, G. (2019). The prototype of regolith sampling tool dedicated to low gravity planetary bodies. In: T. Uhl (Ed.) Advances in Mechanism and Machine Science. IFToMM WC 2019. *Mechanisms and Machine Science*, 73. Springer, Cham.
167. Zhao, Z., Zhao, J., & Liu, H. (2013). Development of a landing mechanism for asteroids with soft surface. *International Journal of Aerospace Engineering*, 873135.
168. Pavone, M., Castillo-Rogez, J. C., Nesnas, I. A., Hoffman, J. A., & Strange, N. J., Spacecraft/rover hybrids for the exploration of small solar system bodies. In *Proceedings of IEEE Aerospace Conference,* Big Sky, MT, USA.
169. Kolvenbach, H., & Voellmy. M. (2019). Design of a Low-Cost, Spherical Air-Bearing Satellite Simulator. In *Proceedings of ASTRA'2019, 15th Symposium on Advanced Space Technologies in Robotics and Automation*, ESTEC, Noordwijk, The Netherlands.
170. Rughani, R., Villafana, L., & Barnhart, D. A. (2019). Swarm RPO and docking simulation on a 3DOF Air Bearing Platform. In *Proceedings of IAC'2019, 70th International Astronautical Congress*, Washington, DC, USA.
171. Papakonstantinou, C., Moraitis, G., Lappas, V., & Kostopoulos, V. (2022). Design of a low-cost air bearing testbed for nano CMG maneuvers. *Aerospace, 9*, 95.
172. Nieto-Peroy, C., Sabatini, M., Palmerini, G., & Jeronimo de Oliveira, É. (2021). A Concurrent testing facility approach to validate small satellite combined operations. *Aerospace, 8*, 361.

Chapter 9
Modular Reconfigurable Multi-robot Systems for Flexible Space Exploration

S. Bartsch, F. Cordes, T. M. Roehr, R. U. Sonsalla, and F. Kirchner

Abstract This chapter discusses the opportunities opened by the utilization of modular reconfigurable multi-robot systems applied in space missions. The involved requirements are considered, solutions already existing and challenges still existing are highlighted. Firstly, different conceivable mission scenarios and application concepts for such systems are presented and the advantages that can be achieved by their use are pointed out. The general design, functionality and required key components, both, in the field of hardware and software are described referring to a concrete example of a system which was developed at DFKI Robotics Innovation Center for planetary surface exploration missions. It is presented how different mission objectives can be achieved by using a reconfigurable multi-robot systems. The procedures and components required to make a multi-robot system feasible are discussed. Planning and control of task distribution and reconfiguration operations of such a complex and flexible system are considered as well. The chapter concludes with a summary and an outlook on remaining challenges.

S. Bartsch (✉) · F. Cordes · T. M. Roehr · R. U. Sonsalla
DFKI Robotics Innovation Center, Robert-Hooke-Str. 1, 28359 Bremen, Germany
e-mail: Sebastian.Bartsch@dfki.de

F. Cordes
e-mail: Florian.Cordes@dfki.de

T. M. Roehr
e-mail: Thomas.Roehr@dfki.de

R. U. Sonsalla
e-mail: Roland.Sonsalla@dfki.de

F. Cordes · T. M. Roehr · R. U. Sonsalla · F. Kirchner
Faculty Mathematics and Computer Science, University of Bremen, Workgroup Robotics, Robert-Hooke-Str. 1, 28359 Bremen, Germany
e-mail: Frank.Kirchner@dfki.de

X. T. Yan and G. Visentin (eds.), *Space Robotics*,
https://doi.org/10.1007/978-3-031-39214-6_9

9.1 Introduction

With increasing objectives, operating time, and complexity of missions and systems for space applications, the achievable benefits of modular and reconfigurable solutions for robotic devices and their control methods are becoming more and more relevant. The use of such systems offers technological, operational, economical and ecological advantages compared with the classical monolithic approaches. This applies to applications in orbital as well as planetary scenarios.

A modular system approach offers the possibility to configure a system for different applications from available standard modules. This can be done already during the design and manufacturing phase on Earth. Development time and costs can be reduced by using existing modules, while availability through mass production of modules is improved. Transporting large structures is simplified by separate transport and subsequent assembly of modules on the operation site. During operation, the lifetime can be extended and functionality can be improved via upgrades made during maintenance work by replacing modules. Systems no longer have to be built for a specific purpose, but can be flexibly modified during operation, for example after completion of a task, in order to be able to fulfill another task. The same applies to the combination of different cooperative systems, which are suitable for the most diverse tasks in different teams.

This chapter gives an overview of the required key technologies and associated issues which must be addressed in order to implement such approaches in future missions.

This section introduces the terminology we use and motivates the topic with exemplary applications and mission concepts. Section two gives an overview on the requirements of modularization and discusses the aspects of interfaces and segmentation into different functional units. Section three discusses the overall system design and required fundamental capabilities of reconfigurable multi-robot systems. The operation, mission planning and adaptive control of such systems is the subject of section four.

9.1.1 Terminology

Several types of systems, combinations thereof and possible ways of their collaboration are discussed in this chapter. This section provides a brief clarification of the terms used to describe system types and modes of operation. The term modular (self-) reconfigurable robot typically refers to a robot system comprising atomic connectible units to form composite robots with a variety of morphologies, sizes and capabilities. The atomic units are typically small homogeneous modules with a common electromechanical interface. Each of these modules can represent a subsystem which provides a well-defined functionality. Often a module comes with one or two Degree of Freedom (DoF) and sensors which are used for self-assembly [16, 20]. A standardized interface allows modules to connect and generate structures to provide, e.g., chain or tree structures [20]. An exhaustive survey on the state of the art of modular (self-) reconfigurable robots is provided in [21]. These types of robots are also referred to as robotic organisms in the following.

If multiple robot systems are combined to achieve a common goal such a system of systems is referred to as a Multi-Robot System (MRS) in the following. When different types of robots are constituting the MRS, the resulting system is said to be a heterogeneous MRS (as opposed to a homogeneous MRS). A reconfigurable MRS is a system consisting of several subsystems that can be connected and disconnected physically or virtually to form new functional units. Also the robots themselves in a reconfigurable multi-robot system can be coupled with each other to form new composite robots. We denote the former atomic agents and the latter composite agents. Creating functional units or composite agents intends to fulfill tasks by exploiting so-called super addition; super addition enables functionality that none of systems offers when operating on its own. When the individual robots of a MRS interact with each other, the behaviour is denoted a cooperative behaviour.

9.1.2 Applications and Mission Concepts

Currently, classical monolithic types of robots are at their limit of solvable applications so that modular (self-) reconfigurable robots as described in Sect. 9.1 are required to achieve the objectives of future space exploration missions. A feasible application could be to build flexible mobile sensor networks using robotic organisms. A scenario for a reconfigurable robotic organism in space exploration context is described in [30] with the goal of establishing a small green house on top of a sandy dune. However, considering long distance traverses or infrastructure setup, currently developed modular (self-)reconfigurable systems are presumably to small in scale. Sensor networks built from robotic organisms could therefore be deployed by bigger scale rover systems in the vicinity of the site of interest: Deployment of a snake-like configuration to reach cavities with small entrances and subsequent re-configuration into a sensor station would be a conceivable application scenario including a "classical" exploration rover and a set of reconfigurable modules.

MRSs built from individual rover systems have been considered for space exploration and infrastructure setup on celestial bodies already two decades ago [29]. Most of the considered exploration scenarios are dedicated to steep cliff negotiation with a robot acting as an anchor and another robot being lowered the steep slope using a tether system, [14, 18]. The individual robots of these MRSs are working in close cooperation, however, the interfaces between the robots are specialized for the respective use case. General modularization is not in the focus of these systems.

An example for a heterogeneous modular reconfigurable MRS for surface exploration is shown in Fig. 9.1. For the depicted scenario, several subsequent missions are assumed, each utilizing the existing modules and enriching the overall system with new capabilities and (sub) systems. A lander unit with four modular interfaces is shown at (A) in the illustration. The lander is equipped with a modular manipulation arm and hosts two energy modules, which can be for example batteries or modular solar panels. A modular radio beacon is connected to the lander for surface communication enhancement. A shuttle rover is transporting a filled sample module to the lander for detailed analysis or return to earth (B). The lander uses its manipulator to take over the sample module from the shuttle rover. Thus, the robots act together resulting in a cooperative behaviour. The rover displayed at (C) is taking soil samples with such a sampling module. The rover at spot (D) in the illustration is making use of a similar arm as that connected to the lander platform: A radio beacon for communication relay and navigation support is deployed on the surface. At position (E) a similar beacon with an additional sensor pack is shown. Long term measurements for example of seismic activities are conducted by this surface deployed payload. The battery might need to be exchanged, this can be accomplished by the rover at (D) which is carrying a full battery pack. A wheeled rover is transporting a legged scout to the rim of a crater (F). The legged scout with its climbing capabilities can reach the crater bottom for scientific investigations. For increased runtime, an additional battery pack is connected to the scout. A tether module is present for connection between wheeled rover and legged scout during descend and ascend in the crater slope.

The illustrated MRS takes aspects of the design from modular (self) reconfigurable systems and combines them with individual heterogeneous robot systems into a heterogeneous modular reconfigurable MRS for exploration. Modularity as aspired in the illustrated exploration scenario facilitates versatility (through reconfiguration possibilities), robustness (replacement of erroneous modules), expandability (both, functionally and spatially) and provides the potential for cost reduction through large scale production and multiple use-cases of available modules, [40].

The application of reconfigurable MRS described here refers only to a planetary surface exploration scenario. However, the use of such systems provides similar advantages in other mission scenarios, such as orbital use cases as described in [13].

The usage of reconfiguration introduces a significant flexibility to any robotic system. The flexibility ranges from hardware to software and resulting in a huge state space for a MRS. Therefore, one of the major challenges lies in the automated

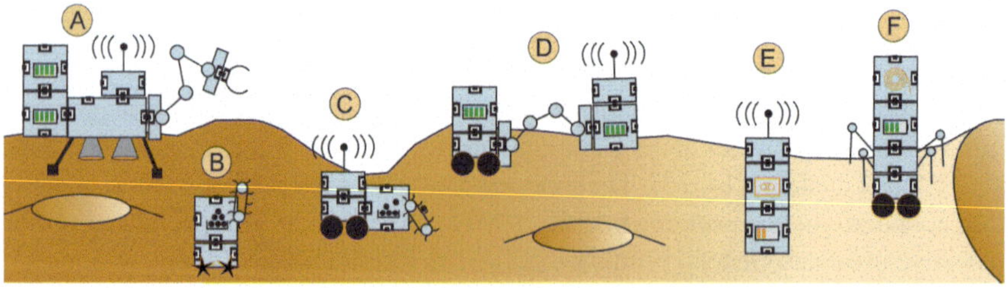

Fig. 9.1 Example for an extensive heterogeneous modular reconfigurable MRS in planetary exploration

exploitation of reconfiguration to improve future robotic missions. We will refer to a modelling and planning approach in Sect. 9.4.

9.2 Modularization

Clearly, modularization generates overheads: Interfaces introduce structures and consequently mass that might be saved without modularity. Mass is an important cost driver in space scenarios, hence, adding mass to a system is only possible when considerable benefits are opened from the additional elements. Additionally, management of configurations is a crucial aspect of a modular system design.

Producing multiple identical modules provides a possible cost reduction effect. Re-assembly of modules can generate new functionalities, hence re-usage of the same modules is possible instead of building, qualifying and flying dedicated robots for several tasks [40]. This can be assured by following a sophisticated standardization philosophy, allowing at least a certain degree of backward compatibility for the individual modules.

9.2.1 Interface Design Considerations

Interfaces play a key role in the context of modular reconfigurable robotic systems and MRS. To allow the handling, attachment and hence, the reconfiguration of the different units, a standardized interface has to be implemented in all modules and systems. The required functional range of the interface depends on the mission and/or system design and may vary between functional units, depending on their functional needs. While mechanical docking or berthing is the core function to allow reconfiguration, the interface can incorporate functions to allow e.g. power, data and thermal transfer or environmental sealing. An overview on possible and recommended functional principles is given in [39] accordingly. In general, two main kinds of docking mechanisms are distinguished: first androgynous mating devices and second, gender mating interfaces.

Androgynous interfaces are characterized by using only one design on either side of the coupling. This allows to dock each individual interface to any other available interface of the same type which provides a maximized flexibility of docking possibilities. Examples for androgynous interfaces in space applications are the NASA Docking System on the ISS, following the International Docking System Standard [12] or the iBoss interfaces for a modular reconfigurable satellite bus as presented in [13, 37]. Advantages of androgynous interfaces are, among others, the possibility to connect any two units and allowing for system level redundancy by role reversion. A drawback of androgynous mating interfaces, which has to be kept in mind with respect to gender mating interfaces, can be seen in potentially extra mass and volume.

This results from the fact, that each interface has to be able to act as active interface part.

In contrast, gender mating interfaces are characterized by introducing a female and a male connector side, which are able to dock to one-another. A classical example of a gender mating interface is given by the Russian Docking System SSVPG4000 used on the ISS or DFKI's Electromechanical Interface (EMI) as described in [8, 38]. In most cases gender mating interfaces consist of an active and a passive side, were the active part encounters the docking mechanism. A benefit of using active and passive gender mating interfaces can be seen in the potential to simplify the passive mate. Simplifying the passive mate to a minimal level allows to design extremely light weight options. As stated in [17] this approach may be needed for weight sensitive vehicles such as Lunar or Mars-bound ascent modules. Such simplified passive mating is e.g. used for the docking interface between the Instrument Deployment Arm and the scientific payloads on NASA's InSight Mars lander [11].

9.2.2 Functional Units

Functional units are to be regarded as expansion units within the framework of the modular robotic system. They can either be stand alone or configurations of various building blocks and/or systems, forming a new functional unit. A generic functional unit is formed by combining a standardized interface with a general module design and adding specific functions. A modular robotic systems as described in Sect. 9.1 requires different building blocks and a reconfiguration architecture, which is formed out of the pool of functional units. While a general modular reconfigurable system can be constructed based on only one functional unit, the system can be expanded in later mission steps by adding additional units as e.g. tools and sensors, which provide new functions. By adding tools to the module pool the system gains the ability of "tool change". Further expanding the base line system, by individual stand-alone robots as well as robotic devices leads to a "system change" capability. In this case a reconfigurable MRS is established.

According to the design approach, the modular units can account for different functions. This allows to build a framework with all kinds of functional units, ranging from simple tools and mechanisms (e.g. a hook or a telescopic mast) to goods (as e.g. energy modules) and electrical units (e.g. for processing, communication or power distribution) up to more complex instruments (as e.g. a camera) and even systems (e.g. manipulation device). Within such a framework of functional units, it is on the one hand possible to extend a system by adding functionality or on the other hand to build up new systems by combining dedicated items accordingly. In this context it is also possible to realize a system-to-system reconfiguration, as the border between system and pure functional unit blurs. One example of such system-to-system reconfiguration is the legged robot CREX acting as a six-fingered gripper device when attached to the manipulator arm of the Sherpa rover, see Fig. 9.3a.

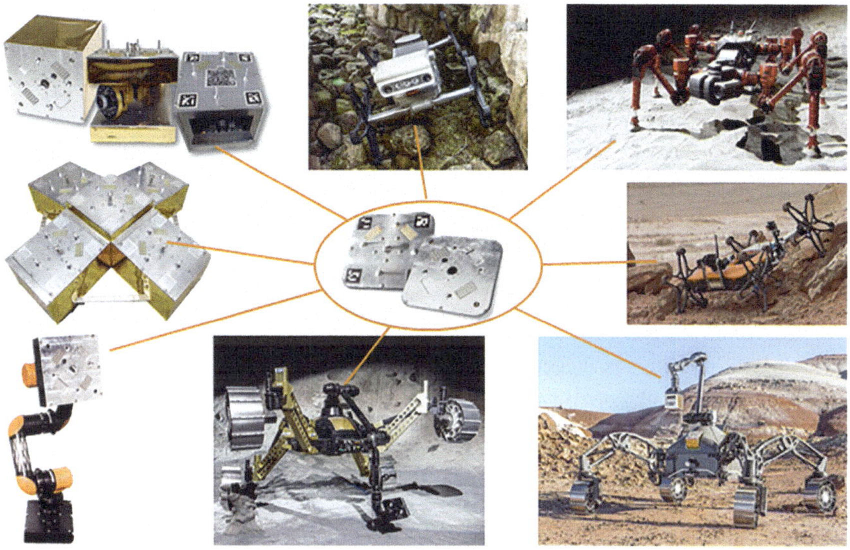

Fig. 9.2 The electromechanical interface (EMI) as central connection of various systems and functional units

Following the approach of building a heterogeneous reconfigurable MRS, a multitude of systems and functional units equipped with the common EMI have been implemented. Figure 9.2 illustrates the EMI and systems as well as functional units providing various functionality that can be connected via the common interface. The following systems with modular interface are depicted in the figure (from centre top, clockwise): legged-wheel robot Asguardv4, six-legged scout CREX, legged-wheel shuttle system CoyoteIII, wheeled-leg exploration rover SherpaTT (equipped with modular items), wheeled-leg exploration rover Sherpa, modular manipulation arm SIMA, a BaseCamp as infrastructure element, several payload items (a battery pack, a camera module and a sampling module). In a classical way the payload items can be seen as the core functional units within the reconfigurable MRS (c.f. [3]). However, as stated above, a clear differentiation between system and functional unit is not possible, as all shown units can potentially be connected to one another and form a new system.

9.3 Reconfigurable System Design

This section discusses several individual and multi-robot systems designed for reconfiguration. The reconfiguration can happen on physical as well as on functional layers. For a single unit system, the reconfigurability can e.g. be represented in reshaping its structure to gain a new set of functions, while a reconfigurable MRS can form new

(a) RIMRES system: six legged scout is connected via common EMI to manipulation arm for deployment.

(b) TransTerrA system: Exploration rover hands a sample container to the shuttle rover for return to lander. Modular arm SIMA is connected to shuttle.

Fig. 9.3 Two examples of heterogeneous modular reconfigurable MRS

functional units by connecting individual agents to one another. A key challenge for designing reconfigurable systems is to balance the overheads introduced by additional functionality against versatility and multipurpose mission needs. Therefore, it is essential to analyse the boundaries of the potential area of operation in order to identify a solid set of requirements for the system design. Even a modular reconfigurable robotic system can only be extended or reshaped and operated within its given physical and functional constraints.

The following examples of reconfigurable MRS highlight the evolution of potential system designs ranging from individual sub-systems and agents to the overall MRS by covering different reference mission scenarios. Furthermore, mobility reconfiguration and manipulation aspects of reconfiguration are considered.

9.3.1 Implemented Systems and Mission Scenarios

A feasibility study of a heterogeneous modular reconfigurable MRS for steep crater exploration at the lunar south pole is presented with the LUNARES MRS [5]. In the scenario, a legged scout robot is transported by an energy-efficient wheeled rover to the crater rim. Modular reconfiguration capabilities of the systems are prepared in the mission structure and realized with mock-ups for a general evaluation of the approach.

The approach is extended by introducing a generalized EMI The RIMRES MRS [25] combines a six-legged walking and climbing scout robot with a hybrid wheeled-leg rover and several modular payload-items. These payload-items are physical implementations of the modules depicted in Fig. 9.1, more details are presented in Sect. 9.2.2. For transporting the legged scout it is connected to the wheeled rover via the EMI and can be deployed either using the active suspension system or the manipulator as shown in Fig. 9.3a. During transport, both systems are mechanically and electrically connected and can share data as well as power via the EMI. The combination of both systems can be considered as a new robot system with different capabilities.

In the follow up MRS TransTerrA [36], a robotic logistics chain for sample return missions is implemented. Modular payload containers are used as sampling and sample storage devices. The sealed samples are transported from the exploration rover to the lander unit using a highly mobile shuttle rover [31]. Individual sequences of the sample return scenario were evaluated in a field deployment in the desert of Utah [32] (cf. Fig. 9.3b). The EMI as central element for reconfiguration and tight coupling of individual subsystems is described in detail in [38]. Both MRSs make use of the EMI as well as immobile payload items for different tasks and capability extension of mobile subsystems.

9.4 Manipulation Capabilities for Reconfiguration

A reconfigurable MRS does not necessarily require manipulation skills for reconfiguration as it is generally the case in modular (self-)reconfigurable systems. Even for MRS built from complex individual robots, manipulation capabilities are not a hard requirement: the coupling between systems can also be established using the kinematic flexibilities of one or both to-be-connected systems, as shown in [5, 25].

However, many applications involve immobile modules, consequently manipulation skills are highly desirable. Here the design of the manipulation device is again depending on the overall MRS design and the desired functional range. Aspects which should be considered during the design phase are e.g. according to the area of operation: work space, payload capacity, docking mechanism and docking routine.

The project RIMRES [25] implements a manipulator arm with an EMI as end effector, so that any compatible module can be attached to the system. Here, attaching a module relies on a visual servoing process which has been implemented in the project TransTerrA [36]. Artificial markers are attached to the modules to facilitate the identification of the relative pose, but in order to achieve the required docking precision, two visual servoing stages are considered. Hence, one set of markers is used to allow for a rough approach bringing the manipulator interface into a known workspace area, while the second set is used to achieve a precise relative pose estimation. The actual docking is then performed by a final blind docking stage which starts from the known relative pose with the manipulator operating in compliance control mode to compensate for minor errors.

9.4.1 Reconfiguration Between Mobility and Manipulation

Mobility reconfiguration can be achieved through several means. Besides of mobility, active locomotion systems can provide manipulation capabilities as well by introducing additional DoF to the platform. This includes individual robot systems as well as multi-robot systems.

In individual systems, one example are wheels that are folded away to change between wheeled and legged locomotion, as presented e.g. with the Roller-Walker robot [10]. An active suspension system for a wheeled rover can be seen as reconfigurable locomotive system as well, examples being ATHLETE [34], MAMMOTH [24] or SherpaTT [6]. Modular (self-)reconfigurable robots can reconfigure to change between various locomotion principles [40].

In a multi-robot system mobility reconfiguration of the overall system can be found for example when looking at a sample return mission. Using the perspective of a sample container it can be seen that it "makes use" of different locomotion platforms: wheeled locomotion (being transported to a crater), legged locomotion (reaching the interior of the steep crater), and legged-wheel locomotion (being transported back to the landing unit).

A functional reconfiguration possibility can be included in the design phase of a system when subsystem components can be used in various ways. One example is the usage of a manipulation arm as a leg or, vice versa, the usage of a leg as a manipulation arm.

In case of the legged scouts Scorpion [5] and CREX [25] one or two front legs are equipped with small gripping devices. Once the system reaches a spot of interest, e.g. at the crater floor, a sample can be picked up with the gripper at the leg now acting as a manipulation arm. Scorpion's sample collecting in an artificial crater is experimentally evaluated in [2].

Using the manipulation arm as a leg for a rover system has been implemented for example in the Sherpa and SherpaTT rovers. This use case has been accounted for already in the design phase of the arm [7]. The rover Sherpa in Fig. 9.2 (bottom centre) is shown with the manipulation arm supporting roughly half of the mass of the 150 kg rover.

A third example for manipulation and locomotion reconfiguration is the modular arm SIMA [31] (cf. Fig. 9.2 (bottom left)). Generally, this arm does not feature its own power source. However, when connected to the CoyoteIII rover it can be docked to a BaseCamp with its manipulation interface. Subsequently, the base interface on CoyoteIII is detached and becomes the new manipulation interface of the arm on a BaseCamp. Hence, the arm moved between two subsystems and the two interfaces changed roles between manipulation end-effector and base anchor point. Furthermore, being connected to a battery module, snake-like locomotion without another mobile system is conceivable for the arm. Thus reconfiguration of multiple types in several use-cases is possible with the SIMA manipulation arm.

9.5 Operation of Reconfigurable Multi-Robot Systems

The ability of reconfiguration provides a significant flexibility to any robotic system. This holds for single as well as for multi-robot systems. An individual robot system's change of the configuration state—e.g. such as the locomotion platform—typically requires the dedicated design of forward and inverse kinematic control approaches as shown in [4]. This can be viewed as low-level reconfiguration planning. The next higher level can be observed in mobile systems, which exploit the reconfigurability of locomotion platform in order to adapt their footprint dynamically as part of path planning or more reactive trajectory control approaches [24].

While these single robots can be part of a multi-robot team and add additional DoFs, a different challenge comes into place with the opportunity to form physical coalitions of robotic agents. Each reconfigurable robot comprises a set of mechanical, or even electromechanical interfaces such as described in [38] in order to establish a connection with another robot. In a simple case this can involve simply restacking and rearranging modules on a robot. However, more complex scenarios require the coordination of two or more mobile systems in order to dynamically form a new system as shown in [25]. A higher level of planning of reconfiguration is therefore required in order to plan robotic missions for reconfigurable multi-robot system.

9.5.1 *Mission Planning for Reconfigurable Multi-robot Systems*

As mentioned, robots in a reconfigurable multi-robot system can be coupled with each other to form new composite agents. The flexibility to couple systems can be seen as an additional DoF. Exploiting this DoF can serve as the basis for new mission designs since hardware resources can be dynamically exchanged and shifted within a reconfigurable multi-robot systems. Such design allows for potentially superadditive capabilities, i.e., designing capabilities neither of the individual robots has. The operation of a reconfigurable multi-robot system is a major opportunity, but a combinatorial challenge at the same time. The flexibility of a reconfigurable multi-robot systems arises from the possibility to form $|P^A| = 2^{|A|}$ distinct agents, including atomic and composite agent. P^A is the powerset for the set of atomic agents A and it includes the empty set, so that for $A = a_0, a_1$ $P^A = \{\{\}, \{a_0\}, \{a_1\}\{a_0, a_1\}\}$. But it is not sufficient to account only for simple agent combinations; agents can comprise multiple physical interfaces for linking, and the compatibility of the interfaces might vary. So on one hand the feasible set of composite agents might be more limited. On the other hand, a single composite agent might be formed by linking through various interfaces, so that different link configurations can establish a composite agent.

This observation has been cast into an organization model [26, 28] which combines existing works of organization theory [9], multiagent domain [23, 35] and knowledge-based reasoning [33]. The model focuses on reconfigurable systems and models agents and interfaces and allows to compute the set of feasible agents based on the set of compatible interfaces. In addition to identifying feasible composite agents, the organization model also enables the identification of suitable agents, where the suitability refers to a particular demanded functionality of the agent.

Based on the organization model a mission planning approach for reconfigurable multi-robot system has been developed. A reconfigurable multi-robot system consists of a set of atomic agents, which can be combined to composite agents, so that each set of atomic agents can form various coalition structures. A mission is represented by a temporal database. This database serves as combined representation of the problem statement and the solution, and defines how the coalition structure evolves over time and how agents distribute over a given set of locations. Planning has to account for the constraints in which way atomic agents can form coalitions. The planning state covers spatio-temporal assignments of agents, and a minimal mission definition requires an initial state and a goal state description. Since a temporal databased is used, intermediate states can equally be states, so that the evolution of a robot missions can be controlled. The planning process will detail the temporal database, and a solution contains a full sequence of operational states, including the transition between coalition structures. While the classical approach for temporal planning requires the application of temporal operators, the developed approach is currently restricted to an implicit set of operators which focus on the reconfiguration and relocation of agents. The mission planning approach uses a constraint-based definition of the problem, and a likewise constraint-based mission planning approach.

9.5.2 Adaptive Control for Reconfigurable Multi-Robot Systems

Planning accounts for reconfigurability of a reconfigurable MRS at a high-level. For the real execution of robotic missions, however, adaptivity has to be accounted for at different level of control. Hence, an operational infrastructure has to be established, which can support reconfigurable MRS.

Reconfigurable multi-robot systems are open-extensible systems in general, but they require a corresponding standardization effort at hardware and software level to achieve true extensibility. Firstly, communication between robots and subsystems to allow for cooperation. Communication in a multi-robot system has to be looked at in two different ways: inter-robot communication and intra-robot communication. While the intra-robot communication can, at least in parts, be robot specific, all communication interfaces which are required to establish inter-robot communication have to be aligned for all robots that shall cooperate. For the robotic teams in TransTerrA [32] and RIMRES [25] the standardization according to FIPA [15] has been advocated and additionally successfully evaluated in the Field Trials of Utah [32]. The implemented FIPA-based communication architecture [27] can support a dynamic distributed team of robots. The standardized communication approach not only allows for an open-extensible communication system, but provides support for the design of custom communication protocols. The communication infrastructure allows to establish a loosely coupled sensor network through the use of the available robotic systems. The approach has been evaluated with a distributed SLAM approach which relies on the continuous sending of sensor information (here limited to point clouds) to a multicast group. A further extension of this communication architecture has been achieved by the use of meshing approaches. For the Field Trip Utah a set of particularly adapted access points in combination with the router operating system OpenWrt [19] has been used in order to establish a communication mesh.

Further, reconfiguration support is required to dynamically account for the addition of functional unit or forming of composite agents. Reconfiguration by attaching or detaching modules will change the morphology of a robot and can therefore require adaptation of the control approach or more general the software structure. Attaching a payload-item to an end effector, e.g., leads to a morphology change. Even more, the attached payload-item acts as new end effector and thus internal software control structures need to adapt accordingly. Firstly, the main rover in [32] uses a visual servoing approach to attach the manipulator to modules. As soon as the end effector attaches a new end effector camera has to be accounted for and used in further activities Secondly, the change of the kinematic structure has to be propagated, so that collision avoidance and blind manipulator activities perform correctly. Thirdly, any calibrated sensor such as a force torque sensor in the manipulator has to consider the additional weight. To facilitate change, one approach is a static, external standardization, e.g., allowing the use of a standardized, single typed functional unit. This approach is clearly too restrictive, so that in the suggested approach the dimensions of each individual payload can be stored in an EEPROM of the respective module. Once it is attached to another module an automatic update of the overall kinematic structure can be triggered. Generally, information about functional unit should come with the units themselves, and apart from outer dimension, transformation frames also capabilities and system properties should be propagated, e.g., in order to feed an existing overall organization model. This approach can lead to a plug-and-play like inclusion of extension modules in software control layers, and industrial applications already show similar configuration approaches [22].

Robot reconfiguration can also be part of an adaptive subsystem, and various examples for an active locomotion systems have been already mentioned [1, 6, 24, 34]. Clearly, the locomotion platform is a key element to any robotic system that operates outdoors and the ability to adapt to a range of terrains types bring clear advantages. Numerous examples in literature proof the benefits of reconfigurable or active locomotion systems. Controlling the outer shape and footprint of a robot allows for a broader operational range of reconfigurable system and the locomotion system can be coupled with path planning; planning algorithm for footprint control in narrow passages is demonstrated in [24].

Modularity and reconfiguration can result in redundancy which increases robustness against failures. Figure 9.4 illustrates a communication reconfiguration for the Sherpa rover. To control the opening and closing of EMI latch two major choices are available for communication: wireless and cable-based. The cable-based choice

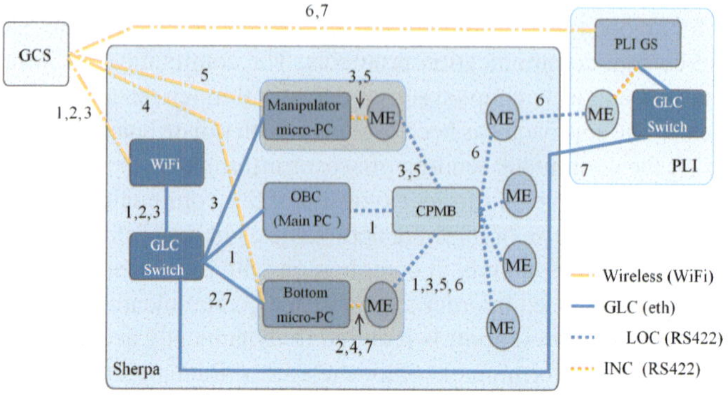

Fig. 9.4 Modular communication bus example: Seven paths for sending command openlatch from GCS to bottom EMI of Sherpa with one connected PLI to a payload bay

comes with further redundancy since commands can be either communicated over Ethernet or RS422. While the redundancies exist an active management of the network configuration might be required, e.g., to avoid package looping in the network. Hence, also the low-level communication systems require to be adaptive and has to dynamically react to changes.

9.6 Conclusions and Outlook

This chapter describes the advantages of modular reconfigurable multi-robot systems for space missions that will become more and more complex in the future. The technologies required to realize such flexibly integrated teams are outlined. It is shown that there is already a multitude of solutions and concepts for associated challenges that have been tested with perfectly harmonized systems. So far, however, these are stand-alone solutions. To be able to make use of the achievable benefits of modularization and reconfiguration on a large industrial scale common standards have to be developed and determined by the community for hardware and software design. This includes the connecting interfaces as well as design rules for the modules up to the representation of the description of their functionality for mission planning.

Acknowledgements The authors would like to acknowledge and express their thanks to the financial support for the projects LUNARES (funded by the German Space Agency (DLR, Grant No. 50RA0706) and the Investment Association Bremen (BIG, Grant No. INNO1036A)), RIMRES (funded by the German Space Agency (DLR,Grant No. 50RA0905)), TransTerrA (funded by the German Space Agency (DLR, Grant No.: 50RA1301)) and SIROM (EU financial support, Horizon 2020 PERASPERA, Grant Agreement Number 730035) in which this topic could be investigated. We would also like to thank all colleagues with whom we have been able to work on these projects in recent years.

References

1. Adachi, H., Koyachi, N., Arai, T., Shimiza, A., & Nogami, Y. (1999). Mechanism and control of a leg-wheel hybrid mobile robot. In *International conference on intelligent robots and systems (IROS'99)* (Vol. 3, pp. 1792–1797).
2. Bartsch, S., Cordes, F., Haase, S., Planthaber, S., Roehr, T. M., & Kirchner, F. (2010). Performance evaluation of an heterogeneous multi-robot system for lunar crater exploration. In: *Proceedings In 10th International Symposium on Artificial Intelligence, Robotics and Automation in Space (iSAIRAS'10)* (pp. 30–37), Sapporo, Japan.
3. Brinkmann, W., Cordes, F., Roehr, T. M., Christensen, L., Stark, T., Sonsalla, R., Szczuka, R., Mulsow, N. A., Bernhard, F., & Kuehn, D. (2018). Modular payload-items for payload-assembly and system enhancement for future planetary missions. In *IEEE Aerospace Conference (AeroConf'18)* (pp. 1–12). Big Sky, MT, USA.
4. Hidalgo, J., & Cordes, F. (2012). Kinematics modeling of a hybrid wheeled-leg planetary rover. In *11th International Symposium on Artificial Intelligence, Robotics and Automation in Space (i-SAIRAS'12)*, Turin, Italy.
5. Cordes, F., Ahrns, I., Bartsch, S., Birnschein, T., Dettmann, A., Estable, S., Haase, S., Hilljegerdes, J., Koebel, D., Planthaber, S., & Roehr, T. M. (2011). LUNARES: Lunar crater exploration with heterogeneous multi robot systems. *Intelligent Service Robotics, 4*(1), 61–89.
6. Cordes, F., Babu, A., & Kirchner, F. (2017). Static force distribution and orientation control for a rover with an actively articulated suspension system. In *IEEE/RSJ International Conference on Intelligent Robots and Systems (IROS'17)* (pp. 5219–5224), Vancouver, Canada, September 2017.
7. Dettmann, A., Roemmermann, M., & Cordes, F. (2011). Evolutionary development of an optimized manipulator arm morphology for manipulation and rover locomotion. In *2011 IEEE International Conference on Robotics and Biomimetics* (pp. 2567–2573), Phuket, Thailand.
8. Dettmann, Z. Wang, W. Wenzel, F. Cordes, and F. Kirchner. (2011). Heterogeneous modules with a homogeneous electromechanical interface in multi-module systems for space exploration. In *IEEE International Conference on Robotics and Automation (ICRA'11)* (pp. 1964–1969), Shanghai, P.R. China.
9. Dignum, V. (Ed.). (2009). *Handbook of research on multi-agent systems: Semantics and dynamics of organizational models*. IGI Global.
10. Endo, G., & Hirose, S. (2011). Study on roller-walker-energy efficiency of roller-walk. In *IEEE International Conference on Robotics and Automation (ICRA'11)* (pp. 5050–5055). Shanghai, China.
11. Fleischner, R. (2013). Insight instrument deployment arm. In *15th European Space Mechanisms & Tribology Symposium (ESMATS 2013)* (pp. 25–27), Noordwijk, the Netherlands, September 2013. ESA.
12. Gerstenmaier, W. H., Krikalev, S., Parker, D., Leclerc, G., & Shirama, R (2016). International docking system standard (IDSS) interface definition document (IDD) Revision E. NASA, Roscosmos, ESA, CSA, Jaxa, October 2016.
13. Goeller, M., Oberlaender, J., Uhl, K., Roennau, A., & Dillmann, R. (2012). Modular robots for on-orbit satellite servicing. In: *2012 IEEE International Conference on Robotics and Biomimetics* (ROBIO) (pp. 2018–2023). IEEE. Guangzhou, China.
14. Huntsberger, T., Stroupe, A., Aghazarian, H., Garrett, M., Younse, P., & Powell, M. (2007). TRESSA: Teamed robots for exploration and science on steep areas. *Journal of Field Robotics, 24*(11–12), 1015–1031.
15. IEEE Foundation for Intelligent Physical Agents. The Foundation for Intelligent Physical Agents: An Overview. http://www.fipa.org, 1 July 2018.
16. Murata, S., Kakomura, K., & Kurokawa, H. (2007). Toward a scalable modular robotic system. *IEEE Robotics and Automation Magazine, 14*(4), 56–63.
17. NASA. (2009). Internationl Docking Standardization.

18. Nesnas, I. A., Matthews, J. B., Abad-Manterola, P., Burdick, J. W., Edlund, J. A., Morrison, J. C., Peters, R. D., Tanner, M. M., Miyake, R. N., Solish, B. S., & Anderson, R. C. (2012). Axel and DuAxel rovers for the sustainable exploration of extreme terrains. *Journal of Field Robotics, 29*(4), 663–685.
19. OpenWRT/LEDE Community. (2018). OpenWRT Homepage. https://openwrt.org, 25 Feburary 2018.
20. Park, M., & Yim, M. (2009). Distributed control and communication fault tolerance for the ckbot. In *ASME/IFToMM International Conference on Reconfigurable Mechanisms and Robots (ReMAR'09)* (pp. 682–688).
21. Parker, L. E., Rus, D., & Sukhatme, G. S. (2016). Multiple mobile robot systems. (Chapter 53, pp. 1335–1379). Springer Handbook of Robotics.
22. Pepperl+Fuchs. (2018). RFID Accelerates Tool Exchange on Mobile Equipment. https://www.pepperl-fuchs.com/usa/en/20042.htm; 3 August 2018.
23. Rahwan, T., Michalak, T. P., Elkind, E., Faliszewski, P., Sroka, J., Wooldridge, M., & Jennings, N. R. (2011). Constrained Coalition Formation. In *25th AAAI Conference on Artificial Intelligence* (pp. 719–725).
24. Reid, W., Fitch, R., Göktoğgan, A. H., & Sukkarieh, S. (2016). Motion planning for reconfigurable mobile robots using hierarchical fast marching trees. In *12th International Workshop on the Algorithmic Foundations of Robotics.*
25. Roehr, T. M., Cordes, F., & Kirchner, F. (2014). Reconfigurable integrated multirobot exploration system (RIMRES): Heterogeneous modular reconfigurable robots for space exploration. *Journal of Field Robotics, Special Issue on Space Robotics, Part 2*, 3–34.
26. Roehr, T. M., & Hartanto, R. (2015). Using an organization-model for PDDL-based planning with a reconfigurable multi-robot system. In *Space Robotics Symposium, Glasgow.*
27. Roehr, T. M., & Herfert, S. (2016). A FIPA-Based Communication Infrastructure for a Reconfigurable Multi-robot System (pp. 665–676). Springer International Publishing, Cham.
28. Roehr, T. M., & Kirchner, F. (2016). Spatio-Temporal Planning for a Reconfigurable Multi-Robot System. In *Proceedings of the 4th Workshop on Planning and Robotics (PlanRob), Ed. by Alberto Finzi and Erez Karpas* (pp. 135–146), London.
29. Schenker, P. S., Pirjanian, P., Balaram, J., Ali, K. S., Trebi-Ollennu, A., Huntsberger, T. L., Aghazarian, H., Kennedy, B. A., Baumgartner, E. T., Iagnemma, K. D., Rzepniewski, & A., Dubowsky. S. (2000). Reconfigurable robots for all terrain exploration. In *Sensor Fusion and Decentralized Control in Robotic Systems III,* volume 4196 of SPIE (pp. 454–468).
30. Shen, W.M., Hou, F., Rubenstein, M., Chiu, H. & Kamimura, A. (2010). Recent progress of superbot. In *ICRA 2010 Workshop "Modular Robots: State of the Art".*
31. Sonsalla, R. U., Akpo, J. B., & Kirchner, F. (2015). Coyote III: Development of a modular and highly mobile micro rover. In *13th Symposium on Advanced Space Technologies in Robotics and Automation (ASTRA'15).*
32. Sonsalla, R., Cordes, F., Christensen, L., Roehr, T.M., Stark, T., Planthaber, S., Maurus, M., Mallwitz, M. & Kirchner, E.A. (2017). Field testing of a cooperative multi-robot sample return mission in mars analogue environment. *In 14th Symposium on Advanced Space Technologies in Robotics and Automation (ASTRA'17), Leiden, The Netherlands.*
33. Tenorth, M., & Beetz, M. (2013). KnowRob: A knowledge processing infrastructure for cognition-enabled robots. *The International Journal of Robotics Research, 32*(5), 566–590.
34. Townsend, J., Biesiadecki, J., & Collins, C. (2010). ATHLETE mobility performance with active terrain compliance. In *2010 IEEE Aerospace Conference (IAC'10)* (pp. 1–7).
35. Ueda, S., Kitaki, M., Iwasaki, A., & Yokoo, M. (2011). Concise characteristic function representations in coalitional games based on agent types. In *IJCAI International Joint Conference on Artificial Intelligence* (pp. 393–399).
36. Sonsalla, R. U., Cordes, F., Christensen, L., Planthaber, S., Albiez, J., Scholz, I., & Kirchner, F. (2014). Towards a heterogeneous modular robotic team in a logistics chain for extended extraterrestial exploration. In *12th International Symposium on Artificial Intelligence, Robotics and Automation in Space (i-SAIRAS'14), Montreal, Canada.*

37. Weise, J., Reimerdes, H., Brieß, K., Adomeit, A., Göller, M., & Dillmann, R. (2012). An Intelligent Building Blocks Concept for On-Orbit-Satellite Servicing. *In Int. Symposium on Artificial Intelligence, Robotics and Automation in Space (i-SAIRAS), Turin, Italy.*
38. Wenzel, W., Cordes, F., & Kirchner, F. (2015). A robust electro-mechanical interface for cooperating heterogeneous multi-robot teams. In *IEEE International Conference on Intelligent Robots and Systems (IROS'15)* (pp. 1732–1737).
39. Yan, X.-T., Brinkmann, W., Palazzetti, R., Melville, C., Li, Y., Bartsch, S., & Kirchner, F. (2018). Integrated mechanical, thermal, data and power transfer interfaces for future space robotics. *Frontiers in Robotics and AI, 5*, 64.
40. Yim, M., White, P., Park, M., & Sastra, J. (2009). Modular self-reconfigurable robots (pp. 5618–5631). Springer New York, New York, NY.

Chapter 10
SIROM an Enabler for Reconfigurable Space Robotics

Javier Vinals, Xiu Tian Yan, Sebastian Bartsch, and Manfred Doermer

Abstract A key to the wider adoption of modular and reconfigurable robotic system is the creation of a standard interface to allow payload modules to be interfaced and changed easily. The payload modules can be found for applications such as on-orbit satellite servicing or planetary exploration. The realization of a standard interface (IF) allows the necessary mechanical interconnections with the possibility of electrical and data connection as well as thermal transfer between these "modular building blocks". Within this context, a standard interface has been developed in the SIROM (Standard Interface for Robotic Manipulation of payloads in future space missions) project and the interface combines the four required functionalities in an integrated and compact form as an integrated solution. With a mass lower than 1.5 kg and having an external diameter of 120 mm and a height of 30 mm, this novel interface permits not only mechanical coupling but also electrical, data and thermal connectivity between so called Active Payload Modules (APMs) as well as other involved operators like the robotic end-effectors. This multi-functional IF is based on an androgynous design to allow replacement and reconfiguration of the individual modules in any combination desired. It consists of the following sub-assemblies: mechanical IF, electrical IF, data IF, thermal IF and IF controller. A clear advantage of SIROM design is that its mechanical IF consists of a latching and guiding systems for misalignment correction capable of withstanding certain robotic arm positioning inaccuracies. For the electrical power connection, SIROM transfers up to 150W electrical power, whereas for the data interface, it provides

J. Vinals (✉)
Sener Aeroespacial, Getxo, Spain
e-mail: javier.vinals@aeroespacial.sener

X. T. Yan
RAS Group, DMEM, University of Strathclyde, Glasgow, Scotland
e-mail: x.yan@strath.ac.uk

S. Bartsch
Airbus, Bremen, Germany
e-mail: Sebastian.Bartsch@dfki.de

M. Doermer
Airbus, Hamburg, Germany

X. T. Yan and G. Visentin (eds.), *Space Robotics*,
https://doi.org/10.1007/978-3-031-39214-6_10

a data transfer rate of 100Mbit/s via SpaceWire and a command communication with a speed up to 1Mbit/s via CAN bus. The thermal IF provides fluidic ports for flow transfer and has the potential to transfer 2500W between APMs provided with the corresponding close-loop heat exchange system. SIROM also exhibits redundant coupling capabilities. It can match and couple another SIROM with the only need to actuate and control just one of them, while the other stays passive. It is provided with main and redundant connectors for thermal, electrical, data and control flow in case of one of the lines fails. All in all, SIROM will allow long duration with no logistic support missions, reparability and reconfiguration of satellites, cost efficiency and simplification of the tool exchange in scientific exploration missions, among others.

10.1 Introduction

Universal and multiple functional interface technologies can become the physical key for flexible and reconfigurable space robots which need to be adaptable for different tasks found in many space missions [1]. The extended use of such space robotic interfaces in a space system will also enable physical flexible configurations and reconfigurations of many space systems [2]. As an example, the use of such interfaces for the flexible reconfiguration of satellites is fundamental to the creation of modular and serviceable satellites in future, in which standard interfaces that groups mainly mechanical, electrical, data and thermal connections are used to both reconfiguration and serviceability. The use of these interfaces can hence provide a critical linking between a target space system such as a satellite and a manipulating space robot.

Based on a study of potential innovative design of a number of single interfaces and current robotic development trends [3], the future flexible space systems will be composed of common building blocks where the autonomous robotic systems will play a significant role as key enablers for on-orbit satellite servicing and planetary exploration missions.

SIROM (Standard Interface for Robotic Manipulation of Payloads in Future Space Missions) project aims to realize a set of integrated and inherently optimised interfaces for mechanical, data, electrical and thermal connectivity that allow reliable, robust and multi-functional coupling of payload to robot manipulators, payload to other payload or client to server. SIROM project is advance the state of the art by investigating multi-functional, well integrated and "Intelligent" IF (mechanical, electrical, data, thermal) to be used as standard building blocks to connect modules within a robotic system and also to connect to the satellite with a servicer.

This standard IF is realised in an integrated form (where mechanical, thermal, electrical, data connections [4] are combined and geometrically optimised for these 4 functionalities). The standard IF will allow the creation of large clusters of modules with high modular and functional optimisation that will be a combination of devices that allow to couple active payload modules (APM) to a manipulator, among themselves and to a spacecraft [5].

This chapter will introduce SIROM concept defining all the interface and modularity concepts with the aim of defining a universal "USB" for space applications looking for modularity and standard ability concepts to be applied to space exploration and servicer missions based on robotics concepts. Finally a vision of potential real applications will be provided.

10.1.1 Challenges in Flexible Space Exploration for Reconfiguration

Times are changing for the way to operate space assets like satellites, satellite clusters, planetary missions or even space stations. The state of the art is to build one of a kind spacecraft which are meant to be launched, operated and never touched again once in orbit and planetary. This process left no room to improvement according to new technologies, adaptation to changing needs or even repair of valuable space infrastructure assets. The economic figures and the ecological responsibility call for a change. This challenge requires new technologies within the field of space robotics that shall enable:

- Planned orbit raising;
- Inspection;
- Deployment/assembly aid;
- Replacement of ORU;
- Planned maintenance;
- Performance enhancement;
- Additional payload;
- Mission evolution;
- Lifetime extension;
- Re/Deorbiting;
- Planetary surface exploration;
- Autonomous navigation;
- Landmark tracking;
- Rendezvous between 2 planetary assets.

Robotic servicing of a cooperative satellite is still an open research area facing many technical challenges. The missing standardization of interfaces poses a great obstacle on the track to serviceable spacecraft. Standardization within robotic applications in space, wide spread for terrestrial applications, is only at its beginnings in the space industry. Building up on a comprehensive technology review of mechanical, electrical and data connections a design of a compact standard IF for orbital servicing and / or planetary applications will be derived.

The actual space infrastructure in operation is based upon one off a kind satellites and cluster of satellites like GPS, GALILEO or Martian Long Range Autonomous Scientist. The future of satellite operations will be asked for the option to enhance satellite capabilities in orbit and planetary, to maintain the operational status and to extend operational times. These measures will secure availability of services. SIROM aim is to demonstrate capability and functionality of a modular standard robotic IF which is needed for future robotic on-orbit servicing & planetary, repair and assembly missions. The results of this development will be used for the design and planning of future European robotic on-orbit servicing and/or planetary missions. SIROM creates a cornerstone for future standard robotic interfaces to be utilized in space for On-Orbit Servicing and Assembly (OOS, OOA) systems required for robotic operations of space elements which need maintenance, repair and/or assembly to extend their life-time, and for planetary missions with long range autonomous capacities.

SIROM aims to study concrete concepts of a standard IF for mechanical connection, transfer of power, data and heat load between APM´s and / robotic gripper which could support European space companies for their space robotics servicing capabilities in orbit and planetary scenarios. Although a great variety of interfaces for data, power exchange and mechanical connection with high TRL or even flight proven variants exist there is technology gap for low profile interfaces actuated and controlled by robots [6]. The project is intended to address the above challenges in both orbital revising and planetary exploration missions.

10.1.1.1 Orbital Track Scenario

End goals for the orbital track: The end goal and demonstration scenarios target orbital operations in close proximity and direct interaction with cooperative structures as shown in Fig. 10.1.

The copyright and credit: ESA – Ducros

Fig. 10.1 Orbital Scenario. The copyright and credit: ESA – D. Ducros

- Robotic in orbit and surface exploration systems require appropriate systems to manipulate objects in order to perform service operations such as maintenance and construction work on technical infrastructure or to carry out scientific investigations and experiments.
- *Expected results*: Open standard architecture of the IF with flexible configuration
- Modular architecture consisting of 'block sets'
- Cost efficient development logic and a consequent oriented design to cost approach

10.1.1.2 Planetary Exploration Track Scenario

The design and operation of planetary robotic systems is extremely challenging and unique, and while a few rovers have been successfully operated, even fewer robotic arms have been designed, flown and operated. Planetary surface robotics systems as exemplified in Fig. 10.2 therefore belong to a class of their own where robustness and minute accuracy must be achieved with limited mass, power and processing capabilities. While in-orbit systems have to contend with a harsh, but repeatable environment, planetary systems have to deal with potentially a highly variable and transient environment with, amongst other, varying temperature and dust level.

Fig. 10.2 ExoMars rover protptype is shown when it is exploring a potenital planetary area (left), and the concpet that ExoMars is collaborating with an orbitor for a common task is also shown (right)

The future of planetary robotic exploration will be calling upon a range of robotic assets including lander and rovers. Increasingly, these systems will need to interact between themselves and with additional payload elements to perform a range of tasks to fulfil the mission objectives [7]. These interactions will involve a range of mechanical, data and power exchanges, be it at lander, mobile platform, robotic arm, or payload level.

10.1.2 Needs to Enable Flexible Reconfiguration

Space robotics is considered one of the most promising approaches for on-orbit servicing (OOS) missions such as docking, berthing, re-fuelling, re-pairing, upgrading, transporting, rescuing and orbital debris removal, and for planetary scenario missions. Many enabling techniques have been developed in the past two decades and several technology demonstration missions have been completed. European countries like Germany (e.g. DEOS) and Italy (e.g. SPIDER) invested to evolve the satellite servicing technology. Also ESA is supporting this strategy by initiating programs like 'CAPTARE' A number of manned on-orbit servicing missions e.g. Hubble servicing were successfully accomplished but unmanned, fully autonomous, servicing missions have not been done yet. Robotic servicing of a cooperative satellite is still an open research area facing many technical challenges.

A reliable and economically viable space servicing business calls for an innovative space servicing architecture to reduce cost; this is one of the SIROM targets. The availability of exchangeable satellite modules opens the perspective for satellites with upgradable modules; e.g. GPS satellites atomic clock module, this would be a valuable option especially for future satellite cluster systems. A continues enhancement of capabilities of a one of a kind satellite (as in the past Hubble) by an update of the sensor suit according to improvements in sensor technology after launch will extend operational capabilities of valuable space infrastructure assets and could extend their lifetime.

Future markets are identified in the area of planned orbit raising to lower launch cost by integrating large orbital assemblies in orbit e.g. as part of post ISS scenario, planned maintenance, planned enhancement. Key factors for a successful evolution of these markets are:

- Reducing mission cost;
- Availability of the enabling technologies;
- Extension of lifetime for satellites by in-orbit servicing measures, which will enlarge the utilization of the satellite;
- Reducing of debris by either satellite refuelling and satellite repair or by actively deorbiting of the satellite and;
- Contributing technical capability for exploration missions.

10.1.3 Methodology to Enable Flexibility

The nature of the flexibility capability presents some challenges that are unique to typical complex systems projects; specifically, the inclusion of a multi-disciplinary IF with a focus on integration of function to a standard. As such, some unique opportunities have arisen to use techniques that are designed to address such an issue. Through using these techniques, a flexible design process is enabled to provide a robust and explorative support for SIROM researchers to develop a flexible mutil-functional interface. These techniques include the following three inter-linked methods.

10.2 Simulation and Multi-perspective Based Validation

In the cases where the potential testing is limited and before parts are bought/manufactured, it is wise to verify and validate them in a virtual setting; traditionally these simulations are known to have inaccuracies and typically do not include detailed factors due to single-perspective modelling techniques. Modelling and testing the thermal transfer, mechanical IF latching, and loads separately do not account for when they are combined in a system.

Multi-perspective modelling techniques alleviate that issue by containing all relevant mechanical, thermal, electrical and environment models in one simulation instance [8]. This will allow to:

- Increase accuracy of representation of real environmental and operational conditions;
- Provides pre-prototype validation;
- Models consolidation to decrease overall modelling time and personnel and increase accuracy;
- Encourages solidification of system requirements and environmental conditions and;
- Enables means of early stage validation, which will allow preliminary calculation errors quicker to catch and fix.

10.3 Virtual Validation Based Methodology

Virtual validation is used as primary design verification method. To determine a set of values, methods and goals for any design project, a design methodology where approach, methods and outcomes will be defined will be considered. This methodology will ensure all of its state-by-stage plans revolve around ensuring the needs of the virtual models are met. This methodology will provide enough evidences that the preliminary system models are in fact part of the conceptual design, allowing for early stage testing.

10.4 Interface Design Principles

Using a combination of the design properties of existing IF designs & literature in the area of IF design needs, it can be established ideal traits that SIROM will develop through key design principles.

Particle mitigation: When designing for both planetary & orbital, IFs can often be populated with sparse amounts of particles or "space dust". Dust tolerance will be an important part of IF design as dust particulates can affect the components performance as well as the success of a connection. Dust can be mitigated in the electronic/data connectors by using a permeable membrane that's penetrable by pins, or spring loading the contacts can help create other contact points. The latching IF requires a more sophisticated solution, such as dimensional tolerance integrated into the geometry.

Rotational symmetry: Increasing the number of acceptable connection orientations will improve the connectivity of the IF overall. Space is not an environment that has a planar constraint, space provides the ability for objects to rotate freely. Having this ability means that orientation of a non-compliant craft can vary wildly relative to the chaser and by having a variety of compatible connection options the odds of a successful coupling is increased. In principle, the more acceptable orientations, the better; a majority of the coupling interfaces surveyed have the capacity to accept at least two orientations (180° symmetry). Introducing rotational symmetry into the design increases the total number of elements used but also increases the internal redundancy of the IF.

Functional element redundancy: Redundancy on a spacecraft is important because maintenance costs are high, especially in a system that is supposed to be able to repair other non-functioning craft. By introducing multiples of each of the function elements, the likelihood of total failure can be significantly reduced. For example, if rotational symmetry of 90° was implemented into a connection IF, elements that are incapable of being centred (contact pins, IR transmit/receive) would be duplicated.

Inclusive design: Inclusion of the 4 IF elements is a paramount to design success. Care to be taken when compromising and optimising each element, especially if requirements contradict each other.

Rigid and mechanical latching: Rigid connections, when handling large payloads and spacecraft, is important as allows for accurate manipulation of the object and helps keep a stronger connection. For manipulator-tool interfaces, this is extremely important as required tool accuracy for OOS may be sub-millimetre.

6-DoF misalignment tolerance: Misalignments are common occurrence in interfacing two interfaces and they can cause an increase in time spent connecting two IFs. By allowing a larger error for alignment for a well-designed interface, the complexity & accuracy of pace manipulators can be reduced, and thus the cost. By ensuring that the two locking IFs have not just translational misalignment tolerance (2 faces are parallel but unaligned) but 6 DoF misalignment tolerances, the probability for a chaser to capture a spacecraft would increase significantly. This is important for orbital depot that may have to repair a satellite that has lost power and is rogue, and since it cannot align itself or control its spin, the manipulator may need connection at potentially random orientation.

Hermaphroditic: Using a manipulator-tool or manipulator-spacecraft means that a male–female connection system is feasible; however there are benefits that a hermaphroditic IF with similar specifications would yield. Firstly, the manufacturing cost per IF would decrease as common component usage doubles which would incur a bulk discount on fabrication. Secondly, the design will be future proof; it will consider that there may be spacecraft elements in the future that could use a similar IF (e.g. manipulator-manipulator docking for a modular arm extension). The final benefit of a hermaphroditic system is the capacity for fail-safe design.

Failsafe docking/undocking: When a docking IF loses the capability to actuate its locking mechanism and the other side can still dock and undock, the design is considered failsafe. This is important in a maintenance manipulator. Additionally, should the manipulator connector fail, it can still dock or undock spacecraft in an emergency by using the other side to release. Failsafe is particularly important as a single side failure could leave a whole system non-functional until repaired, this makes repair easier and also maintains regular functionality.

10.5 Open Niches and Novelty Areas in SIROM

The following open areas are being taken into in the SIROM development:

Defunct APM manipulation and removal: In the event that an APM has failed, it would be imperative that there was some method of removing and replacing it, however if the IF APM side is inoperable then this becomes a challenge. To combat this, SIROM is considering an IF to be designed that can couple and decouple using one side's latch. Being able to connect to a defunct module opens up options for repair or removal; by also opening up the ability for the attaching IF to power the APMs other functions. This method means that in cases of drained power or a power IF malfunction on another side, the module can operate normally.

Total self-alignment: The free-lying nature of spacecraft means that oftentimes the orientation of a spacecraft cannot be guaranteed, or requires expensive correction. In the context of orbital servicing there could also be a rogue target craft with no ability to correct itself. SIROM is considering aligning of the interfaces by simplifying the IF approximation process so that in any angle of approach, the system will self-align to the required angle. By interfering with the craft using geometric thresholds; surfaces that will correct the IF alignment as it draws close, it can be devised a system that does not discriminate its latching effectiveness due to the initial contact angle.

"No-push" mating system: It is expensive to have the target craft push into the IF to activate the latch. SIROM aim is to solve this with inclusion of an independent manipulator grappling the target. By creating an IF geometry that either pulls the target towards the chaser, or latches with torque without pushing the target away, it can be effectively captured free floating craft with little risk of it accelerating away.

This is also vital in the scenario of a defunct, rogue craft as it may not have any means of correcting itself if the latching process transfers translational force.

10.5.1 SIROM Concept

The SIROM interface is a standardized robotic interface consisting of mechanical, electrical data, thermal and control interfaces. It aims connecting payloads to payloads or payloads to structures, with the capability to transfer mechanical loads, data power and heat. With an androgynous design, it can match and couple another SIROM interface on one side. On the other side, it is fixed to an Active Payload Module (APM) [5]. APM is typically a container with integrated subsystem or payload components, which can be used to fulfil a certain task (e.g. special sensor equipment, battery power pack, enhanced computation power). The idea is being capable of constructing a modular and scalable system, based on the association of APMs and spacecraft structure, with mechanical, data, power and thermal coupling (Fig. 10.3).

APMs are integral part of the SIROM development program [5]. During this activity, they are conceptualized, designed and built with the main purpose of providing realistic operational environment for testing and verification of the SIROM interface. This process of test and verification is envisaged to be done through realistic orbital and robotic planetary exploration scenarios, in which APMs play a central

Fig. 10.3 General SIROM/APM architecture

role. Two APMs for each track (two for Orbital and two for Planetary) will be developed for testing the SIROM IF.

The objective for the orbital scenario validation of SIROM interfaces is to demonstrate the successful transfer of an APM from an initial to a final operational location by means of a robot manipulator. The functioning of the interface of the APM with its payload and with the robot end-effector, as well as that of the end-effector with the robot manipulator will be validated. To do so, two different APMs have been developed:

- APM-1 is equipped with a camera and will be commanded to take a picture through SIROM IF.
- APM-2 is not equipped with any payload but it features a SIROM onto which APM-1 will be attached, providing a realistic operation. APM-2 is directly connected to the EGSE / OBC so the communication with APM-1 can be performed through the EE or the APM-2 as long as the APM-1 is connected to the APM-2.

The planetary exploration missions aim at demonstrating an application of battery pack management with a mobile rover and a charging station. In the context, two different APMs are developed, each equipped with SIROM interfaces:

- Primary APM (P-APM) is a transportable solar based battery charging system, equipped with optical sensor/camera for situational awareness.
- Auxiliary APM (A-APM) is a portable (by robotic manipulator) battery pack allowing providing power to rover or to other APMs.

To support the scenarios and the manipulation of APMs with a robotic manipulator, an end-effector is also designed. It consists of an adapter allowing to install the SIROM I/F on the manipulator arm, providing all standardized interfaces to the manipulator to interact with APMs, as well as additional features like force/torque sensors, camera and mechanical misalignment elements.

Finally, an EGSE is also implemented aiming at providing power and data communication interfaces to support integration and testing of SIROM in the different demonstrations.

10.5.2 SIROM Mechanical Connection Description

The SIROM primary objective is the realization of a standard robotic IF consisting of: a mechanical IF, an electrical IF, a data IF, a thermal IF, and an IF controller. All of them will be contained in the SIROM mechanical housing, except the IF controller (and its associated harness) which will be placed inside the APM housing. Following Fig. 10.4 shows the main parts of the SIROM design. A detailed description is depicted in the following paragraphs.

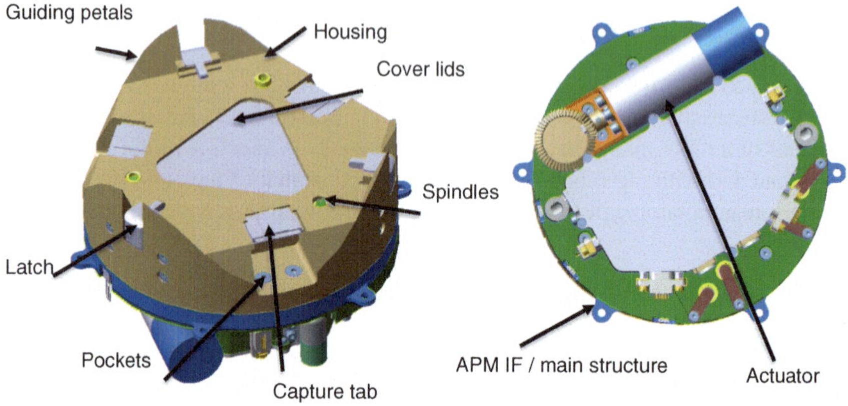

Fig. 10.4 SIROM Main parts

The coordinate system selected for the SIROM envelope is shown in Fig. 10.5. The coordinate centre is on the centre point of the SIROM diameter of SIROM Bottom Plate.

- Axis X goes from the coordinate centre towards one latch, perpendicularly to the androgyny plane of SIROM Connectors Plate.
- Axis Y is the androgyny axis of SIROM Connectors Plate.
- Axis Z goes from SIROM Bottom Plate towards SIROM Contact Plate.

It must be mentioned that androgyny shown in Fig. 10.6 was promoted versus symmetry at SIROM individual level.

10.5.3 SIROM Electric Power Connection Description

Electrical IF (i.e. Power IF) is responsible of providing power to SIROM Controller and APM Controller and it is placed in the Connectors Plate. A power source shall be mounted in the Platform (satellite or rover) or APM, and cables will reach SIROM IF. As the APM may need up to 150 W, SIROM Electrical IF connectors must be compliant with those values. In addition, the Electrical IF will be protected against short circuits.

The Connectors Plate is populated with one male and one female connector achieving an androgynous Electrical IF and redundancy at the same time.

The design of the SIROM electrical interface is based on the custom requirements, which are hereafter described together with the proposed solutions:

- Transfer electrical power in both directions: The SIROM Electrical interface is a bi-directional electronics switch controlled by external command.

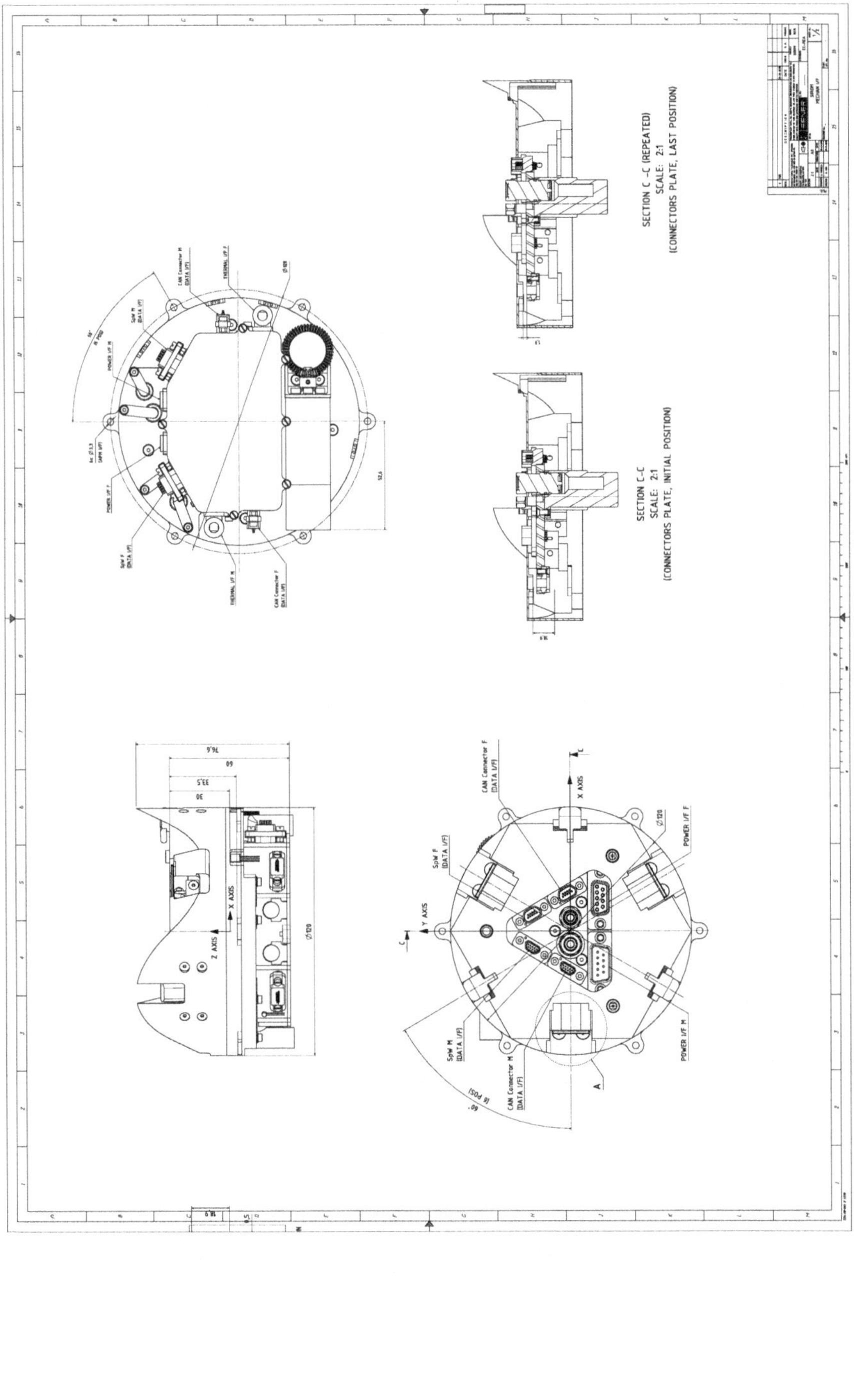

Fig. 10.5 SIROM coordinate system. *Left:* Front view. *Right:* Top view

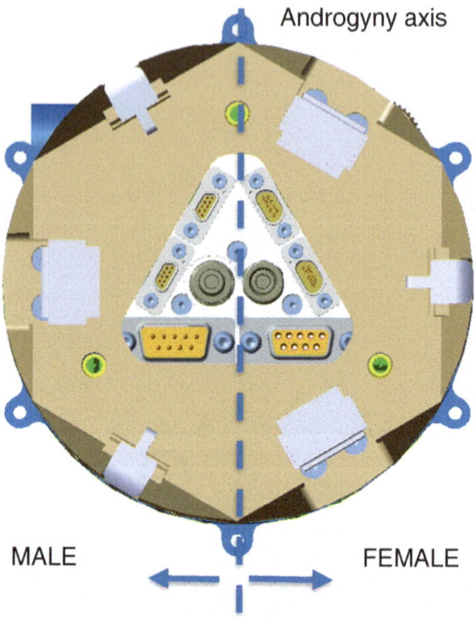

Fig. 10.6 SIROM Male and Female sides wrt Androgyny axis

- Provide on/off switching capabilities: The interface will be commanded by a single logical ON/OFF command at low level (i.e.: 5Vdc).
- Opened and closed 10,000 times: The minimum number of open and close operation shall be 10,000 times.
- No short circuits: The main interface are protected by internal current limiter (Imax) and under voltage (Vmin) circuits (over voltage protection is not implemented, but is highly recommended).
- No surge conditions are expected because no one internal capacitance and power supply are foreseen in the attached payload.
- Electro-magnetic compatible: No one filter is provided for electromagnetic compatibilities because no EMI generator is present in the electrical interface electronics.

10.5.4 SIROM Data Connection Description

The SIROM data interface will support SpaceWire (SpW) interfaces as high-bandwidth data transfer interface and CAN bus interfaces for data transfer and control of SIROM interfaces and APMs.

The SpW data interface will be composed of 2 links (nominal and redundant) supporting full duplex operation and link rates of up to 100Mbps. The SpW data interface and the electrical link characteristics will comply with the European SpaceWire standard ECSS-E-ST-50-12C.

The CAN bus data interface will be composed of 2 CAN buses (nominal and redundant) supporting baud rates of up to 1Mbps. The CAN bus data interface will comply with the European standard ECSS-E-ST-50-15C. The CAN bus interfaces and CANopen protocol are deterministic and provide hard real time guarantees where the delivery time of a message can be determined a priori, at network design phase, using analysis tools.

10.5.5 SIROM Thermal Connection Description

Thermal IF in Fig. 10.7 is responsible of transferring heat flux from APM to APM and vice versa. The system consists on a fluid port able to connect to fluid loops between SIROM A and SIROM B (connectors placed in the Connectors Plate) and 2 heating/cooling elements (one in each APM).

A SIROM thermal IF is mainly composed of:

- Two fluid quick connectors (one male and one female) mounted on the SIROM retractable plate.
- Two flexible lines that allow the required flexibility in the mating and de-mating process of the SIROM.
- A user interface fixed to a SIROM harness.
- Two thermistors (redundant) in order to measure temperature before connection.

The designed SIROM thermal IF will allow multiple possibilities for future implementations. As an example, two concepts are explored:

Close-Loop Fluid Heat Exchanger Module (CL-FHEM) for SIROM: The objective is to provide a maximum heat exchange of 2.5 kW between two APM.

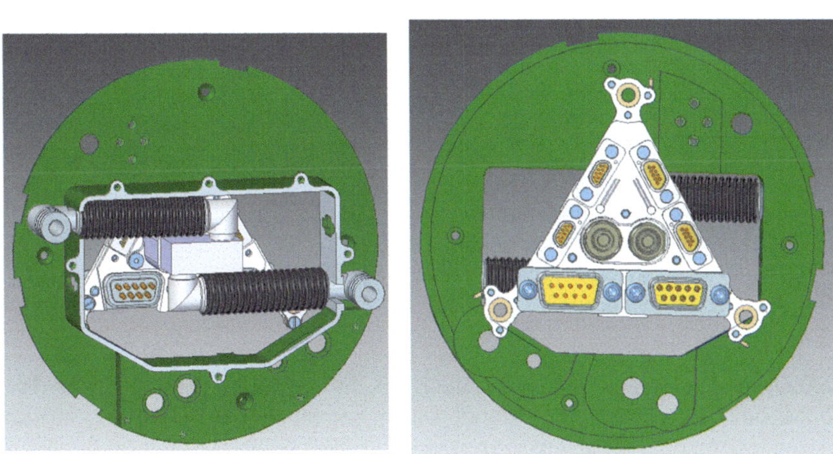

Fig. 10.7 SIROM thermal IF views

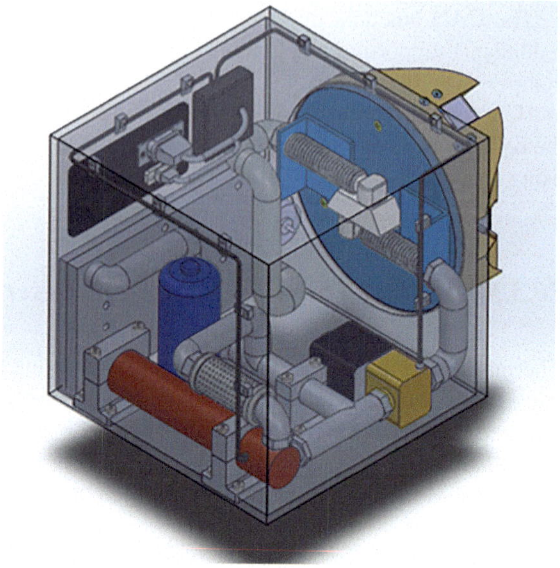

Fig. 10.8 CL-FHEM concept

A modular system including a fluid pumping system between to SIROM IFs and internal conductive thermal IFs to hot and cold sources are provided. With this system, heat can be exchanged up to 2.5 kW between two APMs. Figure 10.8 shows a representation of the concept.

Re-fuelling module concept: A satellite including SIROM thermal IFs can be equipped with a re-fuelling system and a fuel tank to extend operational life of future space missions and in-orbit satellites.

10.5.6 SIROM Controller and Overall Integration

The SIROM I/F controller is designed for orbital and planetary environments. High-level user needs of the controller can be listed as:

- To control the actuators and the switches,
- To monitor the sensors and the switches,
- To execute a control algorithm,
- To execute commands and
- To send telemetry.

SIROM controller is located inside the APM, directly bolted to the structure. The SIROM interface controller consists of 4 main components shown in Fig. 10.9 to achieve these functions, namely:

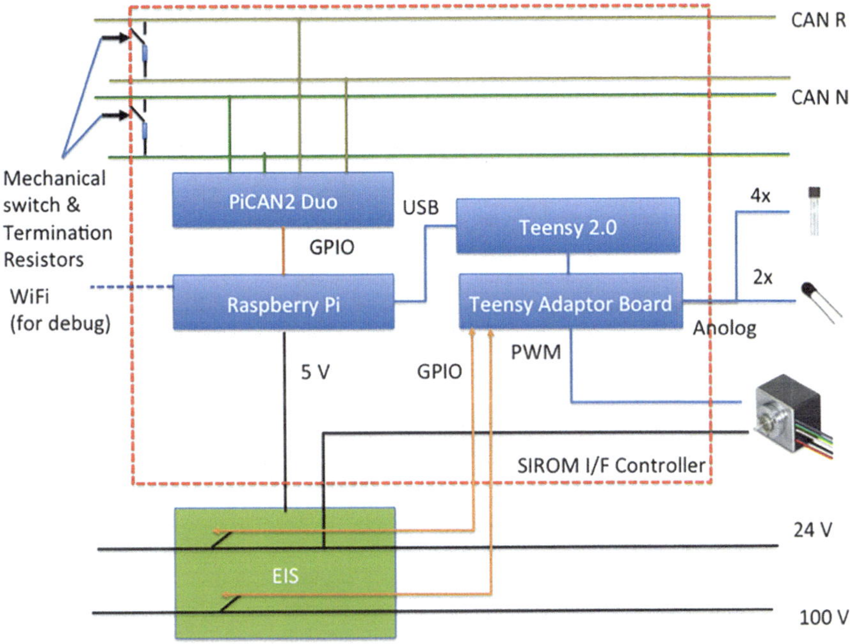

Fig. 10.9 SIROM I/F Controller Architecture

- Raspberry Pi Zero is to control overall operation of the SIROM interface (communication, control and monitoring).
- Teensy 2.0, provides additional I/O interfaces (including analog I/O) and controls I/O operations such as reading the sensors and latching switch status, and controlling the motor.
- Teensy 2.0 adaptor board is a simple board consisting of passive components and connectors for interface and conditioning of the actuator/sensors.
- PiCAN2 Duo is to communicate via CAN bus.

In addition to those components, there are two termination resistors in the system for CAN bus communication. These resistors are connected to the high/low pins of the bus via a mechanical switch. The switch should be off to disable the resistor when an APM is connected to another one. So that, the termination is moved to the next APM. This is shown in Fig. 10.10. The EGSE is one end point of the CAN network with a permanent termination but the other end termination changes depending on the connected APMs. When nothing is connected, the EGSE and the EE forms the basic network with 2 nodes, each terminated (Fig. 10.11).

When two sub-networks are disconnected, all the nodes can be communicated in their sub-networks as their CAN networks are terminated properly. When they are connected together, a mechanical switch disables the resistors of the SIROM's in contact.

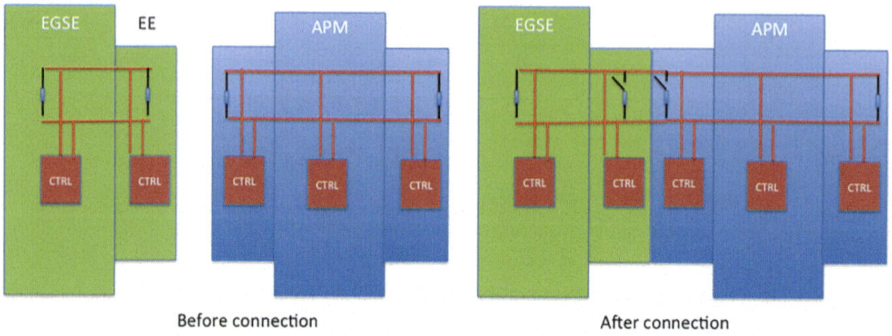

Fig. 10.10 Termination resistors

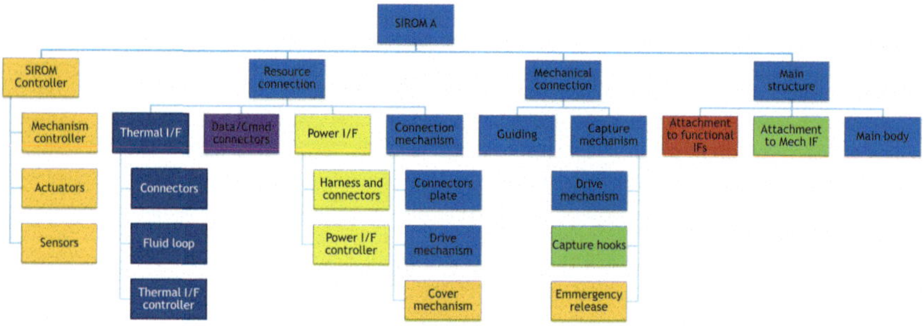

Fig. 10.11 SIROM Modular IF Product tree

10.6 Next Steps and Conclusions

10.6.1 SIROM Integration and Testing

The SIROM equipment will be manufactured, assembled and integrated according to the established PA standards for this kind of equipment of the company which delivers the ready assembled test item. Light weight components are designed and techniques have been considered in order to optimize the weight of the final assemblies [9]. New manufacturing technologies such as additive manufacturing facility has been deployed to make some critical yet difficult to make components.

In general, the equipment performance and compatibility will be verified within the test campaign performed in the OG6 test-field.

There are two reasons to add additional tests on different levels of the equipment integration:

There are requirements which cannot be verified e.g. thermal requirements within the OG6 test-field or it is advantageous to verify them on subsystem level e.g. data rates.

The integrated test set-up of the OG 6 test campaigns is very complex. To assure the performance of SIROM developments there are additional tests foreseen (SIROM orbital & planetary tests) which relies on SIROM equipment only.

This leads to a five level staggered OG5 test campaign:

1. Equipment/subassembly tests performed by equipment responsible
2. SIROM orbital test (orbital test set –up with SIROM equipment only – OG6 not involved) for mechanical verification & SIROM planetary test (SIROM equipment only – OG6 not involved) for data & power verification
3. Orbital test within OG6 test field
4. Planetary test within OG6 test field
5. Thermal tests of thermal IF

10.7 Closing Remarks

The SIROM project will apply a novel approach and consider a number of unique design requirements:

IF standardization and modularization of the different components in an integrated form (where mechanical, thermal, electrical, data connections are combined) or a separated form.

To allow creation of large clusters of modules based on the Standard IF. APMs are considered for demonstration, validation and verification of all properties of the standard IF. APMs will be connected via standard IF to other modules and satellite bus.

An end-effector for a robotic manipulator will be designed according to the layout of the standard IF and coupling functions to be supported.

The Modular IF will take into account long duration missions, no logistics support and missions composed of multiple payloads and architectures.

Expected Impact: Interfaces for modular, maintainable compatible spacecraft (or spacecraft with exchangeable active payload modules) must meet certain criteria in order to enable the exchange of individual modules. The individual modules shall be useable in any combination so the standard IF is required to have an androgynous design. The standard IF design shall have some basic characteristics in order to use it for different applications in space:

- Improve operational capacity
- Reduced logistics with common and modular spares
- Common maintenance standards
- IF architecture flexibility:
  - scalability with low complexity, mass and volume
  - common infrastructure needed to support the modular design
  - internal redundancy
  - compatibility to robot servicing
  - rotation of axis and symmetry

- connection of nearly arbitrary modules without restriction on the relative module orientation
- Mission flexibility (configuration changes)
- Standardizes mechanical, data, electrical, thermal Interfaces
- Keep existing standards where applicable
- Introduce in the design aspects related to interchangeability and interoperability
- Open standard architecture with flexible platform configuration
- Modular architecture consisting of 'block sets'
- Cost efficient development logic and a consequent customer-oriented design to cost approach.

Following the critical design review (CDR), potential future designs/upgrades/amendments of the SIROM interface are envisaged to include the following considerations:

- The use of space qualified components and materials;
- SIROM aiding: Collision avoidance system and Aiding for proximity operations;
- Controller components and mapping to a space qualified version;
- Electrical IF upgrades;
- SpW router;
- Ethernet-based instead of SpW and
- Emergency release system.

SIROM concept has continuity in H2020 SRC in Space Robotic Technologies-Call2 2018 where now the target is to define and develop the proper applications where SIROM interfaces are going to be deployed wither in future space missions or ground based evaluations. The specific objective of this 2018 call is to integrate the previously prepared common building blocks into demonstrators, on ground, towards applications of space robotics in the field of orbital and planetary use (phase 0/A studies). These robotics applications address not only the future needs of exploration and exploitation of space but also potential spin-off and spill-over effects to other areas of robotic activity on Earth, such as agricultural, automotive, mining, nuclear, or underwater. The following provides a summary of these planned research and development for targeted applications demonstrating and using common building blocks, in which SIROM interfaces will pay a critical role:

- Orbital Support Service (OG7): to demonstrate the techniques needed to offer a commercial service to operational satellites. This shall as minimum address robotic deployment and refuelling of satellites in orbit. By means of a general purpose robotic arm, a servicing satellite must be capable to demonstrate release, grasping, berthing and manipulation of a target satellite including services such as refuelling.
- Robotised Assembly Of Large Modular Orbital Structures (OG8): to demonstrate the techniques needed to robotically assemble a large structure on-orbit, comprised of tessellated, hexagonal tiles, which would otherwise not be feasible with a single launch.

- Robotised Reconfiguration Of Satellites (OG9): to demonstrate the techniques needed to realise a satellite-mounted robot system and its related implements that can modify the functionality of a satellite by adding/replacing modules available on-board or provided by another servicing satellite.
- Autonomous Decision Making (OG10) in very long traverses: to demonstrate the techniques needed for realising a planetary rover system with very long traverse capabilities (kilometres a day) by independently taking the decisions required to progress, reduce risks and seize opportunities.
- Exploring Robot-Robot Interaction (OG11) in planetary exploration and exploitation, in two alternative planetary applications: Exploration of difficult sites and Robotised construction.

Acknowledgements The authors would like to acknowledge and express their thanks to all the supporting staff and partners of the SIROM Project: Sener, Airbus DS Ltd., Airbus DS Gmbh, Thales Alenia Space, Leonardo, Strathclyde University, DFKI Gmbh, Teletel, Space Application Services and Mag Soar SL, and the financial support for SIROM as part of the project "PERASPERA" on Space Robotics Technologies from the European Union's Horizon 2020 research and innovation programme under grant agreement No 730035.

References

1. Palazzetti, R., Donaldson, K., Wenzel, W., Bartsch, S., &.Yan, X-T. (2017). Toward a multi-functional interface for future planetary and orbital missions. In *IAC-17.C2.2.7, International Astronautical Congress (IAC)*, 25–29 September, 2017, Australia.
2. Brinkmann, W., Roehr, T. M., Natarjan, S., Cordes, F., Sonsalla, R. U., Szczuka, R., Bartsch, S., & Kirchner, F. (2018). Design and Evaluation of an End-Effector for a Reconfigurable Multi-Robot System for Future Planetary Missions. In *Proceedings of the 2018 IEEE Aerospace Conference.*
3. Yan, X.-T., Brinkmann, W., Palazzetti, R., Melville, C., Li, Y., Bartsch, S., & Kirchner, F. (2018). Integrated mechanical, thermal, data and power transfer interfaces for future space robotics. *Frontiers in Robotics and A, 1*, 5.
4. Wenzel, W., Palazzetti, R., Yan, X-T., & Bartsch, S. (2017). Mechanical, thermal, data and power transfer types for robotic space interfaces for orbital and planetary missions—A technical review. In *14'th ESA Symposium on Advanced Space Technologies in Robotics and Automation (ESA ASTRA 2017)* - Leiden, June 2017, Netherlands.
5. Jankovic M., Wenzel W., Palazzetti R., Bartsch S., & Yan X-T. (2018). Concepts of active payload modules and end-effectors suitable for Standard Interface for Robotic Manipulation of Payloads in Future Space Missions (SIROM) interface. *In 2018 IEEE Aerospace Conference (pp. 1–15).*
6. Vinals, J., Urgoiti, E., Guerra, G., Valiente, I., Esnoz-Larraya, J., Ilzkovitz, M., Cetin, D., Letier, P., Yan, X-T., Henry, G., Quaranta, A., Brinkmann, W., Jankovic, M., Bartsch, S., Kollias, V., Pogkas, N., Fumagalli, A., & Doermer, M. (2018). Systems and Infrastructures to Implement Future Building Blocks in Space Exploration and Development—future space missions with reconfigurable modular payload modules and standard interface—an overview of the SIROM project. *The 69th International Astronautical Congress (IAC 2018), International Astronautical Federation (IAF), 2018.*
7. Roehr, T. M., Cordes, F., & Kirchner, F. (2014). Reconfigurable integrated multirobot exploration system (RIMRES): Heterogeneous Modular Reconfigurable Robots for Space Exploration. *Journal of Field Robotics, 31*(1), 3–34.

8. Vinals, J., Urgoiti, E., Guerra, G., Valiente, I., Esnoz-Larraya, J., Ilzkovitz, M., Franćeski, M., Letier, P., Yan, X-T., Henry, G., Quaranta, A., Cetin, D., Brinkmann, W., Jankovic, M., Bartsch, S., Fumagalli, A., & Doermer, M. (2018). Systems and Infrastructures to Implement Future Building Blocks in Space Exploration and Development - multi-functional interface for payload interconnection of robotic systems in space. In *The 69th International Astronautical Congress (IAC 2018), International Astronautical Federation (IAF), 2018.*
9. McMaster, T., & Yan, X-T (2017). A methodology for design of lightweight parts in harsh environments. In *IAC-17.C2.9.5, International Astronautical Congress (IAC), 25–29 September, 2017, Australia.*

Chapter 11
Coupled Path and Motion Planning for a Rover-Manipulator System

G. J. Paz-Delgado, J. R. Sánchez-Ibáñez, P. Romeo-Manrique, C. J. Pérez-del-Pulgar, and Martin Azkarate

Abstract This paper introduces a motion planning strategy aimed at the coordination of a rover and manipulator. The main purpose is to fetch samples of scientific interest that could be placed on difficult locations, requiring to maximize the workspace of the combined system. In order to validate this strategy, a simulation environment has been built, based on the VORTEX Studio platform. A virtual model of the ExoTeR rover prototype, owned by the European Space Agency, has been used together with the same robot control software. Finally, we show in this paper the benefits of validating the proposed strategy on simulation, prior to its future use on the real experimental rover.

11.1 Introduction

Returning samples from other planets is on the spotlight of space agencies, with the objective of analysing them on Earth [1]. However, it entails a huge series of difficulties to be faced, such as the proper manipulation of the sample. This operation is non-trivial mainly due to the extreme conditions of extraterrestrial surfaces such as those found on Mars [2, 3]. For this reason, unmanned vehicles are suitable to carry out this task, particularly planetary exploration rovers. This kind of systems have successfully performed exploration tasks in previous missions, and hence they are considered as well to be used for sampling tasks together with the use of manipulators. Moreover, since future missions demand higher levels of efficiency, the use of combined path and motion planning algorithms arises as relevant to increase the autonomy of the system and, thus, its efficiency.

G. J. Paz-Delgado (✉) · J. R. Sánchez-Ibáñez · P. Romeo-Manrique · C. J. Pérez-del-Pulgar
University of Málaga, 29070 Andalucía Tech, Spain

C. J. Pérez-del-Pulgar
e-mail: carlosperez@uma.es

M. Azkarate
European Space Agency, ESTEC, Noordwijk, Netherlands
e-mail: martin.azkarate@esa.int

X. T. Yan and G. Visentin (eds.), *Space Robotics*,
https://doi.org/10.1007/978-3-031-39214-6_11

This paper presents the functioning of an algorithm that combines path and motion planning within a simulation environment. This algorithm is aimed at mobile manipulators that must retrieve samples from difficult terrains in planetary surfaces like Mars. Therefore, Sect. 11.1.1 depicts different architectures that have been used to increase the autonomous navigation capabilities in planetary exploration vehicles, and the state of art of motion planning for mobile manipulators is analysed in Sect. 11.1.2. To build up the simulation environment, virtual models of the rover and an uneven terrain are created and detailed in Sect. 11.2. The particular control architecture used and its connection to the simulation platform are introduced in Sect. 11.3. Later on, Sect. 11.4 contains a description about the functioning of the motion planning algorithm. The simulation cases where this algorithm is tested are described in Sect. 11.5. Finally, Sect. 11.6 concludes the paper, illustrating related future research work.

11.1.1 Autonomous Navigation Capabilities on Planetary Exploration

The autonomy software installed on the rovers that have been sent lately to Mars has been progressively improved in the last decades. This is done to compensate the difficulties in the communications between this planet and Earth. Figure 11.1 shows the real scale replicas of some of these rovers. In 1997, the NASA sent to Mars the first rover capable of moving with certain autonomy: Sojourner [4]. This relatively small robot was capable of autonomously performing simple operations like going to a certain waypoint [5]. It achieved a total distance of around 100 m, serving as a proof of concept for sending more robotic mobile explorers to Mars. The immediate next mission was the Mars Exploration Rovers (MERs), in which two twin rovers were sent to different spots of the red planet. One of them was Spirit [6] and the other one was named Opportunity [7]. For this occasion, NASA had augmented the autonomous capabilities of both robots, providing them with a local path planner called GESTALT [8]. During their mission, it was figured out that a local planner was not enough for achieving complete autonomous navigation: the rover could get stuck in different situations involving encounters with rocks. Because of this, a global path planner was also included in the autonomy software. In this way, a long-traverse could be initially planned while the local planner would dynamically produce safe paths avoiding obstacles [9]. This autonomy system was kept for the Mars Science Laboratory (MSL) Curiosity rover in 2011 [3]. Nevertheless, manual commanding from the ground operators is still the most common strategy used to navigate on Mars. Usually, for each Martian day (or *sol*) a series of tasks is previously planned and later sent to these vehicles.

Future missions include Mars Sample Return (MSR) [10]. It consists of taking several tubes with samples inside from the Martian surface and return them to Earth. These tubes are expected to be previously left by the Perseverance NASA rover,

Fig. 11.1 Real scale mock-ups of rovers sent by nasa to Mars. From smaller to bigger size: Sojourner (left below), Opportunity (left above) and Curiosity (right). Credit: NASA / JPL-Caltech

which arrived at Mars in February 2021. Later on, another rover that will be built by ESA, the Sample Fetch Rover (SFR, see Fig. 11.2), will be tasked to retrieve them. For this kind of missions, efficiency is denoted by the number of samples that can be successfully retrieved within a time window, so higher capabilities in both mobility and autonomy are desirable. This implies autonomous mobile manipulation capabilities, which is the problem aimed to be solved in this paper. As a side note, a better understanding of the levels of autonomy of a rover can be achieved thanks to Table 11.1. It contains a division into four levels proposed by the European Cooperation for Space Standardization (ECSS). Reaching the last one, E4, will probably be crucial for complying with the expectations imposed on the SFR, as well as any other future rover with increasingly complex tasks.

11.1.2 Motion Planning for Mobile Manipulators

Planning the motion of a mobile manipulator (mobile base equipped with a robotic arm) is a well-known issue in the robotics literature. The problem could be defined as finding a feasible path of the mobile base and configurations of the manipulator in order to place the end effector in certain poses, to perform a desired task. This entails two main challenges. First, most mobile manipulators are redundant, with 3 DoF for the mobile base and several more for the manipulator, thus there are infinite possible solutions for the problem. Second, mobile bases widely present non-holonomic constraints, which prevent the base to move in certain directions in function of its yaw, hindering the planning process. Additionally, there are several aspects that should always be considered when planning the motion of a robot,

Fig. 11.2 The Sample Fetch Rover (SFR). Credit: ESA

Table 11.1 ECSS levels of autonomy

Level	Description
E1	Direct commanding from ground operators
E2	Execution of pre-planned operations
E3	Reactivity, the vehicle adapts to external events
E4	Goal-oriented, fully autonomous navigation

like collision avoidance, energy consumption, time spent, computational cost, joints limits, task fulfilment precision or uncertainty.

The straightforward solution to tackle the motion planning problem for a mobile manipulator is to decouple the mobile base and the manipulator plans, moving the manipulator exclusively when the base is close to the goal, or when the arm could collide with the environment [11]. Nevertheless, this approach is missing the great advantages that a coupled motion provides, allowing the mobile manipulator to perform more complex tasks, save time and reduce the energy consumption. Thus, most researchers choose dynamic solutions that move both the base and the arm in a coupled way. Three main dynamic approaches are distinguished in the literature, which are defined in the following.

First, the Motion Planning methods based on the mobile Base Path (MP-BP). These methods are characterized by starting planning the bidimensional mobile base path, and afterwards generating the arm motion in the base path vicinity. Regarding this approach, interesting results are presented in [12] with the Optimized Hierarchical Mobile Manipulator Planner (OHMP). It computes a first mobile base bidimensional path using Probabilistic RoadMaps (PRM), and then safe manipulator configurations are found using PRM again in a 2D projection of the 3D space. This method is highly efficient to plan collision-free motions in complex static environments. Conversely, MP-BP methods require to specifically compute the base pose to perform the manipulation task. Hereby, a fast manipulation base pose estimator is proposed in [13], which includes the whole kinematic chain, arm and mobile base, in the inverse kinematics model (IK) of the robot. This way, the computational effort to find a suitable base pose is reduced, since it is found in just a single IK computation. As criteria to choose the best pose, the distance of the joints respect to their limits is considered.

Second, the Motion Planning methods based on the manipulator End Effector Path (MP-EEP). This algorithms are distinguished by starting planning the 3D end effector trajectory of the robotic arm, and then generating the whole robot body configurations to track it properly. A relevant example of MP-EEP is presented in [14], focusing on multi-manipulator systems and mobile manipulators. The motion planner approximates a feasible first solution, uses it as initialization for an optimization algorithm, and then refines the solution with different techniques. In this way, several simulations are carried out planning the motion of a mobile manipulator to open doors and drawers, among others. On the other hand, a fuzzy controller is designed and trained in [15], in order to plan the movements of a mobile manipulator tracking a given end effector path. This method focuses on maximizing the manipulation capabilities of the robotic arm during the thorough operation.

Third, the Motion Planning methods considering the Full Body jointly (MPFB). Although this methods demand bigger computational efforts, they are the most popular due to their versatility: they can be successfully applied to many different robotic fields. MP-FB is based on grouping the complete body of the robot, mobile base and manipulator, in a single model, and generate successive body configurations to perform the desired task. This is clearly stated in [16], where $BI2RRT*$ is presented and used to plan the motion of a mobile manipulator to transport liquids or pull carts. $BI2RRT*$ is an asymptotically optimal motion planner, an improved version of Rapidly-exploring Random Trees (RRT) which is a very popular algorithm mostly used in 2D path planning. Another MP-FB method is proposed in [17], where the priority is to avoid the manipulator singularities and respecting the mechanical limits, including nonholonomic constraints. To ensure collision avoidance, a reactive algorithm modifies the motion of the whole robot when it is moving in the vicinity of obstacles.

Furthermore, MP-FB is broadly achieved using optimization algorithms, which are characterized by finding the best solution considering different criteria, like the aforementioned time spent or energy consumption. For instance, a kinematic planner is presented in [18], which generates whole body optimal motions for greatly redundant mobile manipulators by solving a Constrained Sequential Linear Quadratic Optimal Control problem (Constrained SLQ). Thus, the motion planner tackles optimally the non-holonomic constraints issue of the mobile manipulator. A further effort on optimal motion planning is proposed in [19], where a Model Predictive Control (MPC) control loop is used on a ball-balancing mobile manipulator, which is inherently unstable. SLQ is integrated in the MPC scheme allowing the algorithm to successfully control the robot to track end effector poses or open doors. Finally, it is also fascinating to mention the use of artificial intelligence, specifically deep learning, to improve the performance of optimization based motion planning algorithms, as it is shown in [20]. Here, two different neural networks are used to initialize properly the optimization algorithm, improving noticeably its overall performance in terms of computational cost. Thus, optimal motions are planned for a manipulator spending less computational resources and time.

11.2 Simulation Environment

The proposed planning algorithm is meant to be implemented on a rover platform named ExoTeR (*ExoMars Testing Rover*) [21]. It is a prototype whose kinematic configuration is similar to the one of *ExoMars Rosalind Franklin* rover, that will be sent to Mars as part of the ESA-ExoMars campaign [22]. In this particular case, it is equipped with a manipulator as depicted in Fig. 11.3. Nevertheless, prior to making tests using the real platform, it is desirable to first refine the code via simulation. By using the same control software in both situations, simulated and real, the workflow is more agile and straightforward. Besides, the real platform is not compromised and there is no need for preparing a real setup, which results on saving resources and time. Hence two virtual models have been prepared to run the simulations: the first one is based on ExoTeR, including its manipulator on board, while the second one takes the form of a terrain that resembles the surface of another planet such as Mars.

11.2.1 Coupled Manipulator-Rover System

ExoTeR is a vehicle capable of performing several driving manoeuvres thanks to its kinematic configuration. It can even reconfigure its mode of locomotion thanks to its so called *walking joints* [21], although this feature is not under the scope of this paper. Only *driving* and *steering joints* are here relevant. They are used to make the rover advance and turn, making curved trajectories with variable radii. In case the radius is relatively small, the rover executes a *Spot Turn* manoeuvre, which consists on changing the orientation of the vehicle using zero linear speed, i.e. just rotating without translating. Otherwise, it drives via an *full-Ackermann steering* manoeuvre, which makes use of different values of speed on the wheels, as well as different steering angles according to the location of the centre of rotation.

With regard to the manipulator, it is attached to ExoTeR at its front and has five Degrees of Freedom (DoFs), resulting in a configuration Z-Y-Y-Y–Z. The first three of them are used to determine the position of the end effector, while the other two

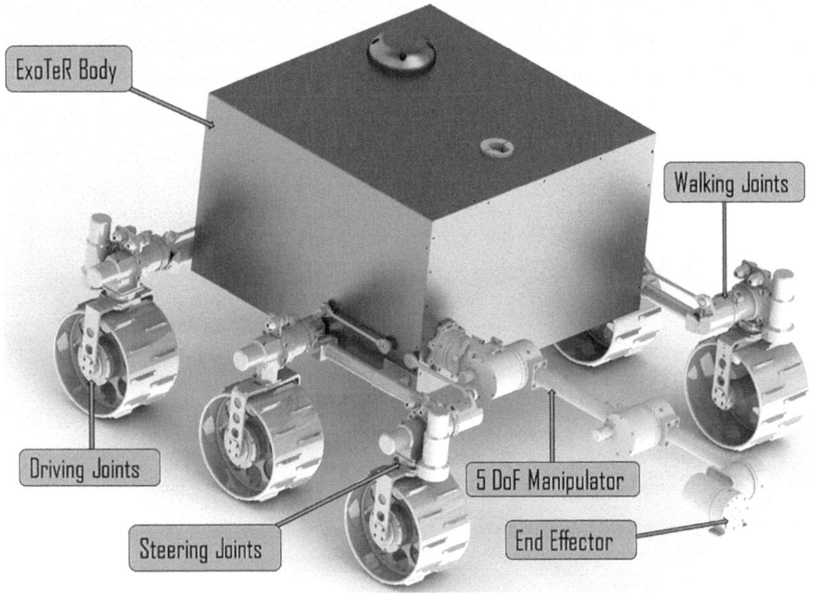

Fig. 11.3 3D model of ExoTeR carrying a 5 DoF manipulator

have more influence on its orientation. Although the number of DoFs introduces a restriction in the orientations the manipulator is capable to reach, it can be compensated by the mobility introduced by the rover locomotion subsystem. On the other hand, the manipulator contains huge gearboxes, which allow the arm to lift up to 2 *kg*, which is a high payload in comparison to the arm weight, 2.4 *kg* and the minuscule joint motors power (6*W*). However, this introduces a drawback in the form of slowness: the maximum speed of each manipulator joint is 0.01 rad/*s*. This fact needs to be acknowledged by the motion planner so as to prepare the manipulator beforehand whenever a sample must be collected.

11.2.2 Planetary Surface

The scene used for the simulation of planetary operations is based on a portion of the Search and Rescue experimental terrain located at the University of Málaga1 (Spain), which is depicted in Fig. 11.4. It is an uneven terrain suitable for carrying out planetary navigation tests, presenting diverse soils, hills, rocks and slopes. By means of photo gravimetry using the Pix4d software2, a virtual model of this terrain is built based on aerial figures that were taken thanks to a camera drone. Besides, a Digital Elevation Map was obtained as output of this software, having a resolution

Fig. 11.4 Overview of the experimental terrain 3d model

of 3 cm. This is considered to be high enough to determine the location of even small obstacles, as well as other morphological properties such as slopes.

11.3 Implementation

A schematic describing the elements that make up the simulation environment and the connections between them is shown in Fig. 11.5. The platform in charge of simulating the behaviour of the vehicle and visualizing the scene elements is VORTEX Studio3. This software has integrated a physics engine that lets model terrains making use of terramechanic parameters like static and dynamic friction, stiffness, damping or slippage [23]. Besides, VORTEX allows the communication with external software to control any virtual robotic systems present in a scene. The simulation environment we show in this paper makes use of this advantage in two ways. On one hand, information is obtained from the virtual scene describing the pose of the robot as well as the state of its joints, i.e. the real speed and torque they are experiencing. This is useful to later estimate the power consumption by the vehicle while driving. On the other hand, commands to each of the motors can be sent in the same way as with the real system, producing motion in each of the active joints. The control software used to implement all navigation layers of the rover is built within the Rock4 (*Robot Construction Kit*) framework. The same layers used for ExoTeR, from path planning to motor commanding, are here connected to VORTEX by means of a bridge programmed using UDP protocol. This bridge is formed by the Rock component that interfaces with the rest of the control software and the UDP module inside VORTEX studio. Once a simulation starts, VORTEX receives from Rock the position and

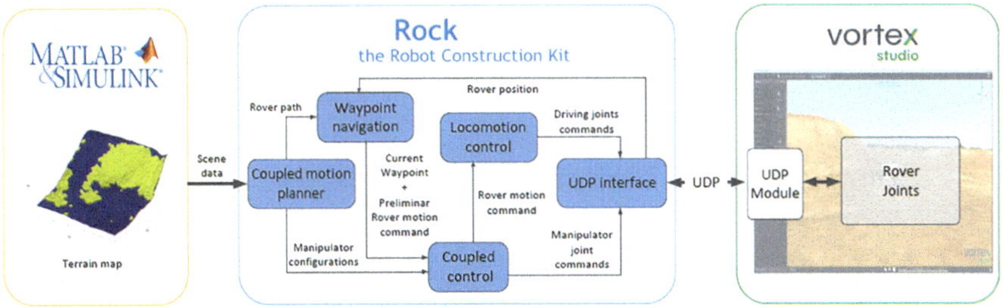

Fig. 11.5 Schematic of the proposed simulation environment

velocity commands corresponding to each active joint, and simulates the motion of the rover on the virtual terrain. It also sends back to Rock information about the position and orientation of the rover body and joints. In this way, Rock is exclusively in charge of the motion control in real time. Each of the Rock components where the control layers are implemented, indicated as well in the schematic of Fig. 11.5, are introduced as follows:

- *Coupled motion planner*. The algorithm proposed, and detailed in Sect. 11.4, is implemented in this component. It makes use of information describing the area surrounding the rover, as well as its initial position, the location of the sample to fetch, and indications of the existing obstacles, their placement and size. All this information is provided by MATLAB scripts that process the DEM information of the experimental terrain. As result, this component provides the path and the manipulator configuration that must be acknowledged by the navigation system to reach the sample.
- *Waypoint navigation*. This component computes translational and rotational speeds of the rover required to follow the path computed. This parameters are computed according to the position of the vehicle, provided by the VORTEX platform, and using the algorithms developed in [24]. For this particular case, the computed values are preliminary and still must be processed by the *Coupled Control* component.
- *Coupled Control*. The resulting configurations of the manipulator computed by the *Coupled motion planner* serve as reference for the manipulator joints to reach them. The maximum velocities and position limits are taken into account during this process. Besides, the speed commands from *Waypoint Navigation* for the driving joints may be altered in this part to coordinate the rover and manipulator motion.
- *Locomotion control*. According to the input rover motion commands, speed and position commands are created for each of the driving and steering joints respectively.
- *UDP interface*. With the commands for the driving joints and the manipulator joints received from *Locomotion control* and *Coupled control* respectively, this component packs the information in a suitable way for UDP communication protocol and sends it to VORTEX studio. This task also receives the current

position and orientation of the rover from VORTEX, and sends it back to the *Waypoint navigation* component.

11.4 Coupled Motion Planning

The proposed algorithm is a solution for the problem of a mobile manipulator that has to reach a sample. This must be done in a safe way, meaning the vehicle has to avoid any encountered obstacles in the form of rocks, dunes or craters. For this purpose, both manipulator and rover base should move in a coordinated way, ensuring none of them are harmed during the process. Assuming the rover is located a few meters from the sample, the objective is to deploy the manipulator and place its end effector in contact with the sample, while at the same time the base is approaching to it. Both base and manipulator coordination is a challenging task when there are redundant degrees of freedom (DoF). Mainly, there are three DoF for the motion of the base (two for the position and one for the orientation) and five more for the motion of the manipulator, so there would be infinite configurations to reach a sample. Considering the available solutions in the literature, as analysed above, the MP-BP methods appear to be the best ones, since they are robust and predictable, prioritizing an optimal safe trajectory for the mobile base before planning the arm motion, allowing to reach the goals even if they are located in hardly-accessible areas.

For that purpose, the first step is to generate a trajectory for the mobile base to follow. This is done by a path planner, which is based in the path planning algorithm developed in [25], which makes use of the Fast Marching Method (FMM) to compute smooth, optimal and continuous paths. For this particular case, it has been modified to ensure the rover is able to reach samples that may be located inside hardly accessible areas. Its functioning is explained in the following.

First, it is necessary to generate a Cost Map that represents the characteristics of the scenario, like obstacles, slopes or the soil properties. This Cost Map will be used by the planner to decide which is the best path to reach the goal. To generate it, a Digital Elevation Map (DEM) of the scenario is required, which is a 2D map that provides the height of the terrain at each spot. The DEM can be analysed to, first of all, differentiate non-accessible areas, like obstacles or big slopes, from the safe ones. To do so, a threshold can be established to limit a maximum allowed slope for the robot to traverse on. Additionally, the roughness of the terrain, which gives information about abrupt changes in the terrain height, can be used to locate obstacles like rocks or craters. Finally, the Obstacles Map is dilated in function of the rover size. This is done to ensure that the whole trajectory is thoroughly safe for the rover to follow, not allowing the rover to drive to close to obstacles. An example of an Obstacles Map is shown in Fig. 11.6.

The generated Obstacles Map altogether with the slope information are used to generate the final Cost Map of the scenario, as it is depicted in Fig. 11.7. The cost of each node represents how difficult and dangerous is for the rover to be placed there. As a result, obstacles areas have the highest costs, and safe areas have the lowest

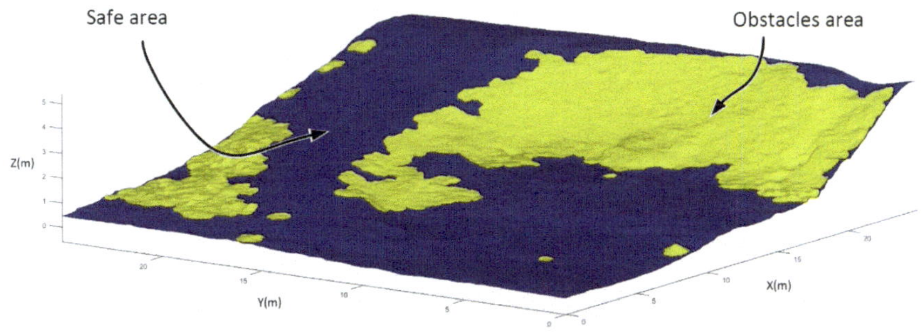

Fig. 11.6 Example of an Obstacles Map, differentiating safe areas from non-traversable ones

ones. Moreover, in order to improve the safety of the sample retrieval operation, repulsive potential fields are created around obstacles, as continuous cost transitions between the high obstacles cost to the low safe areas cost. This has later on an effect on the resulting path, in two ways: first, the computed path gets even further from obstacles. Second, in case the sample is at the border of an obstacle, the path will enter perpendicularly to this edge. This is translated into avoiding high turning maneuvers when the rover is close to the sample, improving the smoothness of the whole path.

This cost map is used by FMM to generate the trajectory, in two steps. First, a wave is expanded on the Cost Map from the goal until reaching the initial rover position. This wave moves faster where the costs are lower, thus, its gradient indicates the optimal direction to reach the goal. Therefore, the second step consists on extracting the path with the Gradient Descent Method, which consists on following the descent direction of the gradient from the initial position of the rover until reaching the goal. An example of a trajectory extracted on the previously shown Cost Map can

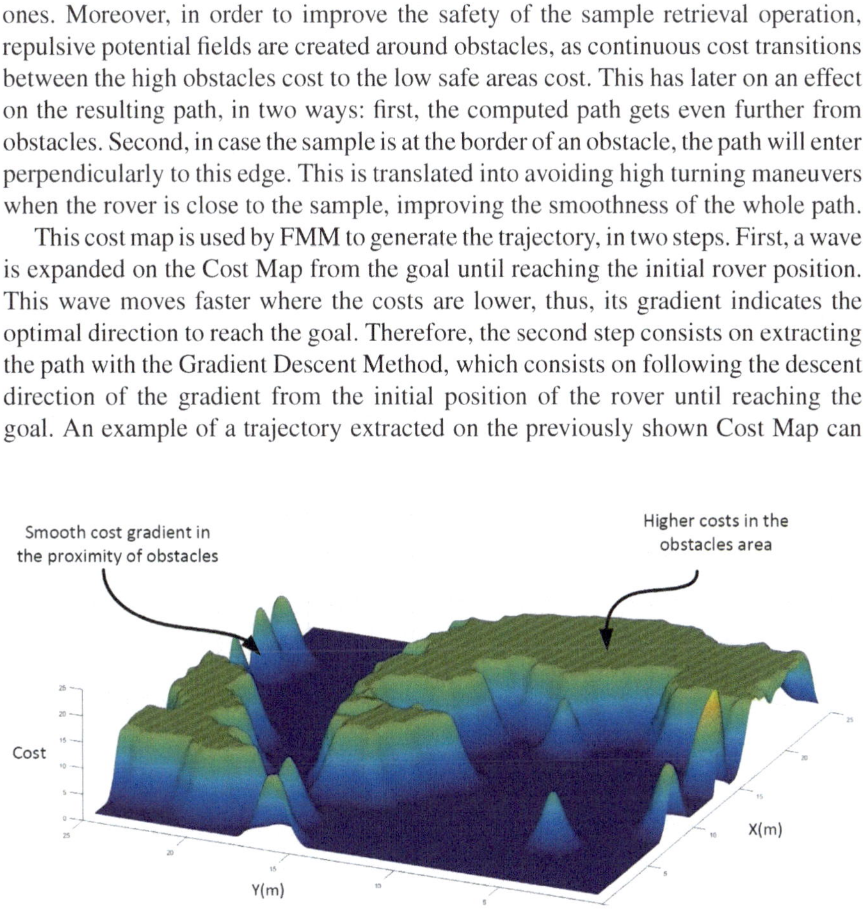

Fig. 11.7 Example of a Cost Map to be used by the path planner, where the higher costs correspond to non-traversable areas

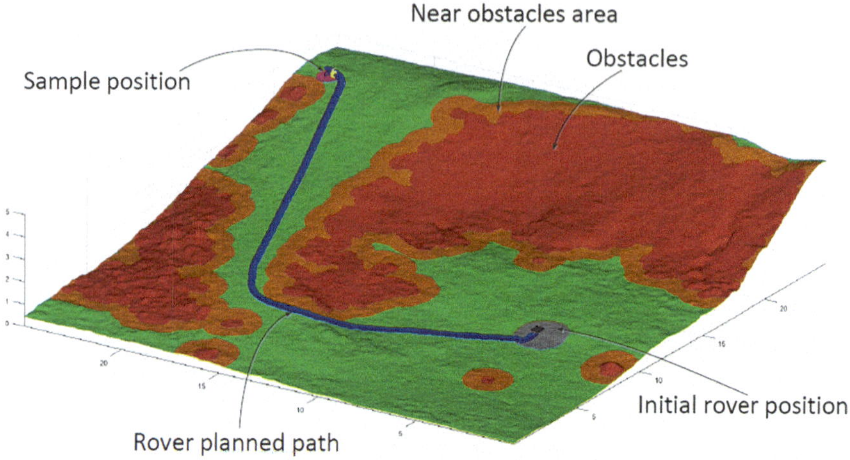

Fig. 11.8 Example of a planned path for the mobile base of the rover, allowing it to safely reach the sample position avoiding all the intermediate hazards

be observed in Fig. 11.8, where the rover path avoids the obstacles areas to reach smoothly the goal, where the sample is located.

Finally, this path is slightly shortened in the proximity of the sample, what is called *Sampling Area*, in order to place the rover in a favourable position for the later manipulation operation. This position is decided upon all feasible waypoints in the section of the path entering the *Sampling Area*, considering different criteria, mainly the arm manipulability and reachability. Moreover, sampling positions that are non-secure are cast aside. In the case all candidate sampling positions are unsafe, the sample is declared as non-reachable.

Once the path of the rover is planned, the second step is to plan the motion of the manipulator as depicted in Fig. 11.9. This process is subdivided into two stages: the early deployment of the manipulator and the planning of the end effector path. The first stage consists on moving the joints of the manipulator from its resting position to another in which the manipulator is deployed and ready to later take the sample.

During this stage it is ensured the rover will continue advancing without colliding with any surrounding obstacle.

The second stage is focused on planning the motion of the manipulator once the rover is a couple of meters away from the sample, moving the arm while the rover is following its respective path to ensure it is prepared to perform the desired task once the rover reaches the final waypoint. For that purpose, a volume surrounding the rover path is delimited in function of the characteristics of the arm, generating a tunnel-like shape that encloses all the possible positions where the arm can be placed at each waypoint of the rover trajectory, as shown in Fig. 11.9. Inside this tunnel, another path is created by means of a 3D version of the Fast Marching Method, which is the one that must be followed by the end effector of the manipulator while the rover is driving. Finally, in order to synchronize both systems, rover and manipulator, the

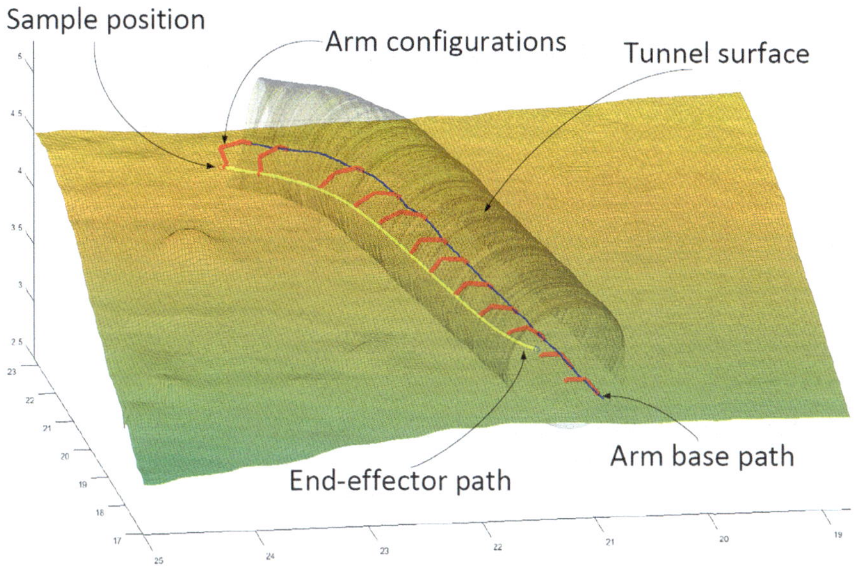

Fig. 11.9 Planned motion for the manipulator, whose movement is constrained by the already drawn rover path, since it generates a tunnel-like volume in its surroundings that encloses all the possible places for the arm to be placed

inverse kinematics model of the manipulator is used to obtain the corresponding arm joints position at each waypoint of the rover trajectory, ensuring both systems are following their corresponding paths in a coordinated manner. In case the manipulator cannot reach some of the planned configurations due to constraints, such as joints limitations or collisions with the rover itself, then the 3d planned path is slightly modified as consequence, resulting on new viable configurations. An example of the resulting manipulator configurations is also shown in Fig. 11.9.

11.5 Experiments

In order to validate the proposed coupled motion planning algorithm before testing it on the real platform, a series of simulations were carried out. Their main purpose was to analyse the performance of the algorithm under different circumstances, i.e. to check whether the coupled system could safely reach a series of samples located on diverse extreme cases. For this reason, five cases were set up as seen in Fig. 11.10. During each simulation execution, the rover always started from the same initial position, having as well the same orientation and being placed on a horizontal plain surface. Then, prior to starting any movement the motion planning algorithm was executed. It was conditioned by the obstacles defined on the map, which were determined under the basis of slope and roughness thresholds. After producing the paths

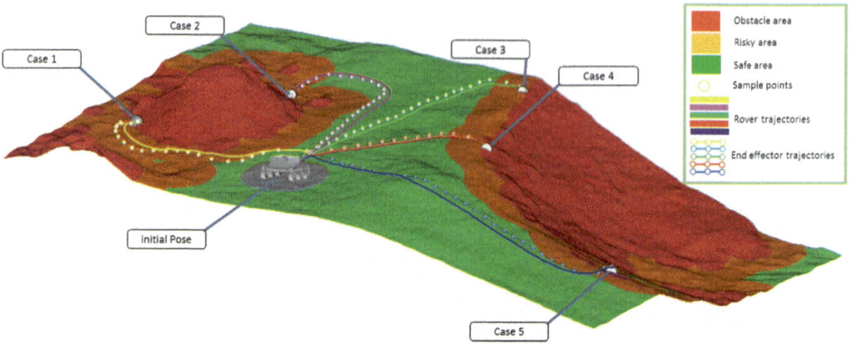

Fig. 11.10 Overview of all sampling cases used for the simulations, including the paths created for the rover base and the end effector

required to guide the rover to the sample, the rover immediately started driving. At the same time, the manipulator joints actuated according to the planning algorithm guidance to make the end effector keep track of the path assigned to it. In this way, the rover deployed the manipulator while it kept advancing towards the sample location, finishing all motion when the end effector was placed on top of the sample. Footage showing the first and fourth cases can be found on a *YouTube video* [26].

With regards to the simulation cases, in the first two of them (Case 1 and 2) the samples were located on the opposite sides of a small hill, as can be checked in Fig. 11.10. There were other elements near them in the form of rocks and slopes that, since they could harm either the rover or the manipulator, were considered as obstacles. In case 1, the rover traversed a narrow corridor placed on *Risky Area* (the area close to obstacles) before having to reorient itself and finally head to the sample. Its final manipulator configuration is depicted in Fig. 11.11(a). Unlike in that case, in the second one the rover was able to follow a more open path prior to reaching the sample, since it went through more *Safe Area*. Later on, the last three cases, from 3 to 5, consisted on taking samples located close to the edge of a cliff, where the rover had to be extremely careful in order not to fall down and collapse. Thus, the approach to the sample was done perpendicularly to this edge. This is observed in case 3: in the last stretch of the path, close to the slope, the rover took a sharp curve and headed to the sample, taking it with a final configuration shown in Fig. 11.11c. The fourth case is very similar to the previous one, except that the sample position was critically inside the slope, resulting in an extreme configuration of the manipulator depicted in Fig. 11.11d. Finally, the last case included a new difficulty as the rover was on a smooth slope while sampling, as shown in Fig. 11.11e.

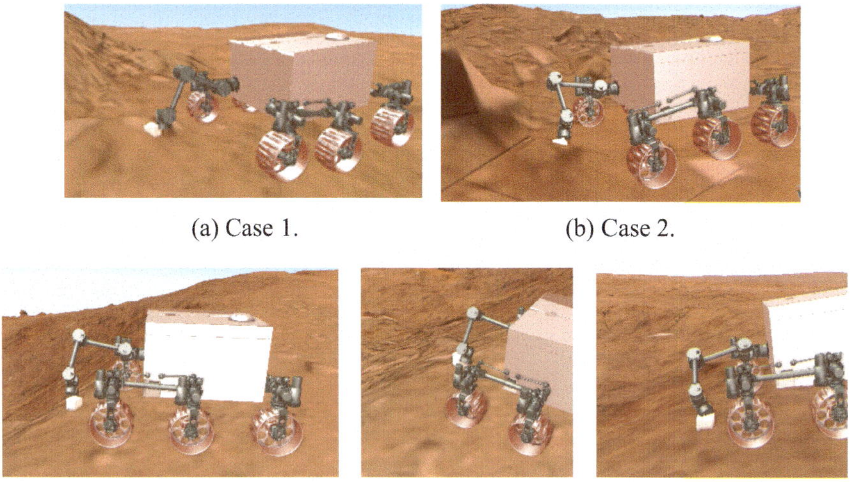

(a) Case 1. (b) Case 2.

(c) Case 3. (d) Case 4. (e) Case 5.

Fig. 11.11 Close-up screen-shots of the virtual rover taking a sample during the simulation cases

11.6 Conclusions

This paper presents the verification and validation of a motion planning algorithm aimed at retrieving samples in planetary exploration missions. This algorithm produces the paths and joint references that must be followed by a rover and a robotic manipulator attached to it. It is intended for future work to carry out tests using this algorithm on the real experimental platform ExoTeR. For this reason, a simulation environment has been built to test and debug such implementation. Within it, the algorithm is programmed using the same software that controls the ExoTeR rover. In this way, the same implementation refined during simulations can be later used in the real platform in a straightforward way. The simulation platform chosen to model the rover and the terrain, VORTEX studio, presents great potential not only for this particular application but also for testing future navigation software, thanks to its capability to handle big scenes and the physics engine it integrates.

With regard to the algorithm here tested on simulations, it has demonstrated to effectively guide the combined system to take samples even when these are placed on uneven terrains. This is translated into augmenting the number of samples that can be retrieved, i.e. to increase the scientific return of the mission. As stated before, is foreseen as future work to carry out experiments using this algorithm on the real rover-manipulator system, even emulating a complete autonomous SFR-like sample retrieval mission. Furthermore, this algorithm is also employed within the U-H2020 project *Autonomous DEcision making in very long traverses*[6] (ADE), serving as a planning module used by a rover to catch any sample of interest that may be found on its way. Additionally, it is expected this algorithm will be refined and improved in several senses. First, more terrain parameters may be considered, not just morphology but soil properties as well. Second, maximizing the performance of alternate locomotion modes to augment the reachability of the rover manipulator. Third, the self-collisions of the manipulator should be considered during the planning stage, in order to directly obtain thoroughly safe and optimal end effector trajectories. Finally, the whole autonomous navigation architecture, including the motion planner, should be tested in a space qualified processor, to confirm its computational cost is low enough, thus the algorithm is viable to be used in future space exploration missions.

Acknowledgements This work was partially supported by the European Space Agency under contract number 4000118072/16/NL/LvH/gp - 2.

References

1. Merlo, A., Larranaga, J., & Falkner, P. (2013). Sample fetching rover (sfr) for msr. In *12th Symposium on advanced space technologies in robotics and automation (ASTRA)* (pp. 1–10). ESA-ESTEC.
2. Woo, M. Y. (2009). Roving on mars. *Engineering and Science, 72*(2), 12–20.
3. Lele, A. (2014). Mars missions: Past, present and future. In *Mission Mars: India's Quest for the Red Planet* (pp. 85–92). Springer.
4. Bajracharya, M., Maimone, M. W., & Helmick, D. (2008). Autonomy for mars rovers: Past, present, and future. *Computer, 41*(12), 2008.
5. Mishkin, A. H., Morrison, J. C., Nguyen, T. T., Stone, H. W., Cooper, B. K., & Wilcox, B. H. (1998). Experiences with operations and autonomy of the mars pathfinder microrover. *In 1998 IEEE aerospace conference proceedings (Cat. No. 98TH8339)* (Vol. 2, pp. 337–351). IEEE.
6. Arvidson, R. E., Bell, J. F., Bellutta, P., Cabrol, N. A., Catalano, J. G., Cohen, J., Crumpler, L. S., Des Marais, D. J., Estlin, T. A., Farrand, W. H., et al. (2010). Spirit mars rover mission: Overview and selected results from the northern home plate winter haven to the side of scamander crater. *Journal of Geophysical Research: Planets, 115*(E7), 2010.
7. Arvidson, R. E., Ashley, J. W., Bell, J. F., Chojnacki, M., Cohen, J., Economou, T. E., Farrand, W. H., Fergason, R., Fleischer, I., Geissler, P., et al. (2011). Opportunity mars rover mission: Overview and selected results from purgatory ripple to traverses to endeavour crater. *Journal of Geophysical Research: Planets, 116*(E7), 2011.
8. Biesiadecki, J. J., & Maimone, M. W. (2006). The mars exploration rover surface mobility flight software driving ambition. In *2006 IEEE Aerospace Conference* (pp. 15). IEEE.
9. Maimone, M. W., Leger, P. C., & Biesiadecki, J. J. (2007). Overview of the mars exploration rovers' autonomous mobility and vision capabilities. In *IEEE international conference on robotics and automation (ICRA) space robotics workshop*.
10. Muirhead, B. K., & Karp, A. (2019). Mars sample return lander mission concepts. In *2019 IEEE Aerospace Conference* (pp. 1–9). IEEE.
11. Pilania, V & Gupta, K. (2018). Mobile manipulator planning under uncertainty in unknown environments. *The International Journal of Robotics Research, 37*(2–3), 316–339, 2.
12. Li, Q., Mu, Y., You, Y., Zhang, Z., & Feng, C. (2020). A hierarchical motion planning for mobile manipulator. *IEEJ Transactions on Electrical and Electronic Engineering, 15*(9), 1390–1399.
13. Lunenburg, J., van de Molengraft, R., & Steinbuch, M. (2013). Minimizing the number of iterations when computing a base pose for manipulation by mobile base inclusion in the inverse kinematics. In *2013 16th International Conference on Advanced Robotics, ICAR 2013*.
14. Kabir, A. M., Kanyuck, A., Malhan, R. K., Shembekar, A. V., Thakar, S., Shah, B. C., & Gupta, S. K. (2019). Generation of synchronized configuration space trajectories of multi-robot systems. In *Proceedings—IEEE International Conference on Robotics and Automation*, volume 2019-May (pp. 8683–8690). Institute of Electrical and Electronics Engineers Inc.

15. Raja, R., & Dutta, A. (2015). Motion planning of a mobile manipulator using fuzzy controller to dexterity measures. In *2015 IEEE Workshop on Computational Intelligence: Theories, Applications and Future Directions, WCI 2015*. Institute of Electrical and Electronics Engineers Inc.
16. Burget, F., Bennewitz, M., & Burgard, W. (2016). BI2RRT∗: An efficient samplingbased path planning framework for task-constrained Mobile manipulation. In *IEEE International Conference on Intelligent Robots and Systems*, volume 2016-Novem, pp. 3714–3721. Institute of Electrical and Electronics Engineers Inc.
17. Pajak, G., & Pajak, I. (2017). Point-to-point collision-free trajectory planning for mobile manipulators. *Journal of Intelligent & Robotic Systems*.
18. Giftthaler, M., Farshidian, F., Sandy, T., Stadelmann, L., & Buchli, J. (2017). Efficient kinematic planning for mobile manipulators with non-holonomic constraints using optimal control. In *Proceedings—IEEE International Conference on Robotics and Automation*, pp. 3411–3417. Institute of Electrical and Electronics Engineers Inc.
19. Minniti, M. V., Farshidian, F., Grandia, R., & Hutter, M. (2019). Whole-Body MPC for a Dynamically Stable Mobile Manipulator. *IEEE Robotics and Automation Letters, 4*(4) 3687–3694.
20. Ichnowski, J., Avigal, Y., Satish, V., & Goldberg, K. (2020). Deep learning can accelerate grasp-optimized motion planning. *Science Robotics, 5*(48), 11
21. Azkarate, M., Zwick, M., Hidalgo-Carrio, J., Nelen, R., Wiese, T., Poulakis, P., Joudrier, L. & Visentin, G. (2015). First Experimental investigations on Wheel- Walking for improving Triple-Bogie rover locomotion performances. In *13th Symposium on 18 Authors Suppressed Due to Excessive Length Advanced Space Technologies in Robotics and Automation (ASTRA)* (pp. 1–6). ESA-ESTEC.
22. Vago, J., Witasse, O., Svedhem, H., Baglioni, P., Haldemann, A., Gianfiglio, G., Blancquaert, T., McCoy, D., & De Groot, R. (2015). ESA ExoMars program: The next step in exploring Mars. *Solar System Research, 49*(7), 518–528.
23. Azimi, A., Holz, D., Kövecses, J., Angeles, J., & Teichmann, M. (2015). A multibody dynamics framework for simulation of rovers on soft terrain. *Journal of Computational and Nonlinear Dynamics, 10*(3):031004.
24. Filip, J., Azkarate, M., & Visentin, G. (2017). Trajectory control for autonomous planetary rovers. In *12th Symposium on Advanced Space Technologies in Robotics and Automation (ASTRA)* (pp. 1–7). ESA-ESTEC, 2013.
25. Sanchez, J.R, Perez-del Pulgar, C.J., & Azkarate, M. (2017). Path planning for reconfigurable rovers in planetary exploration. In *14th Symposium on Advanced Space Technologies in Robotics and Automation (ASTRA)* (pp. 1–7). ESA-ESTEC.
26. https://youtu.be/aqm-CTUp-I

Chapter 12
Rover Autonomy Testing in Simulation and on Breadboard Rovers

Matthias Winter

Abstract To maximise science return, Mars rovers have to traverse long distances through previously unknown hazardous terrain with very limited communication opportunities. For this task, the rovers need highly autonomous Guidance, Navigation and Control (GNC) systems, designed to allow them to safely traverse terrain without real-time assistance from Earth (Winter et al. in ExoMars Rover Vehicle—Detailed description of the GNC System, ASTRA Conference 2015). To demonstrate that these systems will work in a safe and efficient way on Mars, extensive testing is required. This chapter gives a brief overview about the capabilities of GNC systems designed for current and future Mars rovers, including the two Mars Exploration Rovers Opportunity and Spirit (NASA), the Curiosity Rover (NASA) and the ExoMars Rover (ESA). It then takes a detailed look at how these kind of GNC capabilities are tested in simulation, in controlled test facilities and in field trials. It focuses in particular on the main purposes and advantages of each way of testing. As part of a case study, test results from the development and testing of the Mars Exploration Rovers, the Curiosity Rover and the ExoMars Rover are presented. The presented results cover the validation of simulation environments, testing of localisation systems, testing of trajectory and traction control systems, testing of autonomy algorithms and trials of the operational and science procedures.

12.1 GNC Systems of the MER and MSL Rovers

After the early hazard-avoidance capabilities of the Pathfinder microrover [1], the autonomous Mars Exploration Rovers (MER) Opportunity and Spirit and the Mars Science Laboratory (MSL) rover Curiosity featured a much higher level of autonomy, enabling them to drive tens of kilometres across unknown areas on the Martian surface.

There are several driving modes regularly used [2–4]:

M. Winter (✉)
German Aerospace Center (DLR), Robert-Hooke-Str. 7, 28359 Bremen, Germany
e-mail: space-robotics@matthias-winter.eu

X. T. Yan and G. Visentin (eds.), *Space Robotics*,
https://doi.org/10.1007/978-3-031-39214-6_12

1. *Directed drive* executes a pre-defined path without any automated compensation for position and attitude errors caused by wheel slippage and rocks.
2. *Visual Odometry* will additionally stop at regular intervals (commonly about every 10 m) to take images orthogonal to the drive direction and correct the position and attitude to compensate for errors caused by slippage and rocks.
3. *Terrain Assessment* executes a pre-defined path, looking out for geometric hazards. For this, it stops at regular intervals (commonly 0.5–1.5 m [5]), creates a 3D point cloud from stereo images and uses the GESTALT (Grid-based Estimation of Surface Traversability Applied to Local Terrain) algorithms to detect geometric hazards (e.g. rocks, ditches, cliffs). From these outputs, it creates a traversability map. It will stop the rover if the pre-defined path would endanger it. Note that it does not by default use Visual Odometry.
4. *Local Path Selection* selects its own paths from a number of options (see Fig. 12.1) towards a target, utilising either the *Terrain Assessment* mode or the *Visual Odometry* mode or a combination of both.
5. *Global Path Selection* keeps a larger map of previously seen terrain than the *Local Path Selection* mode and can plan arbitrary paths on it. It is useful to recover from complex scenarios for which the Local Path Selection mode does not find a good solution like large obstacles (e.g. rocks larger 1 m, multiple parallel ripples or a crater).

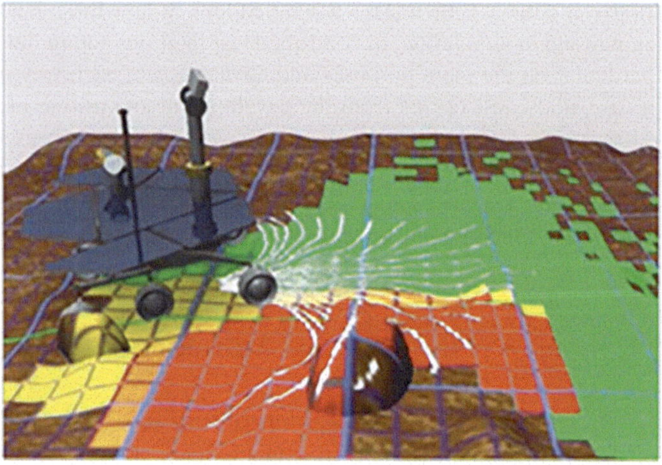

Fig. 12.1 Example of the terrain assessment and path selection by MER ([4], courtesy NASA/JPL-Caltech)

12.2 GNC System of the ExoMars Rover

The ExoMars Rover will be fitted with a GNC system designed to allow it to safely traverse terrain without real-time assistance from Earth. [6]. This GNC system has been designed to have different levels of autonomy. From low to high autonomy, the operator can do the following [7, 8]:

1. Directly drive the rover, for example commanding it to drive straight, to turn on the spot or to stop.
2. Command the rover to follow a pre-defined path: The GNC system drives the rover following the path while compensating for external disturbances (e.g. rocks and slopes) that push the rover away from the path.
3. Command the rover to reach a target: The GNC system autonomously analysis terrain and plans paths towards the target. The paths are then autonomously executed. This is referred to as AutoNav or 'full autonomy'.

It is worth noting that the higher levels of autonomy utilise the functionalities of all lower levels of autonomy. To implement these different autonomy modes, the rover utilises a suite of sensors and actuators, including a movable navigation camera (stereo) mounted on a mast, a static localisation camera (stereo), an IMU and the rover's locomotion subsystem. The overall layout of the ExoMars Rover is shown in Fig. 12.2.

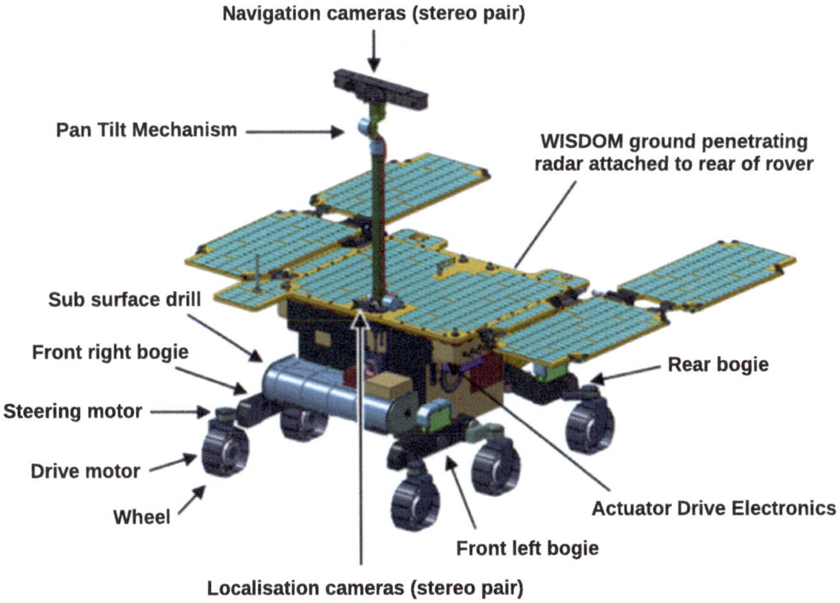

Fig. 12.2 ExoMars Rover ([7], courtesy Airbus)

The ExoMars GNC architecture works with a stop & go approach [7, 8]: Every approximately 2.3 m, the rover stops (from here on referred to as a navigation stop) to perform the following steps:

1. Take three stereo images of the surrounding terrain.
2. Create disparity maps from the stereo images [9].
3. Model the surrounding terrain.
4. Create a local map of the surrounding terrain, which details safe and unsafe areas as well as the terrain difficulty.
5. Merge the data of the local map into a persistent map, which contains the data gathered at previous navigation stops.
6. Plan a new approximately 2.3 m long safe path towards the target.

Then the rover executes the 2.3 m long path with a closed-loop control system, utilising visual odometry, wheel odometry and the IMU [10].

12.3 Rover GNC Testing

To guarantee that an autonomous rover will operate on Mars as expected, a large amount of testing is performed. This involves the testing of each hardware component, many of which are specifically designed or strongly modified to deal with the harsh conditions during biological cleansing (for planetary protection), during launch, in deep-space, during landing and on the Martian surface.

Of particular interest is the testing of the GNC algorithms and software as well as overall system tests. For these, three fundamental kinds of test environments can be used, which will be described in more detail in the following sections:

- Simulators (either pure software or with Hardware-in-the-loop);
- Tests on a representative hardware platform in a controlled environment;
- Field Trials.

12.4 Testing in Simulation

Though the initial effort for setting up a sufficiently representative simulator can be very high, testing in simulation has several strong advantages compared to testing on real hardware:

- *Execution Speed*: Simulators can often achieve execution speeds between tens and thousands of times faster than real-time, depending on the fidelity of the simulator. Moreover, depending on the amount of computing resources available, many instances of the simulator can be run in parallel. Executing large performance test campaigns, as it is common for space projects (e.g. Monte-Carlo campaigns), is only achievable in this environment.

- *Automation*: Large test campaigns with hundreds or thousands of scenarios can be automated in a simulator, allowing a rerun of the campaign with—in a good setup—very limited need for manual inputs/monitoring. Also continuous integration tests can be run during the development phase.
- *Repeatability*: The perfectly controlled nature of a simulator allows rerunning tests with the exact same environment and hardware behaviour, which is impossible to achieve on actual hardware. For example, with actual cameras, it is impossible to recreate exactly the same images because of changed lighting conditions (even with 100% artificial light) and sensor noise. Therefore, in a simulator the behaviour of different GNC software versions can be more accurately compared.
- *Flexibility*: As hardware and environment are simulated, new test scenarios can be set up rapidly. That includes new terrains and changed hardware configurations. For example, the impact of changed camera properties or a changed camera view due to a different rover configuration can be rapidly analysed without the delay introduced by manufacturing hardware.

GNC algorithms and software are commonly tested on several different test benches with a varying degree of model fidelity and a varying degree of incorporation of actual hardware (often prototypes of the flight hardware but in instances also the actual flight hardware).

The main simulator used for testing the algorithmic properties of the GNC software developed for the MER and MSL rovers is called the ROAMS simulator (see [11, 12]). It includes detailed physics-based models of the rover kinematics and dynamics, models of its actuators and sensors (e.g. wheel and steering motors and encoders, IMU, sun sensors and cameras) as well as models of the terrain environment and the terrain/rover interaction (see Fig. 12.3). It also allows for Hardware-in-the-loop testing.

The ROAMS simulator is commonly used for:

- Analyses for system trade-offs
- Development of new rover technologies
- Closed-loop development and test of flight software

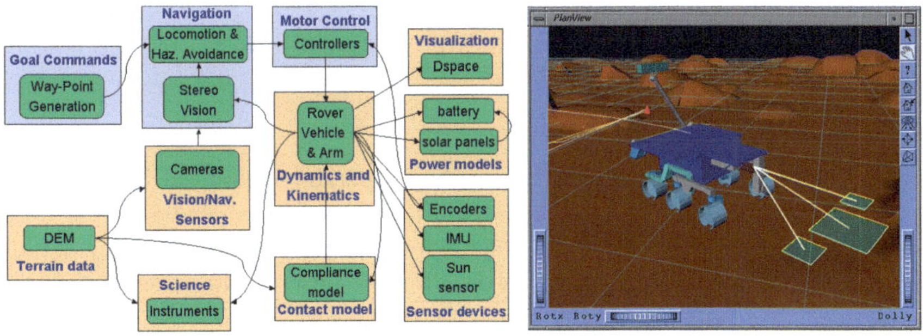

Fig. 12.3 ROAMS components of obstacle detection by Sojourner Rover ([11], courtesy NASA/JPL-Caltech)

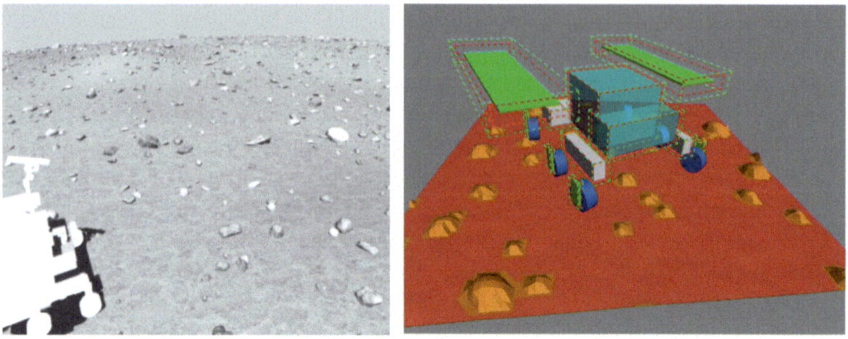

Fig. 12.4 PANGU rendering of Martian terrain and Rover Dynamics Model (from [7], courtesy Airbus)

- As a simulation tool during mission operations (including operator-in-the-loop simulations)

The main simulator for testing the algorithmic properties of the ExoMars GNC algorithms and software is called the Offline-Simulation Environment (OSE). The ExoMars OSE is a high-fidelity simulator with particular focus on the models which influence the sensor and actuator behaviour and the rover dynamics [7]. It includes—among others—the Rover Dynamics Model (RDM [13]), a high-fidelity visualisation environment (PANGU) to render surface images (for both see Fig. 12.4), a high-fidelity camera model, an IMU model and models of the actuators and their sensors.

The OSE is used for several types of testing:

- Impact analysis of changes to the rover architecture (e.g. hardware changes)
- Testing during the development/implementation of GNC algorithms
- Test campaigns for new GNC algorithms
- Test campaigns for the overall GNC system (e.g. Monte-Carlo campaigns)

To illustrate the use cases for simulators, some example results are presented in the following sections.

12.4.1 ROAMS Validation

Figure 12.5 shows a comparison of the slip ratios experienced in ROAMS simulations to the slip ratios of a real rover on the MER tilt table [14]. A total of 400 tests were run, using three different manoeuvres (straight-line, arc drive, turn-in-place) on four different slope angles (0°, 5°, 10°, 20°) with three different directions (up-slope, down-slope, cross-slope).

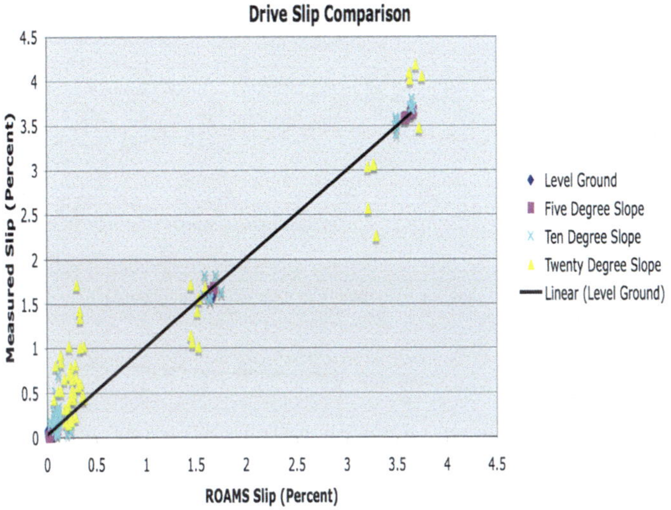

Fig. 12.5 Comparison of slippage in ROAMS simulator to slippage in hardware tests ([14], courtesy NASA/JPL-Caltech)

The results show that the error in the slippage simulation is small for slope angles from zero to 10 degrees but that the error becomes considerably larger for a slope of 20 degrees.

12.4.2 Localisation

Figure 12.6 shows the performance of the ExoMars Rover localisation module with and without slippage correction while the rover is laterally traversing a slope.

The localisation module estimates the rover position and attitude at high frequency while the rover traverses the terrain in order to allow the controller to keep the rover on the path [6]. Visual odometry is used to provide the localisation module with position and attitude estimates obtained by tracking features in consecutive images taken by the localisation camera. These estimates are generated at a relatively low frequency (one sample every 10 s) due to the high computational load of the image processing. Gyro rate measurements and wheel odometry (derived from wheel drive rates and steering angles) are used to propagate the attitude and position of the rover while waiting for the next visual odometry output.

Due to the nature of the Martian surface (sandy slopes and rocks), high slippage is expected while the rover travels. The slippage cannot be detected by wheel odometry. This would lead to a drop in accuracy of the position estimate between visual odometry samples. Figure 12.6 (top image) illustrates this effect while the rover is laterally traversing a 20-degree slope. The up-slope direction is along the positive y-axis. If the slippage remained undetected, the position estimate accuracy

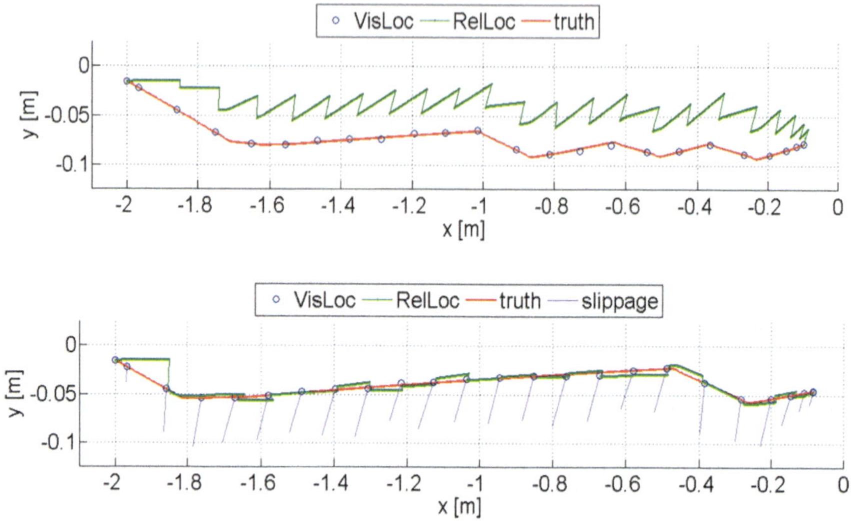

Fig. 12.6 Position estimate without (top) and with (bottom) slippage correction while laterally traversing a slope ([6], courtesy Airbus)

would compromise the capability of the controller to follow the planned path. To avoid this, the localisation module combines visual odometry measurements with gyro and wheel odometry data in order to estimate the average slippage between consecutive visual odometry samples. The slippage estimate is then used to correct the position increments from wheel odometry. Figure 12.6 (bottom image) shows the slippage estimated by the localisation system and the improvement in the position estimate while the rover traverses a slope.

12.4.3 Trajectory Control

Figures 12.7 and 12.8 show an example of the lateral and heading errors for a test case in which the ExoMars Rover traverses laterally across a sandy slope of 15 degrees.

The trajectory control module drives the rover along a path which has been generated either on-board by the path planning algorithms or has been uploaded from ground [6]. To do so, it uses the estimated rover position and attitude provided by the localisation module to calculate the locomotion manoeuvre commands required to follow the planned path closely, utilising closed-loop control.

For path segments, the lateral error from the path and the heading error are maintained within limits by separate controllers. For point turn segments (i.e. heading changes without position change enabled by six-wheel-steering), a single controller acts on the point turn position error. The locomotion manoeuvre control then translates the locomotion manoeuvre command requested by the trajectory control module

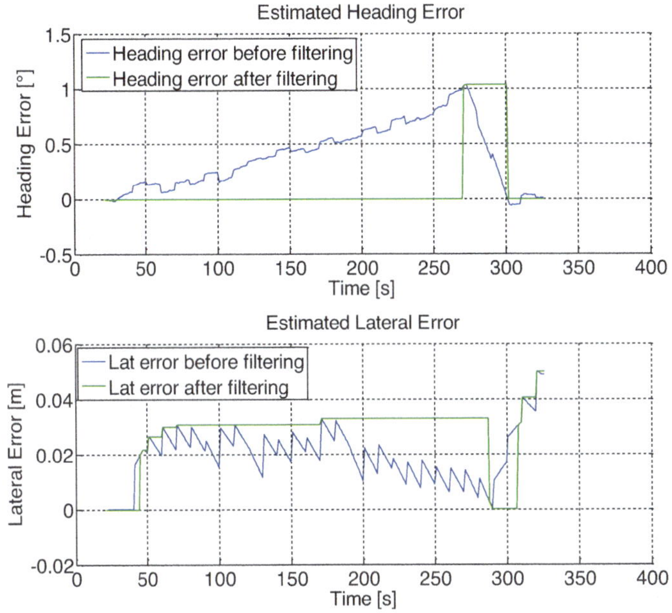

Fig. 12.7 Example of the controller error while laterally traversing a sandy slope of 15 degrees ([6], courtesy Airbus)

at rover body level into wheel level commands. An example of a locomotion manoeuvre command (at body level) is a generic Ackermann command, which defines a rover body speed and a centre of rotation position.

For a test case in which the rover laterally traverses across a sandy slope of 15 degrees, Fig. 12.7 shows the estimated heading and lateral error as well as the action of a filter that removes the saw-tooth pattern in the lateral error input (which originates from the position and attitude update delays of the visual odometry). This filter is one of the features intended to minimise actuation of the steering motors. Figure 12.8 shows the steering commands generated by the controller. The results show that the controllers kept the lateral error and the heading error within the requirements.

12.4.4 AutoNav

AutoNav allows the ExoMars Rover to autonomously plan consecutive 2.3 m long paths to travel to a target position several kilometres away. Considering that commonly the majority of terrain it has to traverse to reach the target position is unknown when starting the traverse, it employs sensible strategies to move around small (e.g. single insurmountable rocks), medium (e.g. gatherings of several insurmountable rocks) and large (e.g. crater) obstacles [7].

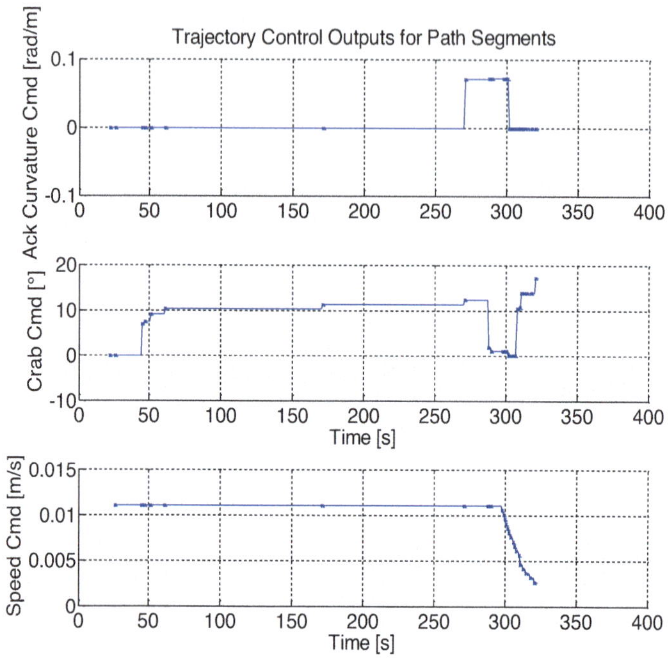

Fig. 12.8 Example of the controller action while laterally traversing a sandy slope of 15 degrees ([6], courtesy Airbus)

The OSE provides the ideal environment to perform detailed comparisons between the terrain models created by the autonomy algorithms (example shown in Fig. 12.9) and the actual terrain, as the actual terrain—due to it simulated nature—is perfectly known. Note that for this purpose, representative stereo-images must be provided to the algorithms. Therefore, a highly representative scene rendering tool (in the ExoMars case: PANGU) and a highly representative camera model is needed.

The fact that the simulated terrain is perfectly known also allows for an easy assessment of the derived terrain attributes used to evaluate if terrain is safe and

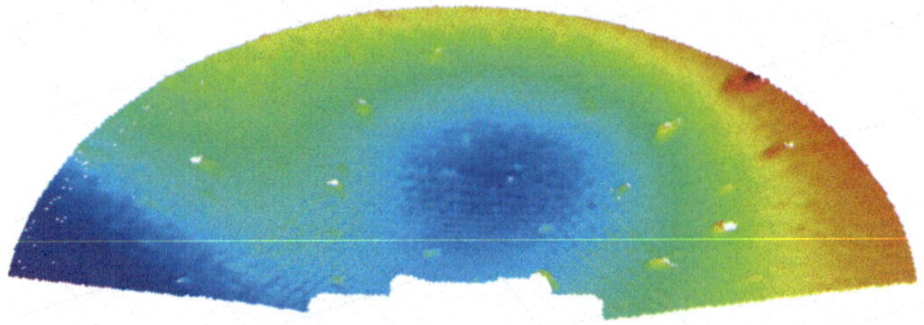

Fig. 12.9 DEM (Digital Elevation Map) from a simulation on the autonomy test terrain. The DEM is approximately 15 m wide and 7 m deep ([7], courtesy Airbus)

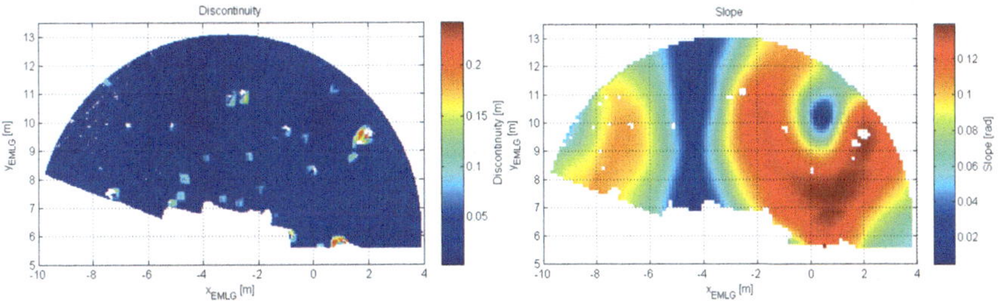

Fig. 12.10 Discontinuity and slope map generated by the AutoNav algorithms for the traversability analysis ([7], courtesy Airbus)

Fig. 12.11 Nominal placement of a virtual rover model by the AutoNav algorithms for the clearance check, visualised in PANGU ([7], courtesy Airbus)

traversable and to assess the difficulty of the terrain. These attributes include the discontinuity of the terrain and the slopes (see Fig. 12.10) but also safety information derived from placing a virtual rover model on the terrain models created by the GNC algorithms (e.g. clearance under the rover bogies, the rover central body and the solar panels, see Fig. 12.11).

The OSE is particularly useful for simulating long traverses (i.e. 40–100 m) to evaluate the long-term behaviour of the full autonomy system (i.e. mapping system, path planning system and closed-loop driving system). An example of a long traverse in progress is shown in Fig. 12.12. It has to be noted that at each point along this traverse, the terrain known to the rover only includes the area up to 6 to 8 m in front of the rover and previously seen areas. This simple traverse shows the autonomy system behaviour when it encounters a dead-end and uses the so far accumulated map to derive a sensible exit and continuation strategy. Hundreds of long traverses can be simulated in a relatively short time-span, exposing the algorithms to a large

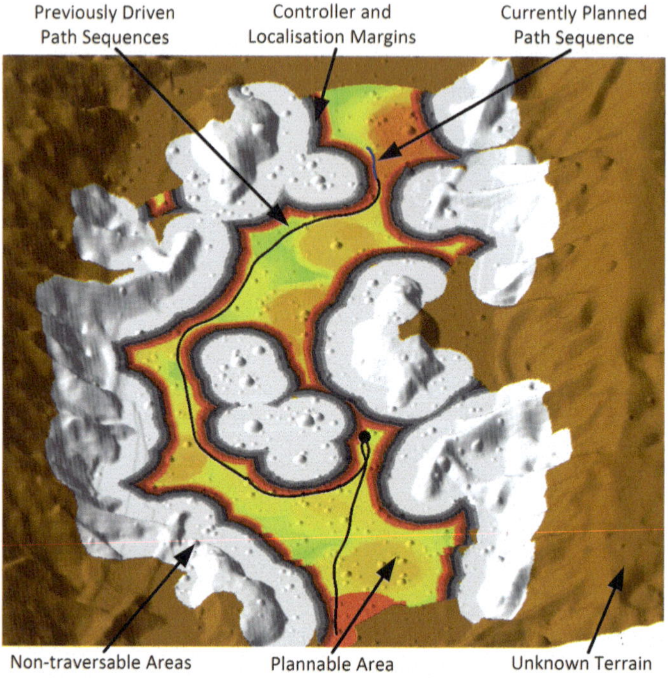

Fig. 12.12 Persistent map generated by ExoMars AutoNav with currently planned path and complete history of previously driven paths, projected on the simulated terrain. Total traverse length at this stage: 43 m ([7], courtesy Airbus)

number of complex scenarios and creating statistics about success rates in terms of reaching the final target and of not endangering the rover (i.e. by grounding the rover bogies or its central body or hitting the solar panels against rocks).

12.5 Tests on Actual Hardware in a Controlled Environment

Testing on actual hardware can give good insights already early during the development of robotic systems, particularly if no high-fidelity simulator is available. But also if high-fidelity simulators are available, it is an important step for gaining confidence in the correct functioning of GNC algorithms and the simulators. Besides showing that the interfaces between hardware and algorithms work, it also confirms that models in the simulator are representative enough to make the simulated robotic system behave in a qualitative and quantitative similar way to what is seen on actual hardware. It can also give the developers a more representative experience of how an operator would actually have to interact with the robotic system (visibility of telemetry, operator command interfaces).

There are several advantages of this kind of testing compared to field trials:

- *Reduced Cost*: Field trials are commonly connected to a number of additional cost factors, which include the cost of the travel to and the accommodation at locations with a representative landscape and related organisational aspects. The locations are often not easily reachable (e.g. Atacama Desert in Chile or El Teide National Park on Tenerife). Moreover, there is the need for an amount of hardware redundancy to avoid having to abort field trials due to hardware issues.
- *Infrastructure*: Test facilities like the Mars Yards at JPL, at CNES in Toulouse or at Airbus in Stevenage provide infrastructure less easily available on field trials: From basic essentials like electricity, internet and accommodation to specialist testing equipment like truth position/attitude reference systems, video systems and systems to measure the terrain.
- *Semi-controlled environment*: In an artificial setup, many conditions can be controlled which are not easily controllable in field trials: terrain layout, weather and lighting conditions (if indoors), sand properties and temperature (if indoors) to name just a few.
- *Flexibility*: If a test facility is easily accessible, it will give a wide range of flexibility to the schedule. That means that testing can be done on a more regular basis (i.e. shorter but more frequent) to see the impact of different design decisions. Also issues with the hardware or software setup do not lead to directly losing time from a fixed test schedule, as the majority of the team can continue to work in different areas while the issue is being fixed.

All of these advantages make these tests a sensible tool during the development of a robotic system, as the robustness of hardware, software and operational procedures is commonly not very high at this stage. They are also a good environment for confidence testing and to increase robustness of a system before starting field trials.

JPL has traditionally performed a lot of testing on real hardware in its test facilitates, which reach from in-door laboratories to a 66 m $\times$ 36 m large Mars Yard. The rovers used include dedicated test rover (e.g. Athena SDM [15], MER and MSL test rovers) and flights spares (e.g. Sojourner flight spare, see Fig. 12.13.)

During the development of the ExoMars Rover GNC algorithms, regular testing of the algorithms on breadboard rovers has been performed in the Mars Yard at Airbus in Stevenage [7]. These breadboard rovers were set up using a mixture of early prototypes of the actual flight hardware and commercial of-the-shelf components [17]. They allow for a rapid integration of the C-coded GNC algorithms, coming from the simulator environment (OSE). The breadboard rovers were designed to have approximately one third of the mass of the ExoMars flight rover but a similar centre of mass position to emulate the dynamics properties the flight rover will have on Mars. The Mars Yard is an indoor test facility about 30 m $\times$ 12 m in size with sand, slopes and rocks, emulating the surface of Mars. The latest GNC breadboard rover and the Mars Yard are depicted in Fig. 12.14.

Fig. 12.13 JPL Mars Yard and test rovers. Front: Sojourner flight spare; Left: MER Project Test Rover; Right: MSL Test Rover ([16] courtesy NASA/JPL-Caltech)

Fig. 12.14 Mars Yard at Airbus in Stevenage with MDM Breadboard Rover ([7], courtesy Airbus)

Besides algorithm tests also open-loop characterisations of the dynamics of the breadboard rovers have been performed [18]. The data from these tests was used for tuning the Rover Dynamics Model (RDM).

To illustrate the use cases for Mars Yard testing, some example results are presented in the following sections.

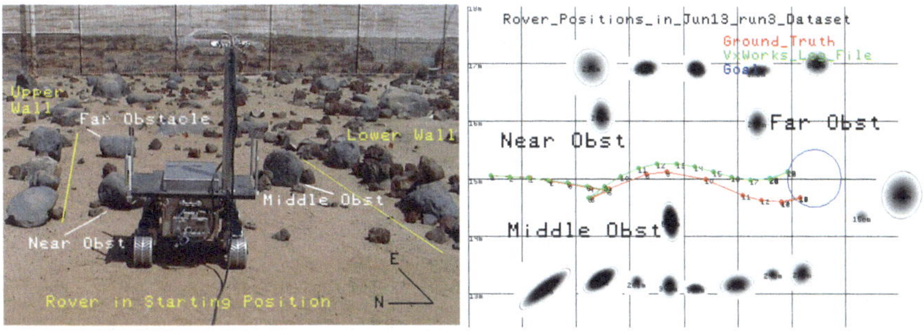

Fig. 12.15 Autonomy test scenario in the JPL Mars Yard ([19], courtesy NASA/JPL-Caltech)

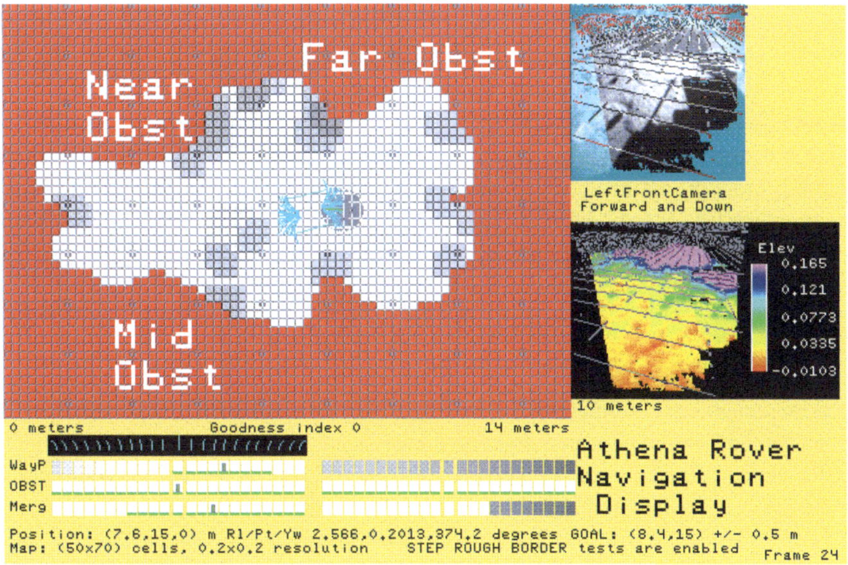

Fig. 12.16 GESTALT diagnostic output ([19], courtesy NASA/JPL-Caltech)

12.5.1 Testing of the GESTALT Autonomy Algorithms

The GESTALT autonomy algorithms are used on the MER and MSL rovers (see Sect. 12.1). They were tested on the Athena SDM test rover before the launch of the MER rovers [19].

Figure 12.15 shows a test scenario set up in a Viking Lander II type terrain in the JPL Mars Yard. It consists of two wall-like line-ups of rocks and three dedicated independent obstacles. The rover is initially positioned at the xy-position 2.4 m/ 15.1 m and is instructed to autonomously reach the xy-position 8.4 m /15 m with an accuracy of $\pm$ 0.5 m while avoiding any obstacles. Figure 12.16 shows the GESTALT diagnostic output when the rover is less than one metre from its final target. On the right, the display shows the view of the left front-facing hazard camera (top) and the corresponding elevation map (bottom). On the left, it displays the traversability map, which has been built up during the traverse. Assumed non-traversable areas are marked in red and traversable areas in grey-scale, with darker grey colours indicating more difficult areas (e.g. rocks). It also displays the considered steering arks, half of them going forwards and half of them going backwards. At the bottom, these arks are ranked according to their goodness values with regards to bringing the rover nearer to its target ('WayP') and their goodness values with regards to avoiding hazards and difficult areas ('OBST'). Both measures are merged together resulting in the merged goodness ('Merg'). The arc with the highest goodness is selected (green vertical line). If several arks jointly have the highest value (horizontal green lines), the one in the middle of the longest adjacent set of best arcs will be chosen.

The true and estimated rover positions during this test are shown in Fig. 12.15 in red and green respectively. The rover successfully reaches its target after eighteen executed paths. It has to be noted though, that the accumulation of localisation error nearly causes the rover to miss the intended target area. This is due to the fact that only wheel odometry and gyro measurements and no visual odometry are used in this test to propagate the rover position. The impact of not using visual odometry becomes more pronounced over longer distances, as can be seen in the 17.5 m long traverse shown in Fig. 12.17.

12.5.2 Traction Control Design Curiosity

After experiencing continuously increasing wheel damage on the MSL rover, JPL developed and deployed a new traction control system to reduce the forces acting on the wheels [20]. This software estimates the wheel speeds needed to avoid wheel slip, using a rover kinematics model fed with the measured rover attitude rates and suspension angles. It does not require any prior knowledge of the terrain and also includes detection and correction algorithms for wheels which have lost contact with the ground.

Figure 12.18 shows a test setup in the JPL Mars Yard with the Scarecrow test rover. The test rover's mobility system is kinematically identical to the one of the Curiosity rover and its overall mass of 318 kg (about a third of the mass of Curiosity) results in similar dynamics properties in Earth's gravity as the Curiosity rover has in Martian gravity.

Figure 12.19 shows the resulting loads when the Scarecrow rover is traversing the test setup in Fig. 12.18, once with the traction control system enabled (right) and once with the traction control system disabled (left). Note that only the first 90 s are shown because after this time the rover would consistently drive off the intended track if the traction control system was disabled. This is due to an increasing heading error in these open-loop tests. This particular test shows a reduction of the maximum load of about 20 percent for the front wheels, 15 percent for the middle wheels and no clear

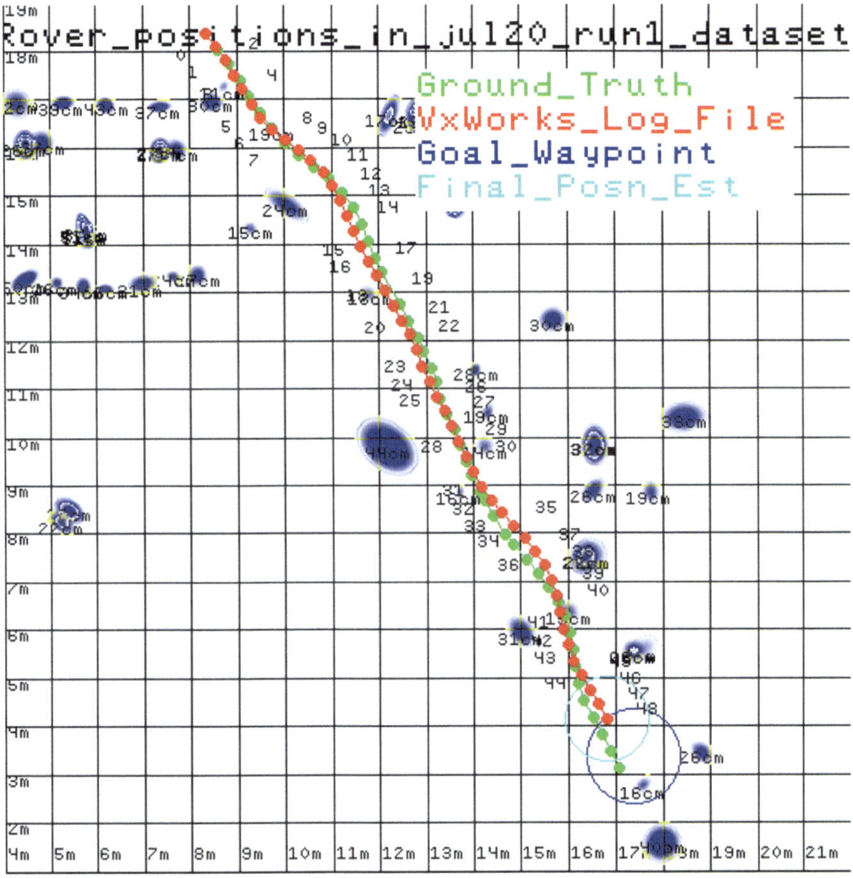

Fig. 12.17 GESTALT autonomous 17.5 m traverse ([19], courtesy NASA/JPL-Caltech)

change for the rear wheels. As the loads on the front wheels and the middle wheels are considerably higher than the loads on the rear wheels, this was a favourable result.

Repetitive tests with a similar setup showed also that the mean resulting loads across all wheels could be reduced by 5.1 percent and that as a further benefit the mean rotational torque across all wheels could be reduced by 15.6 percent, directly resulting in power savings. Additional tests on a loose sand terrain showed that the traction control system has the side-effect of balancing loads between the left and right side of the rover. This leads to a reduction in unintended heading change and wheel slippage on sandy terrain. Figure 12.20 shows a test terrain with non-symmetric sand ripples up to 35 cm high. Without the traction control system, a considerable heading error of up to 12 degrees is introduced (see Fig. 12.21). Using the traction control system, the heading errors are limited to less than 2 degrees.

Fig. 12.18 Traction control test setup with Scarecrow test rover ([20], courtesy NASA/JPL-Caltech)

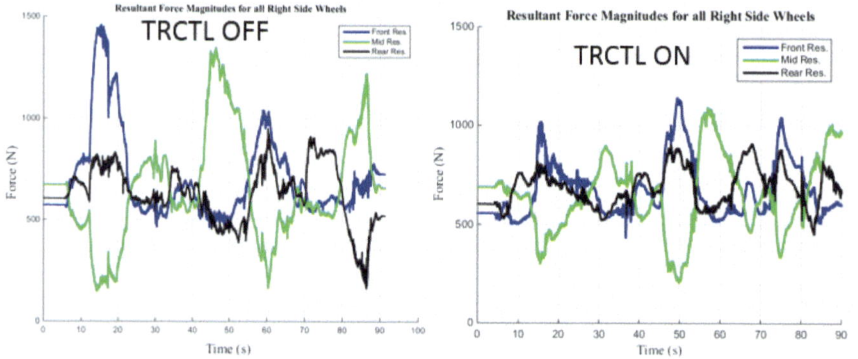

Fig. 12.19 Resulting loads when traversing the test setup shown in Fig. 12.18 ([20], courtesy NASA/JPL-Caltech)

12.5.3 Trajectory Control and Localisation

Figures 12.22, 12.23, 12.24, and 12.25 show the results of a Mars Yard test performed with the MDM ExoMars breadboard rover [10]. The rover was commanded to drive the path shown in Fig. 12.22, utilising the closed-loop trajectory control and localisation system. The path went across a steep slope with a variety of rocks on it. It has to be noted that the scenario shown is actually beyond the limits of what the autonomous system would try to go across and therefore was intended to test out the closed-loop system to its limits. Consequently, while traversing the last fourth of the up-slope

Fig. 12.20 Test setup for the heading test ([20], courtesy NASA/JPL-Caltech)

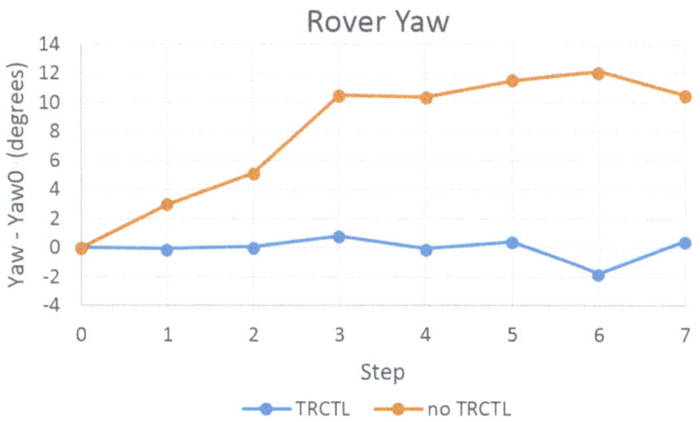

Fig. 12.21 Resulting heading errors when traversing the test setup shown in Fig. 12.20 ([20], courtesy NASA/JPL-Caltech)

path, the longitudinal slippage experienced by the rover increased up to a level at which the rover was not progressing along the commanded path any longer. At this point, path execution was stopped (see Fig. 12.26). Note that if the full autonomous system had been activated, this scenario with one of the wheels considerably dug into the sand would have been avoided by two mechanisms: AutoNav would not have planned a path across these slopes and the slippage monitor would have triggered after 10–20 s of limited progress along the planned path. The test was continued by aborting the path execution and directly executing the subsequent point turn.

Fig. 12.22 Commanded path sequence ([10], courtesy Airbus)

Figure 12.23 shows that during the whole test, the closed-loop system kept the rover within ten to fifteen centimetres of the commanded path and that the accuracy of the localisation system (RelLoc) was continuously in the order of a few centimetres.

The outputs of the trajectory control module during the execution of the second half of the up-slope path are shown in Fig. 12.24. The curvature commands are output to correct the heading of the rover and the crab commands, which command all six wheels of the rover to steer in the same direction, are output to correct for lateral offsets from the commanded path—predominately caused by sliding down the slope [6, 10].

The slippage estimation (Fig. 12.25) shows increasingly large values along the longitudinal axis of the rover while driving up the slope, reaching nearly 100%—which means the rover not moving forward any longer—at the end of the up-slope part. Note that such an amount of slippage would trigger the slippage monitor of the ExoMars Rover FDIR. Additionally, there is a mentionable amount of slippage along the lateral axis due to rocks and the lateral slope component. Also the angular slippage during the point turn is considerable.

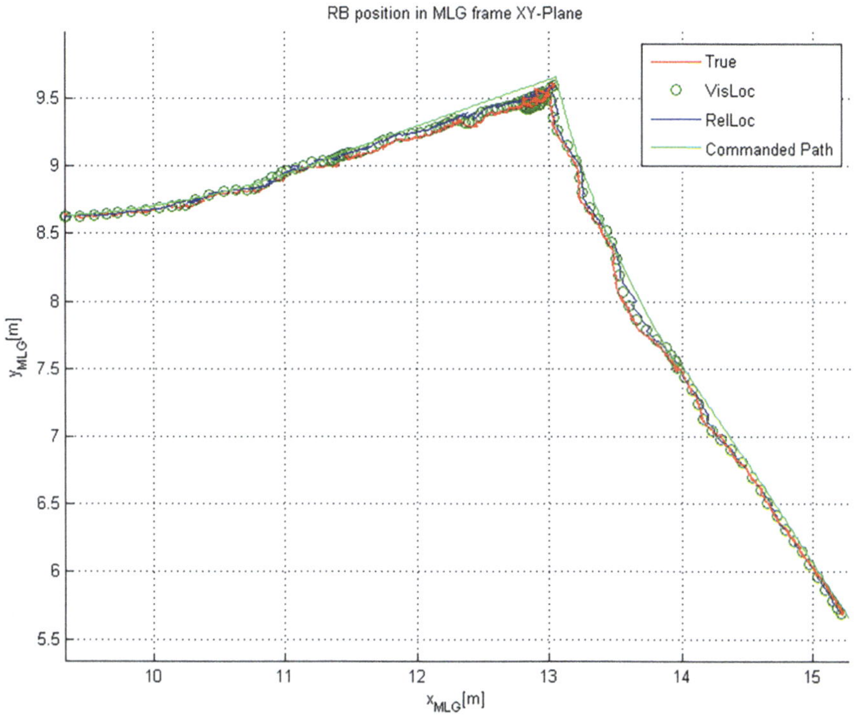

Fig. 12.23 Planned, estimated and true rover position in the horizontal plane ([10], courtesy Airbus)

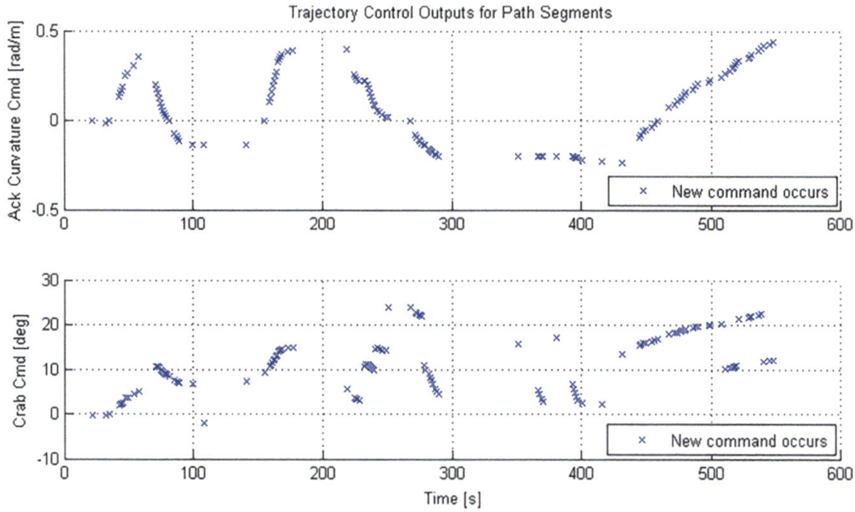

Fig. 12.24 Commands from the trajectory control module during the execution of the second half of the up-slope path (from [10], courtesy Airbus)

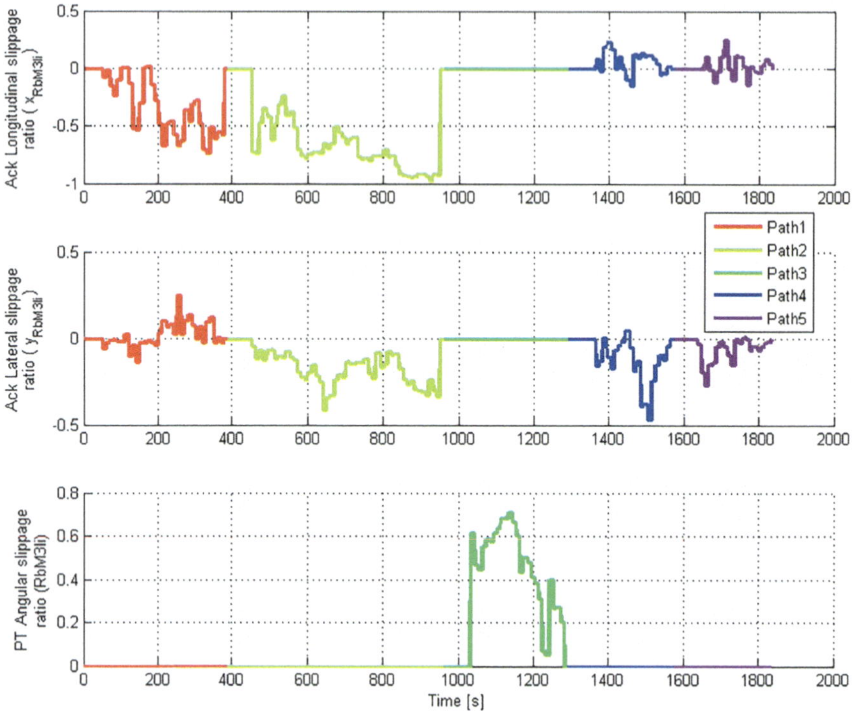

Fig. 12.25 Estimated slippage ratios for the commanded path segments (1 and 2: up-slope, 3: point turn; 4 and 5: down-slope, [10], courtesy Airbus)

Fig. 12.26 MDM rover at the end of the up-slope path ([10], courtesy Airbus)

12.5.4 AutoNav Testing

The fully-autonomous ExoMars GNC system has been tested in the Mars Yard at Airbus in Stevenage. Of particular interest were tests of the perception/terrain-modelling algorithms [9]. The 3D models produced by the rover's GNC algorithms were compared to the measured terrain characteristics of the Mars Yard (e.g. rock heights and distances). These tests showed that the terrain models were an accurate representation of the actual Mars Yard terrain to the accuracy of manual measurement. The tests also confirmed that the test results from the simulator were representative in qualitative ways (e.g. characteristics of terrain models produced) and quantitative ways (e.g. rock height error, resolution switch in disparity map [9], dynamic disparity search range). An example of a terrain model generated in the Mars Yard is shown in Fig. 12.27.

Tests of the mapping and path planning algorithms were performed as well [7]. It is worth noting that the limited size of a Mars Yard together with the large area covered by a single terrain model (up to 15 m wide and 7 m deep) allow only for relatively simple scenarios to be set up, for which the overall rover behaviour needs to be intentionally restricted (e.g. by blocking off some areas using large rocks) to avoid the borders of the Mars Yard having a dominant effect on the rover behaviour. For testing the long-term behaviour and interaction between mapping, path planning and path execution, simulator or field tests can allow for considerably more complex and open scenarios.

The data created during Mars Yard tests was also used to be fed back into the OSE simulator to set up 'open-loop' test scenarios. Though, these test scenarios do inherently not allow the autonomy to make actual decisions (like where the next

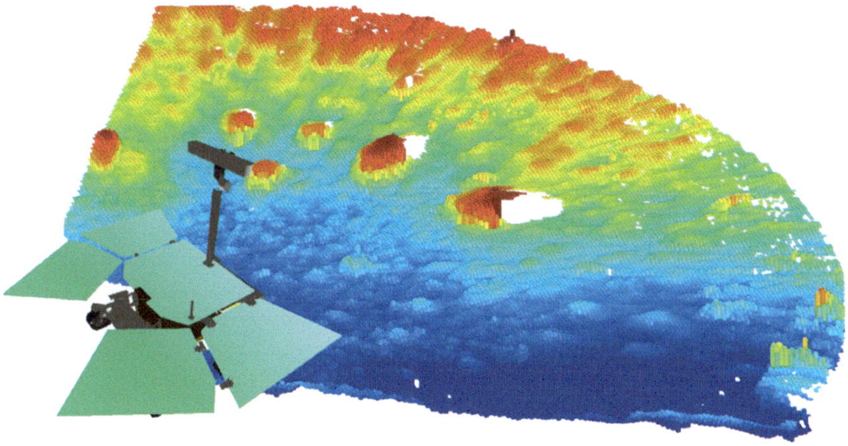

Fig. 12.27 A 3D terrain model from breadboard rover testing. Note the knowledge gaps behind rocks, as these areas are not visible from the perspective of the navigation camera ([9], courtesy Airbus)

navigation stop is going to be), it allows for rapid testing of the mapping and path planning algorithms, using non-simulated images.

12.6 Field Trials

As mentioned at the beginning of Sect. 12.5, the testing on actual hardware is an important step in gaining confidence in the correct functioning of GNC algorithms and simulators. It is often a good idea to perform the initial ones of these tests in a controlled environment (i.e. a lab or Mars Yard). However, after the overall functioning of the GNC system—i.e. hardware and software—has been proven, field trials offer the opportunity to show the robustness of the system and of the operations procedures in a wider variety of larger and longer test scenarios.

Field trials have several advantages compared to testing in dedicated test environments:

- *Size of test scenarios*: The size of lab and Mars Yard test facilities is often very limited, with even the largest outdoor facilities (e.g. CNES Mars Yard or JPL Mars Yard) being considerably less than 100 m in length. By contrast, field trials are often held in areas where the natural terrain imitates the target terrain (e.g. Mars-analogue) for kilometres on end (e.g. Atacama Desert in Chile or the El Teide National Park on Tenerife). This allows for extensive and long test scenarios, even spanning several days or weeks of operations. This is not possible in Mars Yard tests.
- *Variety of test scenarios*: The large size of the available test area also allows to cover a large variety of test scenarios, either as several separate tests in different areas or as continuous test across changing terrain. In controlled environments, for which the area of terrain available is measured in the tens of metres, it requires a lot of effort to set up completely new scenarios, as it involves moving large amounts of soil, rocks and bedrock.
- *Surprises*: Even though the details of a scenario set up in a test facility can be kept secret from the operators, some basic information (e.g. facility size, kind of terrain setup) will always be known to the designers of a GNC system and the operators of a test. Moreover, an artificial environment will always have the disadvantage that it is a designed environment, which means that it only includes things which the designer anticipated (e.g. sand types, rock shapes, colours). A large terrain during a field trial, in which the exact traverse is not known in advance, adds a random factor to the terrain selection, which can expose the system to scenarios not anticipated by the GNC design team, the test planning team and the operator team.

All of these points make field trials one of the ultimate tests for a robotic system, though these tests make only sense after the GNC hardware and software as well as the operations procedures have reached a high level of robustness, as otherwise the effective test time is likely to be very limited. For hardware, this includes a fair amount of system and hardware redundancy (e.g. spare parts).

To illustrate the use cases for field trials, some field trial campaigns in preparation for Martian rover missions will be described in the following sections.

12.6.1 Operations and GNC Testing for the MER Rovers

In the run-up to the MER missions, the FIDO (Field Integrated Design and Operations) rover was deployed into field trial campaigns at Silver Lake (California), Black Rock Summit (Nevada), Soda Mountains (California) and Grey Mountain (Arizona).

The FIDO rover is shown in Fig. 12.28. It is capable of semi-autonomous navigation and of gathering science data from rock and soil science targets. Its instrument suite includes a mast-mounted panoramic science camera, navigational cameras, an infrared point spectrometer and a front mounted robotic arm with several smaller instruments [21]. Field trials were conducted with MER mission scientists and flight operations personnel to test the operations, science and GNC concepts.

The GNC capabilities [21] of the FIDO rover include 3D terrain mapping and hazard-avoidance (using mast-mounted navigation cameras and chassis-mounted front/rear hazard cameras), self-localisation (using visual odometry, wheel odometry, accelerometers, gyros and a sun sensor) and path planning with respect to maps created by the navigation and hazard cameras. To generate ground-truth data, also a differential GPS system was integrated on the rover.

For remote command and control the Web Interface for TeleScience (WITS) was used. This JPL-developed tool (see Fig. 12.29) shows uplink telecommands, downlink telemetry and science data. It also allows 3D pre- and post-visualisation of command sequences.

The MER sensing, control, manipulation and sampling technologies as well as the related remote science operational concepts were tested using an increasingly challenging set of field trials under direction of the MER 2003 science team [21].

The first trials at Silver Lake (California, see Fig. 12.28) in 1999 demonstrated a local sampling loop, including panoramic imaging of the lander area, 3D navigational mapping to ground-designated targets, open-loop traverses to targets, infrared point spectrometer imaging of targets as well as kinematics-referenced 3D visualisation and placement of mast/arm mounted instruments. Moreover, samples were collected and the rover returned to the lander area.

A second, 'blind', field test was performed at Black Rock Summit (Nevada) in 2000 [21, 22]. For this test, the science team was removed from the operations area and instead placed at JPL so that operations was performed only based on data transmitted via satellite communication. Also the prior knowledge of the test site was limited to low-resolution satellite imagery and simulated lander descent imagery. The first action in this scenario was to acquire a full panorama looking out to a distance of 50 to 100 m and correlate this data with the descent data, allowing to situate the rover in a wider terrain context (see Fig. 12.30). The approach taken involved a lot

Fig. 12.28 JPL FIDO rover during field trials at Silver Lake (California, [21], courtesy NASA/JPL-Caltech)

of opportunistic science, in which the operations team frequently stopped the rover to deploy scientific instruments. It has to be noted that in a real Mars scenario this kind of approach is less efficient due to communication limitations like long delays (up to 20 min each way) and limited communication windows. These have not been simulated in the trials at Black Rock Summit. Some data outputs of these trials are shown in Fig. 12.31.

A third set of pre-mission trials was performed at the southern edge of the Soda Mountains in California's Mojave Desert in 2001 [23]. In the scope of these trials, a 40 people operations team simulated 20 days of activities for the first MER mission. These trials were again performed as blind trials but additional limitations were added compared to previous trials like limited uplink/downlink data volume, simulation of communication windows and strict daily mission operations timelines. Additionally, several mission anomalies were staged to test the operations team's handling of unexpected events like temporary loss of communication or telemetry or degraded rover performance due to extreme temperatures. The operations procedures were building up on procedures from the previous field trials: Before the beginning of

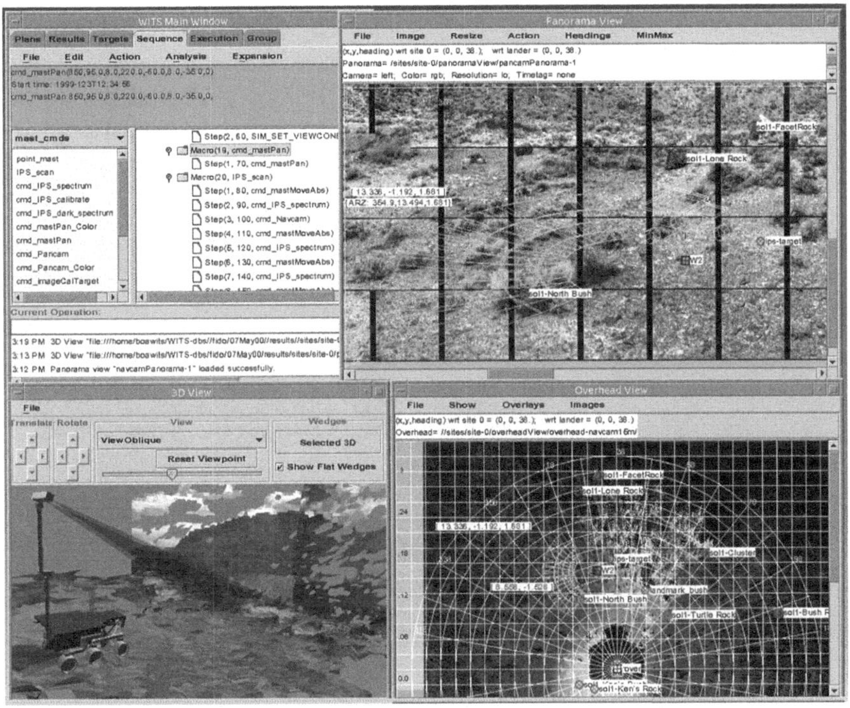

Fig. 12.29 Web Interface for TeleScience (WITS) display, showing command sequence, panoramic view, 3D terrain model and overhead view of surroundings ([21], courtesy NASA/JPL-Caltech)

the mission, the team examined available orbital data and made an initial plan for rover activities. On the first sol, the rover acquired a 360° panorama mosaic. For the second sol, the team examined the panorama data, selected targets and generated the first command sequence, which called for the acquisition of more image data and IR spectra as well as utilising the arm to acquire additional data. Further sequences included traverses to science targets identified in the panorama. The rover executed these autonomously and acquired detailed images and spectra of the target. A total of 135 m distance was traversed during this trial, the longest continuous autonomous traverse being 40 m and the average rover speed during traverses being 60 m an hour.

A final 20-sol dress rehearsal (skipping the first six sols of operations) for the MER missions with many similarities to the third set of field trials was performed in 2002 at Grey Mountain (Arizona). A detailed sol by sol mission log is available on JPL's website [24].

Fig. 12.30 FIDO rover at blind field tests at Black Rock Summit (Nevada). Left: rover after EGRESS from the platform, right: terrain data created from aerial footage ([21], courtesy NASA/JPL-Caltech)

12.6.2 Operations and GNC Testing for the ExoMars Rover

In preparation of the ExoMars Mission, ESA conducted field trials with an ExoMars Rover prototype named 'Charlie' (see Fig. 12.32) in the Tabernas Desert in Spain and the Atacama Desert in Chile with the goal of testing the remote science and operations procedures, the payloads and the GNC hardware and software in a representative scenario [25].

The trials were performed as 'blind' trials, meaning that the operator team was located off-site in Harwell, UK. This way the communication limitations of the ExoMars mission could be strictly enforced (e.g. infrequent communication opportunities, limited data budget). The rover—like for the actual mission—provided telemetry data and executed command sequences uploaded by the operations team. The operations executed included the egress from the landing platform and performing science by autonomously executing traverses towards science targets and autonomously using scientific instruments. All decisions by the operations team were strictly based on data provided by the rover and low-resolution orbital maps of the terrain as are available on a real Mars mission. The payloads operated and tested included the WISDOM ground penetrating radar, the Close-Up Imager (CLUPI), the

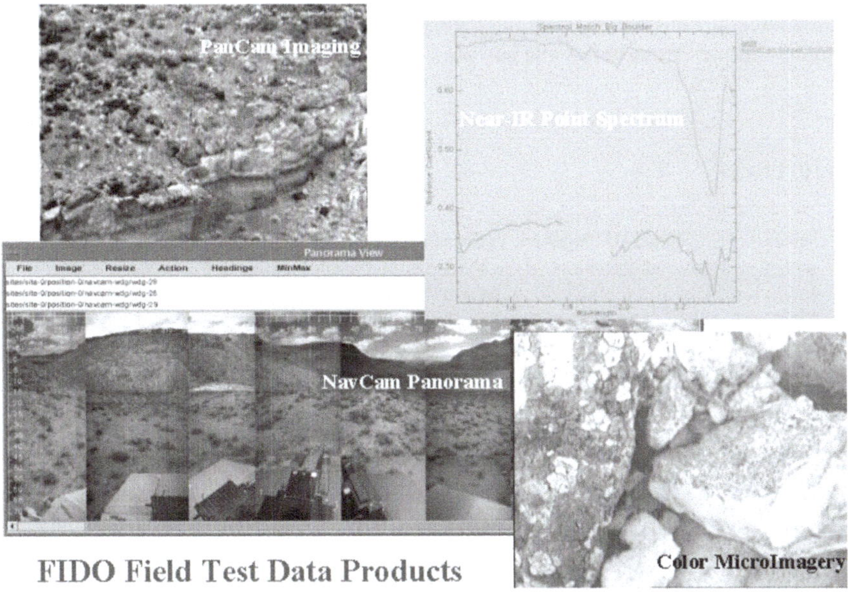

Fig. 12.31 Data outputs from field trials at Black Rock Summit (Nevada). ([21], courtesy NASA/JPL-Caltech)

Fig. 12.32 ExoMars field trials (ExoFiT) in the Tabernas Desert in Spain. Front: rover 'Charlie'; Back: landing platform ([25], courtesy ESA/Airbus)

mast-mounted Panoramic Camera (PanCam) and the ExoMars coring drill, which extracts samples from below the surface.

References

1. Mishkin, A., Morrison, J., Nguyen, T., Stone, H., Cooper, B., & Wilcox, B. (1998). Experiences with operations and autonomy of the Mars pathfinder microrover. In *Proceedings of the 1998 IEEE Aerospace Conference*, Snowmass at Aspen, Colorado.
2. Maimone, M. W., Leger, P. C., & Biesiadecki, J. J. (2007). Overview of the Mars Exploration Rovers Autonomous Mobility and Vision Capabilities. *IEEE International Conference on Robotics and Automation (ICRA) 2007*.
3. Grotzinger, J. P. et al. (2012). Mars science laboratory mission and science investigation. *Space Science Reviews*, September 2012, 170, (1–4), pp 5–56.
4. Bajracharya, M., Maimone, M. W., & Helmick, D. (2008). Autonomy for Mars rovers: Past, present, and future. *Computer, 41*(12), 44–50.
5. Maimone, M. et al. (2014). What drives Curiosity? Robotics Technologies on the Mars Science Rover. In *12th International Symposium on Artificial Intelligence, Robotics and Automation in Space (iSAIRAS),* Montreal.
6. Winter, M., Barclay, C., Pereira, V., Lancaster, R., Caceres, M., McManamon, K., Nye, B. Silva, N., Lachat, D., & Campana, M. (2015). ExoMars Rover Vehicle—Detailed description of the GNC system. In *ASTRA Conference 2015*.
7. Winter, M., Rubio, S., Lancaster R., Barclay, C., Silva, N., Nye, B., & Bora, L. (2017). Detailed description of the high-level autonomy functionalities developed for the ExoMars Rover, In *ASTRA Conference 2017*.
8. Gao, Y. (Editor), Iles, P., Winter, M., Silva, N., Bajpai, A., Muller, J., & Kirchner, F. (2016). Contemporary planetary robotics: an approach toward autonomous systems—Surface Navigation Chapter, John Wiley & Sons, ISBN: 978-3-527-41325-6.
9. McManamon, K., Lancaster, R., & Silva, N. (2013). ExoMars Rover Vehicle perception system architecture and test results. In *ASTRA Conference 2013*.
10. Bora, L., Lancaster, R., Nye, B., Barclay, C., Rubio, S., & Winter, M. (2017). ExoMars rover control, localisation and path planning in a hazardous and high disturbance environment. In *ASTRA Conference 2017*.
11. Jain, A., Guineau, J., Lim, C., Lincoln, W., Pomerantz, M., Sohl, G. T., & Steele, R. (2003). ROAMS: Planetary surface rover simulation environment. In *International Symposium on Artificial Intelligence, Robotics and Automation in Space (i-SAIRAS)*.
12. Jain, A., Balaram, J., Cameron, J., Guineau, J., Lim, C., Pomerantz, M., & Sohl, G. (2004), Recent developments in the ROAMS planetary rover simulation environment, 2004 IEEE Aerospace Conference, Big Sky, MT.
13. Lachat, D., Donchev, A., Kamath, A. K., Menon, P. P., & Watt, M. (2015). Generic rover dynamics model framework for autonomous capability development, verification and validation. In *ASTRA Conference 2015*.
14. Huntsberger, T., Jain, A., Cameron, J., Woodward, G., Myers, D., & Sohl, G. (2008). Characterization of the ROAMS simulation environment for testing rover mobility on sloped terrain. In *On Sloped Terrain: International Symposium on Artificial Intelligence, Robotics and Automation in Space (i-SAIRAS 2008)*.
15. Biesiadecki, J., Maimone, M. W., &. Morrison, J. (2002) The athena SDM rover: A testbed for mars rover mobility, *iSAIRAS Conference*, Montreal, Canada, 2001.
16. Image of three generations of NASA/JPL Martian rovers: https://mars.nasa.gov/resources/3791/three-generations-in-mars-yard-high-viewpoint/
17. Meacham, P., Nuno, S., & Lancaster, R. (2013). The development of the locomotion performance model (LPM) for the ExoMars Rover Vehicle. In ASTRA Conference 2013.

18. Lancaster, R., Silva, N., Davies, A. & Clemmet, J. (2011). ExoMars Rover GNC Design and Development. In *ESA GNC Conference 2011*
19. Goldberg, S. B., Maimone, M. W. & Matthies, L. (2002). Stereo vision and rover navigation software for planetary exploration. *IEEE Aerospace conference proceedings 2002*, volume 5, Big Sky, Montana, USA, 2002.
20. Toupet, O., Biesiadecki, J., Rankin, A., Steffy, A., Meirion-Griffith, G., Levine D., Schadegg, M., & Maimone, M. (2018, March). Traction control design and integration onboard the Mars Science Laboratory Curiosity Rover. *IEEE Aerospace Conference 2018.*
21. Schenker, P. S. et al. (2002, June). FIDO: A field integrated design & operations rover for Mars surface exploration. In *6th Intonational Symposium on AI, Robotics and Automation in Space, Montreal*, Canada, Paper No. AM012, 2001.
22. Arvidson, R. et al. (2002). FIDO prototype Mars rover field trials, Black Rock Summit, Nevada, as test of the ability of mobility systems to conduct field science. *Journal of Geophysical Research: Planets, 107*(E11), FIDO-2.
23. Tunstel, E. et al. (2002). FIDO rover field trials as rehearsal for the NASA 2003 Mars Exploration Rovers mission. In *Proceedings of 9th International Symposium on Robotics & Applications,* 5th World Automation Congress, Orlando, 2002.
24. NASA/JPL website: FIDO Field Test sol-by-sol mission log https://mars.jpl.nasa.gov/mer/fido/index.html (accessed December 2018)
25. ESA Website: ExoFiT Field Trials http://www.esa.int/spaceinimages/Images/2018/11/ExoFitness (accessed December 2018)

Chapter 13
Design and Validation of Orbital Robotic Missions

Marco De Stefano, Máximo A. Roa, Ribin Balachandran, Alessandro M.Giordano, Hrishk Mishra, Nassir Oumer, Bernhard Brunner, Martin Lingenauber, Martin Stelzer, and Roberto Lampariello

Abstract A servicer spacecraft equipped with robotic manipulators is a key technology for space debris removal and on-orbit servicing missions. The servicer is a composite system consisting of robotic manipulators, on-board cameras, proprioceptive and exteroceptive sensors, and the spacecraft actuation system. This hybrid nature of sensing and actuation on the spacecraft and manipulator subsystems poses several challenges to achieve the task of the robotic mission. This chapter outlines the challenges in such robotic missions and provides hardware design and control strategies, with a key focus on on-ground validation. Firstly, the mission aspects of the robotic capture of a satellite are introduced while considering the e.Deorbit mission study. The on-ground verification and validation of the control algorithms under micro-gravity condition is a necessity before the space mission commissioning. To this end, a state-of-art on-ground robotic facility, namely OOS-SIM, is described and used later for validation. From a control perspective, strategies for the orbital robot including vision-based semi-autonomy, shared control and teleoperation are reviewed. In particular, these controllers are designed to accomplish the mission task while considering the dynamic interaction between the spacecraft and the manipulator, the heterogeneity of the actuators available for control, and the lack of direct measurements of the complete state. In this context, the combined impedance control of the robotic manipulator and the satellite is specifically designed to be compliant against interaction with space targets. Finally, an end-to-end validation of a space-mission for on-orbit servicing is presented involving both on-board and on-ground segments, with realistic space communication characteristics.

M. De Stefano · M. A. Roa (✉) · R. Balachandran · A. M.Giordano · H. Mishra · N. Oumer · B. Brunner · M. Lingenauber · M. Stelzer · R. Lampariello
German Aerospace Center (DLR), 82234 Wessling, Germany
e-mail: Maximo.Roa@dlr.de

M. De Stefano
e-mail: marco.destefano@dlr.de

X. T. Yan and G. Visentin (eds.), *Space Robotics*,
https://doi.org/10.1007/978-3-031-39214-6_13

13.1 Introduction

Since the launch of the first satellite, the Sputnik 1 in 1957, the number of satellites in orbit has substantially increased. During the last decade, about 100 satellites were launched every year, and more than 29,000 objects greater than 10 cm are orbiting the low Earth orbit now [1]. The objects in orbit include not only satellites, but also space debris, a specific type of non-functional human-made space objects that orbit around the Earth. Examples of space debris are the discarded upper stages of launch vehicles, inactive satellites, and fragments produced by on-orbit collisions. The high density of space debris and the high velocity associated with the low orbit can generate mutual collisions leading to a chain reaction. This cascade effect is defined in literature as *Kessler syndrome* [2]. One of the major examples of such effect is the unintentional collision occurred in 2009 between the defunct Russian satellite Kosmos 2251 and the operational satellite Iridium 33. The collision generated more than 200,000 fragments greater than one centimetre, and the debris around the Earth increased drastically [3]. Furthermore, in recent years on-orbit failures have exceeded the launch failures, leading to an increase in space debris and losses of billions of dollars [4]. These failures are mainly related to failures in thruster systems or star tracking sensors, problems in the apogee motors or deployment of solar arrays, or wrong orbit location [5]. Besides that, every satellite eventually runs out of fuel and thus, must be decommissioned even if it is still functional.

Mitigation of space debris and resolution of satellite failures have driven the main space agencies to define a recovery plan, thus introducing the concept of On-Orbit Servicing (OOS) [6]. The term OOS refers to interventions on a defective system in orbit, including assembly, refuelling and repairing. The goal of an OOS mission is to extend the lifetime of satellites after failures and to actively remove space debris with a controlled re-entry in the Earth's atmosphere. It is notable that such tasks have motivated the development of new technologies in space, including unmanned missions. Among all, robotics emerges as a promising technology to perform such servicing tasks.

The assembly and maintenance in orbit using robotic technologies began in 1981, when the Space Shuttle manipulator, the *Canadarm*, was used for the first time. Several manned missions have followed for the capture and repair of malfunctioning satellites, including the repairing of the Hubble Space Telescope. Also the use of the manipulator onboard the International Space Station (ISS), the *Canadarm 2*, represented a milestone for the support of the ISS extra-vehicular activities. In addition to the robotic servicing capabilities, that are bound to the now decommissioned Space Shuttle or to the Canadarm 2, unmanned mission demonstrators were flown in orbit. The most relevant missions and demonstrations include ROTEX (the Robot Technology Experiment) developed by the German Aerospace Center (DLR) [7], the Japanese Engineering Test Satellite VII (ETS-VII) [8, 9], the German Robotic Component Verification experiment aboard the ISS (ROKVISS) [10], and the Demonstration of Autonomous Rendezvous Technology (DART) [11]. A particular milestone was the success of the DARPA Orbital Express mission in 2007 [12], whose goal was the demonstration of autonomous rendezvous and docking operations, including maintenance activities like refuelling on a *partially-cooperative* satellite. However, space debris and defective satellites are usually non-cooperative, i.e. they provide no support for their attitude and orbit control. Furthermore, defunct satellites might not have a docking port or retro reflectors used for vision-based navigation. This factor represents a challenge for a control system due to the lack of information about the state of the target.

Currently, many space projects focus on the capture of a non-cooperative spacecraft using a robotic manipulator. An overview of these space robotic projects is presented in Sect. 13.2. The on-ground validation of the algorithms and controls prior to the launch plays a key role for the success of an OOS mission. For supporting this on-ground testing, a robotic facility has been developed at DLR, as described in Sect. 13.3. Novel nonlinear controllers based on compliant behaviour for a robotic manipulator in an OOS mission have been designed by DLR and are reported in Sect. 13.4, including an end-to-end simulation summarized in Sect. 13.5.

13.2 Robotic Design for On-Orbit Servicing and De-Orbiting Missions

Typically, an OOS mission requires the use of a manipulator arm operating on a controlled satellite base. The arm performs multiple tasks, including the approach to a non-cooperative target satellite, followed by grasping and stabilization of the target. Later, the docking manoeuvre begins and allows the connection of the two spacecraft through a dedicated interface (e.g. the apogee motor), hence an OOS task or a de-orbiting manoeuvre can be performed on the combined structure.

Several missions have been proposed for performing OOS, among them, the German Orbital Servicing Mission study (DEOS - Deutsche Orbital Servicing Mission). DEOS investigated technologies to perform satellite rendezvous and close proximity operations, as well as to capture a tumbling and uncooperative target satellite with a dexterous manipulator in autonomous or in teleoperation mode [13]. Although this particular project did not continue beyond the preliminary design phase, the work on the robotic capture system was used as the technological heritage for the e.Deorbit [14] and COMRADE (COntrol and Management of Robotics Active DEbris removal) [15] mission studies, both projects commissioned by the European Space Agency (ESA). The most challenging task considered in e.Deorbit and COMRADE is the removal of Envisat, an eight-ton Earth-monitoring satellite that is defective and free-tumbling in space since 2012. It is recognized as the biggest debris in low Earth orbit.

A robotic capture design was developed within the e.Deorbit project for the capture of Envisat based on a chaser satellite equipped with a seven degrees-of-freedom (DoF) dexterous robotic manipulator, holding a dedicated linear two-bracket gripper [16]. The grasping point was chosen to be on the Launch Adapter Ring (LAR) of

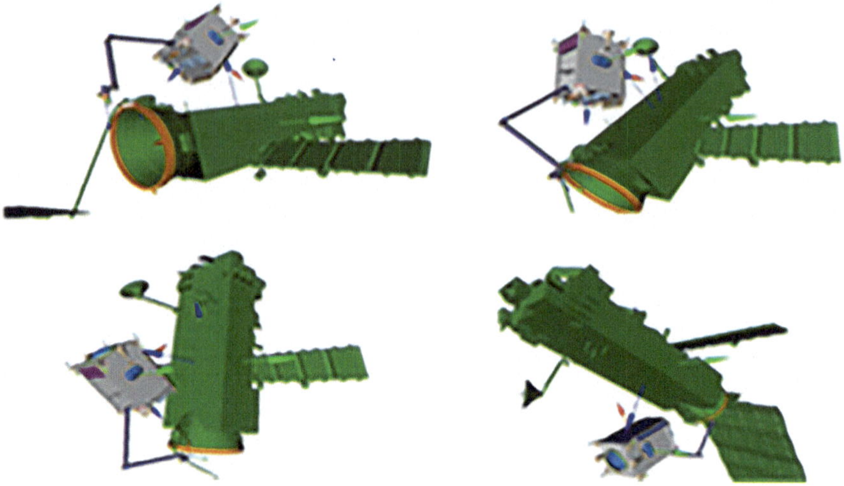

Fig. 13.1 E.Deorbit scenario: servicer satellite equipped with a robotic arm during the tracking of Envisat. Envisat tumbles about its intermediate axis with an angular velocity of 5 deg/s

Envisat, which provides a very solid, stiff and well-defined structure. The robotic operation consists of two major tasks: first, the approach of the manipulator towards the LAR, and second, the subsequent grasping and stabilization phase of the arm. The latter creates a connection with the Envisat, which can be then controlled for the subsequent de-orbiting manoeuvre.

Figure 13.1 shows the synchronized flight between a chaser satellite equipped with a manipulator arm and the Envisat satellite. Envisat tumbles around the y-axis of its body frame, and the chaser manipulator approaches it from the top side. The main advantage of choosing this position is the absence of obstacles at this side of the ring, such as the antenna and connectors that were previously used to connect Envisat with the launcher vehicle, thus offering greater robustness for the grasping manoeuvre.

The robotic arm concept for capturing the target is shown in Fig. 13.2. The arm features seven DoF, due to its seven revolute joints (see [16] for details about the design), with a total stretched length of 4.3 m. This chosen kinematical structure is redundant (of redundancy degree one) with respect to the grasping task, which endows the arm with greater dexterity and robustness to singularities. The linear-driven gripper was designed to achieve full form- and force-closure on the standardized LAR. The arm joint design is based on the heritage from the third generation of the Light-Weight Robot technology [17] and from the ROKVISS mission [10].

In ROKVISS, two robot joints were space qualified during a five year period of operation outside the Russian service module of the ISS. Each joint features integrated position as well as torque sensors, operating in classical position or active impedance control modes. The latter allows minimizing the possibly detrimental effect of collisions with the target during the grasping phase.

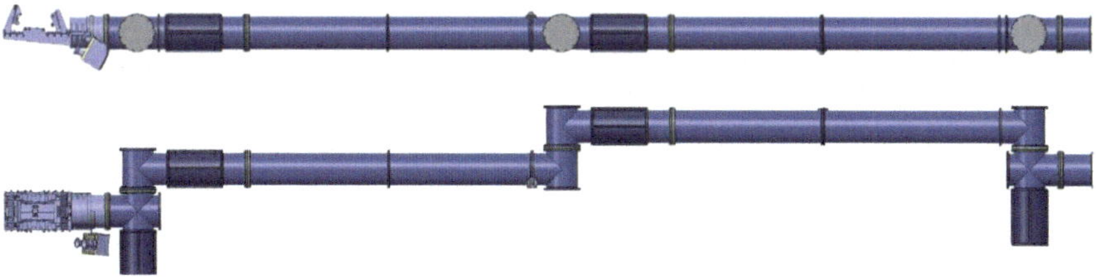

Fig. 13.2 Robotic arm in stretched configuration with built-in gripper and stereo camera system at the end-effector

In order to verify the task-specific performance of the chosen manipulator length and configuration, the kinematics of the manipulator was validated and analysed using the method of the reachability and capability map [18]. With this technique, the workspace for the end-effector is discretized into voxels (3D translation), and each voxel has an associated structure discretizing the pointing orientation and rotation around it (3D orientation), thus accounting for all six dimensions of the end-effector pose. Such a model provides a global overview of the workspace and the capabilities of the robotic arm in positioning and orienting the end-effector. A kinematic performance measure called reachability index is then derived from the workspace representation as a percentage of reachable rotations within each voxel.

The workspace models are generated accounting for self-collisions of the robot and collisions of the robot with the chaser body. This is important to make sure that the workspace model properly reflects the overall system capabilities. Figure 13.3 depicts the arm delivery point pose (left) and the cross-section of the capability map with the chaser satellite (right). The capability map represents in a colour code the dexterity of the robot in each particular region of the workspace; blue regions have high dexterity, while red regions are regions with poor arm dexterity.

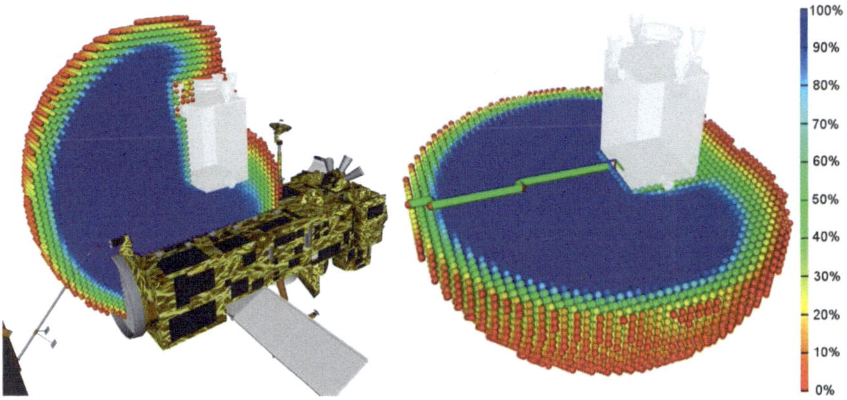

Fig. 13.3 Capability map of the manipulator arm for the e.Deorbit mission

The robotic manipulator features a stereo camera system mounted laterally on its end-effector, including an integrated illumination system. Both during grasping and fixation tasks, the image processing algorithm estimates the relative roto-translation (6 DoF pose parameters) from the camera to the target. The performance of the visual tracking is very important, as it is the key contributor to the positioning accuracy of the robotic arm. Using these images, a visual tracking algorithm [19] detects the visible adapter ring edges and matches them to an internal model. The measured tracking error between the cameras and the grasping point on the LAR is the input to a visual servoing system, which is necessary for achieving an accurate relative positioning of the gripper. Initially, the servoing starts with a predicted pose that must be provided using robot kinematics and trajectory planning. This prediction is refined by minimizing a cost function (visual matching error between the end effector and the target) through a local optimization algorithm.

The orbital environment conditions were simulated with the ASTOS simulation tool [16]. This allows uploading CAD models of the robot and of Envisat, as well as positioning Sun and Earth in relation to the two satellites in a realistic fashion. An example of a camera image rendered with ASTOS for the e.Deorbit scenario is shown in Fig. 13.4. The estimated pose is exploited by the visual servoing algorithm, which autonomously drives the manipulator to the final grasping point.

The main control modes of a manipulator that can be used in an OOS robotic space mission can be classified in general as follows [20],

- Semi-autonomy: a reference trajectory is planned based on the motion of the target satellite, derived from the estimated target pose. The trajectory is executed on the robotic manipulator with a local controller that takes into account disturbances and prediction errors. The semi-autonomy mode considers that the grasping point is given (e.g. preselected by the operator).
- Teleoperation: the manipulator arm is controlled remotely by a human operator using a haptic interface, which delivers to the operator visual and force feedback

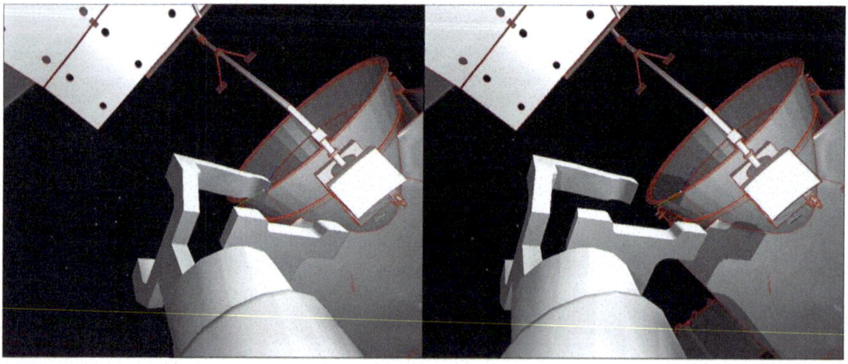

Fig. 13.4 E.Deorbit scenario: example of end-effector camera image rendered by ASTOS

from the real environment. This feedback increases the feeling of being directly operating on the remote environment.

- Shared control: the command input to the manipulator (e.g. torque) is provided either by teleoperation or by shared autonomy, depending mainly on the performance of the visual servoing component: when it degrades, the human should take over to guarantee successful execution of the tasks.

In order to assess the success of the mission, the validation of the robot control algorithms for the de-orbiting task (of Envisat in this case) is a key factor. A ground-based simulator is used in this validation process in order to test the different control modes and strategies, as explained in the following section.

13.3 On-Ground Validation: The DLR OOS-SIM

For performing on-ground experimental validation of robotic control algorithms for on-orbit tasks, including grasping and servicing of a tumbling satellite, DLR developed the OOS-SIM [20]. This facility exploits the hardware-in-the-loop (HIL) simulation technology, in which a computer-based dynamic model of the satellite system is used to command the robots in order to reproduce the desired motion under microgravity conditions.

The OOS-SIM, shown in Fig. 13.5, uses two 6-DoF industrial robots (KUKA KR-120) to simulate the gravity-free conditions of the satellites. These robots can carry payloads up to 120 kg. The robot on the right-hand side carries the client satellite, e.g. Envisat, and provides the dynamics of the free-tumbling satellite that requires servicing. This robot is equipped with a force-torque sensor (FTS) at the end effector, which is exploited for measuring external interaction forces. The robot on the left-hand side carries the servicer satellite, which integrates a docking interface and OOS-like elements such as a refuelling interface. It is also equipped with a FTS at the end effector. To perform the operations, the servicer carries a 7-DoF Lightweight Robot (DLR-KUKA LBR) equipped with a stereo-camera and a three-fingers robotic gripper from Robotiq. The LBR has position and torque sensors at all the joints. The sensors used in the OOS-SIM facility are summarized in Table 13.1, including their corresponding interface, sampling time, and functional description.

Servicer and client robots operate in a common workspace for testing tracking, grasping and stabilization of the free-tumbling client satellite through the servicer manipulator. The common workspace for the simulator is obtained by analysing the setup with the capability maps. For the OOS-SIM case, the robots are treated as two independent serial kinematic chains, and their individual workspaces are intersected, as shown in Fig. 13.6. This provides an estimation of the workspace available for simulating on ground different operations for the close-range scenario of a robotic space mission.

The software architecture of the OOS-SIM is designed to separate elements that correspond to the space dynamic simulation from those that do not need to be

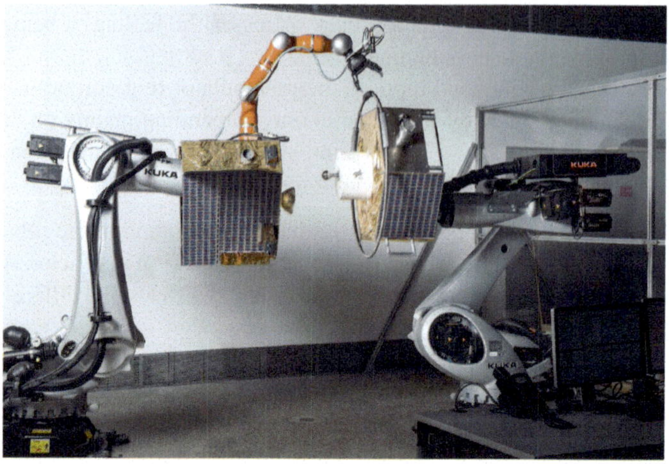

Fig. 13.5 The OOS-SIM facility: servicer robot (left) and client robot (right)

Table 13.1 Sensors and interfaces in the OOS-SIM facility

	Sensor	Interface	Sampling Time	Function
Client FTS	FTS160	Ethernet	4ms	Physical interaction
Servicer FTS	FTS160	Sercos	1ms	Physical interaction
Cameras	GigE	Ethernet	1ms	Visual servoing
IMU (Inertial Measurement Unit)	Xsense MTI	Sercos	1ms	Star field simulation

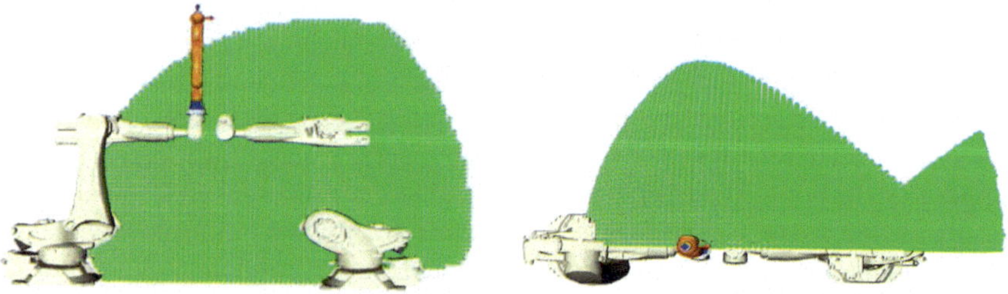

Fig. 13.6 Workspace available for simulation of space robotic operations with the OOS-SIM: sideview (left) and top-view (right)

simulated and can be taken as in the real mission, e.g. the space robot control. The schematic of the communication between hardware and sensors is shown in Fig. 13.7. A real-time computer (RT-C1) performs the computation of the model-based dynamics for the servicer and client. This computer runs with a sampling time

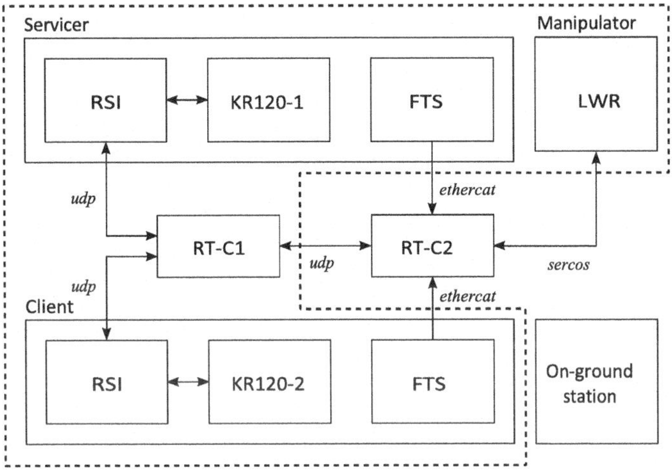

Fig. 13.7 Overview of the system communication in the OOS-SIM. The dashed line divides simulated and mission-like elements

of 4 ms. A second real-time computer (RT-C2) runs at 1 ms and controls the manipulation arm (LBR) with a Sercos interface, which allows real-time communication. The two real-time computers communicate through UDP-protocol. The communication with the industrial robots (KR120-1 and KR120-2) is established through the Robot Sensor Interface (RSI). Additionally, the RSI is synchronized to RT-C1 in order to preserve real-time determinism.

The industrial robots can be controlled only by means of position commands, and this fact imposes an admittance causality on the controller architecture for rendering the dynamics of a satellite. A generic architecture for controlling the industrial robots for satellite dynamics simulation is shown in Fig. 13.8. The input to the simulated dynamics (Sat.Dyn.) is the measured forces and torques, expressed in the wrench vector $\boldsymbol{F}_e$ from a FTS. Hence, the satellite dynamics can be computed according to the external physical interaction. The block L represents the integral process and the transformation required to compute a desired homogeneous matrix $\boldsymbol{H}_d$ commanded to the robot through the inverse kinematics. In particular, by double integrating the simulated dynamic model, a relative position can be defined, and using the Euler-Rodrigues formulation, a relative orientation matrix can be found. As a result, the robot moves in Cartesian space simulating the satellite motion [21].

However, the use of industrial robots for dynamics simulation represents a challenging task. The discrete-time integration of the model-based dynamics and the presence of time delay after a hardware contact for the corresponding simulation-driven reaction is recognized as a limiting factor [1]. In particular, the time delay in the control loop might lead the system to become unstable [22], and a non-consistent dynamics integration can lead to an unrealistic motion [23].

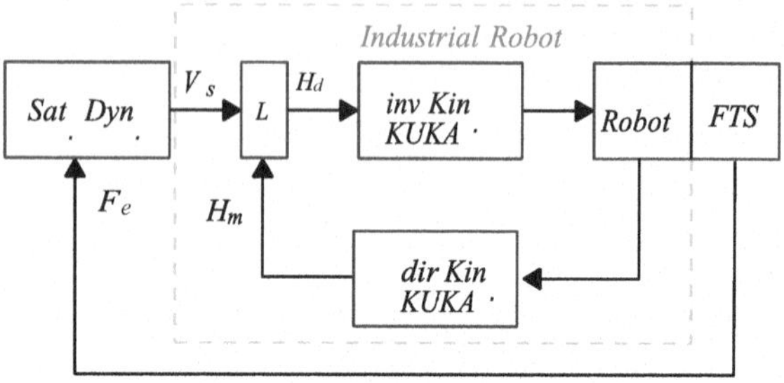

Fig. 13.8 Control architecture of the industrial robot for reproducing satellite dynamics

For the time delay problem, the approach employed in the OOS-SIM facility is based on the time-domain passivity approach [24], which shows good performances when compared with common energy-based methods [25]. Forces and energy observers were designed to monitor the activity of the system [21]. Therefore, the exact amount of energy that could destabilize the system is known, and a time-varying element is designed to dissipate it, namely the Passivity Control (PC). The PC takes the form of a dissipative element in a series or parallel configuration, and it restores the energy balance for guaranteeing the passivity of the system. In order to preserve the performance, an optimal damping design solution has been found based on an optimization problem resolved in [26].

The energy generation that results from the standard Euler integration process has been also identified in the OOS-SIM [27]. This energy affects the discrete time dynamics, which deviates from the ideal one, resulting in position drifts or stability issues. The approach to solve the problem is to design an additional energy observer in order to reshape the simulated dynamics and to preserve its physical energy properties [28]. A unified architecture is designed, which combines the time delay compensation and the passive integrator, and acts as a passivity layer between the real hardware and the mathematical model. Time delay effects are resolved with a passivity control (PC) which acts in impedance mode (force correction), and the discrete integration effects are solved with a second PC in admittance mode (velocity correction) [29].

The OOS-SIM is also equipped with a sun simulator in order to emulate on-ground the sunlight conditions as in orbit. It consists of a strong spotlight on a pan-tilt-unit, a linear axis with a semicircular shape to move the spotlight around the servicer and client robots, and a system of black Molton screens to block any external light and to provide a space-like background. The spotlight provides approximately the same spectrum and colour temperature as sunlight above the atmosphere. It is operated with a short arc metal halide lamp, which provides a maximum total output of 30,000 lumen at 1400 W with a Colour Rendering Index Ra (CRI Ra) of 95 at 6000 K colour temperature, in comparison to approximately 5800 K colour temperature of sunlight in Earth orbit. Due to the lamp's high CRI Ra, close to the maximum value of 100 for natural light, the colours of illuminated objects are revealed as if they would be illuminated with natural sunlight. The opening angle of the light beam can be changed between approximately 10° and 45°, thus allowing to either fully illuminate the workspace around the client or to provide a high intensity light on a

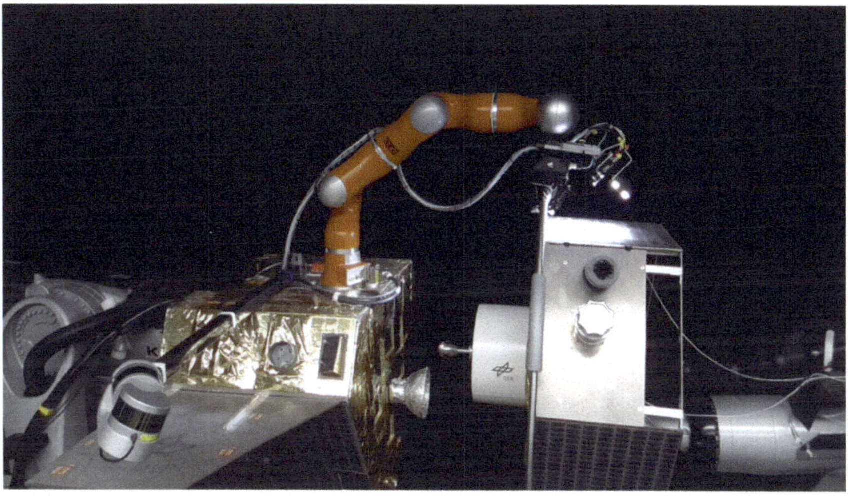

Fig. 13.9 The OOS-SIM facility in space light condition. A LED light mounted at the end-effector illuminates the target, and a spotlight simulates the sun illumination

dedicated target area. By moving the spotlight on the semi-circular linear rail around the satellites, it is possible to illuminate the scene from all positions from behind the servicer and behind the client satellite, i.e. with an azimuth angle between the sun vector and the flight vector of 0° to 180°. Thanks to the spotlight's pan-tilt unit, the light can always be directed to the same target area on the client, independent of the spotlight's position on the linear rail. In order to provide constant illumination for the camera systems, even when the satellites are simulated to be in the shadows, a LED matrix is attached to the end-effector of the robotic arm, next to the stereo camera system. It illuminates the volume in front of the cameras during shadow or eclipse conditions as shown in Fig. 13.9.

The effects of the atmosphere, such as light scattering, are still present to some extent since the OOS-SIM is not located in a thermal-vacuum chamber. Nevertheless, due to the strong light and the sharply formed beam, the desired challenges for vision systems, such as extremely bright regions, strong reflections on shiny surfaces or hard shadows with sharp edges, are present in the set-up. Therefore, it is possible to perform validation tests of the vision algorithms and control in the development stage with near-realistic illumination conditions as in orbit.

13.4 Control Strategies

The design of controllers for orbital robots presents several challenges from the point of view of dynamics, sensing, and actuation. Some of the major challenges are the dynamic interaction between the spacecraft and the arm, the heterogeneity of the actuators available for control, and the lack of direct measurements of the complete state. Furthermore, a compliance capability for the manipulator is required in applications where external interaction is involved, such as in grasping a target satellite. In contrast to classical position control, Cartesian impedance control offers more robust interactions [20]. In this section, different controllers developed for the space manipulator and validated on the OOS-SIM facility are presented.

13.4.1 Semi-Autonomy

The autonomous control system on the servicer relies primarily on measurements obtained from the exteroceptive and proprioceptive sensors located on the servicer for rendezvous and approach towards a non-cooperative and tumbling satellite. The exteroceptive measurements include the estimated relative pose (attitude and position) of the grasping frame located on the target satellite. This pose is obtained using processing algorithms on the exteroceptive sensor data. Furthermore, LIDAR-based measurements and star/sun trackers are commonly employed for rendezvous missions. Alternatively, due to its small form and cost factors, an optical camera-based solution is of increasing interest for an OOS mission.

The tracking of the target is usually performed with a model-based tracking, which exploits the knowledge of the CAD model of the target (client satellite). It relies on tracking pre-determined geometric features of the target object (e.g. edges and contours) using camera images obtained online during the approach and rendezvous phases [30]. The camera specifications, including field of view, depth of field, focal length and image resolution, are selected depending on the distance to the target. The distance between target and servicer limits the expected accuracy of orientation and position estimation. This factor affects the robot control performance, which requires millimetre accuracy for safe and reliable operation of the desired task. Tracking the target at very close-range poses difficulties because of a harsh lighting conditions and occlusions, as well as feature invisibility, while tracking the target at far-range does not provide the required accuracy for manipulation. The lighting conditions in space are extremely challenging for camera-based tracking due to the direct sunlight and reflective characteristics of a client satellite multilayer insulation (MLI). Moreover, a complex tumbling motion of the target poses additional difficulties for the visual tracking.

Traditional tracking algorithms can be robustified to deal with the challenging conditions of tracking an object in space. Methods based on edge detection have proven useful in this sense [31]. First, model contours are sampled and re-projected at an initially known pose, followed by a frame-to-frame optimization of the rigid body pose. The re-projection error is minimized to match sampled visible contours and image edges using a Levenberg–Marquardt algorithm. The use of multiple features of the object as references can improve estimation, for example during visual servoing and close-range approach where unnecessary features can be removed for the specific

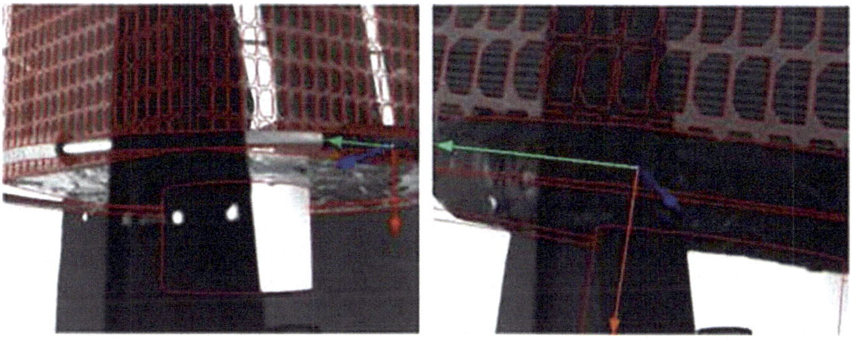

Fig. 13.10 Tracking a target satellite: pose estimation for very close range approach phase with overlay of the model contours (in red) and image edges. The perfect overlapping of image edges and model contours indicates a precise orientation and position estimation

task. Furthermore, motion constraints can be incorporated into the tracking to tackle relatively large target motions and improve tracking robustness. Thus, a prediction scheme is integrated into the optimization in order to align the image edges and model contours. Assuming a constant frame-to-frame motion, an Extended Kalman filter (EKF) is used to reduce the search space during minimization of the cost function. This method has been experimentally demonstrated for the tracking of a tumbling target satellite with the OOS-SIM facility [32]. An example of visual tracking is shown in Fig. 13.10, where the re-projection of the visible model edges at the estimated pose are indicated in red.

In a first approach, a cascade control consisting of a visual tracker, motion planner, an EKF, and a feedback controller were used to approach, track, grasp and stabilize the tumbling target by means of a kinematically redundant robot manipulator, validated on the OOS-SIM facility in [32]. The nonlinear stability analysis of such an interconnected system was verified in [33] using vector Lyapunov functions, specifically while considering an observer, kinematic visual servo and a PD + controller.

With the development of combined control strategies that employ full actuation capability of the servicer [15], an Inertial Measurement Unit (IMU) sensor is necessary on the servicer for agile motions. However, factors like the time-varying nature of the control task, sensor noise and low sampling rates negatively affect the control performances. Hence, a state-reconstruction is required, which is a key element for a successful autonomous robotic capture of a tumbling satellite. The estimated motion states provide the necessary feedback/feed-forward terms for the time-varying Cartesian control required for a mission like the capture of Envisat in [15]. Within this context, a nonlinear geometric observer is needed to estimate the time-varying inertial motion states of the client satellite using relative exteroceptive measurements, as presented in [34]. A coupled EKF, which combined sensor measurements from a LIDAR, camera, IMU and star/sun tracker to estimate the inertial motion states of the servicer platform and the tumbling satellite was validated in [35] for a fully-actuated approach strategy. The EKF has demonstrated to be capable of providing robustness against sensor noise, outliers, and sensor outage, which improves controller performance.

The controllers of the manipulator developed for the semi-autonomy mode are designed to have a compliant behaviour, which facilitates the adaptation of the robotic arm to uncertainties in the task and environment. This can be achieved by controlling the manipulator in torque mode. In early control concepts, attention was given to the possibility of completely turning off the spacecraft's actuators and controlling only the manipulator by adopting the so-called free-floating control strategies [36] under the assumption of zero initial momentum. In particular, a transposed-Jacobian method was proposed in [37] for the torque-based control of the end-effector; this method was also experimentally tested on the OOS-SIM [38].

The limitations of the free-floating control approach arise when any voluntary or accidental contact happens, as it induces an inertial drift in the free-floating system with respect to the given target location [39, 40]. This risky situation can only be recovered by using a fully-actuated strategy that utilizes the satellite's external actuators to stop the drift and restore the proper robot-target location. These strategies are also considered in space projects for the de-orbiting of Envisat (e.g. [15, 41]).

Fully-actuated strategies were derived firstly in [42] to simultaneously regulate the poses of the spacecraft and the end-effector of the manipulator with respect to the inertial frame, and the tracking problem of the target was resolved in [43] with a nonlinear controller. The fully-actuated strategies make intense use of thrusters due to the task of actively controlling the pose of the spacecraft. This factor leads to high fuel consumption, plume impingement, and limitation of the manoeuvring speed due to thrusters' saturation and low bandwidth.

A new class of momentum-based whole-body controllers was recently developed to mitigate these limitations and reduce the thrusters usage. The controllers replace the task of regulating the pose of spacecraft with the so-called momentum dumping and whole-body Center-of-Mass (CoM) control tasks (see [40, 44]). The controllers make use of triangular actuation decompositions to avoid direct coupling of the end-effector task into the thrusters. Therefore, they do not enforce requirements on the pose of the spacecraft, and completely decouple the robot control from the thrusters. In this way, thrusters are ideally not activated during robot manoeuvres, and are activated automatically only to stabilize the system after contact, limiting their usage to a minimum. For space missions requiring control of the spacecraft orientation, momentum-based whole-body controllers were proposed in [45], where a partial decoupling is employed to avoid activation of the thrusters in translation during robot manoeuvres, though the thrusters are still used for rotations.

A comparison between the aforementioned strategies was performed in order to evaluate the fuel consumption while considering ideal thrusters model. The comparison considers a space robot composed of an LBR mounted on a 150 kg spacecraft, for the following strategies:

1. Fully-actuated control (coordinated control [42, 43]): the spacecraft translation and rotation are controlled using the actuators.

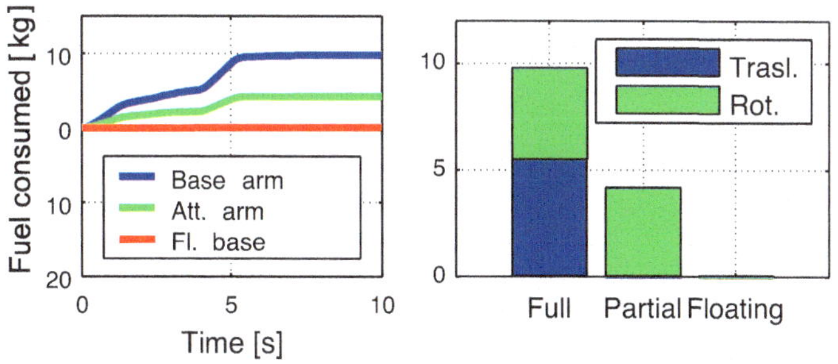

Fig. 13.11 Fuel consumption for different control strategies for a robotic manipulator in space

2. Partially-actuated control (whole-body coordinated control [45]): the spacecraft attitude is controlled and the translation is left free. The space robot's CoM is controlled instead.
3. Free-Floating control (whole-body coordinated control [44]): both the spacecraft translation and rotation are not controlled. The space robot's CoM and the angular momentum are controlled instead.

Figure 13.11 shows the results of the comparison. During contact-free manoeuvres of the arm, the fuel consumption can be reduced to zero when using the floating-base whole-body control in [44], i.e. the motions of the fully-actuated robot do not consume fuel. In scenarios where the attitude control is mandatory and only thrusters are available (e.g. for the de-orbiting of Envisat), the whole-body strategy with partial base control [45] can be used as an intermediate solution to save at least the considerable amount of fuel required for translation. Note that this zero fuel consumption is only ideal, in practice the limit cycle induced by real discrete thrusters consumes a minimal amount of fuel.

Space missions that require a fully-actuated control strategy impose constraints on the frequency of control available for the manipulator and the GNC (Guidance Navigation and Control) system. Typically, the control of the spacecraft base runs at low rate (between 1 and 10 Hz) and the controller of the manipulator at a higher rate (usually 1000 Hz), since the latter has to interact with the environment [41]. The discrete nature of the controller and the difference in the actuation frequencies must be taken into account during the design of the controller because they can jeopardize the stability of the system [22]. The effect of low actuation of the base controller was first analysed in [46] for the regulation case, and a novel control strategy for a space manipulator operating on a controlled base was developed in [43]. Besides guaranteeing trajectory tracking for the manipulator in Cartesian space and the regulation for the base, the multi-rate nature of the controllers was also addressed from an energetic perspective. In particular, the effects due to the different sampling rates of the manipulator and the base controllers were identified, isolated and addressed through energy-based control strategy. A passivity-based controller with a time-varying damping was designed for controlling the base and the manipulator to avoid instability issues. The approach has been experimentally validated in the OOS-SIM facility, as presented in [43].

13.4.2 Teleoperation

The semi-autonomous system explained in the previous section relies highly on perception of the target satellite, which is affected by internal and external factors. In particular, for the vision-based system, internal factors include camera noise and its time-varying nature, and communication failures. In adverse cases, the robot may negatively affect its perception due to occlusions/image losses that result from its own motion. Eye-in-hand cameras are affected by close-range vision degradation, or from the target going out of the image [47]. Among external factors, sensing capabilities are affected by illumination changes, reflections and shadows on the tracked target [48]. These factors might lead to a failure in the task execution by an autonomous system, and a human-in-the-loop supervision is always preferred due to the critical nature of the orbital robotic missions.

In teleoperation mode, a bilateral teleoperation scheme is exploited wherein a human operator, using a Human–Machine-Interface (HMI) device (*leader*), controls the robotic manipulator on the servicer satellite (*follower*) to perform tasks on the target satellite. The forces felt by the follower while interacting with the target are reflected back to the human operator by closing the force feedback loop, making it more intuitive to the operator to implement the task at hand [49]. The presence of time delay in the communication channel between the human operator and the follower in a bilateral teleoperation scheme can render the system unstable, and even if stabilized, can affect the performance of the operator as a trade-off for stability [50]. Teleoperation for OOS mission can be achieved by using a 4-channel bilateral controller, which ensures stability while exploiting the time-domain passivity approach [24].

Although the time delay for a direct link between the human operator and the robotic manipulator for a Low-Earth Orbit (LEO) servicer satellite is 20–30 ms, the operation windows are extremely low (maximum 10 min) [51], which is not enough to complete practical servicing tasks. The operation windows can be significantly improved if the signals are routed through multiple satellites in the geosynchronous Earth orbits (GEO) (see Fig. 13.12, left), but this increases the communication delay between the operator and the follower manipulator.

To validate orbital teleoperation tasks in LEO satellites with long operation windows, experiments were conducted with the ASTRA GEO satellite acting as a relay system for the signals from the operator and the OOS-SIM manipulator [52], where the round-trip delay between the leader and follower was 539.44 ms with standard deviation of 3.11 ms and packet loss of 2.12%. Figure 13.12 shows the targeted scenario of controlling the manipulator on the LEO servicer through the GEO relay satellite (on the left), and the validation scenario using the OOS-SIM facility (on the right). It was presented in [52] that grasping and stabilization of the free-floating

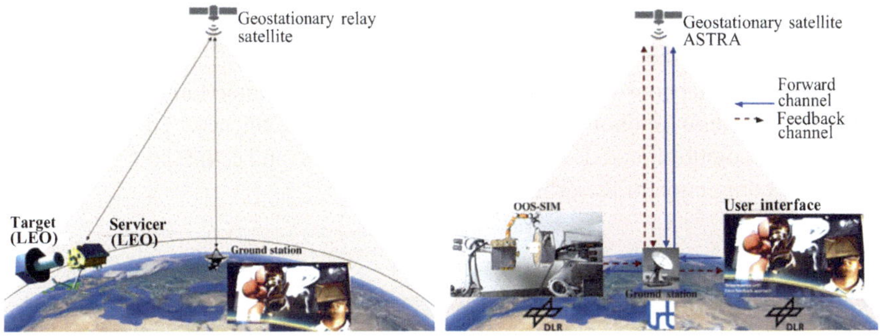

Fig. 13.12 Target scenario (left) and validation scenario of teleoperation with OOS-SIM (right)

OOS-SIM target using the servicer manipulator with teleoperation is feasible even under large time delays of about 540 ms.

In spite of the successful validation of the teleoperation system, it was observed that there were significantly high levels of physical and cognitive loads demanded from the operator during the operation of grasping the free-floating target satellite, especially in presence of high delays.

13.4.3 Shared Control

The two control strategies presented earlier, namely, semi-autonomy and teleoperation, have both benefits and demerits [48]. In a shared control approach, the benefits of both strategies are combined in a mixed initiative form [53]. This means that the final wrench $\boldsymbol{F}_s$ (force and torque) commanded to the manipulator on the servicer satellite is a weighted sum of the wrenches from the autonomous system, $\boldsymbol{F}_a$ and teleoperation controller, $\boldsymbol{F}_t$, as follows:

$$F_s = \alpha(t)F_t + \underbrace{(1-\alpha(t))}_{\beta(t)}\boldsymbol{F}_a, \tag{13.1}$$

The weights α and β are called authority allocation factors, which can be fixed or adaptive over time [54].

In OOS-SIM, an adaptive approach was validated wherein the control authority shifts from semi-autonomy to teleoperation. The values of α and β vary in such a way that when the manipulator is in fully autonomous mode, $\beta = 1$ and $\alpha = 0$. On the other hand, when the manipulator is in pure teleoperation mode, $\beta = 0$ and $\alpha = 1$. As soon as the vision-based pose estimation of the tracked target satellite degrades, the control is transferred to the human operator by decreasing β and simultaneously increasing α. The degradation happens mainly due to increase in noise of the camera measurements, which arises either due to changes in lighting conditions or during the final grasping phase where the eye-in-hand camera is in close proximity to the target [47]. Since the authority shift from semi-autonomy to teleoperation happens only during these cases, human intervention is necessary only when the semi-autonomous system fails to complete the task. This reduces the physical and cognitive load demanded from the operator, and at the same time allows successful completion of the orbital robotic tasks.

13.5 Validation of Close-Range Mission

On-ground validation of a robotic mission for OOS includes also the operations between ground and space segment. For the ground segment, the existing legacy of satellite operations is extended for an OOS mission, where the manipulator is considered to be a payload. Specifically, the actions commanded on ground (telecommands, TC) as well as the data received from the on-board system (telemetry, TM) are used in an on-ground infrastructure. Hence, an on-board system (e.g. the control of a space robot) can be validated for its readiness in a real mission scenario by closing the entire end-to-end loop [55].

For implementing the end-to-end simulation, two hardware-in-the loop simulators at DLR are used to represent the space segment: the EPOS (European Proximity Operations Simulator) 2.0 facility [56], for simulating the close range rendezvous phase, and the OOS-SIM facility for simulating the capture. The OOS-SIM facility, presented in Sec. 3, represents the on-board system (satellite with robotic payload) that performs the task of tracking and grasping the target satellite. Both facilities are shown in Fig. 13.13.

The combination of both hardware-in-the-loop simulators allows to reproduce the rendezvous phase of two satellites and the grasping of the target under realistic physics during contacts. However, a complex software system is needed in order to simulate the satellites behaviour. The satellite simulator contains different subsystems such as dynamics model, actuator model for the servicing satellite, basic

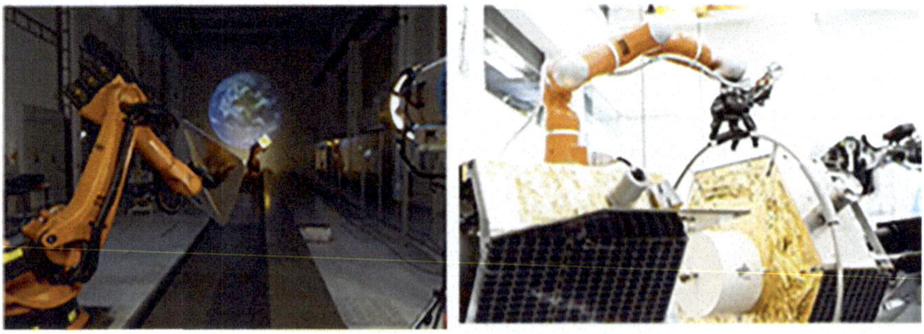

Fig. 13.13 Hardware-in-the-loop simulators at DLR: EPOS facility for simulating rendezvous (left), and OOS-SIM facility for simulating the robotic capture of the target (right)

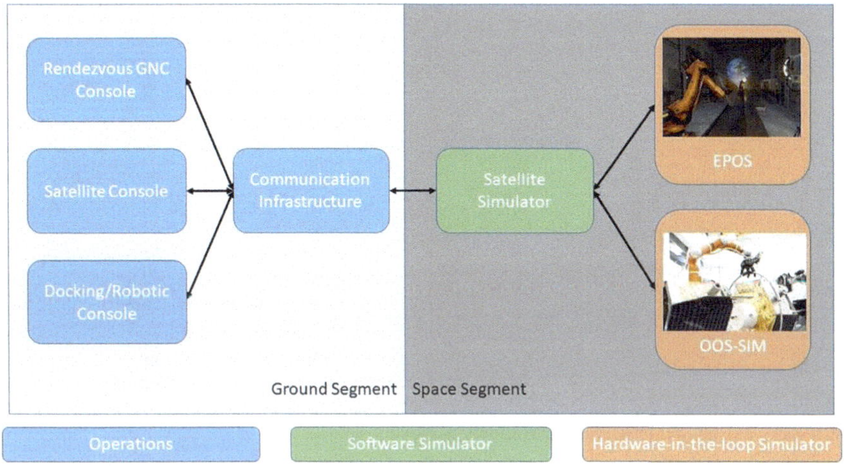

Fig. 13.14 Subsystems of an on-orbit-servicing mission simulation system

functionalities for power and thermal subsystems, and the complete on-board data handling system (telecommand and telemetry).

An overview of the end-to-end simulation concept is given in Fig. 13.14, and the validation results are reported in [55]. The figure illustrates the communication infrastructure, which plays a major role in the simulator set-up by connecting the ground consoles with the satellite simulator. In particular, the telemetry generated on-board the servicing spacecraft is transmitted via the communication infrastructure, which simulates both the communication path (including representative jitter and delays) and the connection of the antennas at the ground station to the consoles in the control room.

The robot control system is part of the satellite on-board controller. The robot controller communicates with the robotics console via the uplink and downlink stream. The main tasks of the robot control system include the processing of the telecommand and telemetry data, the decomposition of high-level commands into low-level control commands for the manipulators and/or available sensors (action interpreter), and the core control functions to operate the space robot in different control modes, e.g. position or torque control mode. Figure 13.15 shows the overall architecture of the robotic control system.

Regarding the autonomy aspect, high-level commands (received from ground) must be interpreted and decomposed into smaller units for the execution on the real-time motion controller. The autonomous part of the robot control system is complemented by the visual tracking system, which was described in Sec. 4.1, as well as a supervisor block observing the current execution state. To handle uncertainties or not predefined actions, the telepresence approach has been also included in the end-to-end simulation such that the operator on ground can control the manipulators on-board, as described in the previous section.

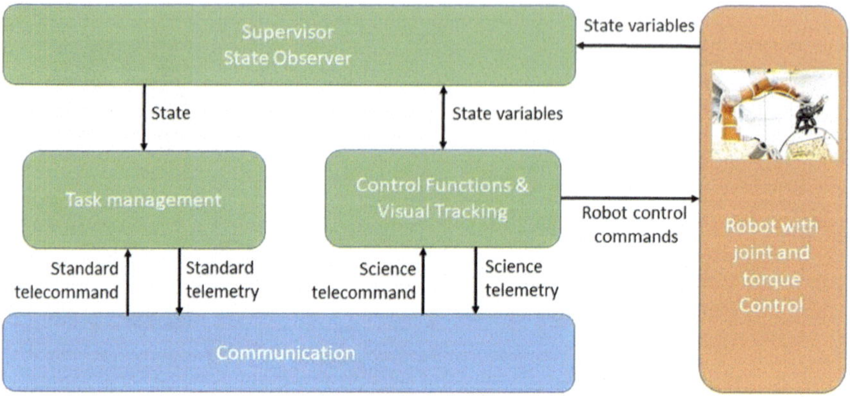

Fig. 13.15 Functional blocks of the robotics payload system

13.6 Summary and Outlook

Facilities such as the OOS-SIM represent a new class of hardware-in-the loop simulators that allow the development and testing of space robotics control algorithms, mainly related to On-Orbit Servicing operations. The facility can be potentially used for end-to-end tests, to verify in a micro-gravity environment a complete robotic OOS mission. Different types of OOS missions can be tested on the facility, including refuelling and repair of non-operative satellites, as well as de-orbiting of orbital debris and dead satellites such as Envisat. Recently, the OOS-SIM has been employed to validate the control strategies for the approach and capture of the Envisat satellite (by means of a redundant manipulator) within the COMRADE project [15] and the final set-up is shown in Fig. 13.16. To accommodate its large motion on the OOS-SIM facility, a novel relative dynamics formulation has been developed for testing the approach and capture phase on ground, and the results are reported in [57].

Research is now focused mainly on the development and improvement of control systems for the robotic payload, the visual tracking behaviour, as well as the different modes for controlling the servicer satellite and the robotic manipulator. Furthermore, the sun simulator will be exploited to validate the vision system for OOS robotic missions under realistic illumination conditions. Future developments will deal also with user studies to gain statistical insight on the human factor when dealing with latencies and specific on-orbit servicing tasks during teleoperation.

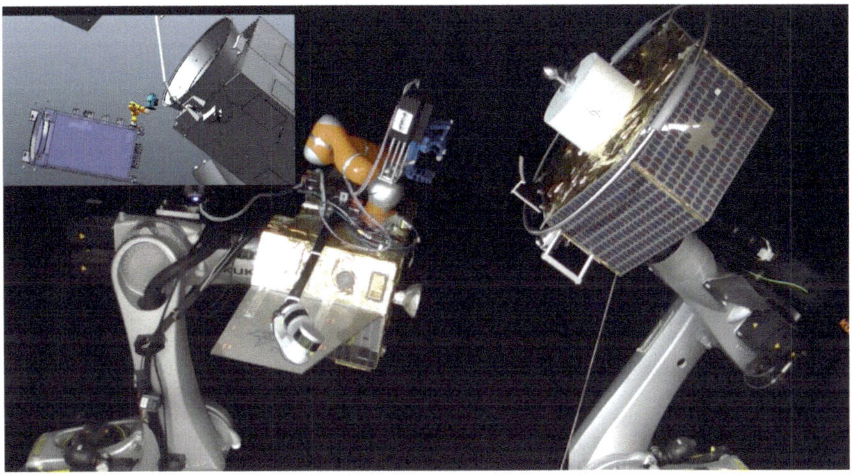

Fig. 13.16 OOS-SIM facility employed for simulating the approach and grasping of Envisat. The client robot is equipped with the LAR profile, and the servicer has an LBR robot with a gripper developed by PIAP. Top-left: simulated space scenario for the capture of Envisat

References

1. Flores-Abad, A., Ma, O., Pham, K., & Ulrich, S. (2014). A review of space robotics technologies for on-orbit servicing. *Progress in aerospace sciences, 68*, 1–26.
2. Kessler, D. J., & Cour-Palais, B. G. (1978). Collision frequency of artificial satellites: The creation of a debris belt. *Journal of Geophysical Research: Space Physics, 83*(A6), 2637–2646.
3. Liou J.-C. (2011). An active debris removal parametric study for LEO environment remediation. *Advances in Space Research, 47*(11), 1865–1876.
4. Sullivan, R., & Akin, D. L. (2001). A survey of serviceable spacecraft failures. In *AIAA Space, 2001*.
5. Ellery, A., Kreisel, J., & Sommer, B. (2008). The case for robotic on-orbit servicing of spacecraft: Spacecraft reliability is a myth. *Acta Astronautica, 63*(5–6), 632–648.
6. Miller, R., Minsky, M., & Smith, D. (1982). Space applications of automation, robotics and machine intelligence systems (ARAMIS). *Executive summary report NASA-CR, 162079*.
7. Hirzinger, G., Brunner, B., Dietrich, J., & Heindl, J. (1993). Sensor-based space robotics-ROTEX and its telerobotic features. *IEEE Trans. Robotics and Automation*, 9(5), pp. 649–663, Oct 1993.
8. Oda, M., Kibe, K., & Yamagata. F. (1996). ETS-VII, space robot in-orbit experiment satellite. In *Proc. IEEE Int. Conf. Robotics and Automation - ICRA*, pp. 739–744.
9. Yoshida. K. (2003). Engineering test satellite VII flight experiments for space robot dynamics and control: Theories on laboratory test beds ten years ago, now in orbit. *Int. J. Robotics Research* , *22*(5), 321–335.
10. Albu-Schaffer, A., Bertleff, W., Rebele, B., Schafer, B., Landzettel, K. & Hirzinger, G. (2006). ROKVISS - robotics component verification on ISS current experimental results on parameter identification. In *Proc. IEEE Int. Conf. Robotics and Automation - ICRA*, pp. 3879 –3885.
11. Howard, R. T., Johnston, A. S., Bryan, T. C., & Book, M. L.R. (2004). Advanced video guidance sensor (AVGS) development testing. In *Spaceborne Sensors*, vol. 5418, pp. 50–61. Int. Society for Optics and Photonics.

12. Shoemaker, J., & Wright, M. (2004). Orbital express space operations architecture program. In *Spacecraft Platforms and Infrastructure*, vol. 5419, pp. 57–66. Int. Society for Optics and Photonics.
13. Rank, P., Mühlbauer, Q., Naumann, W., & Landzettel, K. (2011). The DEOS automation and robotics payload. In *Symp. on Advanced Space Technologies in Robotics and Automation - ASTRA*.
14. Telaar, J., Ahrns, I., Estable, S., Rackl, W., De Stefano, M., Lampariello, R., Santos, N., Serra, P., Canetri, M., Ankersen, F. & Gil-Fernandez, J. (2017). GNC architecture for the e.Deorbit mission. *Europ. Conf. for Aeronautics and Space Science - EUCASS*.
15. Colmenarejo, P., Branco, J., Santos, N., Serra, P., Telaar, J., Strauch, H., Fruhnert, M., Giordano, A.M., De Stefano, M., Ott, C., Reiner, M., Henry, D., Jaworski, J., Papadopoulos, E., Visentin, G., Ankersen, F., & Gil-Fernández, J. (2018). Methods and outcomes of the COMRADE project - Design of robust Combined control for robotic spacecraft and manipulator in servicing missions: comparison between between Hinf and nonlinear Lyapunov-based approaches. In *Proc. Int. Astronautical Congress - IAC*.
16. Jaekel, S., Lampariello, R., Rackl, W., De Stefano, M., Oumer, N., Giordano, A. M., Porges, O., Pietras, M., Brunner, B., Ratti, J., Muehlbauer, Q., Thiel, M., Estable, S., Biesbroek, R., & Albu-Schaeffer, A. (2018). Design and Operational Elements of the Robotic Subsystem for the e.Deorbit Debris Removal Mission. *Frontiers in robotics and AI, 5*, 100.
17. Hirzinger, G., Sporer, N., Albu-Schäffer, A.O., Hähnle, M., Krenn, R., Pascucci, A., & Schedl, M. (2002). DLR's torque-controlled light weight robot III-are we reaching the technological limits now? *In Proc. IEEE Int. Conf. on Robotics and Automation - ICRA*, pp. 1710–1716.
18. Porges, O., Stouraitis, T., Borst, C., & Roa, M. A. (2014). Reachability and capability analysis for manipulation tasks. In *ROBOT2013: First Iberian Robotics Conference*, pp. 703–718.
19. Panin, G. (2011). Model-based visual tracking: the OpenTL framework. John Wiley & Sons, 2011.
20. Artigas, J., De Stefano, M., Rackl, W., Lampariello, R., Brunner, B., Bertleff, W., Burger, R., Porges, O., Giordano, A., Borst, C. & Albu-Schaeffer, A. (2015). The OOS-SIM: An on-ground simulation facility for on-orbit servicing robotic operations. In *Proc. IEEE Int. Conf. Robotics and Automation - ICRA*, pp. 2854–2860.
21. M. De Stefano, J. Artigas, W. Rackl, and A. Albu-Schäffer. Passivity of virtual free-floating dynamics rendered on robotic facilities. In *Proc. IEEE Int. Conf. Robotics and Automation - ICRA*, pp. 781–788.
22. Diolaiti, N., Niemeyer, G., Barbagli, F., & Salisbury, J. K. (20016). Stability of haptic rendering: Discretization, quantization, time delay, and coulomb effects. *IEEE Trans. Robotics, 22*(2), 256–268, April 2006.
23. De Stefano, M. (2019). Energy-based control for simulation of multi-body dynamics using robotic facilities. In *Ph.D. thesis, Univ. of Modena and Reggio Emilia, Italy*, March 2019.
24. Hannaford, B., & Ryu, J. H. (2001). Time domain passivity control of haptic interfaces. In *Proc. IEEE Int. Conf. Robotics and Automation - ICRA*, pp. 1863–1869.
25. De Stefano, M., Vezzadini, L., & Secchi, C. (2019). Time-delay compensation using energy tank for satellite dynamics simulators. In *Proc. IEEE/RSJ Int. Conf. on Intelligent Robots and Systems - IROS*, pp. 521–528.
26. De Stefano, M., Artigas, J., & Secchi, C. (2016). An optimized passivity-based method for simulating satellite dynamics on a position controlled robot in presence of latencies. In *Proc. IEEE/RSJ Int. Conf. on Intelligent Robots and Systems - IROS*, pp. 5419–5426.
27. De Stefano, M., Balachandran, R., Artigas, J., & Secchi, C. (2017). Reproducing physical dynamics with hardware-in-the-loop simulators: A passive and explicit discrete integrator. In *Proc. IEEE Int. Conf. Robotics and Automation - ICRA*, pp. 5899–5906.
28. De Stefano, M., Artigas, J., & Secchi, C. (2017). A passive integration strategy for rendering rotational rigid-body dynamics on a robotic simulator. In *Proc. IEEE/RSJ Int. Conf. on Intelligent Robots and Systems - IROS*, pp. 2806–2812.
29. De Stefano, M., Balachandran, R., & Secchi, C. (2020). A passivity-based approach for simulating satellite dynamics with robots: Discrete-time integration and time-delay compensation. *IEEE Trans. on Robotics, 36*(1), 189–203.

30. Drummund, T., & Cippola, R. (2002). Real time visual tracking of complex structures. *IEEE Trans. Pattern Analysis and Machine Intelligence, 24*(7), 932–946, July 2002.
31. Oumer, N. W., Panin, G., MÃijlbauer, Q., & Tseneklidou, A. (2015). Vision-based localization for on-orbit servicing of a partially cooperative satellite. *Acta Astronautica, 117*, 19–37.
32. Lampariello, R., Mishra, H., Oumer, N., Schmidt, P., De Stefano, M., & Albu-Schäffer, A. (2018). Tracking control for the grasping of a tumbling satellite with a free-floating robot. *IEEE Robotics and Automation Letters, 3*(4), 3638–3645, Oct 2018.
33. Mishra, H., De Stefano, M., Giordano, A. M., & Ott, C. (2018). Tracking control with robotic systems for a moving target: A vector Lyapunov function approach. In *Proc. IFAC Symp. on Robot Control - SYROCO*, pp. 471–478.
34. Mishra, H., De Stefano, M., Giordano, A. M., & Ott, C. (2019). A nonlinear observer for free-floating target motion using only pose measurements. In *Proc. American Control Conference - ACC*, pp. 1114–1121.
35. Gallardo, A. P., Mishra, H., Giordano, A. M., & Lampariello, R. (2019). Robust estimation of motionstates for free-floating tumbling target capture. In *Proc. IEEE Aerospace Conf.*, pp. 1–11.
36. Umetani, Y., & Yoshida, K. (1987). Continuous path control of space manipulators mounted on OMV. *Acta Astronautica, 15*(12), 981–986.
37. Masutani, Y., Miyazaki, F., & Arimoto. S. (1989). Sensory feedback control for space manipulators. In *Proc. IEEE Int. Conf. Robotics and Automation - ICRA*, pp. 1346–1351.
38. De Stefano, M., Artigas, J., Giordano, A., Lampariello, R., & Albu-Schäffer, A. (2015). On- ground experimental verification of a torque controlled free-floating robot. In *Symp. on Advanced Space Technologies in Robotics and Automation - ASTRA*.
39. Giordano, M., Garofalo, G., De Stefano, M., Ott, C., & Albu-Schäffer, A. (2016). Dynamics and control of a free-floating space robot in presence of nonzero linear and angular momenta. In *Proc. IEEE Conf. on Decision and Control*, pp. 7527–7534.
40. Giordano, M., Garofalo, G., & Albu-Schäffer, A. (2017). Momentum dumping for space robots. In *IEEE Ann. Conf. on Decision and Control - CDC*, pp. 5243–5248.
41. Telaar, J., Estable, S., De Stefano, M., Rackl, W., Lampariello, R., Ankersen, F., & Fernandez, J. G. (2017). Coupled control of chaser platform and robot arm for the e.deorbit mission. In *Proc. Int. ESA Conf. on Guidance Navigation and Control Systems - GNC*.
42. Papadopoulos, E., & Dubowsky, S. (1991). Coordinated manipulator/spacecraft motion control for space robotic systems. In *Proc. IEEE Int. Conf. Robotics and Automation - ICRA*, pp. 1696–1701.
43. De Stefano, M., Mishra, H., Balachandran, R., Lampariello, R., Ott, C., & Secchi, C. (2019). Multi-rate tracking control for a space robot on a controlled satellite: A passivity-based strategy. *IEEE Robotics and Automation Letters, 4*(2), 1319–1326, April 2019.
44. Giordano, M., Calzolari, D., & Albu-Schäffer, A. (2018). Workspace fixation for free-floating space robot operations. In *Proc. IEEE Int. Conf. Robotics and Automation - ICRA*, pp. 889–896.
45. Giordano, M., Ott, C., & Albu-Schäffer, A. (2019). Coordinated control of spacecraft's attitude and end-effector for space robots. *IEEE Robotics and Automation Letters, 4*(2), 2108–2115.
46. De Stefano, M., Balachandran, R., Giordano, A. M., Ott, C., & Secchi, C. (2018). An energy-based approach for the multi-rate control of a manipulator on an actuated base. In *Proc. IEEE Int. Conf. Robotics and Automation - ICRA*, pp. 1072–1077.
47. Forsyth, D., & Ponce, J. (2002). Computer vision: a modern approach. Prentice Hall, 2002.
48. Schmidt, P. W., Balachandran, R. & Artigas, J. (2016). Shared control for robotic on-orbit servicing. In *Robotics Science and Systems - RSS*.
49. Hokayem, P. F., & Spong, M. W. (2006). (2006) Bilateral teleoperation: An historical survey. *Automatica, 42*(12), 2035–2057.
50. Anderson, J. & Spong, M.W. (1989). Bilateral control of teleoperators with time delay. *IEEE Trans. on Automatic Control, 34*(5), 494–501.
51. Artigas, J., Balachandran, R., Riecke, C., Stelzer, M., Weber, B., Ryu, J.H. & Albu-Schaeffer, A. (2016). KONTUR-2: force-feedback teleoperation from the International Space Station. In *Proc. IEEE Int. Conf. Robotics and Automation - ICRA*, pp. 1166–1173.

52. Artigas, J., Balachandran, R., De Stefano, M., Panzirsch, M., Lampariello, R., Albu-Schaeffer, A., Harder, J. & Letschnik, J. (2016). Teleoperation for on-orbit servicing missions through the ASTRA geostationary satellite. In *IEEE Aerospace Conf.*, pp. 1–12.
53. Musić, S., & Hirche, S. (2017). Control sharing in human-robot team interaction. *Annual Reviews in Control, 44*, 342–354.
54. Parasuraman, R. & Riley, V. (1997). Humans and automation: Use, misuse, disuse, abuse. *Human factors, 39*(2), 230–253.
55. Benninghoff, H., Rems, F., Risse, E., Brunner, B., Stelzer, M., Krenn, R., Reiner, M., Stangl, C. and Gnat, M. (2018). End-to-end simulation and verification of GNC and robotic systems considering both space segment and ground segment. *CEAS Space Journal, 10*(4), 535–553.
56. Boge, T. Wimmer, Ma, O., & Tzschichholz, T. (2010). EPOS - using robotics for RvD simulation of on-orbit servicing missions. In *AIAA Modeling and Simulation Technologies Conf.*
57. De Stefano, M., Mishra, H., Giordano, A. M., Lampariello, R., & Ott, C. (2021). A relative dynamics formulation for hardware- in-the-loop simulation of on-orbit robotic missions. *IEEE Robotics and Automation Letters, 6*(2), 3569–3576.

Chapter 14
Hybrid Testbeds with a Focus on Sensors in Space Robotics Applications

Jörn Thieling and Jürgen Roßmann

Abstract Environment sensors, such as camera and lidar, are important technologies for increasing automation in space robotics applications, but they also provide a significant difficulty in terms of system verification, validation and optimization. Often, required system tests in dynamic operational environments are either less realistic (e.g. downscaled mock-ups in laboratory environments) or enabled too late (i.e. during the real mission in space). To allow for early and continuous system tests during development, we propose the use of Digital Twins and Hybrid Testbeds (HTBs), which enable a seamless transition from a first digital prototype to the final real system. To this end, real and simulated components are combined based on availability or necessity to allow for simulation-based testing throughout the development process. In this article, we introduce the overall concept of HTBs as well as selected functionalities and applications of HTBs. This includes essentials of physical sensor simulation using ray tracing and decomposition of simulation tasks based on data parallelism. As a result, not only reliable but also real-time-capable HTBs are realized, which are applied for open-loop as well as closed-loop testing of real software and hardware components. We present the benefits of HTBs applied for the development, testing and optimization of a lidar sensor that is used for automation of Rendezvous- and Docking (RvD) emphasizing the importance of HTBs for already completed as well as currently planned missions, such as Mars Sample Return.

J. Thieling (✉) · J. Roßmann
Institute for Man-Machine Interaction, RWTH Aachen University, 52074 Aachen, Germany
e-mail: thieling@mmi.rwth-aachen.de

J. Roßmann
e-mail: rossmann@mmi.rwth-aachen.de

X. T. Yan and G. Visentin (eds.), *Space Robotics*,
https://doi.org/10.1007/978-3-031-39214-6_14

14.1 Introduction

Highly automated and autonomous systems from the space robotics domain and beyond are equipped with various environment sensors to perceive and interpret their operational environment. Nowadays they are no longer limited to raw measurements of distances, intensities, angles and velocities but also come with comprehensive algorithms that internally process the raw sensor data to reliable encoded sensor data (e.g. with reduced noise and removed outliers). Moreover, they often additionally provide information-enriched sensor data that already includes interpretations of the perceived environment as higher-level features. For instance, modern camera systems often come with online calibration and self-diagnosis (e.g. glare and soil detection to classify the validity of image regions) while lidar sensors for space applications may provide algorithms for docking port detection and pose estimation. Both sensor types and the reliable and information-enriched data, which they provide, are fundamental to enable comprehensive cycles of Sense-Think-Act. This holds not only for space robotics applications but also for the probably most prominent application of autonomous vehicles from the automotive domain as well as for increasing automation in production plants from the industrial domain (compare Fig. 14.1).

However, independent of the field of application and the specific sensor types and sensor versions used, the development of such sensors and algorithms is facing various challenges (in terms of verification, validation and optimization of all system components), which increasingly require the complementary use of advanced 3D simulations to be efficiently solved. The creation of Digital Twins and their execution by means of 3D simulation allow for an efficient and systematic investigation of component and overall system functionality with a sufficient test coverage—thanks to safe and reproducible virtual 3D environments.

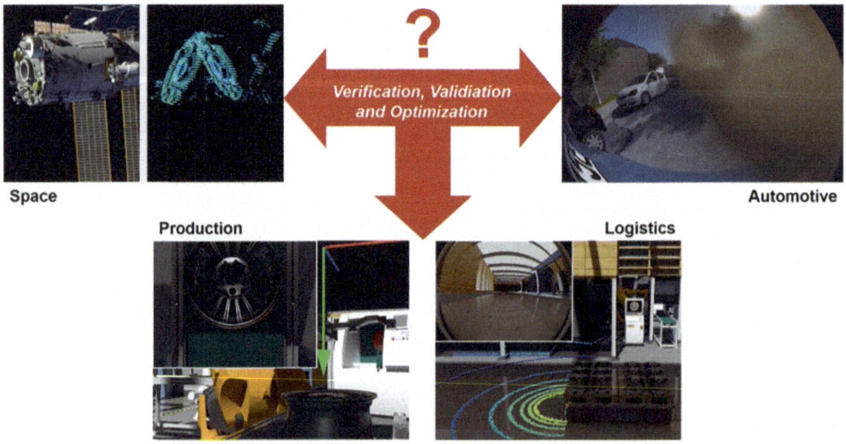

Fig. 14.1 Overview of domains, which are facing challenges in the development of sensor-based systems with regard to their verification, validation and optimization during development

In the following sections, we firstly take up the most relevant challenges that arise in the development of sensor-based systems. Here we also include current approaches, which target to solve these challenges from the simulation perspective but still do not meet the needs for flexible, continuous and realistic testing during development. Consequently, we then introduce our solution and overall concept of Hybrid Testbeds (HTBs) addressing their multidimensional scalability. Afterwards, we highlight underlying essentials of physical sensor simulation and decomposition of simulation tasks that ensure the reliability as well as real time capability of our HTBs. Finally, we present the benefits of our HTBs achieved for their application in open-loop and closed-loop testing. In this context, we present the adaptability of HTBs to support the development of a lidar sensor used for automation of Rendezvous- and Docking (RvD) in various space missions. This is also the case for the currently planned Mars Sample Return mission, which we address at the end of this article by presenting the benefits of our HTBs in supporting the verification, validation and optimization of sensor and system configuration for this application.

14.2 Challenges in the Development of Sensor-Based Space Systems

As introduced above, using environment sensors for increasing automation of various (space robotic) systems and applications leads to the fact that **things are turning from complicated to complex**. On the one hand, the interactions between those sensors and their dynamic environment include numerous intrinsic and extrinsic influences and disturbances, which further correlate with each other. On the other hand, the scenario-specific behaviour of Artificial Intelligence (AI) or classic perception algorithms is hard to predict when it comes to variant-rich environments and a multitude of possible disturbances (e.g. varying illumination, dust, motion etc.). The increasing use of neural networks opens up new approaches of machine learning in the area of AI, but at the same time confronts developers with the challenge to explain and verify the learned behaviour for all possible scenarios the system may face.

Especially in (but not limited to) the space sector, the development of sensor-based systems, which are equipped with environment sensors of various types and versions, leads to the following major challenges during system development:

- First and foremost, relevant reference data from space is very limited and mostly only available when the overall system or space mission is running or completed. Thus, the optimal configuration of environment sensors and the training of neural networks requiring such reference and training data becomes unfeasible or even impossible.
- Second, identifying key performance indicators (KPIs) of the developed software and hardware components requires close-to-reality testing that is very limited on Earth when solely using (downscaled) mock-ups in laboratory environments.

Especially testing these components as integrated part of the resulting overall system is very limited without using simulation as a complement.
- Third, performing manual system tests in laboratory environments is neither efficient nor sufficient to achieve a sufficient test coverage for a functional validation of the resulting overall system—i.e. testing all scenarios that the environment sensors and space system may face.

Taking a look at the state of the art reveals various (sensor) models and simulation methods that try to overcome these challenges by the complementary use of simulation technology. Approaches are ranging from models describing the sensor-intrinsic components [1, 2] to some focusing on the sensor-environment interaction [3, 4] and further include material models [5] and simulation methods like rasterization [6] and ray tracing [7]. All of such approaches aim to enable early and reliable system tests, which are carried out with high efficiency, systematism and reproducibility to provide an insight into the functionality and applicability of individual system components at an early stage of development. Unfortunately, the found approaches are often very sensor- or application-specific and focus on a single task within the development process instead of supporting the development continuously—i.e. adapting to the varying needs during development.

Depending on the use case, sensor models are built from an idealized, phenomenological or physical point of view [8], but are mostly designed to be used with a single simulation method (e.g. rasterization or ray tracing). This leads to less flexibility and missing interchangeability of simulated with simulated or simulated with real components during development. In general, simulation of environment sensors in particular is a fundamental limitation in all fields, which means that existing approaches are either not realistic or require too much computational effort [9] resulting in the exclusion of real-time applications.

The state of the art also provides approaches that deal with simulation-based development as well as virtual integration and validation by means of X-in-the-Loop (XiL) [10]. However, using limited models and simulations as introduced above, the solutions are again either limited to a specific sensor version, development step or field of application. This is why different domains (such as automotive, industry and space) neither benefit from each other nor harmonize with each other—even though many comparable challenges exist in the context of environment sensing and AI.

14.3 Needs for Flexible, Continuous and Realistic Testing of Space Systems During Development

Summarizing the aforementioned challenges, there is a need for a framework that continuously supports the entire development process, with all of its varying requirements on the simulation and virtual integration of developed system components. To this end, Fig. 14.2 depicts a typical development process visualized as the well-known V-model. This starts with requirement definition and different design phases

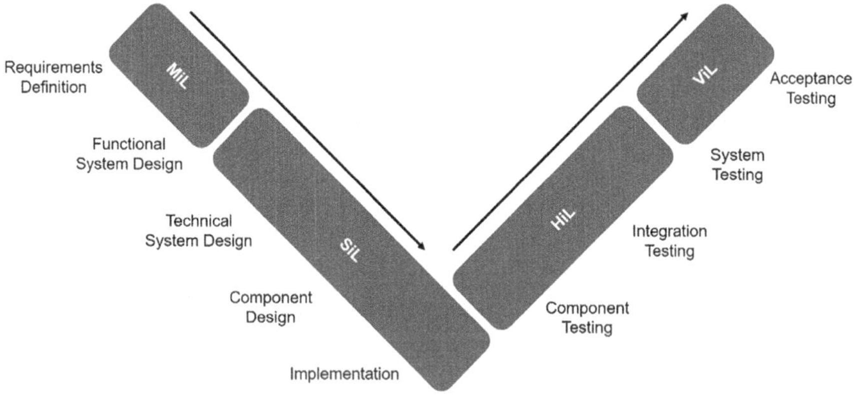

Fig. 14.2 Use of X-in-the-loop (XiL) structured along the V-model of a typical development process showing hard boundaries between the different forms of XiL—where X can stand for Model (MiL), Software (SiL), Hardware (HiL) or Vehicle (ViL). Based on [11]

on the left, goes down to the implementation phase on the bottom followed by the integration and testing of the overall system to the upper right.

Furthermore, Fig. 14.2 also shows the use of X-in-the-Loop (XiL) methods during system development, which acronyms are indicated in each box along the development process. These XiL methods comprise different integration and validation techniques, all of which virtually integrate a developed component or overall system into its planned operational environment to assess its behaviour for various simulated test scenarios. Here, the X can stand for Model, Software, Hardware or Vehicle, with each of these types (and their use in the development process) putting distinct demands on the simulation's realism, abstraction level and real-time capability. The common purpose of all XiL setups is to create a simulated system and/or operational environment that is sufficiently realistic from the view of the component or system put in the "loop" (the X in XiL).

While XiL methods are well known from the automotive sector, within the space domain one might link these methods with Ground Support Equipment (GSE) that allows for testing the developed software and hardware components on Earth—i.e. before the actual space mission starts. Especially SiL and HiL are the primary use cases of such GSE.

In order to realize different kinds of XiL presented in Fig. 14.2, it needs a framework for flexible, continuous and realistic testing during development. While current approaches just focus a single step or task within system development, our goal is to ensure consistency, reduce redundancy and avoid divergence of the used (sensor) models and simulation methods that are need within each of the depicted development steps. As tasks and goals vary during development, so do the requirements for the models and simulations that are needed for XiL.

Typically, at the beginning of the development process, very first digital prototypes need to build within simulation using primarily idealized or phenomenological models to design and verify system or components at a very early stage. Once first software or hardware components are implemented, they need to be tested, verified and optimized using increasingly realistic (i.e. physical) models and simulations. With ongoing development, more and more real-world components become available, which need to be virtually integrated into the overall system or the planned operational environment for continuous testing and early adjustments to keep iteration loops small. Especially when it comes to closed-loop testing in SiL or HiL setups, real-time capability becomes very important to make the software or hardware believe that the simulated environment is real and not a slow "fake". This application in closed-loop testing is very demanding for current simulations, since often real-time capability comes with reduced realism of simulation.

This has to be faced with implementing a sophisticated decomposition of simulation tasks that allows for distributing the computational effort over all available computational resources to achieve real-time capability without reducing the simulation quality. Finally, in the case of ViL—or Final-System-in-the-Loop (FiL) as we proposed in [8]—the final real system is implemented and solely the operational environment is still kept in simulation in order to test the developed real-world overall system (including all sensors, algorithms and controls) within a virtual space mission before sending it to space in reality. Figure 14.3 shows the approach of FiL (or rather ViL) for the automotive domain where the real-world vehicle is driving through a systematic and reproducible virtual test scenario in order to verify and validate all system components as integrated part of the overall system (i.e. automated vehicle). Exchanging the depicted vehicle with a real-world spacecraft and the depicted environment with a space environment, this approach of FiL is also very beneficial for space domain—allowing for a final functional validation of the overall system without the need to build real-world mock-ups of the planned environment.

The basis for this and our work in HTBs builds the concept of Digital Twins [12], which are 1:1 replicas of real systems in the digital world comprising all relevant system components, including their functions and interfaces, and brought to life through the use of 3D simulations.

14.4 Overall Concept of Hybrid Testbeds (HTBs)

To meet the needs introduced before, we propose the use of HTBs, which include realistic and flexible (i.e. generalized and modularized) models, simulations and XiL methods that support the complete development process by adapting to all its varying requirements and challenges. The key to this is the **multidimensional scalability** of our sensor models and simulation methods, which we presented in [8] and visualize in Fig. 14.4.

Figure 14.4 outlines the desired seamless transitions at the level of actors, methods, models and processes. For this purpose, the horizontal axis runs along the development process. The vertical axis, in turn, structures the modules of our HTBs into layers

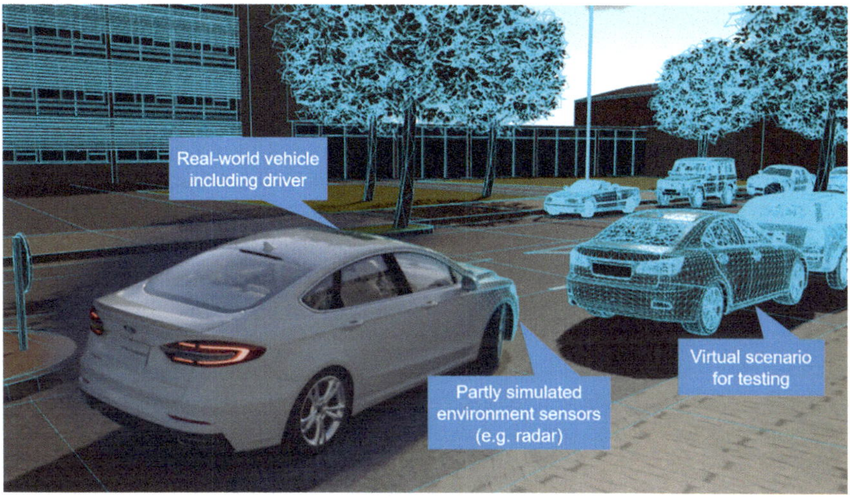

Fig. 14.3 Vehicle-in-the-Loop (ViL)—as special case of Final-System-in-the-Loop (FiL)—used within the automotive domain to verify and validate the developed driving functions as part of the real-world vehicle (i.e. overall system) driving within a safe and reproducible virtual environment

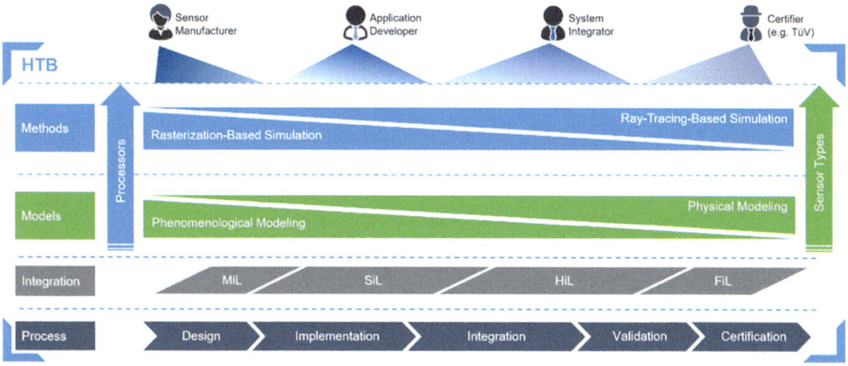

Fig. 14.4 Multidimensional scalability of HTBs to allow for seamless transitions within system development (development process plotted in horizontal direction)—compare use of XiL in V-model from Fig. 8.2

of simulation methods, underlying sensor models and XiL methods for their integration into the development process. During system development, the final integration layer ensures digital prototypes are gradually replaced by real prototypes, regardless of whether the component is software or hardware. In this way, HTBs enable continuous testing of the developed system components along the complete development process—starting with purely simulated prototypes (the Digital Twins) through partly simulated and partly real prototypes (the Hybrid Twins) to the completion of the final real system (the Real Twin). In contrast to the state of the art, the design, testing and integration during development is done with soft transitions between the used models, simulations and integration methods. **This makes the Digital Twin fade to its Real Twin seamlessly, having application-specific configurations of Hybrid Twins in between.**

Thus, the multidimensional scalability depicted in Fig. 14.4 comprises three major scalability axes: (1) simulation depth, (2) sensor types and (3) processors, which our sensor models and simulation methods achieve by using a physical, generalized and decoupled approach presented in [8]. For instance, generalization and harmonization allow single models and approaches to be reconfigured and recombined to satisfy the needs of several applications and domains. A lidar simulation built for satellite servicing in orbit, for example, can readily be converted to simulate a lidar used for autonomous driving on Earth or even Mars. To achieve physical simulations of all relevant sensor types with feasible computation time, our sensors scale their computational workload across all available processors (i.e. multiple cores of a GPU and multiple GPUs within a server).

In summary, HTBs provide a flexible software and hardware environment, which allows modelling and simulation of all relevant components (e.g. different sensor types, data processing steps and dynamic effects) integrated in the overall system and its planned operational environment. By combining simulated and real components, HTBs not only enable an early functional validation by means of Digital Twins, but also allow seamless transfer from an initial digital prototype to the final real system with continuous testing using Hybrid Twins during development. As a result, HTBs provide not only very detailed (i.e. physical) simulations to achieve reliable simulation results, but also very fast and efficiently distributed simulations to allow for real-time applications—both of which are addressed in the following sections.

14.5 Physical Sensor Simulation as Key to Reliable HTBs

As introduced in the previous sections, physically correct simulations are the key to reliable HTBs enabling simulation-based development and validation of sensor-based systems. This includes not only physical models of sensor-intrinsic components (e.g. lens, detector, signal processor etc.) but also models for simulating the interaction between sensor and environment (i.e. the propagation of light or other electromagnetic and acoustic waves used for environmental sensing). Typically, the simulation of those interactions takes the most demanding part of sensor simulation and is realized either by rasterization or ray tracing. These two approaches are well-known concepts in the field of computer graphics where they are used for generating synthetic images of virtual 3D scenes. Going beyond this original application, rasterization and ray tracing are also usable for the simulation of all other environment sensors (e.g. lidar, radar and ultrasonic sensors). In this context, Table 14.1 compares both methods regarding their simulation capabilities.

Table 14.1 Comparison of rasterization-based and ray-tracing-based simulation methods regarding simulation capabilities in the context of environment sensors

Simulation capabilities	Rasterization-based	Ray-tracing-based
Sample pattern	Limited	Full
Beam divergence	Full	Full
Multipath propagation and return	None	Full
Material properties	Limited	Full
Surface scattering	Limited	Full
Caustics	Limited	Full
Computational effort	Low	High
Detection of small targets	Good	Limited

Comparing both approaches in terms of simulation capabilities, as listed in Table 13.1, reveals that benefits of rasterization—which firstly projects 3D geometries onto a 2D plane followed by sampling this plane appropriately to the sensor's sample pattern (e.g. a lidar's scan pattern)—are mainly located in computational effort and the detection of small targets. This comes at the cost of an increasing gap between simulation and reality that does not converge to real behaviours, even if spending much computation time, especially because the effect of multipath propagation will always be missing. This also leads to limited material properties, which—in case of rasterization—do not support mirror-like reflection or volumetric effects such as diffraction of glass. Thus, ray-tracing-based methods are preferred for close-to-reality simulations of sensor-environment interactions. A qualitative comparison of simulated camera images generated by rasterization versus ray tracing is found in Fig. 14.5, showing clear benefits of ray tracing with regard to physical correctness of direct and indirect illumination of the ISS.

Since physical models of a specific real-world effect may not always be available right from the beginning of the development process, our HTBs focus on seamlessly fading between different level of details provided by those models. This is achieved by putting real-world structures in the centre instead of a specific simulation method. In other words, no matter if models are idealized, phenomenological or physical, they are always fused together using natural (i.e. physical) interfaces. This ensures an easy exchange of models with different level of detail as soon as they are available or required within system development. In this regard, Fig. 14.6 provides an overview and classification of different model types by naming typical subjects addressed by each. This starts with idealized models providing less detail at less effort (bottom left) and ends with physical models being the most detailed but at the same time most effortful models regarding modelling and simulation (top right).

The physical material description plays an essential role for physical sensor-environment interactions. Here, next to the absorption coefficientα, the index of refraction n and the normal vector $\underline{n} \in \mathbb{R}^3$, the Bidirectional Scattering Distribution Function (BSDF) takes a major part in describing a physically correct propagation of the respective information carrier used for environmental sensing—be it light,

Fig. 14.5 Comparison of simulated camera images generated by rasterization (left) and ray tracing (right) for the very same scenario clearly show benefits of ray tracing in terms of physically correctness. Ray tracing results include reflections visible on the solar panels, indirect illumination by self-reflections of the ISS as well as ambient light from the atmosphere (leading to blue illumination visible at down facing surfaces of the ISS)

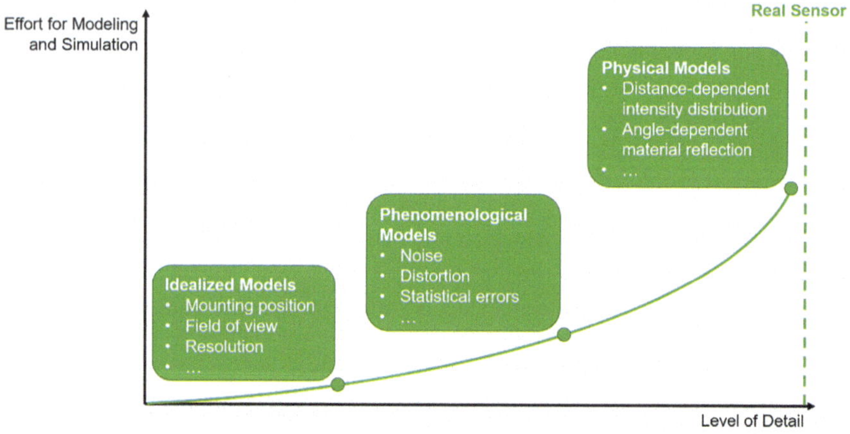

Fig. 14.6 Classification of different modeling methodologies (i.e. model types) regarding their level of detail and effort to be spent for modeling and simulation. Structured in idealized, phenomenological and physical models naming representative subjects of each model type within the appropriate boxes

infrared, radar or ultrasonic waves. The information-carrier-specific BSDFs comprise reflection as well as transmission functions to model all possible scattering properties of a given material. Thus, the BSDF includes the Bidirectional Reflectance Distribution Function (BRDF) and Bidirectional Transmission Distribution Function (BTDF) as depicted in Fig. 14.7 top and bottom row. Note that BSDFs are always described from the perspective of the particular sensor type or information carrier respectively, since **each sensor sees the same material but perceives it in a different way**. To

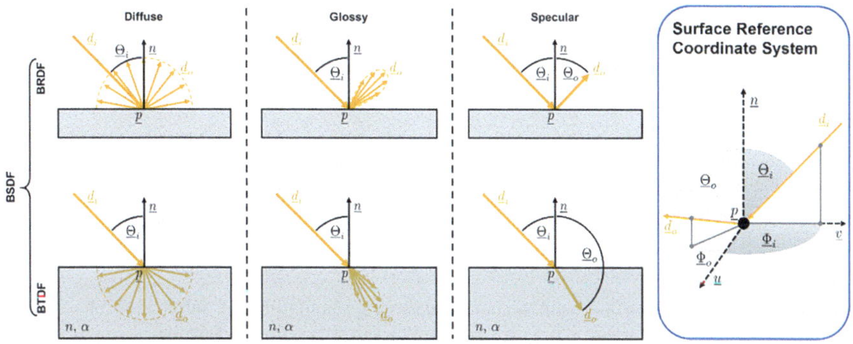

Fig. 14.7 Building blocks for diffuse, glossy and specular reflection/transmission that are combined to mimic even complex scattering properties of arbitrary materials (left) defined within the surface reference coordinate system depicted on the right

this end, the BSDF is a sensor-type-specific function of wavelength λ and is defined over all solid angles (Θ, Φ) of the surface reference coordinate system—with surface point $\underline{p}$, normal $\underline{n}$, tangent $\underline{u}$ and bitangent $\underline{v}$.

Combining the depicted building blocks from Fig. 14.8 (left), arbitrarily complex scattering characteristics of given materials are described to achieve a highly detailed physical simulation in terms of sensor-environment interaction. Typically, when it comes to ray tracing, one might first think about idealized reflections at the objects' surfaces that are traced as depicted in Fig. 14.8. However, this idealized approach only holds for mirror-like materials which just take a small part in real-world applications.

Thus, instead of tracing only idealized reflections, our sensor simulation respects the full physical material description introduced above to adequately mimic real-world scattering. In order to enhance the realism of the simulated sensor-environment interactions in terms of material behaviour and raw sensor measurements (i.e. multiple returns reflected from different distances/objects), we take material- and sensor-type-specific BSDFs at each hit point between the ray and a virtual 3D object into account. In contrast to the idealized ray tracing approach (from Fig. 14.8),

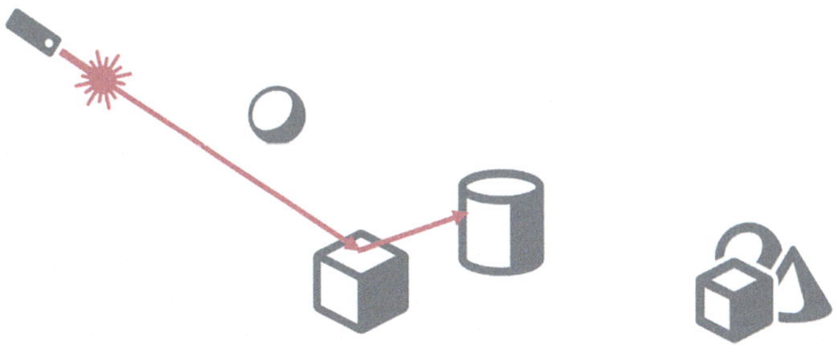

Fig. 14.8 Idealized ray tracing with idealized reflections that only apply to mirror-like materials but can not mimic indirect illumination caused by diffuse or glossy reflections

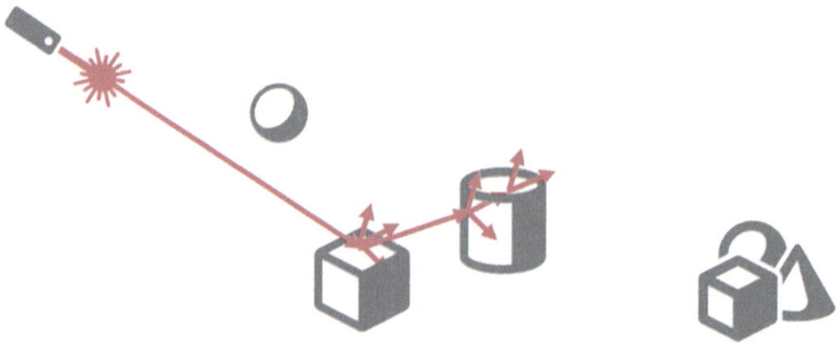

Fig. 14.9 Physical ray tracing with complex scattering simulating the interaction at surfaces respecting full physical material properties including arbitrarily complex BSDFs

which either returns at the first hit point (in case of diffuse or retroreflective materials) or perfectly reflects (in case of mirror-like materials), our approach samples the arbitrarily complex BSDF at each hit point allowing for multiple reflections and transmissions even for complex materials (compare Fig. 14.9).

This leads to samples representing information carriers (light etc.) travelling through the scene along paths that are no ideal reflections but close-to-reality approximations of real distributions in terms of material-specific scattering into multiple directions. Since real-world scattering is not digital but analog, BSDFs are defined as a continuous function describing the angle-dependent distribution of incoming information carriers (light etc.) over all outgoing angles. Thus, simulating their propagation has to be implemented by sampling the given BSDF multiple times that leads to multiple sample rays approximating the underlying BSDF as depicted in Fig. 14.10.

Here, we are using algorithms of importance sampling in order to prefer samples that have a major influence on the simulation result. In other words, our simulations do **not only trace much rays, but they preferably trace the right rays**. For instance, as depicted in Fig. 14.10 (right), directions in which the incoming light is primarily reflected are sampled with a higher probability, leading to a higher number of samples shot into this direction. Thus, directions which are less or not relevant for BSDF evaluation are ignored. This leads to fast convergence of simulated results to real-world behaviour.

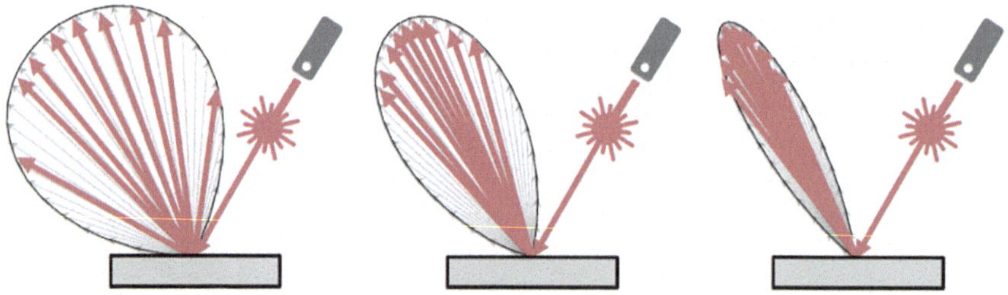

Fig. 14.10 Importance sampling of three differently wide opened BSDFs (visualized as lobes) with a finite number of rays at each hit point of the objects' surfaces

14.6 Decomposition of Simulation Tasks as Key to Real-Time HTBs

With regard to the realization Hybrid Twins—consisting of partly simulated and partly real components—another essential role takes the scalability of sensor models across available processors (i.e. achieving parallel computations). Only the decomposition of simulation tasks at a very low level allows real-time-capable simulations by parallelization of the resulting computational effort. Thus, our implementations with reusable building blocks are always parallelized from the ground up by describing work packages as functions over the sensor-centric dimensions (azimuth, elevation and time; see [8]) and thus decoupling the computations as far as possible. In detail, the computational effort required for the simulation is parallelized over thousands of cores of a Graphics Processing Unit (GPU) and all available GPUs of a server, which guarantees the generation of simulated sensor data with feasible computational time—without necessarily reducing simulation depth. For instance, simulating lidar sensors sampling the environment at frequencies of up to one million points per second, all samples/points (each including multiple subsamples and reflections/transmissions; as introduced in the previous section) are simulated in parallel. Thus, dynamic scan patterns of arbitrary lidar sensors are efficiently simulated allowing for real-time applications also for complex scenes and interactions (compare Fig. 14.11).

Consequently, all compute-intense physical simulations are designed with respect to scalability, to allow for highly detailed but at the same time online simulation runs. Since single-threaded CPUs are very limited in terms of computational capabilities, our implementation supports simulations on highly parallelized GPUs or server clusters. Therefore, the focus lies on an appropriate design for the anyway

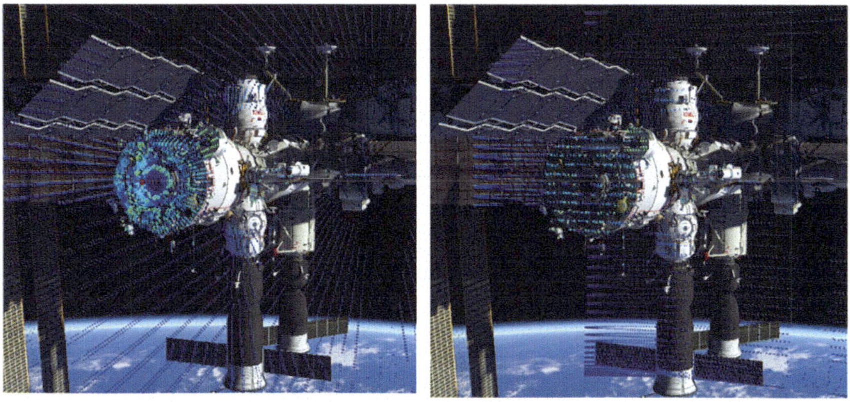

Fig. 14.11 Simulated lidar measurements visualized as point clouds for two differently scanning lidars. Independent of the specific scan pattern and internal processing steps used by each sensor, computations are always decomposed along the sensor-centric dimensions azimuth, elevation and time—being extended by the dimension of triangles when it comes to the simulation of sensor-environment interactions (see [8]). 3D model of ISS provided by NASA

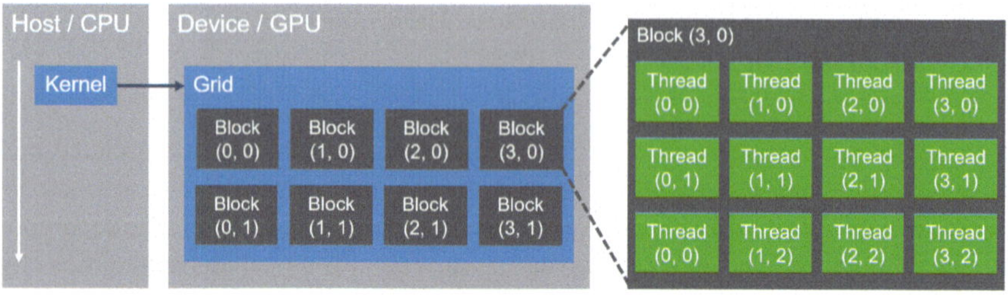

Fig. 14.12 CUDA thread hierarchy consisting of grids and blocks, each of which can be declared by the programmer as one-, two- or three-dimensional. Based on [13]

required discretization of the analog real world and the decomposition of physical processes on very low levels of abstraction. In order to achieve maximally decoupled computations, we further decompose the functional description on subsample- and triangle-level so that—in theory—computations can be done for every pair of subsample and triangle in parallel.

Having a look from the implementation perspective, brings us to the Compute Unified Device Architecture (CUDA) developed by NVIDIA [13]. CUDA is a programming model that allows for a GPU-accelerated processing of tasks by beneficially using data parallelism, such as the data parallelism that we achieve with the aforementioned decomposition of our simulation models.

In detail, implementing our models and simulations in CUDA allows us to parallelize computations from CPU to GPU by executing them on a multi-dimensional grid where each block within the grid comprises multiple threads that process the given simulation (sub-) tasks with high parallelism as depicted in Fig. 14.12. Depending on the GPU architecture used, this typically comprises more than 4,000 CUDA cores for floating point and integer operations as well as further RT cores specifically designed to speed up ray-tracing-specific operations, such as ray-triangle intersections and Bounding Volume Hierarchy (BVH) traversals (see [14]).

As a result, our simulations achieve real-time-capable computation times as depicted in Figs. 14.13 and 14.14. Both performance measurements were made for a quite complex scenario targeting on high simulation quality. In detail, a virtual Rendezvous- and Docking (RvD) with the International Space Station (ISS) was simulated, as shown in Fig. 14.11, where the 3D model provided by NASA comprises 4,586,592 triangles (including various texture and material properties). The simulated lidar sensor itself was parametrized to measure 921,600 samples per launch and the simulation was parallelized on a single NVIDIA RTX 6000 GPU with 4,608 CUDA cores, 72 RT cores and 24 GB of RAM.

Figure 14.13 shows measured computation times achieved with the ray-tracing-based (physical) lidar simulation of our HTBs. The plot shows times needed for the simulation of sensor-environment interaction (*launch*), intrinsic components (*intrinsics*), further internal processing units (*processingSteps*) and the final copy of simulation results from GPU back to the CPU (*finalToDynStruct*). Note that all sensor

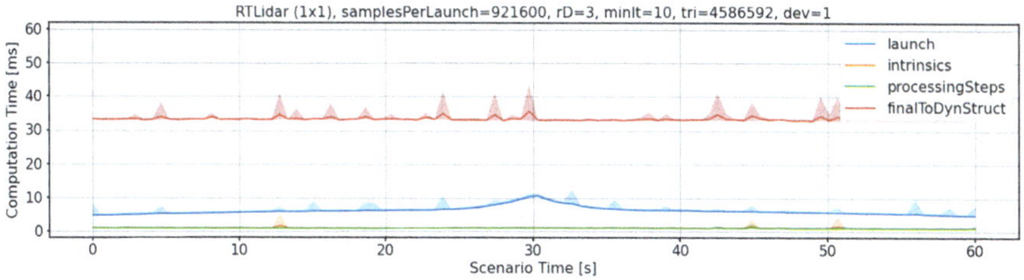

Fig. 14.13 Computation times needed to generate 921,600 lidar measurements plotted over scenario time showing slight variations dependent on the complexity of objects that are within the field of view at the specific point in time. All GPU-based simulation tasks (*launch, intrinsics* and *processingSteps*) are solved in less computation time—thanks to high parallelism—while copying simulation results from GPU to CPU (*finalToDynStruct*) is noticeable high, even though still less than 40 ms

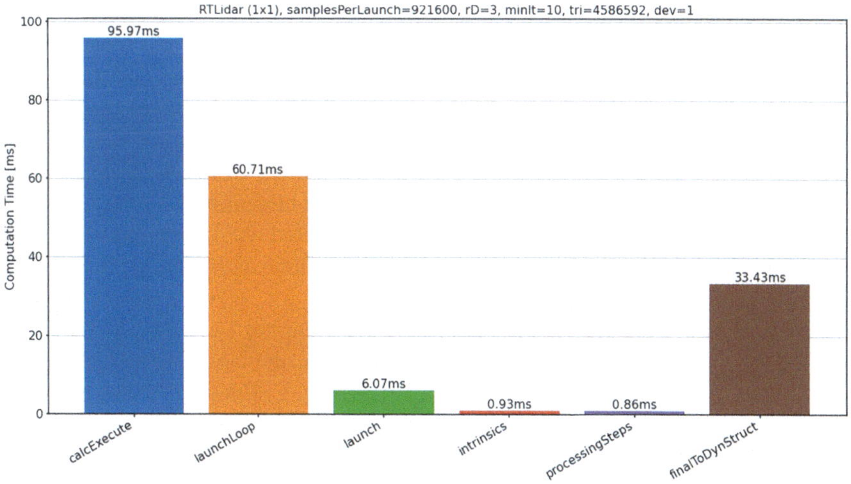

Fig. 14.14 Computation times of physical lidar simulation averaged over the simulated RvD scenario with the ISS as described for Fig. 14.13. This time additional computation times are depicted showing the time needed for using ten iterations for sensor-environment interaction (*launchLoop*) and the resulting overall computation time (*calcExecute*), which summarizes all averaged times needed for simulation and copy tasks in this scenario. Note that *launch* is included in *launchLoop* and not counting twice in this plot

simulation tasks are solved with high parallelism on the GPU leading to less computation time. Only the final copy of simulation results from GPU to CPU is done by the CPU which takes a noticeable amount of time (compare *finalToDynStruct*)—even though simulating the interaction between 921,600 lidar rays and 4,586,592 triangles with up to three reflections/transmissions (i.e. recursions) has a much higher complexity than just copying result data.

In summary, our sensor simulation has proven to simulate 117.25 million samples per second for the described RvD scenario. This is 2,931 times faster than a typical real space lidar performs (assuming 40 kHz measurement rate). Even when running ten launch loops (i.e. computing ten subsamples per sample) and taking final result copy into account, our simulation achieves an average computation time of less than 100 ms for this quite complex RvD scenario (see *calcExecute* in Fig. 14.14). This equals around twelve million samples simulated per second. In contrast, even a real-world high-resolution multi-layer lidar sensor from the automotive sector equipped with e.g. 64 rotating laser diodes (such as Ouster OS1-64) generates "only" about three million samples/points per second.

14.7 Seamless Transitions Between Simulation and Reality Using HTBs

Once having not only reliable but also real-time HTBs, as introduced before, Digital and Real Twins are synchronized to efficiently realize systematic and reproducible system tests. Based on work presented in [15] we synchronized our laboratory (in spatial and temporal domain) in order to verify, validate and optimize environment sensors. One configuration of this laboratory used for optimization of a lidar flight software (including a pose estimation algorithm) is depicted in Fig. 14.15. Here, reference scenarios with the real RVS3000(-3D) lidar manufactured by Jena-Optronik and real mock-ups (e.g. ISS in 1:100 scale) are conducted within our real laboratory (Fig. 14.15 left), while the planning and evaluation is done within the virtual laboratory (Fig. 14.15 right). Doing so, all relevant system states, lidar measurements and processing results are intuitively visualized using the Digital Twin. This includes the joint angles of all axes of the real Kuka robot (synchronized via the Fast Robot Interface) as well as point clouds and pose estimation results of the real RVS3000(-3D) lidar (synchronized via SpaceWire interface).

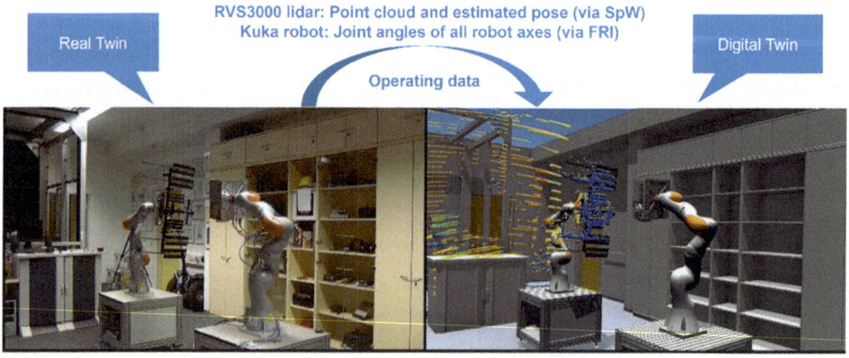

Fig. 14.15 Spatial and temporal synchronization of real twin (left) and digital twin (right) including operating data of lidar sensor and Kuka robots. Here, two Kuka robots realize systematic and reproducible motions between lidar and ISS mockup with 1:100 scale

As a result, the best of both worlds is fused—i.e. real hardware, real measurements and real processing results (for a maximum of realism) with virtual representations, visualizations and ground truth (for best systematism, reproducibility and evaluation possibilities). Here, "maximum of realism" applies for the developed components (e.g. lidar hardware), but may not hold for real-world mock-ups that often need to be miniaturized for laboratory. For instance, when using a downscaled mock-up of the ISS as depicted in Fig. 14.15, physical effects caused by scanning over the objects' edges become prominent within the lidar measurements which do not mimic real measurements from space. Further, due to the miniaturization, relative motions need to be very precise and smooth, since one centimetre within the mock-up represents one meter of the real operational environment. Thus, we use two Kuka robots which manipulate and measure the actual poses of both carried objects (i.e. lidar and ISS) very precisely and can reproduce user-defined trajectories with high accuracy. For the latter, a first calibration of the carried weights is conducted using the Kuka software that results are afterwards respected by the Kuka's control to achieve high reproducibility in terms of pose accuracy.

The benefits of our HTBs become visible, when we no longer want to use real mock-ups (e.g. due to their lag in realism) but virtual mock-ups instead. For this purpose, the physical sensor simulation comes into play which provides close-to-reality sensor data. In order to parametrize our generalized sensor models, real-world experiments are conducted. This includes the measurement of material- and sensor-type-specific material properties (such as the BSDF as presented in the previous sections of this article). Further, application-specific comparisons of the simulated and real scenario are performed to verify that given models and simulation methods are suitable for the current use case.

In [16] we presented a methodology that combines a laboratory-based with a scenario-based approach to verify and validate such models and simulation methods from an intrinsic, extrinsic and application-specific perspective. As a result, **the Digital Twin no longer just visualizes operating data from its Real Twin but generates close-to-reality sensor data using sensor simulation**. Thanks to our hybrid laboratory with its synchronization of the spatial and temporal domain, both worlds can be directly compared even on the level of sensor data generated by the Digital Twin and Real Twin respectively.

Figure 14.16 shows such a comparison for an RvD scenario where the lidar is approaching a 1:1-scale mock-up of the International Docking Adapter (IDA) within the real and virtual laboratory. Sensor data generated by each of the Twins are visualized as color-coded point clouds below and as overlay on each camera image.

Consequentially, Jena-Optronik plans to use the presented laboratory setup as well as our HTBs for further reference measurements, which shall be conducted in collaboration with the Institute for Man–Machine Interaction of RWTH Aachen.

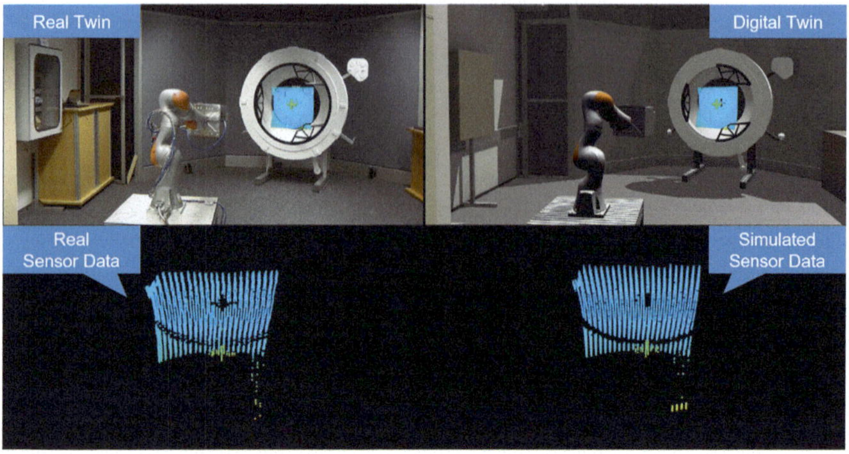

Fig. 14.16 Hybrid laboratory with real (left) and digital twin (right) both generating sensor data for the very same scenario. This allows for a direct comparison of both worlds based on sensor data that is perfectly synchronized in spatial and temporal domain

14.8 Closed-Loop Testing of Space Systems on Earth Using HTBs

Open-loop testing evaluates the sensing and perception part of an automated system. In our case, we target to support the development process of the RVS3000(-3D) lidar that is used for automation of RvD scenarios. In this context, one of the largest challenges is to verify and validate the lidar's flight software (i.e. target detection and tracking as well as pose estimation between target and sensor/chaser). Typically, real mock-ups are used to allow the evaluation of planned scenarios on Earth—if necessary, in a scaled form (see ISS mock-up with 1:100 scale from Fig. 14.17left).

However, those tests are hardly reproducible and only partly systematic without having HTBs, such as the hybrid laboratory introduced in the previous section, which is also used in Fig. 14.17. On the left, the real image is fused with the real point cloud to have clear insight on which parts of the mock-up are perceived and which objects of the laboratory are still falsifying the experiment. Furthermore, the results of the pose estimation are visualized as coordinate system and as bounding box aligned properly to the real ISS.

Moreover, these valuable pieces of information are not only visualized to allow for an intuitive insight, but are also matched to the available ground truth. This is shown on the right of Fig. 14.17. Here, the key performance indicators (KPIs) of the lidar's flight software are calculated based on the ground truth given by the real axes angles that are applied to the virtual robot—i.e. the estimated pose is compared to the actual pose between lidar and ISS. Thanks to this, very precise error metrics are calculated (Fig. 14.17 right). Here, even smallest periodic peaks of one centimetre and below can be identified and resolved to optimize the pose estimation of the lidar

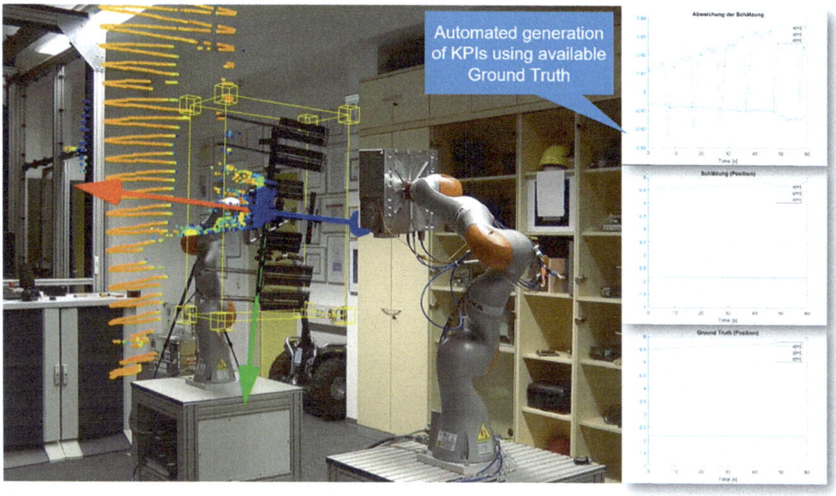

Fig. 14.17 Hybrid laboratory showing real sensor and ISS mockup with additional information of estimated pose and bounding box of the target fused within the image

with very high accuracy and systematism. This could hardly have been achieved without the use of our HTBs.

These results underline the benefits of using HTBs within the development process. Even though measurement campaigns with the lidar were extensively conducted in the past, those small effects could not be detected, caused by their low scale that is hardly noticeable without having a hybrid laboratory fusing the Real and Digital Twin.

Moreover, using HTBs within system development eliminates the need for real mock-ups and real laboratories even when it comes to open-loop testing of developed hardware components. For instance, instead of testing the lidar's hardware with real sensor data (e.g. generated by scanning some downscaled mock-ups), the HTBs' physical sensor simulation generate simulated sensor data including much more details on the planned target and operational environment (assumed that 3D models and physical knowledge on the environment etc. is given). Thus, as a complement to Figs. 14.17 and 14.18 shows the appropriately simulated RvD scenario where the simulated RVS3000(-3D) lidar is scanning the ISS to generate close-to-reality sensor data for open-loop testing of pose estimation etc. To this end, a stimulation unit (StimUnit) provided by Jena-Optronik is used to inject sensor data into the real lidar hardware making all internal and subsequent components believe that they are on the real mission. Details on the StimUnit (before known as IF-box) are found in [17]. Since open-loop testing excludes decision and control, for this use case, sensor data does not have to be generated in real-time but can be generated and recorded with less than real-time for a later replay in real-time. This will change when it comes to closed-loop testing discussed in the following section.

Fig. 14.18 Automated Rendezvous- and Docking (RvD) with the virtual ISS. Simulated RVS3000(-3D) lidar sensor mounted on a spacecraft (left) scanning the ISS for distance measurement and pose estimation. Sensor data generated by simulated lidar visualized as color-coded point cloud. 3D model of ISS provided by NASA

14.9 Closed-Loop Testing of Space Systems on Earth Using HTBs

The previously presented open-loop testing aims to evaluate the KPIs of the environment sensing and perception algorithms, which are the first step to a full automation of space systems. The second step is decision taking and control, which is addressed by closed-loop testing discussed in the following section.

In our case, we focus on a Guidance, Navigation and Control (GNC) system, which uses the pose estimate provided by the lidar to take control of a spacecraft with the goal of automating a complete RvD manoeuvre. In this way, unmanned missions are realized even in distant areas of space where, due to the communication speed limited by the speed of light, manual intervention from Earth is only possible with a latency of several minutes.

The difficulty in developing a GNC is its validation as an integrated part of the overall system/spacecraft. This corresponds to the investigation of the fully closed loop including the sensors, the perception algorithms and the control of the thrusters, which in turn result in the movement of the spacecraft and its sensors—and so closing the loop. Even if all hardware and software components have already been fully developed and integrated, such experiments are only possible to a limited extent on Earth, so that the functionality of the overall system will first be revealed in the

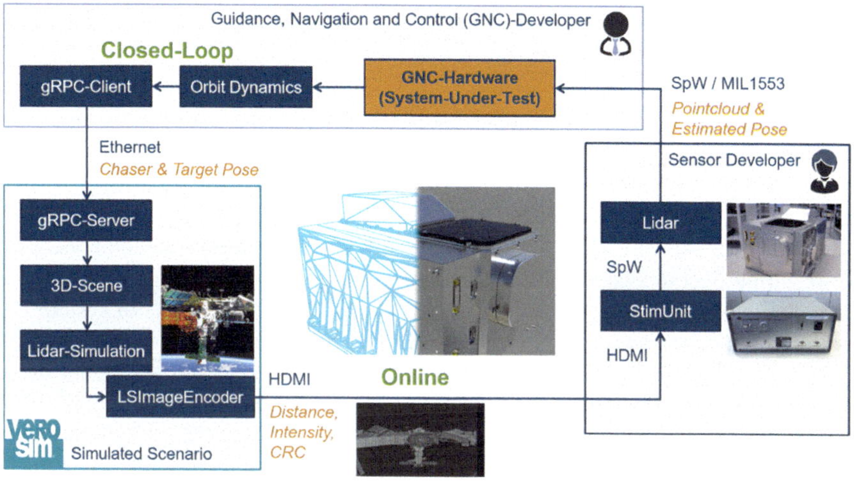

Fig. 14.19 Hybrid closed-loop setup for testing real software and hardware components (right and top) within a simulated operational environment (left)

final mission. This is particularly problematic in view of the lack of possibility for manual intervention.

Here, HTBs support the development by simulating the complete operational environment and stimulating the developed hardware and software components with simulated sensor data realizing XiL setups. A key function of simulation for this application is the real-time generation of simulated sensor data, which is directly injected into the real hardware (i.e. online). In contrast to open-loop testing, closed-loop testing also respects the control variables of GNC to close the control loop as depicted in Fig. 14.19.

In this figure, the GNC communicates as System-Under-Test with the real lidar (top right), which in turn is stimulated with simulated sensor data coming from the simulated scenario (bottom left). In terms of a closed-loop setup, the GNC control variables effecting the orbit dynamics are continuously respected by updating the simulated 3D scene (top left).

Crucial for this setup is the HTBs' scalability over processors to enable real-time applications by massive parallelization—for which the decomposition of simulation tasks presented in the previous sections gains significant importance. To this end, all major simulation tasks needed for sensor simulation are running on thousands of cores of a single GPU (see previous section on real-time HTBs) or even multiple GPUs within a server (see results presented in [8]). Since even copying simulation results from GPU to specific data structures on CPU can lead to noteworthy increase in overall computation time (see performance measurements discussed above), this time, simulated sensor data is directly injected from GPU via HDMI to the StimUnit, which in turn feeds the real RVS3000(-3D) lidar (see Fig. 14.19). As a result, pose estimation made by the lidar and attitude control made by the GNC are directly effecting the simulated scenario leading to online updates of simulated sensor data injected into the real lidar.

Thanks to the modular configuration of HTBs regarding simulated (sensor) components, the setup depicted in Fig. 14.19 can also be easily adapted so that lidar components becoming part of the simulated scenario (left). This enables the successive exchange of simulated and real parts of the control loop, which enables not only the final validation of the already developed components, but also the continuous validation of those still under development—always as an integrated component of the resulting overall system. Hence, the hybrid setup depicted in Fig. 14.19 allows for a continuous verification, validation and optimization of the overall system on Earth right from the beginning of development.

The closed-loop testing setup with HTBs outlined here is already being used intensively by Jena-Optronik for the validation of their sensors and algorithms during development and will also be made available to their customers in a next joint step.

14.10 Application of HTBs in Mars Sample Return Mission

As space exploration has increased, so has international interest in returning samples from other planets (e.g. Mars) to Earth. A very prominent and forward-looking example is the Mars Sample Return (MSR) mission outlined by ESA in collaboration with NASA (see Fig. 14.20). Here, in particular, ESA's sub-mission Earth Return Orbiter (ERO), in which a sample container launched into orbit is captured by an automated spacecraft and brought back to Earth, poses new challenges to the planning and functional validation of such missions. Due to the large distance to Earth and the associated latency of about three to nearly 20 min (depending on the constellation of the planets), an intervention to control the spacecraft from Earth is not reasonably possible. This makes the simulation-based development and virtual validation of sensor-based systems—such as the lidar-based RvD introduced in the previous sections—becoming even more important.

However, this special and at the same time very relevant case of RvD brings current sensor simulations to their limits, since they are mostly based on a discretization of the dynamic measurement process occurring continuously in the real world. As a consequence, especially sensor simulations for scenarios with small and dynamic targets over large distances lead to a strong bias between simulation and reality. Therefore, HTBs provide a sophisticated physical sensor simulation and combine different methods for lidar simulation that ensures both, the detection of small objects over large distances despite discrete simulation steps (i.e. finite number of subsamples to approximate the lidar beam) as well as simulated sensor data that includes all relevant dynamic effects. The latter include, for example, the blurring of the target object (here: orbiting sample container) within the sensor data (i.e. 3D point cloud) due to the relative motion to the sensor.

The simulation of such dynamic physical effect is especially enabled by a parametric pose description that allows the interpolation of objects' poses by respecting

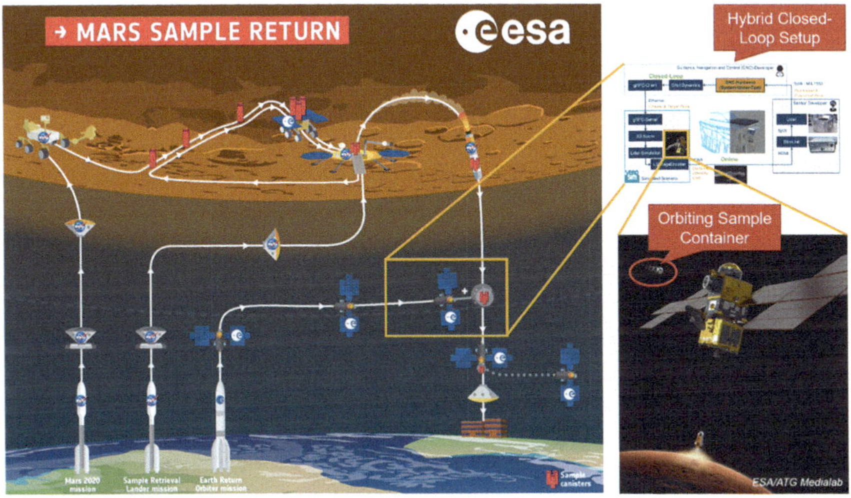

Fig. 14.20 Overview of mars sample return mission (left) and hybrid closed-loop setup to be used within sub-mission Earth Return Orbiter (right). Based on [18]

their velocity vectors even within a single simulation step—making simulations very performant. Without having this parametric pose description and applying velocity-based interpolation within a single simulation step, the simulation tasks would tend to be solved iteratively (in order to simulate dynamic effects)—losing benefits of GPUs and parallelization. Thus, HTBs provide not only reliable but also real-time results that include even such dynamic physical effects in sensor data when applied for closed-loop testing as presented above.

Due to the aforementioned benefits of HTBs, Jena-Optronik also will use our HTBs (including the closed-loop testing setup from above) to verify and validate their sensors and algorithms for the ERO mission. HTBs will especially support the planning and development phase by enabling comprehensive system tests before the developed system is sent to space.

14.11 Conclusion and Future Work

The aim of this article was to assess the benefits of using our novel approach of Hybrid Testbeds (HTBs) for continuous testing of sensor-based systems during development. To this end, we firstly introduced current challenges within system development requiring simulation as a complement to real-world mock-ups and laboratories. However, available approaches for the simulation of environment sensors has been shown to be less realistic or require too much computational effort that excludes them from real-time applications.

To address these limitations and the existing needs for flexible, continuous and realistic testing of sensor-based systems during development, we presented our concept of HTBs. We introduced the benefits of their multidimensional scalability as well as major advances achieved in the physical simulation of environment sensors as part of our HTBs.

To this end, we presented scalable and physical models and simulation methods ensuring reliable simulation results, as well as the implemented decomposition of simulation tasks that is fundamental for real-time applications—namely closed-loop testing of developed software and hardware by means of XiL. Related performance measurements confirmed that even physical (lidar) sensor simulations in complex environments (such as the presented RvD manoeuvre with the virtual ISS) can be simulated in real-time by using our proposed decomposition and parallelization across thousands of cores and multiple GPUs.

Taken together, these results show that HTBs are reliable as well as real-time capable to continuously support the development process ensuring a seamless transition from a first digital prototype to the finale real system. This makes the Digital Twin fade to its Real Twin seamlessly, having application-specific configurations of Hybrid Twins in between.

In this context, we presented various use cases and examples in which our HTBs support the development process in terms of continuous open-loop and closed-loop testing of sensor-based systems. In particular, we presented their use for the increasing automation of RvD missions using pose estimation provided by Jena-Optronik's RVS3000(-3D) lidar. Here, we also presented the currently planned Mars Sample Return (MSR) mission for which our HTBs will be used in order to verify, validate and optimize the RVS3000(-3D) flight software (incl. target tracking and pose estimation) before they are being sent to space as part of the Earth Return Orbiter (ERO) submission.

The trend towards increasingly detailed simulations is generally associated with an increase in effort for modelling and simulation. The presented HTBs already mitigate this through reusable physical models and highly parallelized simulations. However, the initial calibration and validation of those models and simulations is still associated with a high effort, and even powerful simulation hardware comes up against hardware limits above a certain degree of parallelization. Here, the research field of Artificial Intelligence (AI) could bring beneficial approaches. Based on real reference data, AI could learn physical relationships and parameters that are difficult to model analytically (e.g. models for mapping the atmosphere of other planets), or whose simulation would require much computations. By using models trained in advance, an AI-based simulation could not only increase realism, but also reduce the need of manual modelling. Possible approaches are AI-based error models that learn from real sensor data to refine the simulated sensor data in terms of realism (as presented in [19]), or AI-based model parameters that automatically learn the required input variables (e.g. variant-rich material textures of the environment) in the sense of parameter identification. In summary, AI thus appears to be not only a key to complex data processing and analyzation, but is also potentially of great use in the generation and execution of Digital Twins within our HTBs. Research on this topic has already been started within the project "ViTOS-3", supported by the German Aerospace Center (DLR) with funds of the German Federal Ministry of Economics and Technology (BMWi).

Acknowledgements This work shows results of the projects "ViTOS-II" and "ViTOS-3", supported by the German Aerospace Center (DLR) with funds of the German Federal Ministry of Economics and Technology (BMWi), support code 50 RA 1810 and 50 RA 2120. Special thanks also go to our industrial partner Jena-Optronik, who have supported the in-depth practical relevance of this work through valuable discussions and research assignments.

References

1. Kannala, J., & Brandt, S. (2006). A generic camera model and calibration method for conventional, wide-angle, and fish-eye lenses. In *IEEE transactions on pattern analysis and machine intelligence 28.8* (pp. 1335–1340). issn: 0162–8828. https://doi.org/10.1109/TPAMI.2006.153.
2. Fasogbon, P., & Aksu, E. (2019). Calibration of fisheye camera using entrance pupil. In *2019 IEEE international conference on image processing (ICIP)* (pp. 469–473). isbn: 978–1–5386–6249–6. https://doi.org/10.1109/ICIP.2019.8803832.
3. Mohammadikaji, M., Bergmann, S., Beyerer, J., Burke, J., & Dachsbacher, C. (2020). Sensor-realistic simulations for evaluation and planning of optical measurement systems with an application to laser triangulation. *IEEE Sensors Journal, 20*(10), 5336–5349. https://doi.org/10.1109/JSEN.2020.2971683
4. Chipengo, U. (2018). Full physics simulation study of guardrail radar-returns for 77 GHz automotive radar systems. *IEEE Access, 6*, 70053–70060. https://doi.org/10.1109/ACCESS.2018.2881101
5. NVIDIA Corporation. (2020). Material definition language: technical introduction. Online: https://mdlhandbook.com/.
6. Laine, S., & Karras, T. (2011). High-performance software rasterization on GPUs. In *Proceedings of the ACM SIGGRAPH symposium on high performance graphics—HPG'11* (pp 79–88). isbn: 9781450308960. https://doi.org/10.1145/2018323.2018337.
7. Haines, E. (2019). Ray tracing gems. In E. Haines & T. Akenine-Möller (Eds.), Berkeley, CA: Apress. ISBN: 9781484244265. https://doi.org/10.1007/978-1-4842-4427-2.
8. Thieling, J., & Rossmann, J. (2021). Scalable sensor models and simulation methods for seamless transitions within system development: From first digital prototype to final real system. *IEEE Systems Journal, 15*(3), 3273–3282. https://doi.org/10.1109/JSYST.2020.3006739
9. Hakuli, S., & Krug, M. (2015). Virtuelle Integration. In: H. Winner, S. Hakuli, F. Lotz et al. (Eds.), *Handbuch Fahrerassistenzsysteme: Grundlagen, Komponenten und Systeme für aktive Sicherheit und Komfort*. Wiesbaden: Springer Fachmedien Wiesbaden (pp. 125–138). isbn: 978–3–658–05734–3. https://doi.org/10.1007/978-3-658-05734-3_8.
10. Rosique, F., Navarro, P. J., Fernández, C., & Padilla, A. (2019). A systematic review of perception system and simulators for autonomous vehicles research. *Sensors, 19*(3), 648.
11. Hakuli, S., & Krug, M. (2016). Virtual integration in the development process of ADAS. In H. Winner, S. Hakuli, F. Lotz et al. (Eds.), *Handbook of driver assistance systems* (pp. 159–176). Cham: Springer International Publishing. isbn: 978–3–319–12352–3. https://doi.org/10.1007/978-3-319-12352-3_8
12. Schluse, M., Osterloh, T., & Rossmann, J. (2019). Development of modular spacecraft with experimentable digital twins. In *The workshop on simulation for European space programmes (SESP), ESA/ESTEC* (pp. 1–13).
13. Cheng, J., Grossman, M., & McKercher, T. (2014). Professional CUDA C Programming 1st. Indianapolis: Wrox. isbn: 978–1–118–73927–3

14. Kilgariff, E., & Moreton, H. (2018): NVIDIA turing architecture in-depth. In *NVIDIA Developer Blog*. Online: https://developer.nvidia.com/blog/nvidia-turing-architecture-in-depth/.
15. Sondermann, B., Atorf, L., Grinshpun, G., Priggemeyer, M., Thieling, J., Rast, M., & Rossmann, J. (2017). A virtual testbed for optical sensors in robotic space systems–VITOS. In *Proceedings of the 14th symposium on advanced space technologies in robotics and automation* (pp. 1–8)
16. Thieling, J., & Roßmann, J. (2021). Modulare Validierung simulierter Sensorsysteme für interagierende Digitale Zwillinge. In H. Proff (Eds.), *Making connected mobility work* (pp. 223–240). Springer Fachmedien Wiesbaden.
17. Dochow, S., Heilmann, C., Kolb, F., Linhart, B., Schmitt, C., Schwarz, M., & Windmüller M. (2018). The RVS3000 and RVS3000–3D lidar sensors—test results and development outlook. In *Advances in the astronautical sciences AAS/AIAA guidance, navigation and control,* 164. AAS 18–064.
18. European Space Agency—Science and Exploration (2022). Mars Sample Return. https://www.esa.int/Science_Exploration/Human_and_Robotic_Exploration/Exploration/Mars_sample_return.
19. Thieling, J., Elspas, P., & Roßmann, J. (2019). Neural networks for end-to-end refinement of simulated sensor data for automotive applications. In *2019 IEEE International systems conference (SysCon)* (pp. 1–8). IEEE.J. https://doi.org/10.1109/SYSCON.2019.8836918.

Chapter 15
In-Space Assembly Operations

Maximo A. Roa, Korbinian Nottensteiner, Ismael Rodriguez, and Gerhard Grunwald

Abstract In the last two decades, robotized On-Orbit Servicing (OOS) missions are increasingly gaining importance. OOS addresses the maintenance of satellites in orbit, including inspection, repairing, refueling and assembly. On-orbit robotic assembly can potentially reduce costs and allow the construction of large structures directly in space. This chapter provides an overview of existing or emerging robotic technologies for space-born assembly, including missions that use robotics arms, and development of standard interconnects that combine mechanical latching with electronic and power connectivity. Given the high delays in communication and the repetitive and demanding tasks required, teleoperating a robotic assembly system from Earth is not straightforward. Therefore, the robotic assembly system should behave autonomously. This kind of autonomous assembly system has been explored for ground applications, as illustrated with an example developed at the Institute of Robotics and Mechatronics in the German Aerospace Center—DLR.

15.1 Introduction

Large scale space mission scenarios for the Post-ISS era such as the Moon Village (ESA), Mars exploration (NASA), and Orbital Hub in LEO (DLR) [1], require the availability of dexterous robot systems capable of performing complex assembly tasks. In contrast to the construction of the International Space Station (ISS), which was performed by astronauts and took over a decade, new orbital structures are expected to be assembled by robotic systems. In the last two decades, robotized On-Orbit Servicing (OOS) missions have increasingly gained importance. OOS addresses the maintenance of space systems in orbit, including repairing and refuelling, using technologies that can be extended to on-orbit robotic assembly. In-space assembly can enable the construction of much larger orbital structures, for

M. A. Roa (✉) · K. Nottensteiner · I. Rodriguez · G. Grunwald
Institute of Robotics and Mechatronics, German Aerospace Research Center (DLR), 82234 Wessling, Germany
e-mail: maximo.roa@dlr.de

X. T. Yan and G. Visentin (eds.), *Space Robotics*,
https://doi.org/10.1007/978-3-031-39214-6_15

instance for setting up more sensitive radar imaging or for obtaining much higher communication speeds by assembling bigger and more sensitive antennas.

In the future, robotic construction capabilities could also be applied to create space infrastructure, standing structures such as refuelling depots, in-space manufacturing facilities, space-tourism complexes, and asteroid mining stations [2]. Current manufacturing and technological limitations are evident in the construction of antennas and mirrors that have to be deployed from a single launch with a single satellite. However, a dexterous and capable space robot could assemble pieces coming in multiple launches to construct antennas and other structures of enormous size, thanks partially to the expected reduction of launch costs from private companies such as SpaceX.

Operations required for servicing, maintenance and assembly missions, such as autonomous rendezvous, docking and undocking, are relatively common nowadays, but autonomous on-orbit robotic servicing is currently under development [3]. Robot-based assembly in the absence of gravity addresses fundamental technical questions that do not exist for terrestrial applications. The robot is mounted on an actively regulated platform, and the motion of the robot has dynamic effects on the platform itself. The dynamic effects depend on the mass of the objects and the velocity of the motion. When the robot works on a platform different to its own floating base, the required physical contact affects both platforms. The contact can also affect the perception system, as minor changes in position might lead to significant visual occlusions. These effects must be considered for planning motions of the robotic manipulators.

Some approaches for robotic on-orbit servicing and assembly are being developed, but they are mostly at conceptual level [4]. Current missions such as NASA's OSAM-1 (Restore-L) [5] and DARPA's MRV (RSGS) [6] are expected to demonstrate the enhancement of functionalities well beyond servicing existing satellites. It is expected that within five years some systems in orbit will be able to demonstrate the autonomy, dexterity, and delicacy needed to work on orbital assembly. This chapter provides an overview of robotic technologies aiming to provide such capabilities (Sect. 15.2).

Main in-space assembly technologies used so far by astronauts include the deployment of truss and beam sub-assemblies, with revolute joints or special latches for easy snap-on, and recently, in-space manufacturing using additive manufacturing techniques [7]. Independently of the technology used for the mechanical frame, one of the main challenges for assembling a large functional structure is the combination of power and data connectors together with the mechanical latch. Possible alternatives for solving this issue include magnetic latching and reversible joints that allow disassembly for repairing or for module replacement. A brief overview of standard interconnects that allow full functionality for modular assemblies is presented in Sect. 15.3.

While teleoperated or partially assisted assembly operations are possible on ground, autonomous assembly of structures has been recently proven feasible through the combination of adaptable perception, integrated assembly and grasp planning, and compliant control of the robotic manipulators [8]. A description of such ground-based demonstrator and its main technological features is presented in Sect. 15.4. We conclude the chapter with a discussion on some open topics in the field of autonomous on-orbit robotic assembly (Sect. 15.5).

15.2 Current Technologies

In-space assembly has been studied for almost four decades now [9] due to the great potential applications for performing tasks such as assembling systems larger than current cargo areas in commercial transportation vehicles, and can also be applied to more commercially interesting activities such as repair, refuel and upgrade of current satellites, although the economic benefit of those operations is controversial [10, 11]. Possible applications of in-space assembly include the construction of artificial gravity or asteroid redirect vehicles, space transportation hubs, space telescopes, solar electric power and propulsion, sun shields, and atmospheric decelerators [7]. An analysis of the potential applications shows that the demanded cross-cutting (non application-dependent) capabilities are robotic assembly, standardized interfaces, modular design with high stiffness, deployable subsystems, and docking and berthing. This chapter is focused on the topics of robotic assembly and standardized interfaces, which are more relevant for roboticists working on space applications. The design of deployable structures with high stiffness corresponds to the subfield of space architecture; autonomous docking and berthing are manoeuvres that have been tested in multiple occasions and are currently being applied to on-orbit servicing operations [3].

Several on-orbit servicing missions have been carried out, and have created the technological basis for on-orbit assembly. The first robotic on-orbit servicing mission was performed by NASDA (now JAXA, Japan) with the ETS-VII (KIKU-7) mission [12]. It used the first satellite equipped with a robotic arm, a 6 DoF (degrees of freedom), 2 m long manipulator, which was employed mainly to perform experiments of rendezvous and docking between a chaser (Hikoboshi) and a target (Orihime), and also to prove the feasibility of doing an ORU (Orbital Replacement Unit) exchange. The experiments were performed in 1999 [13]. The next relevant mission was Orbital Express, a joint effort of DARPA and Boeing. Experiments took place in 2007 [14], and included rendezvous, docking, fluid transfer and ORU transfer between two satellites, a servicing satellite equipped with a 6 DoF arm (ASTRO), and a serviceable satellite (NEXTSat) [15].

Different technologies have been proposed for robotic assembly in space, although real experiments have been performed only at the level of servicing and maintenance tasks, mainly with the Canadarm2 and Dextre manipulator (Fig. 15.1). Dextre, also known as the Special Purpose Dexterous Manipulator (SPDM), was assembled in 2008 on-board the ISS. It was built by MacDonald, Dettwiler and Associates (MDA), and maintained by the Canadian Space Agency (CSA). The original goal of the robot was to relieve astronauts from external servicing and maintenance tasks on the ISS, including removal and installation of batteries, opening and closing covers, or reconnecting cables [16]. It has also been recently used for removing cargo payload from supply ships like the Dragon capsules. Dextre is a dual arm manipulator, each arm with 7 joints and a length of 3.35 m, and a total mass of about 1600 kg [17]. It has a power fixture at the "wrist", which can be grasped by the Canadarm2 for repositioning at different work sites, or to be attached directly to the ISS. For performing operations, Dextre moves one arm at a time, teleoperated either from the ISS or from a ground control centre. In 2009 NASA initiated the Robotic Refueling Mission (RRM), aiming to use Dextre for performing robotic servicing tasks in orbit, including refuelling and maintenance [18]. The first phase was developed between 2012 and 2013, and employed four different tools: a wire cutter, a multifunction tool, a safety cap removal tool, and an EVR nozzle tool. It proved that the robot was capable of performing on-orbit refuelling of existing satellites. The second phase was launched in 2013 to further mimic servicing tasks, which were performed between 2015 and 2016. The tasks included the installation of two task boards onto an existing module, the visual inspection of Canadarm2 using a specialized tool, and assessment of performance of a solar cell. The mission ended in March 2017 with the removal of the RRM payload.

In 2011, Robonaut 2 (Fig. 15.1) became the first humanoid robot in Space [19]. It was co-developed by NASA and General Motors (GM) to serve as an assistant to human workers [20]. The initial Robonaut included only the torso, but in 2014 the mobility platform for Robonaut was delivered, to augment the robot with two legs for manoeuvring inside the ISS. The robot has human-like arms and hands,

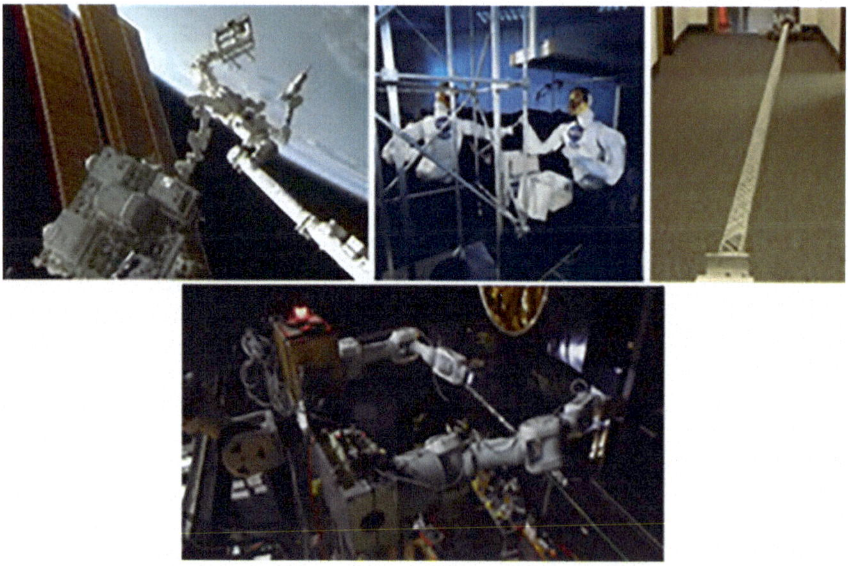

Fig. 15.1 Current robotic demonstrations of in-space assembly. From left to right: Dextre approaching the RRM module (Courtesy of NASA), Robonaut assembling truss structures on ground (Courtesy of NASA), Trusselator for SpiderFab (Courtesy of Tethers Unlimited), and initial ground demonstration of Phoenix with the FREND arms (Courtesy of DARPA)

which enables it to use the same tools as the crew members. On-ground tests have demonstrated the ability of Robonaut to perform different tasks, including assembly of truss structures [21]. On-orbit uses of Robonaut were limited to cleaning and monitoring tasks inside the ISS, although the robot had the potential for becoming a useful assistant for instance in EVA tasks. The robot was brought back to Earth in 2018.

Traditionally, satellites are fabricated and tested on ground, and then launched as payload on rockets. This demands a large effort and cost in the preparation of the system to survive the launch process, while the size of the satellite is limited by the cargo capabilities of current spacecrafts. To survive these restrictions, Tethers Unlimited has proposed the SpiderFab architecture [22], a satellite "chriysalis", combining manufacturing techniques with robotic assembly technologies (Fig. 15.2). This effort was supported through a NASA Innovative Advanced Concepts (NIAC) grant [4]. The architectural concept included:

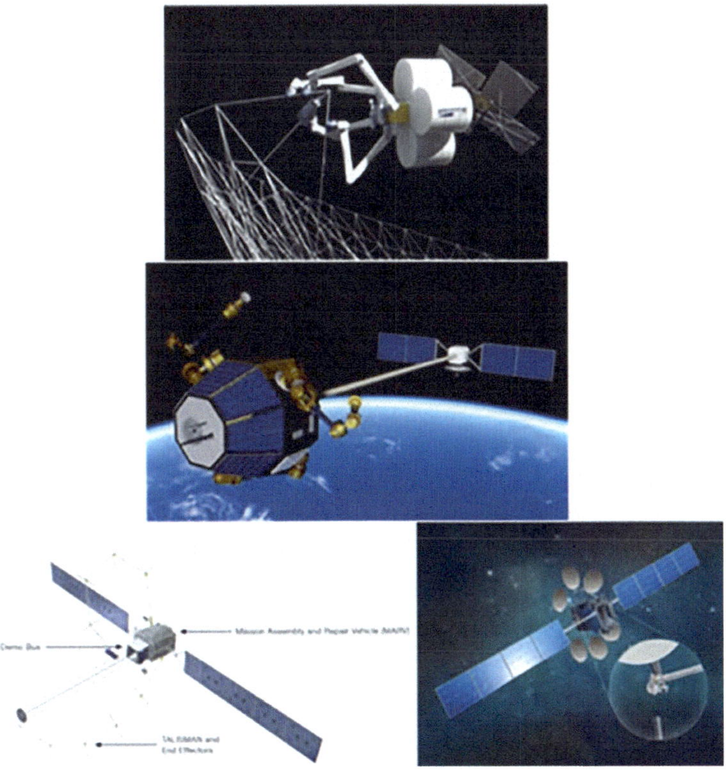

Fig. 15.2 Different robotic concepts for in-space assembly. From left to right: SpiderFab (Courtesy of Tethers Unlimited), Archinaut (Courtesy of Made in Space), CIRAS (Courtesy of Orbital ATK), and Dragonfly (Courtesy of Space Systems Loral)

- Techniques for processing suitable materials to create structures: while most rapid prototyping techniques on ground rely on gravity to facilitate bonding of the materials, the microgravity environment in space calls for a different approach. However, the lack of gravity also means that structures can be fabricated in any direction without the distortions induced by gravity. The system proposes the combined use of Fused Filament Fabrication to 3D print high-performance reinforced polymers. An initial test of the manufacturing of supporting structures was performed with the "Trusselator" (Fig. 15.1), for creating large truss structures by using a continuous fibre-reinforced thermoplastic yarn [23]. This technology was selected for a follow-up project, the MakerSat, aiming to endow small satellites with the ability to grow their size in space.
- Mechanisms for mobility and manipulation of tools and materials: involves the use of multiple dexterous robot arms to position the fabrication heads, position the robot along the constructed structure, and position the structural elements for assembly. As initial concept, the "Kraken" 7 DoF, 1 m long arm was developed for nanosatellite applications.
- Methods for assembly and joining structures: bonding can be accomplished in multiple ways, including welding, mechanical fasteners, adhesives, and snap-on latches. Fusion bonding can also be exploited by combining heat and pressure to fuse thermoplastic materials.
- Thermal control: one big challenge is the control of temperatures and heat gradients on the manufactured parts, especially considering the influence of the sun according to the position on orbit. The materials must be processed such that they cool and solidify in a manner that does not induce undesired stresses in the structure.
- Metrology to enable closed-loop fabrication: required to measure the overall shape of the structure, and to enable accurate positioning of the fabrication heads.
- Methods for integrating functional structures: the concept of SpiderFab was focused on creating support structures, but it required also suitable mechanisms to integrate functional elements that are produced on ground, such as solar cells, antenna panels, sensors, wiring, reflective membranes, and payload packages.

In 2012 DARPA started the Phoenix program, aiming to demonstrate the capabilities of a multi-arm satellite with tool changing capabilities (Fig. 15.1). The short-term mission goals were repairing of satellites, such as fixing a component or deploying an antenna. Medium-term goals included performing upgrades on orbit, and the long-term goal was the on-orbit construction of a geostationary satellite capable of servicing an operational spacecraft [24]. The arms for the satellite were developed under the FREND program (Front-End Robotics Enabling Near-Term Demonstration) for grappling and repositioning Resident Space Objects (RSOs); they have 7 DoF, a weight of 78 kg, and a total length of 2.5 m [25]. As part of the Phoenix program the arms were upgraded, new components like tools and tool changer were designed, and novel algorithms for control, teleoperation, mission sequencing and fault detection were also expected. A video with some of the on-ground results of the project is available in [26]. Two parallel missions under development and with similar goals are NASA's OSAM-1 (formerly known as Restore-L) and DARPA's MRV (formerly known as RSGS). Restore-L (Refuel Government-owned client satellite in Low Earth Orbit) aims to create a spacecraft that will extend satellites' lifespans by refuelling and relocation in LEO, mostly through teleoperation for critical actions [5]. RSGS (Robotic Servicing of Geosynchronous Satellites) aims for replacing and repairing faulty hardware, relying mostly on autonomous operations [6].

At the end of 2015, NASA announced new Public–Private Partnerships to advance "Tipping Point", Emerging Space Capabilities. Three projects were selected to explore the future of robotic in-space manufacturing through manipulation of structural trusses: Archinaut, CIRAS and Dragonfly (Fig. 15.2). Phases A/B were developed during two years,.

The first project, Archinaut (officially called Versatile In-Space Robotic Precision Manufacturing and Assembly System) was developed by Made In Space Inc. for construction, assembly and integratation of large structures in space [27]. It is constructed around a space-capable manufacturing unit already tested on the ISS, and capable of producing structures larger than itself. The setup for testing additive manufacturing of structures is called ESAMM (Extended Structure Additive Manufacturing Machine), and was created to test manufacturing in simulated space conditions (thermal-vacuum). To provide the assembly robotic component, the 3D printer was complemented with three manipulator arms built by Oceaneering Space Systems, which help to conform the "Ulisses" spacecraft for in-space manufacturing. Northrop Grumman provided systems engineering, control electronics, software, testing and assistance with Archinaut's space station interface.

The second project, CIRAS (Commercial Infrastructure for Robotic Assembly and Services), was carried out by Orbital ATK [28]. It aimed to attach and detach a solar array, demonstrating assembly and reuse capabilities on an integrated ground test using an air-bearing testbed. The system used two upgraded 15 meter TALISMAN (Tension Actuated Long Reach In-Space Manipulator) arm together with precision EBEAM welding using reversible quick-connects, and a special jigging precision assembly robot. The project built upon existing capabilities of the Lightweight Surface Manipulator System (LSMS) [29].

The third project, Dragonfly, or On-Orbit Robotic Installation and Reconfiguration of Large Solid RF Reflectors [30, 31], was developed by Space Systems Loral, and aimed to modify existing antenna and robotic equipment to perform a flight demonstration of antenna assembly for a GEO satellite. It used existing assembly interfaces with robot manipulators, to demonstrate stowage techniques for larger than traditional solid reflectors. Tethers Unlimited was a partner in this project, contributing with the in-space truss manufacturing facility developed with SpiderFab.

Northrop Grumman has recently demonstrated the functionality of two Mission Extension Vehicles (MEV): MEV-1 in 2020, and MEV-2 in 2021, both successfully attaching to Intelsat satellites for performing satellite servicing and repositioning [32]. NASA has continued the developments of Restore-L under the OSAM-1 program (On-orbit Servicing, Assembly, and Manufacturing), for performing on-orbit refueling of satellites [33]. The Archinaut developments are nowadays called

Fig. 15.3 OOS-SIM, an on-orbit servicing simulator developed at DLR

OSAM-2; the mission will build two beams and deploy a solar array using robotic manipulation [34]. Both missions are expected to be launched not earlier than 2025.

Concepts for an on-orbit servicing mission that can be potentially expanded for on-orbit fabrication have also been developed at DLR, with the DEOS (Deutsche Orbitale Servicing Mission) study [35]. As part of these activities, DLR developed the OOS-SIM, an On-Orbit Servicing simulator (Fig. 15.3) that replicates conditions of operations in micro-gravity environments. The setup is used for testing for instance combined platform-robot control schemes [36, 37]

15.3 Connectivity and Interfaces

A standard interface is defined as "a combination of devices that allow to couple active payload modules (APMs) to a manipulator, amongst themselves, and to spacecraft. It shall allow transferring of mechanical loads, electrical signals and data, as well as thermal flux between the coupled modules" [38]. A recent review on standard interfaces can be found in [39]. This section focuses mainly on detachable interfaces for orbital robotics; several of them have already been implemented in space missions (Fig. 15.4). Robotic arms often possess an End Effector (EE) that interfaces with a grapple fixture. Examples are the Shuttle Remote Manipulator System (SRMS), and the latching end effector (LEE) on the Canadarm2 (also known as Space Station Remote Manipulator System, SSRMS). The JEMRMS (Japanese Experiment Module Remote Manipulator System) on the ISS Japanese experimental module (JEM), has a similar mechanism to interface with several kinds of common grapple fixtures placed on various components of ISS modules. Their ability to grasp and move massive objects in space environment has been significantly demonstrated so far in different ISS projects. Also, the ORU/Tool Changeout Mechanism (OTCM) of the SPDM can attach at the tip several tools to conduct various tasks. A typical interface such as LEE is a detachable electro-mechanical interface (DEMI) designed to possess gender, i.e. an EE side and a grapple fixture side. All the interfaces possess active mechanisms on the arm side, an exclusive actuator used only for the docking operation.

Besides those already implemented interfaces, there are several interfaces developed on ground aiming for future implementation. As part of a space-borne manipulator, JAXA has developed an EE and grapple fixtures aiming to be used for construction of large space structures (Fig. 15.5). However, as the majority of interfaces in space, it is designed to provide only electro-mechanical connection between components. Another example is the Compact Tool Exchange Device (CTED, shown in Fig. 15.5), developed for connecting DexArm (ESA) and Dexhand (DLR) [40]; it was initially developed as part of the EUROBOT project. The RSGS interface was developed by MDA and adopted to interface the FREND arm and different EEs. Also MDA has developed a passive interface mechanism (PIF,) for the Next Generation Canadarm (NGC) project of CSA, to interface mechanically between the NGC arm and tool tip. Another interface named DWIM (Fig. 15.5) was developed by

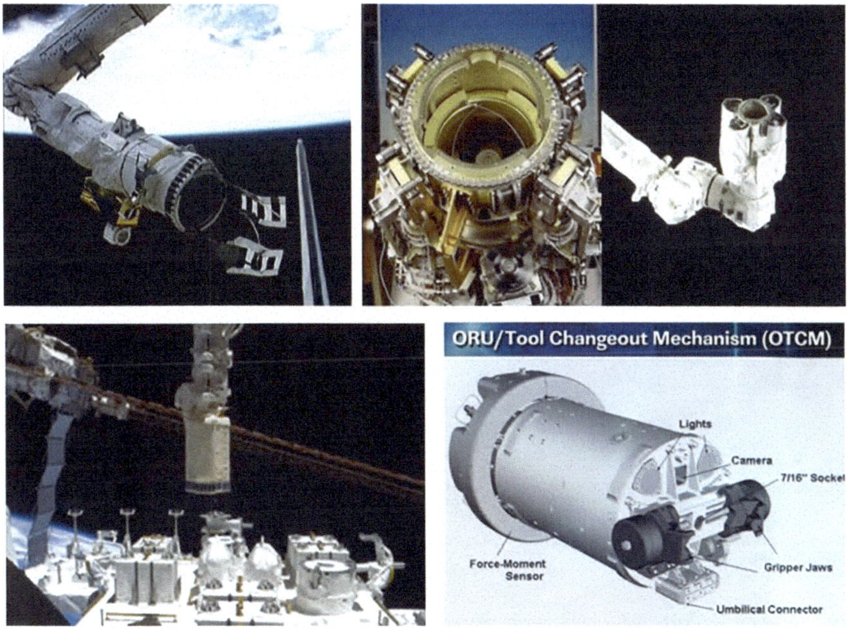

Fig. 15.4 Existing standard interfaces: top left: EE on SRMS (Courtesy of NASA), top right: LEE on SSRMS (Courtesy of NASA), bottom left: JEM EE (Courtesy of JAXA), bottom right: OTCM (Courtesy of NASA)

Tokyo Tech for a manipulator of an astronaut-supporting robot. While DWIM and PIF are detachable mechanical interfaces (DMI), the other three (JAXA EE, CTED, RSGS) are DEMI. Also, all interfaces possess gender. In addition, all of them are active, except for the MDA PIF (has no actuator on both sides). A recent proposal that combines mechanical coupling, power, heat and data transfer in one interface is the iSSI (Intelligent Space System Interface) [41], which provides a flat mating surface with an optional form fit to provide guidance during the mating process. More recently, the PERASPERA consortium supported the development of two novel standard interfaces: SIROM and HOTDOCK (Fig. 15.6), combining mechanical coupling, power, heat and data transfer in one interface. HOTDOCK features an androgynous and 90-degree symmetrical design [42]. It provides a form-fit that enables simultaneous connection of up to three orthogonally mounted interfaces. Its central connection plate is equipped with spring-loaded pogo pin connectors to provide electrical power as well as bi-directional data transfer between connected subsystems. A fluidic transfer capability can be optionally mounted for thermal cooling. SIROM shares some features with HOTDOCK, including the androgynous form fit and pogo pin electrical connectors. Its capture mechanism is based on three capture hooks evenly distributed at 120 deg on the mating surface, complemented with guiding petals to correct misalignments during approximation [43]. This mechanism allows the capture of the interface before the contact between the surfaces happens. A benchmarking approach to compare the functionality of the three latter interfaces (iSSI, HOTDOCK and SIROM) has been recently proposed [44], and the three interfaces are currently evolving toward full space qualification.

On ground applications, typical industrial manipulators use adapters or quick tool changers that enable quick change of end effectors. For instance, the SWS series from Schunk provides electro-mechanical interfaces with diameters from 40 mm to several hundred mm. A common model is the SWS-060 Schunk adapter, which is DEMI, and possesses gender to be able to use an active mechanism on the arm side that adopts pneumatic actuation to dock/undock the tool [45].

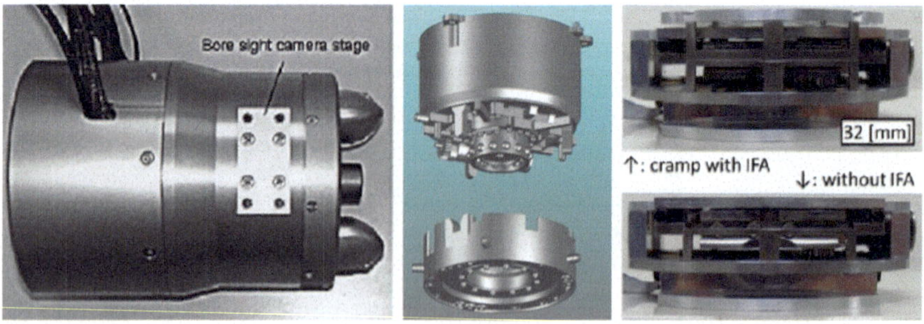

Fig. 15.5 Different standard interfaces: JAXA EE (Courtesy of JAXA), CTED (Courtesy of ESA), and DWIM (Courtesy of Tokyo Tech)

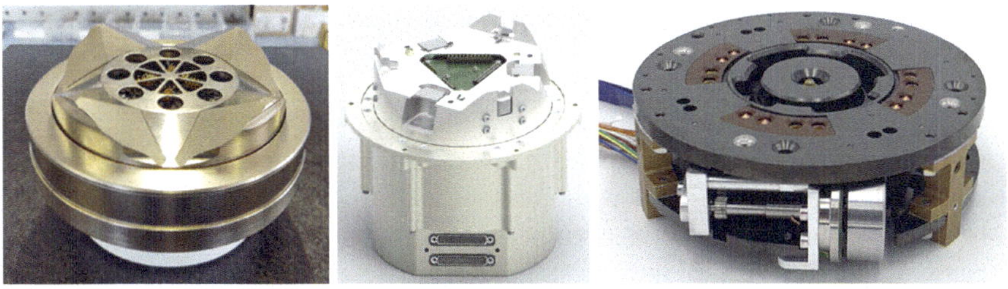

Fig 15.6 European standard interfaces. From left to right: HOTDOCK (courtesy of Space Application Services), SIROM (courtesy of SENER) and iSSI (courtesy of iBOSS)

15.4 Autonomous Robotic Assembly

Different research projects investigate new challenging applications in the field of on-orbit assembly (OOA), including for instance the construction of ultra-large structures such as telescopes, as in the PULSAR project (Prototype of an Ultra Large Structure Assembly Robot, [46]), as well as modular reconfigurable satellites, as in the project MOSAR (MOdular Spacecraft Assembly and Reconfiguration, [47]), or even concepts of an orbital factory mounted on the ISS, as in the PERIOD project (PERASPERA In-Orbit Demonstration) [48] (Fig. 15.7). . These assembly applications are intensively looking into autonomous robotic systems. Currently, the control of space robotic systems usually relies on teleoperation, which requires an appropriate feedback channel for the operator, typically affected by substantial time delays. The concept of shared autonomy increases the dexterity of such systems and reduces the effort for the operators in complex tasks [49]. Nevertheless, teleoperation approaches only have limited use when it comes to the assembly of complex structures. Because of the fine granularity of assembly tasks, classical teleoperation becomes unfeasible as it consumes substantial amounts of time for the synchronization of operator commands and manipulator actions. Therefore, a robotic assembly system should be capable of performing a sequence of operations or even the complete assembly task autonomously. This level of autonomy has been reached for robotic assembly system on Earth as described in the following.

Fig. 15.7 European projects considering robotic assembly in space. From left to right: artistic concepts for PULSAR, MOSAR and PERIOD

15.4.1 Development of Autonomous Robotic Assembly Systems

Traditional robots in industrial applications are expected to manufacture the same product in large quantities and with constant quality. Such systems are expected to be mainly robust and precise, ideally without the need for human intervention. Because of the trend toward mass customization, assembly systems are expected to also be flexible and adaptable to different products, again with little human effort. Autonomous robotic assembly systems can provide capabilities to plan and execute necessary tasks for the production of unique products.

Since the development of the SHAKEY robot from 1966 until 1972 [50], the concept of autonomy has been increasingly exploited in robotics. One of the first autonomous robotic systems developed for assembly tasks was the Karlsruhe Autonomous Mobile Robot (KAMRO) [51, 52]. The KAMRO System is a mobile two-arm robot based on a function-oriented planning, which means that the system interprets unexpected situations instead of just reacting to them as in behaviour-oriented systems. In the experiments, the KAMRO system was able to perform an assembly plan autonomously. At the beginning the KAMRO system navigates through a room collecting parts while avoiding obstacles, and then assembles them. The final assembly is a mechanical structure that requires peg-in-hole skills plus bimanual manipulation to tighten bolts. Since those developments, autonomous assembly has evolved with different techniques for planning and decision making. One of the examples with most impact in the media has been the IkeaBot [53], where an array of heterogeneous robots was able to assemble an IKEA chair using off-the-shelf hardware. More examples can be found in different literature reviews [54–56].

The evolution of autonomous assembly processes brought the need of more specific representations beyond the ones used in the KAMRO system, where only simple precedence relations were used. Better representations were developed, not only of the product to be built, but also for the steps required to construct such product and the constraints raised during the process. The KAMRO system used precedence graphs, other widely adopted representation are the AND/OR graphs [57]. Some object-oriented product models included relations between subassemblies, components and features of these components [58]. This allowed the shift from not only modelling geometrical features, but also adding more information related to the function of each component. Already in [59], authors replaced the model used in KAMRO for a model that separates the assembly process into tasks, instead of seeing the process as a complete unit. A task description was included, plus a distribution and allocation process for these tasks in order to coordinate all the system components. A new paradigm for ontology-based assembly design was introduced in [60]. In this new framework, previous assembly relation models are enhanced in order to provide interpretable knowledge shared between different machines.

Another research area in the autonomous assembly field is the planning and scheduling of tasks. Different complexity measurements were already introduced in the 90s [61] in order to find solutions that were more suitable to the desired problem. These measurements include for example number of non-linear steps, number of re-orientations, and number of assembly steps required. Based on different measurements, the systems have increased their planning capabilities focusing on assembly task allocation and assembly task execution. A first approach to this problem was presented in [62], introducing the multi-manipulator systems (MMS). MMS is an architecture that could handle complex interactions between agents. The MMS also includes a distributed trajectory planning, giving to the system better flexibility, scalability, reconfigurability, and fault-tolerance.

The real execution of an assembly task is strongly affected by uncertainties and unforeseen events. Therefore, robotic assembly systems need to provide robust execution strategies. Key features of such strategies are compliance and force control, which help to close the gap between planning and execution. Early works show how concepts of mechanical compliance [63, 64] and control algorithms using force feedback can precisely mate parts with small clearances [65, 66]. Today, various kinds of force control methods for robotic assembly exist [67], and the main challenge nowadays is to ease the programming of novel tasks and to increase the reusability of implemented solutions in various situations. Especially in terrestrial scenarios, programming by demonstration and kinesthetic guiding, e.g. [68, 69], are explored to teach new robotic skills. However, these approaches typically require direct interaction between human and robot, which makes them difficult to use in space scenarios. Human robot interaction in simulated virtual environments could make this teaching and learning approaches feasible [70]. Similarly, such virtual environments might be required to make use of artificial intelligence methods, including deep-reinforcement learning [71] or self-supervised learning [72] methods for assembly skill acquisition, as in space applications online learning is hardly possible. While those new learning-based approaches show promising results, existing approaches using predefined strategies [73] or planners for contact-rich tasks [74] in combination with compliant control can already provide sufficient reusability and robustness in multiple space applications.

In recent years new techniques have been introduced in the assembly process to facilitate the use of robotics systems. A relevant technique is learning from demonstration (LfD). LfD has been used to teach robots easily how to implement tasks autonomously, specially to learn manipulation behaviours from observing the motion executed by human demonstrators [75]. Also, new interfaces have been created in order to increase the usability of the systems. The click and assemble robotic assembly (CARA) system [76], has a simple but effective way to introduce the product to assemble. It receives as an input a CAD model, and no further action is required by the user. The system autonomously performs the planning phase and generates an execution plan for the robotic setup. In [77] a similar process is performed, where a Relational Assembly Model is derived directly from the CAD data. This model generates the tasks to be completed in order to perform the assembly, based on the spatial relations between the parts. These tasks are then decomposed in skill primitives, based on the start and goal conditions. Some systems [78, 79] have taken similar approaches by generating user-friendly interfaces that enhance the user experience and give more information to the system about the relations between the parts of the assembly. An associated problem is transferring the already acquired knowledge into future uses of the system, in order to minimize re-planning as much as possible. For instance, MARVIN [80] is a cognitive system that does learning and transfer of sensorimotor level and planning level information. Another example of this knowledge re-use is our demonstrator for autonomous assembly, described below.

15.4.2 Case Study: Autonomous Assembly of Structures

A robotic system for autonomous assembly of aluminum structures was developed for a terrestrial use case at DLR [8]. However, the concepts used to solve this assembly task can be potentially transferred to systems in space applications. The structures are made up of a modular building kit typically used in industrial facilities for the setup of production equipment, composed by aluminum profiles of various lengths, and angle bracket connectors. The robotic assembly system is able to automatically decompose a given assembly specification, i.e. a list of parts and their relative configuration, into a sequence of tasks, which are appropriately translated into robotic skills. Figure 15.8 shows this workflow, starting with an assembly planning unit that recognizes the tasks required for the assembly of the structure. Then, a module is in charge of the task-to-skill mapping, and finally another module schedules and executes the necessary skills.

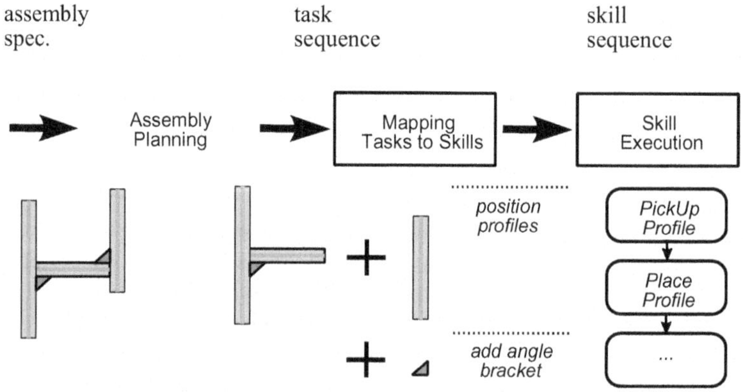

Fig. 15.8 Workflow of the robotic assembly system. An assembly specification is automatically decomposed into a task sequence and then mapped to appropriate robotic skills. These skills are then executed in the robotic setup

15.4.2.1 Robotic Platform

The initial demonstration of autonomous assembly was performed with a single arm and a number of specific fixturing units conveniently located in the robot work cell [8]. Later, the setup was extended with a second robot arm (Fig. 15.9) to increase flexibility of the system by completely removing the required fixtures. This extra arm enables the assembly of more complex structures by the implementation of multi-arm collaboration and manipulation. The actual system is composed by two KUKA LBR iiwa R800 manipulators, each arm with 7 DoF and equipped with torque sensors at each joint. These manipulators are the industrial version of the light-weight robot arm (LBR) technology developed at DLR [81]. The robotic components of the LBR were verified for space applications with experiments on the ISS in the ROKVISS mission [82]. In most of the autonomous execution of assembly tasks, the LBR's sensitivity and the impedance control mode are required for robust and stable fine manipulation.

Although humanoid robots have already been used for simple space operations [19], the general dexterity and sensing capabilities for autonomous execution of complex assembly tasks are not yet available. In our system, we combine general dexterous manipulators with specialized tools and grippers (as seen in Fig. 15.10) suitable for the target building kit in order to facilitate the reliable execution of the tasks. The design effort for application-specific tools and grippers is justified by the low amount of distinctive tasks to solve in space compared to general robotic assistants, as demonstrated for instance with the Robotic Refueling Mission (RRM) for the ISS Dextre robot [83]. For our demonstrator, we adapted an industrial Schunk

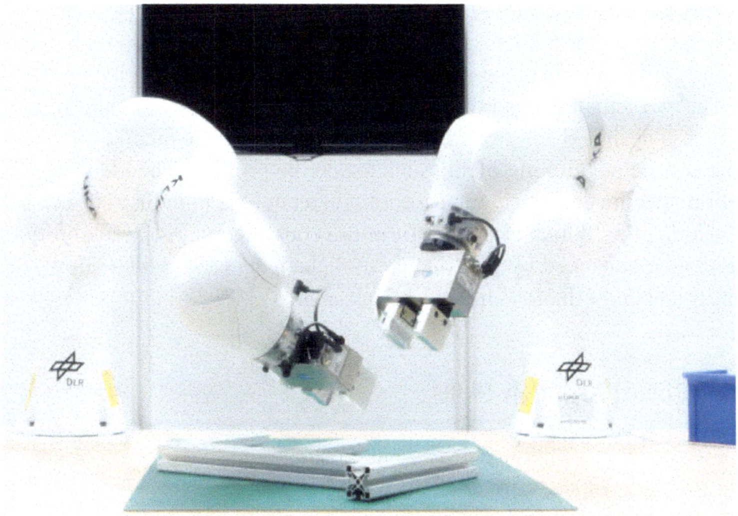

Fig. 15.9 Two light-weight robot arms (KUKA LBR iiwa) collaborate in the assembly of aluminum structures

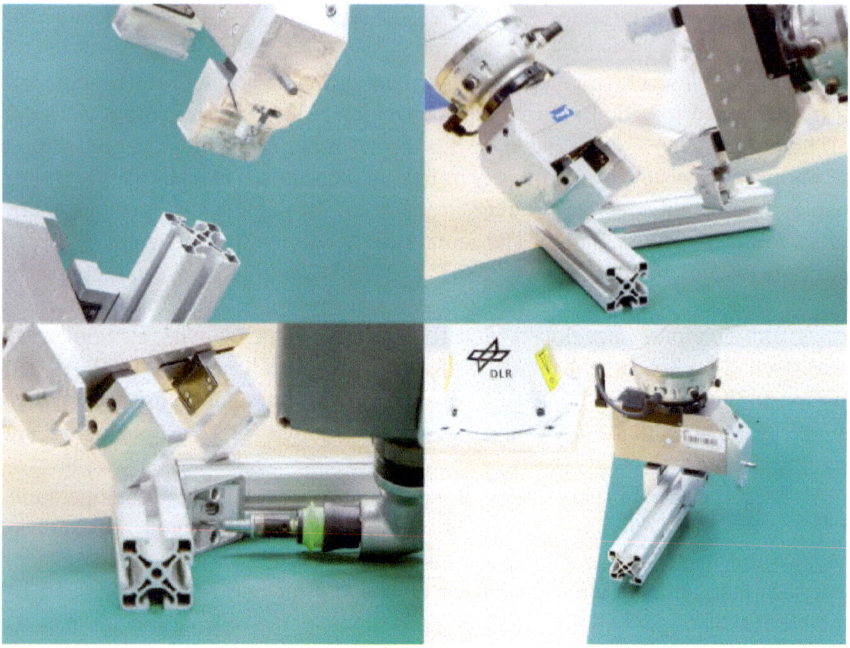

Fig. 15.10 Realization of various subtasks. Clockwise: slot nut insertion, placement of an angle bracket, placement of a profile, and fastening the angle bracket with screws using the screwdriver

WSG 50 gripper, where specific fingertips were designed for the parts and tools of the building kit. The modified geometry of the fingers allows form-closure grasps on the parts and the required tools (screwdriver). In this way, grasping uncertainties are significantly reduced, and grasp stability can be guaranteed directly by the mechanical design. The two actuated fingers are mainly used for power grasps, while an additional third finger is designed for grasping angle brackets and slot nuts; it also contains a small pin for the positioning of slot nuts inside the profiles. The idea of generating application-specific end-effectors is a common technique in industrial scenarios and is seen also in other domains such as furniture construction [84]. Flexibility of such systems can be enhanced by introducing tool changers [85], which allows to select the appropriate end-effector for a given task.

15.4.2.2 Robust Assembly Skills

In space assemblies, as in on-Earth assemblies, modularity is an important requirement for the reuse of software and hardware components. Nowadays, users ask the system to be able to generate personalized products, therefore the systems must be able to cope with variation of goals using modular production techniques. With our robotic platform, an infinite number of structures can be theoretically constructed.

The building set is composed by square-shaped profiles that are customizable in length, and connections that can be freely placed along the four sides of the profiles. A single connection between two profiles is made with one or two angle brackets, each of them fastened with screws and nuts that are inserted in the slots of the profiles. The subtasks required by the system during the construction of an aluminum structure are [8]:

- insert slot nuts in the profiles;
- position profiles;
- add angle bracket;
- add screws.

All subtasks require robust assembly strategies for their execution in the robotic setup. These strategies are implemented as a list of basic skills that encapsulate the robot capabilities in a parameterizable and reusable way. Skills can be adapted to the current task, and the desired behaviour can be guaranteed to solve a given task through a certification process [86]. Open-loop robustness on a higher level is achieved here thanks to the sensitivity of the robot arm and the impedance controller. A robust strategy for such insertion tasks can be chosen considering the passive alignment properties and the compliance of the robot in the contact [73]. Furthermore, observation algorithms based on joint torque measurements are developed to reduce uncertainties and monitor and adapt the executions online [87, 88].

15.4.2.3 Planning for Autonomous Execution

Given the wide variety of products that come from the modularity of the domain, it might not be possible to engineer and pre-compute every assembly sequence. Also, the system needs to adapt the execution according to the present state of the environment, which calls for an autonomous planning tool. In order to achieve the required autonomy, our system uses a set of dedicated planners located in the planning unit. The outcome of such units is used to select and parametrize the skills for the present tasks. Therefore, an assembly sequence planning, a grasp planning, a workspace analyzer and a motion planning component are tightly connected through a runtime world model for symbol grounding according to the present state of the assembly, as depicted in Fig. 15.11. Grasps are generated and checked for feasibility for all subassemblies [89]. Data from the planners is used to parametrize skills in order to find the appropriate configurations, e.g. a specific grasp is selected amongst the available grasps in the database according to reachability criteria and available paths provided by the motion planner [8]. The motion planner [90] synchronizes its internal geometrical representation with the runtime world model, where all objects are registered with type and pose, and are also labelled semantically to connect with the planning units and skills. The sequence planning [91] takes into consideration not only the constraints relevant to the product to be built, but also the skill set of the robotic agents. A solution is generated after sequential feasibility checks in different levels of detail. Solutions that are not feasible are then used to prune the

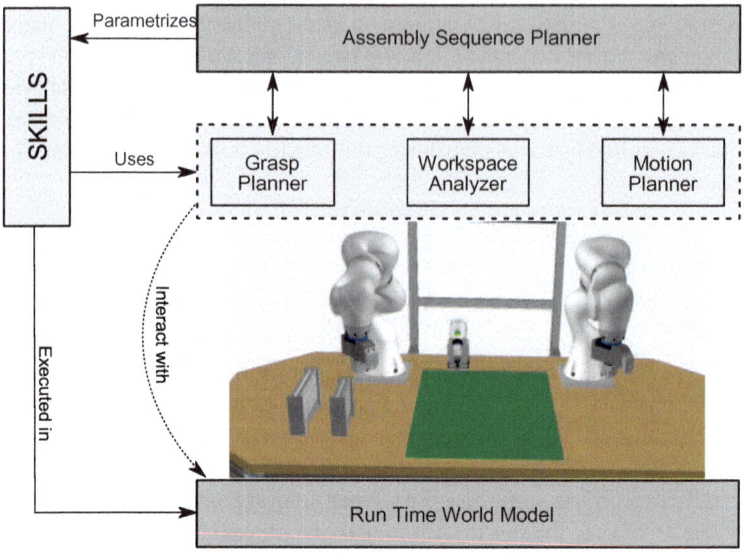

Fig. 15.11 Components of the planning unit for the autonomous execution of assembly skills. The sequence planner, the grasp planner and the motion planner interact with a runtime world model to select and parameterize an appropriate sequence of skills

search algorithm by avoiding the same errors that already occured. An extension to this algorithm has been developed to minimize the need of human expert knowledge [92]. In this extension, the knowledge on how to build a certain assembly is reused to avoid extra planning in future assemblies. The method is based on matching key features between different assemblies and representing the relevant constraints in a simple semantic way.

So far, the system is capable of automatically planning sequences of tasks and skills for the construction of planar aluminum structures by only providing a specification of the desired assembly goal. Ongoing work aims to add more flexibility to the system by using advanced reasoning algorithms for assembling more complex structures based on an expanded set of connectors.

15.5 Final Discussion

New technologies such as additive manufacturing of components have been recently introduced for space applications. In a medium term, however, the construction of large-scale structures will need to rely on robotic manipulators for performing delicate on-orbit assembly operations. Several challenges remain to prove the feasibility of this ambitious idea, but initial tests on ground have verified the capabilities and versatility of autonomous assembly using robotic manipulators. These terrestrial applications typically rely on the use of fixtures that provide mechanical support during the assembly operation. However, the use of fixtures in space can be limited due to the dynamic effects of zero gravity conditions. Feeding of parts for the assembly, size of the parts to be handled, feasibility and limits of usage of one or two arms for the operations, or use of assembly kits are still open questions in the field. The extension and testing on relevant environments of zero-gravity autonomous robotic assembly is a current topic of interest at DLR.

Acknowledgement This work was partially supported by the European Union's Horizon 2020 research and innovation programme under grant agreement No. 821858, project PULSAR.

References

1. Garcia, A., Lamb, A., Sleptsov, A., Moreno, C., Victorova, M., Glazkova, N. et al. (2016). Post-ISS plans: What should be done? *REACH—Reviews in Human Space Exploration, 1*, 63–73.
2. Barnhart, D., Will, P., Sullivan, B., Hunter, R., & Hill, L. (2016). Creating a sustainable assembly architecture for next-gen space: The Phoenix effect. In *National space symposium*.
3. Artigas, J., de Stefano, M., Rackl, W., Lampariello, R., Brunner, B., Bertleff, W., et al. (2015). The OOS-SIM: An on-ground simulation facility for on-orbit servicing operations. *Proc. IEEE Int. Conf. on robotics and automation (ICRA)* (pp. 2854–2860).
4. Hoyt, R., Cushing, J., & Slostad, J. (2013). SpiderFab (TM): Process for on-orbit construction of kilometer-scale apertures. In *NASA Innovative Advanced Concepts (NIAC)*, Report No. NNX12AR13G.
5. NASA: Restore-L robotic servicing mission. URL https://svs.gsfc.nasa.gov/12023/
6. DARPA: *RSGS: robotic servicing of geosynchronous satellites*. http://www.darpa.mil/program/robotic-servicing-of-geosynchronous-satellites
7. Belvin, K., Doggett, B., Watson, J., Dorsey, J., Warren, J., Jones, T., et al. (2016). In-space structural assembly: Applications and technology. In *3rd AIAA spacecraft structures conference*.
8. Nottensteiner, K., Bodenmueller, T., Kassecker, M., Roa, M.A., Stemmer, A., Stouraitis, T., et al. (2016). A complete automated chain for flexible assembly using recognition, planning and sensor-based execution. In *47st Int. Symp. on Robotics (ISR)*, pp. 1–8.
9. Akin, D., Minsky, M., Thiel, E., & Kurtzman, C. (1983). Space applications of automation, robotics and machine intelligence systems (ARAMIS)—phase II. In *NASA Contractor report* (p. 3734).
10. Krishna, A., Skenjana, M., Do, V., Yoshida, R., Wang, Q., & Rizzo, F. (2016). In-orbit maintenance: the future of the satellite industry. In *MCTP project*, Toulouse Business School.
11. Purves, L. (2002). A method for estimating costs and benefits of space assembly and servicing by astronauts and robots. In *Proc. IEEE Aerospace conference*.
12. Oda, M., Kibe, K., & Yamagata, F. (1996). ETS-VII, space robot in-orbit experiment satellite. In *Proc. IEEE Int. Conf. on robotics and automation (ICRA)* (pp. 739–744).
13. Imaida, T., Yokokohji, Y., Doi, T., Oda, M., & Yoshikawa, T. (2004). Ground–space bilateral teleoperation of ETS-VII robot arm by direct bilateral coupling under 7-s time delay condition. *IEEE Trans. Robotics and Automation, 20(3)*, 499–511.
14. Friend, R. (2008). Orbital Express program summary and mission overview. In *Proc. SPIE, sensors and systems for space applications II*.
15. Whelan, D., Adler, E., Wilson, S., & Roesler, G. (2000). DARPA Orbital Express program: effecting a revolution in space-based systems. In *Proc. SPIE Small Payloads in Space*.

16. Coleshill, E., Oshinowo, L., Rembala, R., Bina, B., Rey, D., & Sindelar, S. (2009). Dextre: Improving maintenance operations on the International Space Station. *Acta Astronautica, 64,* 869–874.
17. CSA: *Dextre, the International Space Station's robotic handyman.* http://www.asc-csa.gc.ca/eng/iss/dextre/default.asp
18. NASA: RRM: Robotic Refueling Mission. URL. https://sspd.gsfc.nasa.gov/robotic_refueling_mission.html
19. Diftler, M., Mehling, J., Abdallah, M., Radford, N., Bridgwater, L., Sanders, A., Askew, R., Linn, D., Yamokoski, J., Permenter, F., Hargrave, B., Platt, R., Savely, R., & Ambrose, R. (2011). Robonaut 2—the first humanoid robot in space. *Proc. IEEE Int. Conf. on robotics and automation (ICRA),* pp. 2178–2183.
20. Bluethmann, W., Ambrose, R., Diftler, M., Askew, S., Huber, E., Goza, M., et al. (2003). Robonaut: A robot designed to work with humans in space. *Autonomous Robots, 14,* 179–197.
21. Diftler, M., Doggett, W., Mehling, J., & King, B. (2006). Reconfiguration of EVA modular truss assemblies using an anthropomorphic robot. In *Proc. AIP Conference* (pp. 992–999).
22. Hoyt, R., Cushing, J., Slostad, J., Jimmerson, G., Moser, T., Kirkos, G., et al. (2013). SpiderFab: An architecture for self-fabricating space systems. In *AIAA SPACE Conference.*
23. Hoyt, R., Cushing, J., Slostad, J., & Jimmerson, G. (2014). TRUSSELATOR: On-orbit fabrication of high-performance composite truss structures. In *AIAA SPACE Conference.*
24. Henshaw, C. (2014). The DARPA Phoenix spacecraft servicing program: Overview and plans for risk reduction. In *Proc. i-SAIRAS Conference*
25. Debus, T., & Dougherty, S. (2009). Overview and performance of the front-end robotics enabling near term demonstration (FREND) robotic arm. In *AIAA SPACE Conference*
26. DARPA: *Phoenix program demonstration of latest advances.* https://www.youtube.com/watch?v=QQ0mSNsGlcQ
27. Made in Space: Project ARCHINAUT. URL https://redwirespace.com/newsroom/2019-2-13-archinaut-the-in-space-robotic-manufacture-and-assembly-technology/
28. Orbital ATK: CIRAS, the commercial infrastructure for robotic assembly and services. URL https://www.nasa.gov/missions/tech-demonstration/orbital-atk-supports-ground-testing-on-ciras-at-nasas-langley-research-center/
29. Dorsey, J., Doggett, W., Hafley, R., Komendera, E., Correll, N., & King, B. (2012). An efficient and versatile means for assembling and manufacturing systems in space. In *AIAA Space Conference.*
30. Lymer, J., Hanson, M., Tadros, A., Boccio, J., Hollestein, B., Emerick, K., et al. (2016). Commercial applications of in-space assembly. In *AIAA SPACE conference.*
31. Space Systems Loral: Project Dragonfly. URL https://www.nasa.gov/technology/nasas-dragonfly-project-demonstrates-robotic-satellite-assembly-critical-to-future-space-infrastructure-development/
32. Northrop Grumman: SpaceLogistics. https://www.northropgrumman.com/space/space-logistics-services/.
33. NASA: OSAM-1 mission. https://nexis.gsfc.nasa.gov/osam-1.html.
34. NASA: OSAM-2 mission. https://www.nasa.gov/mission_pages/tdm/osam-2.html.
35. Reintsema, D., Thaeter, J., Rahtke, A., Naumann, W., Rank, P., & Sommer, J. (2010). DEOS—the German robotics approach to secure and de-orbit malfunctioned satellites from low Earth orbits. In: *Proc. i-SAIRAS conference.*
36. De Stefano, M., Mishra, H., Giordano, A., Lampariello, R., Ott, C. (2021) A relative dynamics formulation for hardware- in-the-loop simulation of on-orbit robotic missions. *IEEE Robotics and Automation Letters, 6*(2), 3569–3576.
37. Mishra, H., De Stefano, M., Giordano, A., Lampariello, R., Ott, C. (2020) A geometric controller for fully-actuated robotic capture of a tumbling target. In *Proc. American Control Conference (ACC),* pp. 2150–2157.
38. European Commision. (2016). Guidelines for strategic research cluster on space robotics technologies. *Horizon 2020 Space Call 2016, Final Version.*

39. Wenzel, W., Palazzetti, R., Tan, X., & Bartsch, S. (2017). Mechanical, thermal, data and power transfer types for robotic space interfaces for orbital and planetary missions-a technical review. In *Proc. ASTRA*.
40. Wedler, A., Chalon, M., Baumann, A., Bertleff, W., Beyer, A., Burger, R., et al. (2011). DLR's space qualifiable multi-fingered hand DEXHAND. In *Proc. ASTRA*.
41. Weise, J., Briess, K., Adomeit, A., Remierdes, H., Goller, M., & Dillmann, R. (2012). An intelligent building blocks concept for on-orbit satellite servicing. In *Proc. i-SAIRAS Conf*erence
42. Letier, P., Siedel, T., Deremetz, M., Pavlovskis, E., Lietaer, B., Nottensteiner, K., Roa, M., Sanchez, J., Corella, J., Gancet, J. (2020) HOTDOCK: Design and validation of a new generation of standard robotic interface for on-orbit servicing. In *Proc. Int. Astronautical Congress (IAC)*.
43. Diaz, M., Bedialauneta, P., Vinals, J., Gala, J., Guerra, G. (2021) Standard interface for robotic manipulation (SIROM): Current state and future developments for one of the main building blocks for space robotics advancement. In *Proc. Int. Astronautical Congress (IAC)*.
44. Brinkmanna, W., Yuksel, M., Jankovic, M., Wirkus, M., Saffer, J., Soto, I., Gancet, J., Letier, P., Schervan, T., Kreiseld, J., Estable, S., Kirchner, F. (2022) A standard interconnect benchmark for a european in-orbit services, manufacturing and assembly (ISMA) demonstrator. In *Proc. Int. Astronautical Congress (IAC)*.
45. Schunk: SWS, schnellwechselsysteme. URL https://schunk.com/at/de/automatisierungstechnik/wechselsysteme/c/PUB_8306
46. Roa, M., Koch, C.E., Rognant, M., Ummel, A., Letier, P., Turetta, A., Lopez, P., Trinh, S., Rodriguez, I., Nottensteiner, K., Rouvinet, J., Bissonnette, V., Grunwald, G., Germa, T. (2022) PULSAR: Testing the technologies for on-orbit assembly of a large telescope. In *Proc. Symp. on Advanced Space Technologies in Robotics and Automation (ASTRA)*.
47. Letier, P., Yan, X., Deremetz, M., Bianco, A., Grunwald, G., Roa, M., Krenn, R., Munoz, M., Dissaux, P., Garcia, J., Ruiz, R., Filippis, L., Porceluzzi, G., Post, M., Walshe, M., Perryman, P. (2019) MOSAR: MOdular Spacecraft Assembly and Reconfiguration demonstrator. In *Proc. Symp. on Advanced Space Technologies in Robotics and Automation (ASTRA)*.
48. Bartscha, S., Estable, S., Vogel, T., Manza, M., Aridon, G., Caujolle, R., Colmenero, F., Alonso, M., Esquer, M., Silveira, D., Serra, C., Jankovic, M., Brinkmann, W., Soto, I., Guerra, G., Gancet, J., Letier, P., Barrio, A., Shilton, M., Allouis, E., Ordoubadian, B., Bertels, E., Chamos, A. (2022) In-orbit demonstration of isma robotic capabilities to pave the way for a new generation of space systems. In *Proc. Int. Astronautical Congress (IAC)*.
49. Birkenkampf, P., Leidner, D., & Borst, C. (2014). A knowledge-driven shared autonomy human-robot interface for tablet computers. In *IEEE-RAS Int. Conf. humanoid robots*, pp. 152–159.
50. Nilsson, N. J. (1984). Shakey the robot. Tech. rep., SRI International.
51. Hormann, A., & Rembold, U. (1991). Development of an advanced robot for autonomous assembly. In *Proc. IEEE Int. Conf. on robotics and automation*, pp. 2452–2457.
52. Lueth, T., & Rembold, U. (1994). Extensive manipulation capabilities and reliable behavior at autonomous robot assembly. In *Proc. IEEE Int. Conf. on robotics and automation (ICRA)*
53. Suarez-Ruiz, F., Zhou, X., & Pham, Q. C. (2018). Can robots assemble an IKEA chair? *Science Robotics, 3*(17), 2018.
54. Andersen, A. L., Brunoe, T. D., & Nielsen, K. (2015). Reconfigurable manufacturing on multiple levels: Literature review and research directions. In S. Umeda, M. Nakano, H. Mizuyama, N. Hibino, D. Kiritsis, & G. von Cieminski (Eds.), *Advances in production management systems: innovative production management towards sustainable growth*, Vol. 459, pp. 266–273. Springer.
55. Fulea, M., Popescu, S., Brad, E., Mocan, B., & Murar, M. (2015). A literature survey on reconfigurable industrial robotic work cells. *Applied Mechanics and Materials, 762*, 233–241.
56. Setchi, R., & Lagos, N. (2004). Reconfigurability and reconfigurable manufacturing systems state-of the-art review. In *IEEE Int. Conf. on Industrial Informatics*, pp. 529–535
57. De Mello, L. H., & Sanderson, A. C. (1990). AND/OR graph representation of assembly plans. *IEEE Trans. on robotics and automation, 6*(2), 188–199.

58. van Holland, W., & Bronsvoort, W. F. (2000). Assembly features in modeling and planning. *Robotics and Computer-Integrated Manufacturing* 16(4), 277–294.
59. Lueth, T., & Laengle, T. (1994). Task description, decomposition, and allocation in a distributed autonomous multi-agent robot system. In *Proc. IEEE/RSJ Int. Conf. on Intelligent Robots and Systems (IROS)*, pp. 1516–1523.
60. Kim, K. Y., Manley, D. G., & Yang, H. (2006). Ontology-based assembly design and information sharing for collaborative product development. *Computer-Aided Design, 38*(12), 1233–1250.
61. Goldwasser, M., Latombe, J. C., & Motwani, R. (1996). Complexity measures for assembly sequences. In *Proc. IEEE Int. Conf. on robotics and automation* (ICRA), pp. 1851–1857.
62. Fraile, J. C., Paredis, C., Cheng-Hua, W., & Khosla, P. (1999). Agent-based planning and control of a multi-manipulator assembly system. In *Proc. IEEE Int. Conf. on Robotics and Automation (ICRA)*, pp. 1219–1225.
63. Drake, S. H. (1978). Using compliance in lieu of sensory feedback for automatic assembly. *Ph.D. thesis, Department of Mechanical Engineering, Massachusetts Institute of Technology.*
64. Whitney, D., & Nevins, J. L. (1979). What is the remote center compliance (RCC) and what can it do. In *Proc. Int. symposium on industrial robots (ISIR).*
65. Goto, T., Takeyasu, K., & Inoyama, T. (1980). Control algorithm for precision insert operation robots. *IEEE Trans. on Systems, Man, and Cybernetics, 1*(10).
66. Inoue, H. (1974). Force feedback in precise assembly tasks. *Massachusetts Institute of Technology Artificial Intelligence Laboratory AI Memo 308.*
67. Marvel, J., & Falco, J. (2012). Best practices and performance metrics using force control for robotic assembly. US Department of Commerce, National Institute of Standards and Technology.
68. De Chambrier, G. P. L. (2016). Learning search strategies from human demonstrations. *Ph.D. thesis,* École Polytechnique Fédérale de Lausanne.
69. Kramberger, A., Gams, A., Nemec, B., Chrysostomou, D., Madsen, O., & Ude, A. (2017). Generalization of orientation trajectories and force-torque profiles for robotic assembly. *Robotics and Autonomous Systems, 98*, 333–346.
70. Scherzinger, S., Roennau, A., & Dillmann, R. (2019). Contact skill imitation learning for robot independent assembly programming. In *Proc. IEEE/RSJ Int. Conf. on intelligent robots and systems (IROS)*, pp. 4309–4316.
71. Inoue, T., De Magistris, G., Munawar, A., Yokoya, T., & Tachibana, R. (2017). Deep reinforcement learning for high precision assembly tasks. In *Proc. IEEE/RSJ Int. Conf. on intelligent robots and systems (IROS).*
72. Lee, M. A., Zhu, Y., Srinivasan, K., Shah, P., Savarese, S., Fei-Fei, L., et al. (2019). Making sense of vision and touch: Self-supervised learning of multimodal representations for contact-rich tasks. In *Proc. IEEE Int. Conf. on Robotics and Automation (ICRA)*, pp. 8943–8950.
73. Stemmer, A., Albu-Schaffer, A., & Hirzinger, G. (2007). An analytical method for the planning of robust¨ assembly tasks of complex shaped planar parts. In *Proc. IEEE Int. Conf. on Robotics and Automation (ICRA)*, pp. 317–323.
74. Wirnshofer, F., Schmitt, P. S., Feiten, W., Wichert, G. V., & Burgard, W. (2018). Robust, compliant assembly via optimal belief space planning. Proc. *IEEE Int. Conf. on Robotics and Automation (ICRA)*, pp. 1–5.
75. Zhu, Z., & Hu, H., (2018). Robot learning from demonstration in robotic assembly: a survey. *Robotics, 7*(2), 17.
76. Fakhurldeen, H., Dailami, F., & Pipe, A. G. (2019). CARA system architecture-a click and assemble robotic assembly system. In *Proc. IEEE Int. Conf. on robotics and automation (ICRA)*, pp. 5830–5836.
77. Mosemann, H., & Wahl, F.M. (2001). Automatic decomposition of planned assembly sequences into skill primitives. *IEEE Trans. on Robotics and Automation, 17*(5), 709–718.
78. Paxton, C., Hundt, A., Jonathan, F., Guerin, K., & Hager, G. D. (2017). Costar: Instructing collaborative robots with behavior trees and vision. In *Proc. IEEE Int. Conf. on robotics and automation (ICRA)*, pp. 564–571.

79. Thomas, G., Chien, M., Tamar, A., Ojea, J. A., & Abbeel, P. (2018). Learning robotic assembly from CAD. In *Proc. IEEE Int. Conf. on robotics and automation (ICRA)*, pp. 1–9.
80. Savarimuthu, T. R., Buch, A. G., Schlette, C., Wantia, N., Roßmann, J., Martínez, D., et al. (2017). Teaching a robot the semantics of assembly tasks. *IEEE Trans. on Systems, Man, and Cybernetics: Systems, 48*(5), 670–692.
81. Albu-Schaffer, A., Haddadin, S., Ott, C., Stemmer, A., Wimb ock, T., & Hirzinger, G. (2007). The DLR lightweight robot: design and control concepts for robots in human environments. *Industrial Robot: An Int. Journal, 34*(5), 376–385.
82. Albu-Schaffer, A., Bertleff, W., Rebele, B., Schafer, B., Landzettel, K., & Hirzinger, G. (2006). ROKVISS—robotics component verification on ISS, current experimental results on parameter identification. In *Proc. IEEE Int. Conf. on robotics and automation (ICRA)*, pp. 3879–3885.
83. NASA: Robotic refueling mission (2013). https://svs.gsfc.nasa.gov/10988/
84. Knepper, R. A., Layton, T., Romanishin, J., & Rus, D. (2013). Ikeabot: An autonomous multi-robot coordinated furniture assembly system. In *Proc. IEEE Int. Conf. on robotics and automation* (ICRA), pp. 855–862.
85. Lehmann, C., Staedter, J. P., & Waechter, K. (2016). A variant-flexible assembly cell for hydraulic valve sections using a sensitive lightweight robot. In *Proc. 47st Int. Symp. on Robotics* (ISR), pp. 1–7.
86. Steinmetz, F., & Weitschat, R. (2016). Skill parametrization approaches and skill architecture for human-robot interaction. In *Proc. IEEE Int. Conf. automation science and engineering (CASE)*.
87. Nottensteiner, K., Sagardia, M., Stemmer, A., & Borst, C. (2016). Narrow passage sampling in the observation of robotic assembly tasks. In *Proc. IEEE Int. Conf. on robotics and automation (ICRA)*, pp. 130–137.
88. Nottensteiner, K., Stulp, F., & Albu-Schaffer, A. (2020). Robust, locally guided peg-in-hole using¨ impedance-controlled robots. In *Proc. IEEE Int. Conf. on Robotics and Automation (ICRA)*.
89. Thomas, U., Stouraitis, T., & Roa, M. (2015). Flexible assembly through integrated assembly sequence planning and grasp planning. In *Proc. IEEE Int. Conf. automation science and engineering* (CASE), pp. 586–592.
90. Lehner, P., & Albu-Schaffer, A. (2017). Repetition sampling for efficiently planning similar constrained manipulation tasks. In *Proc. IEEE/RSJ Int. Conf. on intelligent robots and systems (IROS)*, pp. 2851–2856.
91. Rodrıguez, I., Nottensteiner, K., Leidner, D., Kaßecker, M., Stulp, F., & Albu-Schäffer, A. (2019). Iteratively refined feasibility checks in robotic assembly sequence planning. *IEEE Robotics and Automation Letters, 4*(2), 1416–1423.
92. Rodríguez, I., Nottensteiner, K., Leidner, D., Durner, M., Stulp, F., & Albu-Schäffer, A. (2020). Pattern recognition for knowledge transfer in robotic assembly sequence planning. *IEEE Robotics and Automation Letters, 5*(2), 3666–3673.

Chapter 16
PULSAR: Verification of Technologies for In-Orbit Assembly of a Large Telescope

Maximo A. Roa, Christian Ernst Siegfried Koch, Mathieu Rognant, Antoine Ummel, Pierre Letier, Alessio Turetta, Pablo Lopez Negro, Souriya Trinh, Ismael Rodríguez, Korbinian Nottensteiner, Julien Rouvinet, Vincent Bissonnette, Gerhard Grunwald, and Thierry Germa

Abstract The EU project PULSAR (Prototype of an Ultra Large Structure Assembly Robot) carried out a feasibility analysis for a potential mission that could demonstrate robotic technologies for autonomous assembly of a large space telescope. The project performed the analysis using two hardware demonstrators, one devoted to show the assembly of five segmented mirror tiles using a robotic manipulator, and another one showing extended mobility for assembling a large structure in low gravity (underwater) conditions. The hardware demonstrators were complemented with a simulation analysis to demonstrate the operation of a fully integrated system and to address the mission challenges, especially in the field of attitude and orbital control. The techniques developed in the project support the path toward In-Space Servicing, Assembly and Manufacturing (ISAM).

M. A. Roa (✉) · I. Rodríguez · K. Nottensteiner · G. Grunwald
German Aerospace Center (DLR), 82234 Wessling, Germany
e-mail: maximo.roa@dlr.de

C. E. Siegfried Koch
German Research Center for Artificial Intelligence (DFKI), 28359 Bremen, Germany

M. Rognant
ONERA—The French Aerospace Lab, 31055 Toulouse, France

A. Ummel · J. Rouvinet
CSEM S.A, 2000 Neuchatel, Switzerland

P. Letier
Space Applications Services NV/SA, 1932 Sint-Stevens-Woluwe, Belgium

A. Turetta
Graal Tech, 16152 Genova, Italy

P. Lopez Negro
Thales Alenia Space, 06150 Cannes, France

S. Trinh · V. Bissonnette · T. Germa
Magellium, 31520 Ramonville Saint-Agne, France

X. T. Yan and G. Visentin (eds.), *Space Robotics*,
https://doi.org/10.1007/978-3-031-39214-6_16

16.1 Introduction

Space Industry is attracting unprecedented levels of attention in the last few years. Favoured by the progressive reduction of traditional entry barriers (technology complexity, high investments) and stimulated by an ever increasing demand of new, more performing services, a proliferation of innovative ideas is emerging [1]. One of the key aspects enabling the current golden era is the improvement of launch capabilities: today's launchers allow a drastic reduction of the cost per launched kilogram, mostly thanks to the introduction of reusable rockets [2]. While some of the new launchers multiply the payload capacity (in mass) compared to its predecessors, still size and accommodation limitations impose severe constraints in the design of spacecrafts. These constraints become important limiting factors in some particular missions requiring very large structures. Indeed, the need of enhanced services of space infrastructure combined with the renewed interest in science and space exploration allows to identify a number of use cases requiring large-size structures and/or appendages [3]. Potential applications include large space telescopes (Fig. 16.1), large antenna reflectors, orbital platforms, very large solar arrays, deployable radiators, and thermal shields [4].

These large space structures do not easily fit inside a launcher fairing, and usually require some kind of mechanism for deployment, or a specific design allowing unfolding when the structure shape allows for it. This was the approach taken with the James Web Space Telescope (Fig. 16.2), launched in 2022, which used a deployable primary mirror splitted in three segments [5]. However, novel assembly technologies are required for substantially increasing the size of structures to be deployed in

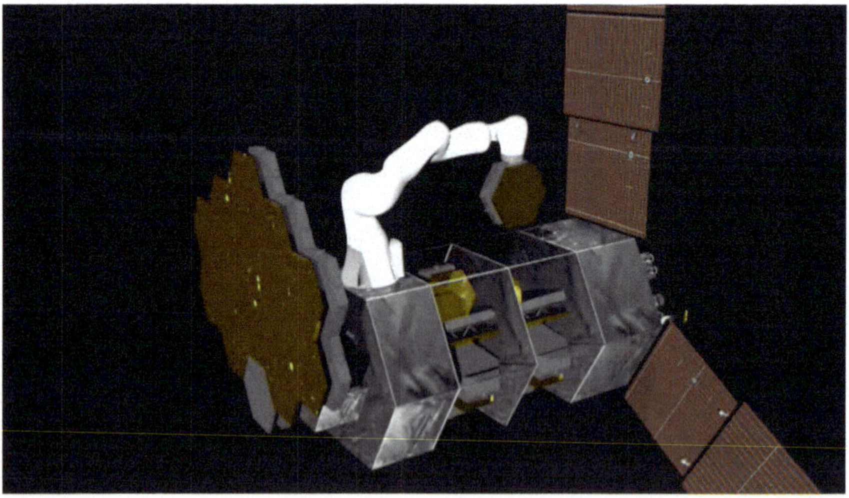

Fig. 16.1 Artistic concept of the PULSAR project for assembling a large space telescope

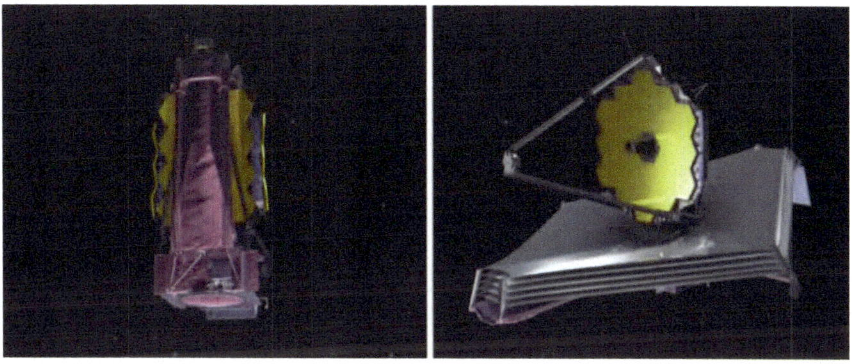

Fig. 16.2 Deployment of the James Webb Space Telescope. Left: folded configuration, for transportation in the cargo area. Right: fully deployed configuration in space (Courtesy of NASA)

space. In order to make the solutions scalable and able to cope with complex geometries and high performance requirements, the use of robotic manipulators to perform assembly in-orbit is a subject of increasing interest among governments, academic and industrial institutions [6].

In this context, prominent industrial actors in the European Union have identified the need of fostering Europe's capabilities in space robotics and in-orbit assembly. The European Union funded, through the Space Robotic Technologies program within Horizon 2020[1], the project PULSAR (Prototype of an Ultra Large Structure Assembly Robot)[2]. This project aimed to develop and demonstrate the technologies that will allow the in-orbit precise assembly of a segmented mirror using an autonomous robotic system. PULSAR was organized in three demonstration tracks to address the major challenges of in-space autonomous assembly of telescopes:

- dPAMT, demonstrator of Precise Assembly of Mirror Tiles: addressed the challenge of assembling a mirror composed by multiple hexagonal segmented mirror tiles. This requires a robotic system that provides high assembly accuracy and adaptability to cope with the accumulated assembly errors.
- dLSAFFE, demonstrator of Large Structure Assembly in Free Floating Environment: addressed the challenge of assembling a large structure in a free-floating environment, in this case, underwater. This requires advanced mobility to overcome the limits of robotic arm range, adaptability to cope with accumulated assembly errors, and optimal attitude and position control systems.
- dISAS, demonstrator of In-Space Assembly in Simulation, addressed the challenge of simulating in the most realistic conditions possible the autonomous deployment of a large structure in space while ensuring the stability and the safety of the spacecraft. It was used to simulate a fully representative scenario

[1] https://cordis.europa.eu/project/id/640026

[2] https://cordis.europa.eu/project/id/821858

with respect to the spacecraft Attitude and Orbit Control System (AOCS) considerations, including for instance the effect of the arm motion on the spacecraft attitude.

This chapter summarizes the main results achieved within the project PULSAR based on the three demonstrators above described.

16.2 Demonstrator of Precise Assembly of Mirror Tiles (dPAMT)

The main objectives of the dPAMT demonstrator were to prove the feasibility of performing autonomous assembly of Segmented Mirror Tiles (SMTs) for a space telescope using standard connection interfaces and a suitable Robotic Assembly System (RAS), to verify the assembly precision using an external measurement device, and command calibration motions of the optical surfaces to correct alignment errors induced by the assembly process.

The primary mirror to be assembled consisted of a fixed, central tile, and five SMTs that had to be properly placed by the robotic system in an appropriate assembly sequence (Fig. 16.3). Two of the SMTs were active, in the sense that they had a mobile stage with a positioning mechanism that could adjust a circular mirror mounted at the centre of the tile [7] (Fig. 16.4). The adjustment range was $\pm$ 3 mm in position, with 1 μm resolution and 5 μm repeatability, and $\pm$ 1° in tip/tilt angle, with 4μrad resolution and 20μrad repeatability. The mirror was representative of the telescope optical surface, and it was used as metrological target for the demonstration. The SMTs also contain the supporting structure for several Standard Interconnects (SIs), HOTDOCK in this case (Fig. 16.3). The HOTDOCKs provided mechanical, electrical and data connection between the tiles, and also served as means for manipulating the SMTs with the robotic system [8].

The demonstrator was implemented using a KUKA KMR iiwa robotic system. Its main components are a mobile platform and a robotic manipulator mounted on top of it. The mobile platform provides the robot navigation capabilities, and carries the SMT container and the on-board control computer. Furthermore, it carries a camera mounted on a pole, with a pan-tilt unit, to acquire the required visual information from the assembly (Fig. 16.5). The robotic manipulator is a KUKA LBR iiwa 14, which is a seven degrees-of-freedom (DoF) lightweight robot that can be controlled in position or impedance mode, as required to fulfill the delicate assembly tasks performed during the demonstration. The robot was endowed with an active HOTDOCK as end effector, to retrieve and manipulate the SMTs.

The sequence of steps to assemble the structure was determined using a hybrid planning framework [9]. This planner is structured in two main parts, a physical layer and a logical layer (Fig. 16.6). Its main objective is to generate the best possible assembly sequence using as cost function the torque demands on the robot manipulator due to the weight of the SMTs under ground (1g) conditions. The physical

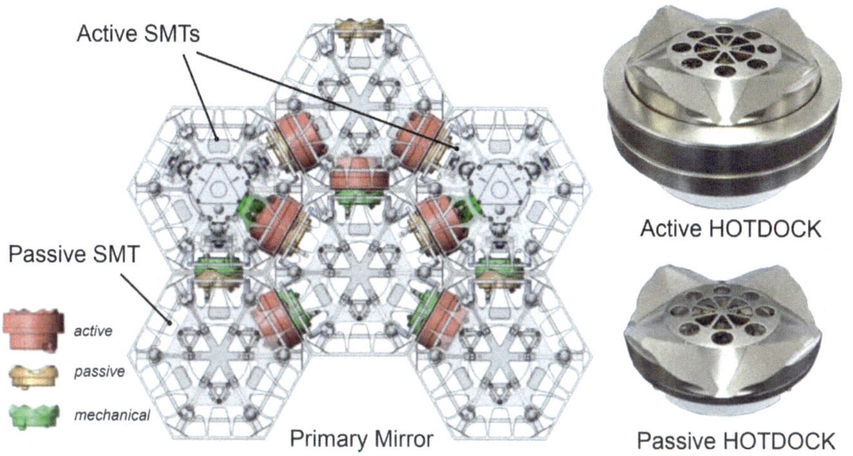

Fig. 16.3 Primary mirror in the dPAMT demonstrator. Left: components of the assembly. Right: HOTDOCK standard interconnect, passive and active version

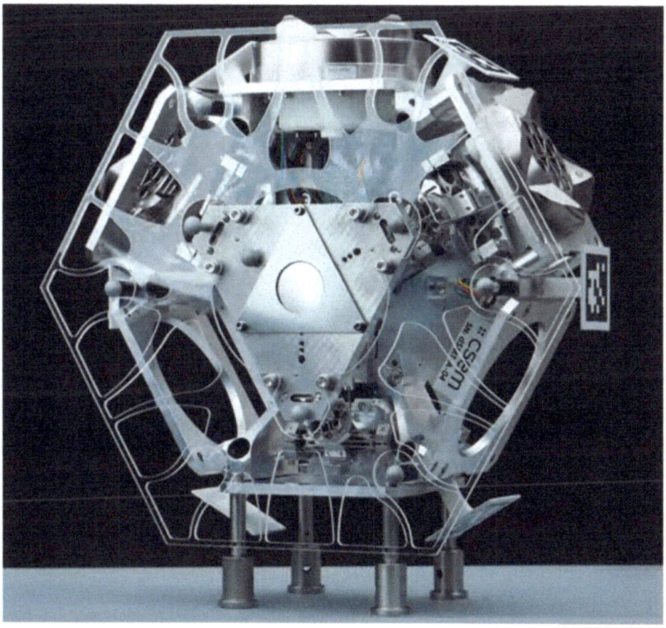

Fig. 16.4 Active segmented mirror tile of dPAMT

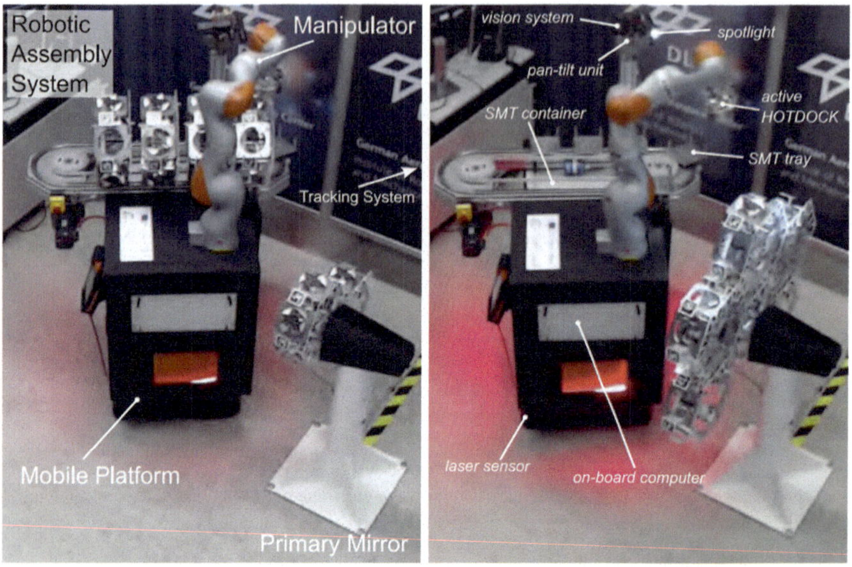

Fig. 16.5 Components of the dPAMT demonstrator

layer computes the trajectories to place each tile minimizing the joint torques; this optimization is achieved using Stochastic Trajectory Optimization for Motion Planning (STOMP) [10]. In the logical layer, the assembly sequence is represented as a graph, where the nodes are sub-assemblies in the process, and the edges contain the information given by the physical layer on the largest required joint torques to assemble the corresponding tile. Once the graph is generated, the planner searches for the sequence that minimizes the overall cost. Furthermore, the search takes into account the connectivity between the tiles in order to ensure mechanical stability of the assembled structure and to guarantee control access to all active SMTs.

Suitable robotic manipulation skills are assigned to each assembly step. These skills include different compliant assembly strategies, depending on the degree of uncertainty in the estimated pose of the HOTDOCK interface. For navigation, the robot base is localized relative to the primary mirror using two laser sensors located on the mobile platform. Additionally, a perception module implemented the localization of the tiles using AprilTags [23] as visual (fiducial) markers (Fig. 16.7).

The demonstrator successfully showed the assembly of five SMTs. The HOTDOCK connectors allowed single, double and triple matings during the assembly process thanks to their form fitting and passive guidance features, which allowed deviations up to 15 mm in position, and up to 10° in orientation (Fig. 16.8). The visual localization successfully reconstructed the relative location of the tiles during the assembly process. A tolerance threshold on their relative pose is defined based on the maximum misalignment allowed by the HOTDOCK specifications, in order to ensure their proper latching. Also, using an external VICON tracking system for verification, the relative position of the active SMTs could be reconstructed, and

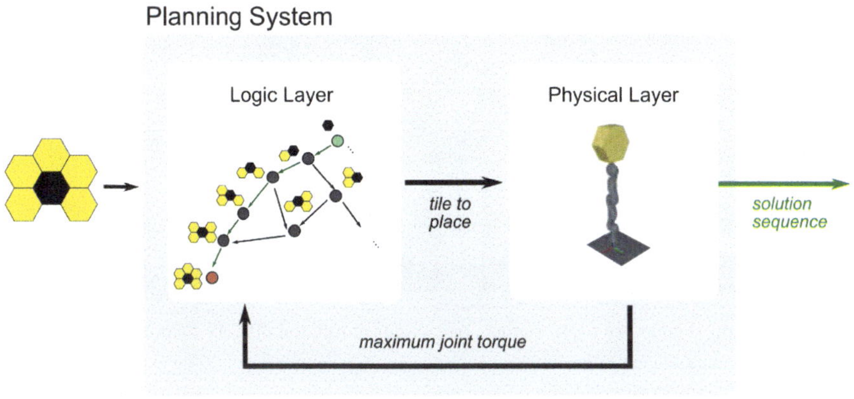

Fig. 16.6 Planning of the assembly sequence for dPAMT using a hybrid planning framework

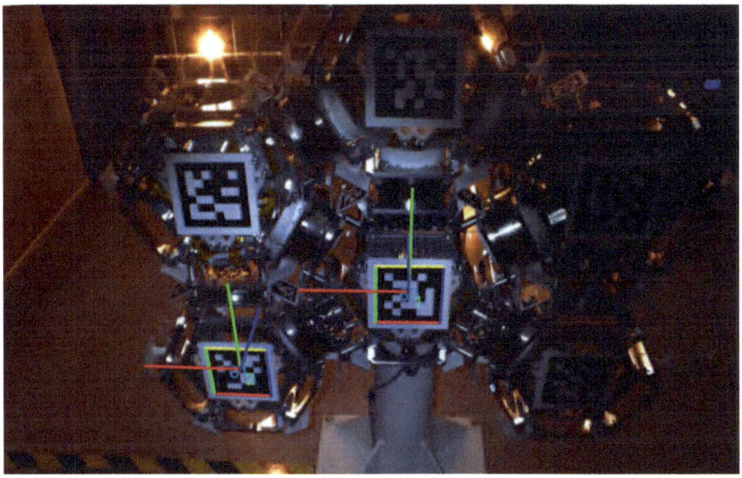

Fig. 16.7 Visual reconstruction of the assembly using AprilTags. The image shows the alignment verification of SMT2 (bottom left) with respect to SMT1 (central tile)

suitable adjustment commands were successfully transmitted and executed by the active tiles. The demo was successful, even though it had a primary limitation: gravity, as the structural capacity of the robotic arm limited the achievable size of the assembled structure. The tiles had a diameter of 30 cm, while the tiles on the real scenario would have a diameter of 1.3 m. The next demonstrator, dLSAFFE, worked with 1:1 tiles in low gravity conditions.

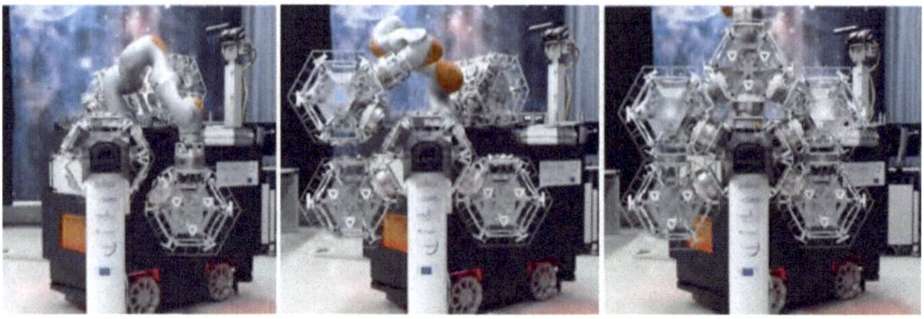

Fig. 16.8 Single, double and triple mating connections during the dPAMT demonstration

16.3 Demonstrator of Large Structure Assembly in Free Floating Environment (dLSAFFE)

The main objective of dLSAFFE was to demonstrate the extended mobility required to assemble very large structures that exceed the workspace of the robotic manipulator, focusing mainly on motion planning and control. The demonstration was set in an underwater zero-buoyancy environment to lift size and weight constraints, as well as to partially simulate the free-floating dynamics. Using components adapted for underwater deployment, the robotic assembly system and segmented mirror tiles were represented with realistic proportions and workspace [11].

The robotic assembly system consists of a manipulator arm with six DoF situated on a linear rail system acting as prismatic joint. The rail increases the workspace of the manipulator and allows traversing the distance between storage compartment and assembly target. However, even with the rail in place, the large mirror structure to be assembled exceeds the workspace of the robotic assembly system. To overcome the workspace limit and facilitate the assembly of much larger structures, an extended mobility concept was introduced. Instead of assembling the mirror tile by tile (individually), several tiles are pre-assembled, forming a subassembly. This subassembly can be manipulated as a unit and assembled to its final position on the mirror structure (Fig. 16.9).

For dLSAFFE, the robotic assembly system and mirror tiles were designed at a near one-to-one scale to accurately represent the workspace: the hexagonal mirror tiles measure approximately 1.3 m in diameter with a thickness of 0.4 m; the lightweight manipulator arm has a reach of 2 m while weighing only 20 kg in air. As required by the extended mobility concept, multiple mirror tiles are manipulated simultaneously. Weight constraints were lifted by setting the demonstration in a zero-buoyancy underwater test environment. The assembly system, mirror tiles and standard interconnects were adapted towards underwater usage (Fig. 16.10).

The robotic assembly system consists of a six DoF manipulator arm mounted on a linear rail. The linear rail allows vertical movement of 1.2 m, and it uses a spindle drive mechanism for actuation. The GraalTech Underwater Modular Arm was used as baseline to create the customized manipulator for dLSAFFE. The manipulator can

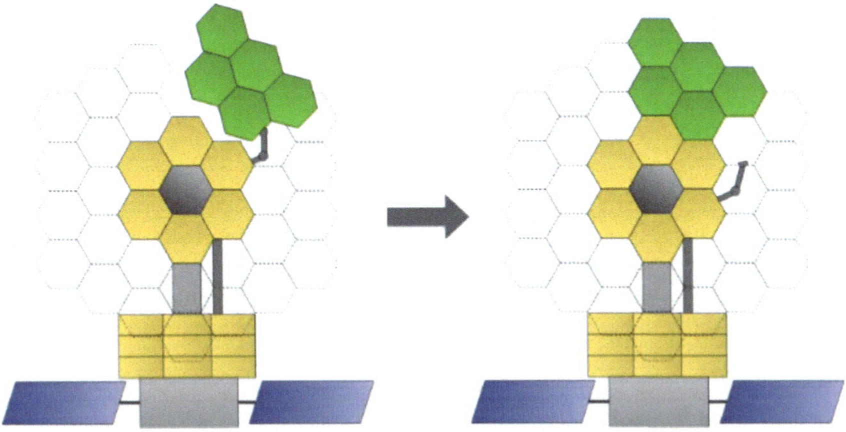

Fig. 16.9 The outermost ring of the mirror structure lies outside the workspace of the robotic assembly system. To overcome this limit and facilitate the assembly, a unit consisting of five mirror tiles (green) is pre-assembled before being mounted to its final position on the outer ring

Fig. 16.10 dLSAFFE underwater spacecraft mock-up in the test facility. The robotic assembly system manipulates a subassembly consisting of three mirror tiles. Two tiles of the inner ring of the mirror structure are already assembled to the stationary central tile. Without mirror tiles, the mock-up is 3.5 m tall

be controlled in joint position and velocity mode. The segmented mirror tiles were designed considering weight and drag minimization, to reduce undesired effects on the demo. Instead of a solid body, the tiles consist of a hexagonal frame of carbon fibre beams. Weight and buoyancy of the tiles are carefully calibrated such that the tiles are very slightly positively buoyant, with as little restoring torque as possible.

The mirror tiles are connected to each other via standard interconnects based on HOTDOCK, adapted for underwater usage. The interfaces are simplified and only provide mechanical connection, while power and data transmission were omitted. To avoid movable components, the mechanical connection is provided using permanent magnets. An active standard interconnect is also used for grasping the tiles by means of an electro-magnetic version of the interface, used as end-effector on the manipulator.

To a large extent, the software was shared between PULSAR demonstrators. In the context of dLSAFFE, motion planning and control software were implemented, while higher level execution functions were developed and tested with dPAMT and dISAS. A simplified view of the software architecture is given in Fig. 16.11. In particular, the motion planning component generates trajectories for the robotic assembly system to accomplish a desired task, e.g., to reach a target position, while satisfying constraints such as joint limits and collision avoidance. This was based on a C++ library targeted towards the ROCK framework[3]. The library provides interfaces to different state-of-the-art planning [12–14], kinematics [15, 16] and collision detection algorithms [17]. In PULSAR, the library and the dLSAFFE software were adapted and integrated within the ESROCOS framework [18].

dLSAFFE successfully demonstrated planning and execution of constrained manipulation of mirror tiles. The demo showed the assembly of the outer ring, i.e., the extended mobility function, where multiple mirror tiles were manipulated simultaneously, as shown in Fig. 16.12.

16.4 Demonstrator of In-Space Assembly in Simulation (dISAS)

The dISAS demonstrator was conceived for testing the complete assembly of the primary mirror with the highest accuracy possible via simulation, and to provide a set of tools for the specification, execution and visualization of large structures assembly via simulation, in order to validate their feasibility. The complete spacecraft is modelled and simulated accurately using a combination of the Webots simulator[4], which uses ODE[5] (Open Dynamics Engine) as physics engine, and dedicated controllers to deal with flexible elements such as the sunshields. Taking into account the physics of the flexible elements required a dedicated development, to be able to

[3] https://www.rock-robotics.org/

[4] https://cyberbotics.com/

[5] https://www.ode.org/

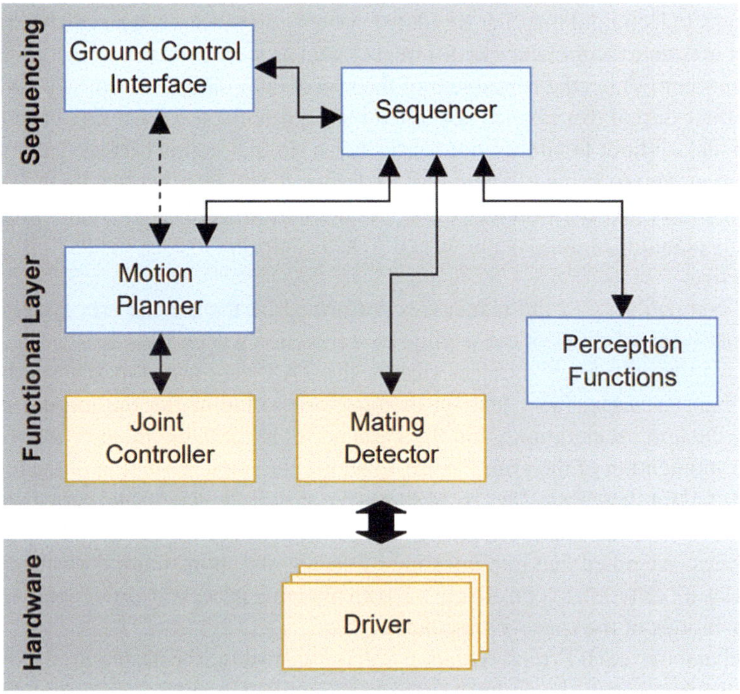

Fig. 16.11 Simplified software architecture for dLSAFFE. Components in blue are shared among demonstrators, while yellow components are specific to dLSAFFE

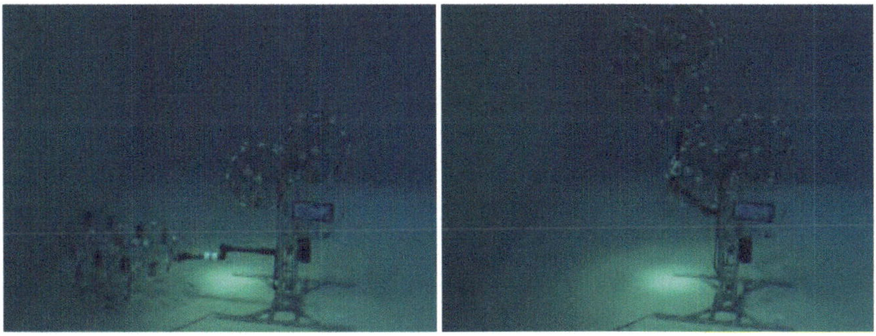

Fig. 16.12 dLSAFFE demonstrating the manipulation of a subassembly with three mirror tiles

realistically simulate the effect of the flexible appendages on the whole spacecraft. Resulting forces and torques are then computed and applied at each simulation time step on the different spacecraft anchor points. The simulated system has twenty-two flexible modes: the first four modes of each of the two solar panels, and the first three modes of the four support beams of the sun shield.

Thanks to its modularity, the simulated system complexity can be easily adjusted in order to isolate the influence of different system properties on the correct execution of the mission. Thus, the robustness of the spacecraft controller to inertia variation and to the external disturbance generated by the robotic arm have first been tested on a model without flexibility nor sensor noise. In that context, $H\infty$ synthesis [3] has been proven to be an efficient tool to design controllers that handle the system dynamic evolution (inertia and flexibility) and sensor noise. The simulator integrated assembly planning, motion planning and AOCS controller to consider the full system perspective.

The compatibility of the maneuvres performed by the robotic arm system with the actuation capabilities of the satellite base (reaction wheels) has also been considered. Two sets of reference trajectories for the RAS were tested to achieve the full deployment scenario: (a) a first set of trajectories minimizing the torques generated by the arm accelerations, and (b) a second set minimizing the derivative of the angular momentum of the arm. Fig. 16.13 shows the implementation of the attitude control used for these tests. Fig. 16.14 shows the attitude deviation and period without communication link to ground during the full deployment for the two tested sets of trajectories. We noted that minimization of the angular momentum derivative is the most relevant criterion to consider in the path planning process in order not to exceed the capabilities of the spacecraft actuators.

A behaviour tree (BT) representation [19] is used to describe the required sequence of actions to perform the complete mirror assembly. Complex actions are decomposed into simpler ones, until atomic operations are described, which allows better reusability and composability. Figure 16.15 shows a simplified BT representation of the current scenario, together with the corresponding simulation state in Webots. The BT formalism allows to easily define the different sets of actions to achieve the full assembly sequence, with possible recovery cases if needed. It is similar to finite state machines, but with better maintainability (nodes can be designed independent from each other in BT, allowing adding or removing new nodes without the necessity to

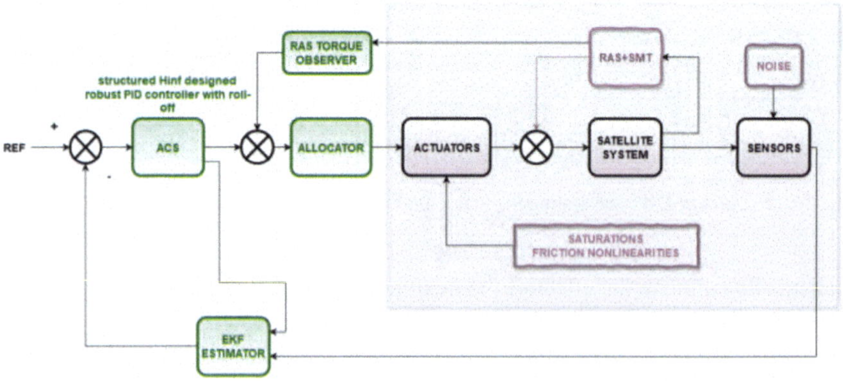

Fig. 16.13 dISAS final implementation of the attitude controllers

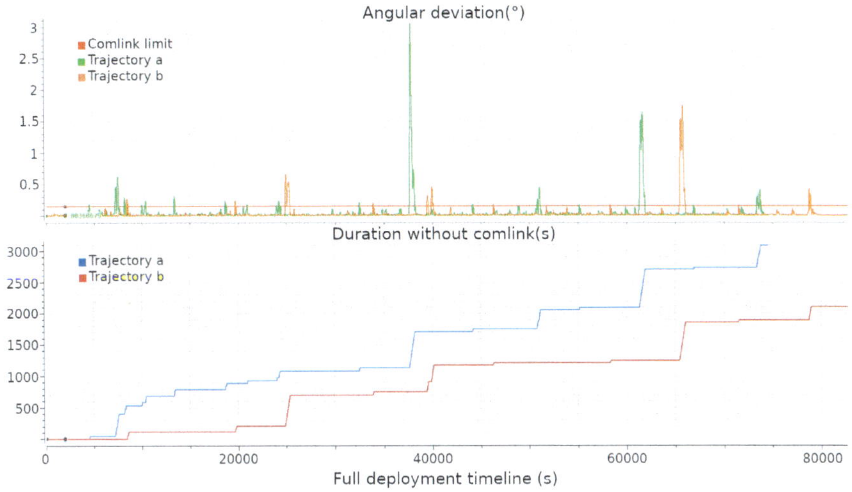

Fig. 16.14 Attitude deviation (deg) of the spacecraft and duration (s) without comlink during the full deployment with two sets of trajectories

change other parts of the model), scalability (possibility to decompose BT into small subtrees), and reusability (since nodes are independent in BT, the subtrees are also independent).

A perception capability using fiducial marker detection and object pose estimation was also implemented and tested. For the implementation in simulation and the real execution in hardware in the dPAMT demonstrator, each tile has the same fiducial marker configuration (one big marker at the back, and two small markers around

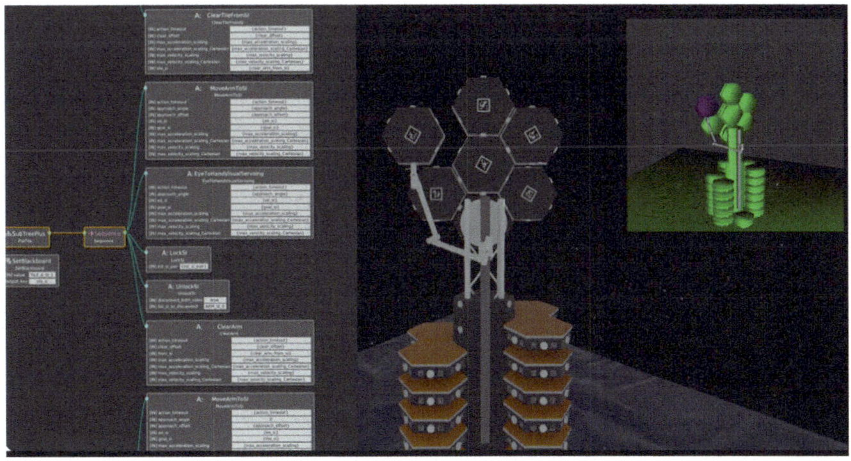

Fig. 16.15 Behaviour tree (simplified representation) for the primary mirror assembly (left); current assembly state in Webots (right)

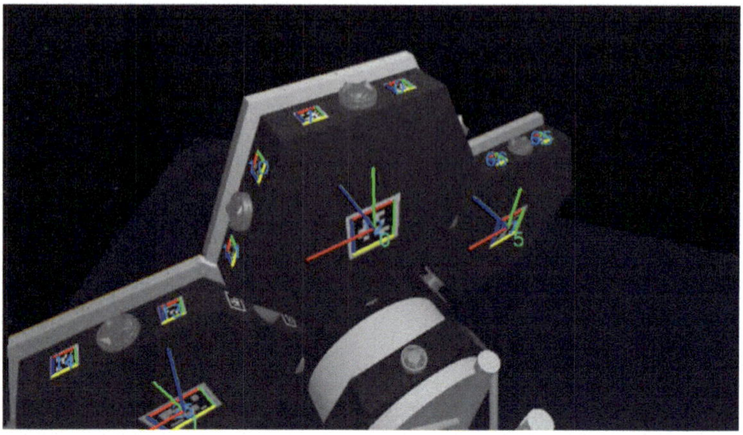

Fig. 16.16 Tile identification and localization using AprilTag fiducial marker detection

each SI), but with different marker patterns in order to uniquely identify and localize the different tiles. Figure 16.16 shows an example of successful identification and pose estimation by the perception module. The small fiducial markers located near the SI are used to precisely localize the tile in the camera frame for picking tasks, when the robotic arm is close to it. Such a perception approach allows to use vision to autonomously perform tasks such as tile picking or tile assembly. With an embedded camera on the robot end-effector (eye-in-hand camera), the tile grasping operation has been successfully tested in simulation. At the beginning of the operation (Fig. 16.17), the two markers can be simultaneously detected. From their precise corners coordinates extraction and the model of the markers, it is possible with a calibrated camera to accurately compute the pose of the tile with respect to the camera. A visual servoing control scheme then takes care of commanding the robot in order to reach the desired end-effector pose [24]. This can be visualized with the yellow frames in Fig. 16.17, which represent the pose of the tile in the camera frame at the desired pose. Thus, converging toward the goal pose brings the current tile frame (represented by the RGB coordinates frame) closer to the target (yellow) frame. At convergence of the visual servoing control law, the robot end-effector reaches the desired pose with SIs correctly mated. This desired end-effector pose corresponds to a specific pose of the tile with respect to the camera frame (Fig. 16.18). Advanced approaches for visual servoing exist in literature. In particular, by selecting and combining the appropriate 2D and 3D features, other system behaviors can be achieved, such as simultaneously ensuring the global convergence of the system while maintaining the visual features in the field of view of the camera. An appropriate control law for joint limit avoidance is another illustration of advanced approaches, and exploited in the present case thanks to the joint redundancy of the Robotic Arm System in dISAS.

dISAS can be used to model realistic sensors and actuators. It demonstrates that core components (AOCS, RAS motion planner, perception functionalities) are able to interface seamlessly with real and simulated hardware. The simulator also allows

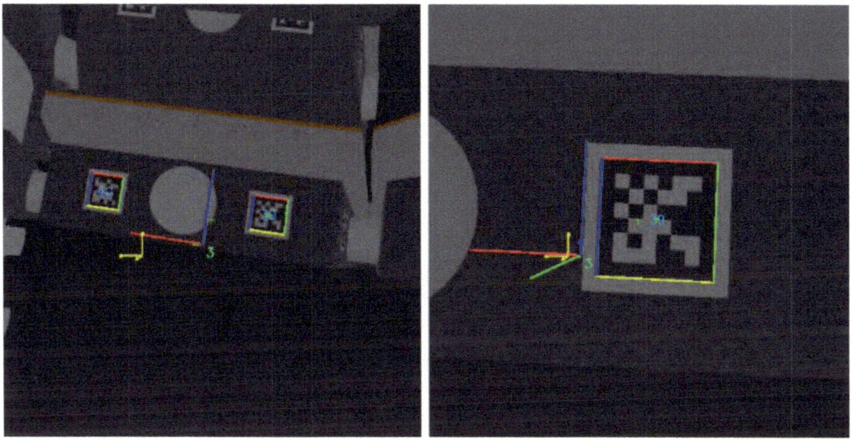

Fig. 16.17 Tile pose computation at the beginning of the operation (left), and close to convergence (right)

the modelling of component flexibility and related disturbances. Multiple tests have been performed to assess the validity of the different modelled sensors. Custom sensor models are implemented in order to output realistic values from perfect data, including also the possibility to use custom sensor noise models. The repeatability of the different implemented instruments (gyroscope, star tracker, reaction wheel) has been validated thanks to this demonstrator.

The simulation has been an invaluable tool to prototype an implementation of perception functions for assembly tasks [20]. It enabled rapid testing of, for example, different sensor specifications, sensor positioning, or illumination conditions, at various assembly steps. It allowed to simulate the full assembly process for a large telescope, including effects of the arm motion on the AOCS of the spacecraft (Fig. 16.19). It also allowed us to gain interesting insights on the limitations of such functions in an operational context, as well as give clues towards the trade-offs to be made in order to use these types of capabilities in a real mission.

16.5 Final Discussion

The European project PULSAR was introduced, which aimed to provide a first experimental verification for low-level technologies that need to be further developed for in-space autonomous assembly of complex structures such as large telescopes. This goal was achieved through three different demonstrators, based on a mobile robotic manipulator (for testing autonomous assembly and optical verification of the telescope), an underwater platform (for testing assembly in a low gravity environment),

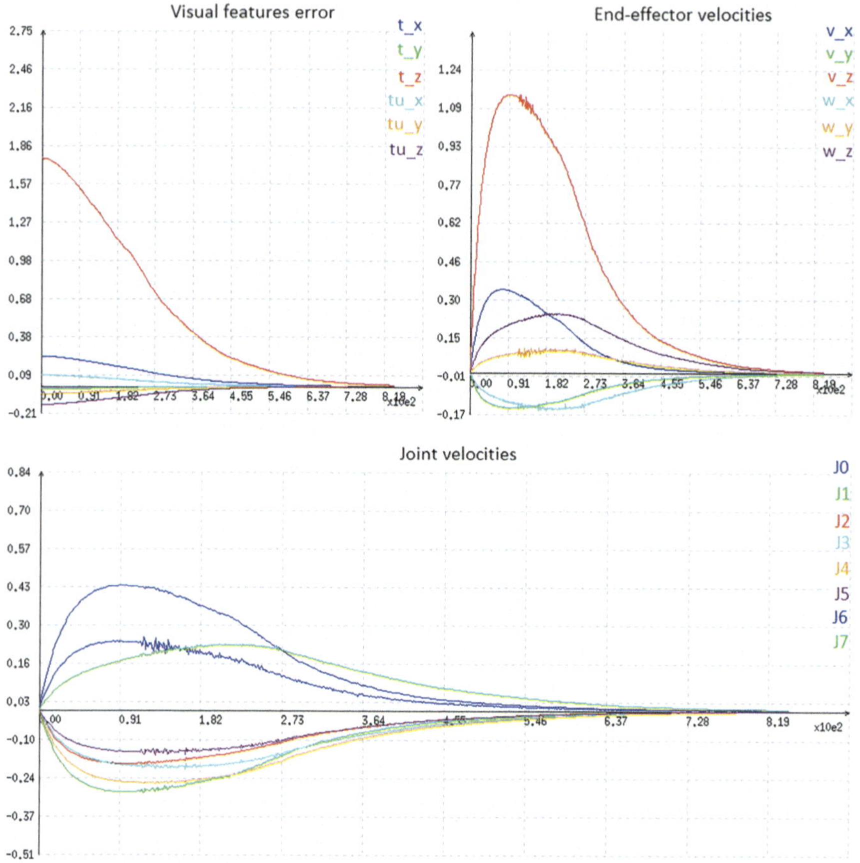

Fig. 16.18 Evolution of the visual features error during the servoing operation (top left), with the correct regularization to zero of the error function, and evolution of the end-effector velocities (top right). The corresponding commanded joint velocities are displayed at the bottom

and a simulation-based approach for testing a full mission. The final demonstrations were successfully performed in 2021 (links to videos of the demonstrations are provided in the Appendix).

The dPAMT demonstrator proved the feasibility of autonomous assembly of the primary mirror, using a combination of adaptable perception, integrated assembly and motion planning, and compliant control of the manipulators. HOTDOCK was effectively employed as standard interconnect that enabled the assembly of multiple components. The SMTs provided a modular approach for constructing a large mirror, facilitating the manipulation and providing high motion accuracy and adjustment of the optical surfaces, as required for a space telescope. The knowledge gained in this demonstrator has strong synergies with other ground-based applications, where collaborative robots are used for assembling different structures, for instance, using modular construction kits [21].

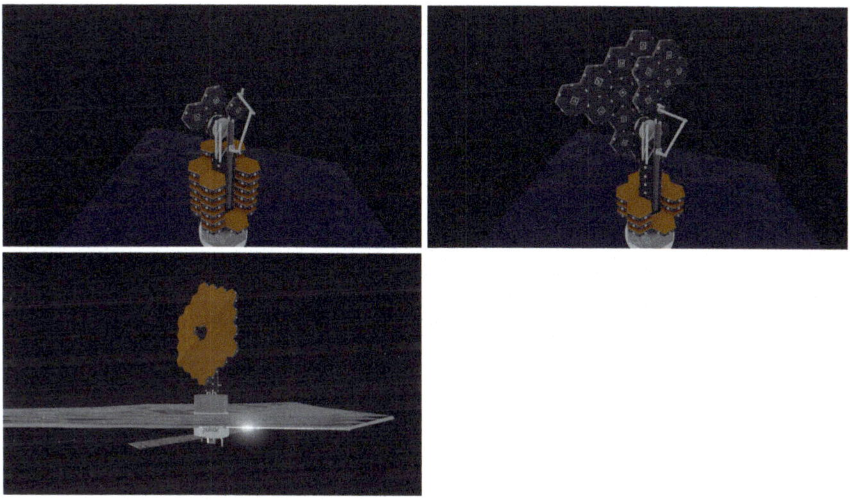

Fig. 16.19 Large telescope assembly using the dISAS simulator, visualized at different steps of the assembly process

The dLSAFFE demonstrator worked in a zero-buoyancy facility, and used an underwater robotic manipulator of representative size mounted on a linear guide. This partially lifted the constraints on the robotic arm workspace, and provided the required mobility to simulate retrieval and assembly of real-size segmented mirror tiles, considering partially the free-floating dynamics and focusing on motion planning and control. The demonstrator was limited by the water drag, which damps structural flexibility effects, and only considered the mobility of the robot but not the integration with the AOCS. However, the strategy to work with sub-assemblies was verified as a promising venue to work with large size structures.

The dISAS demonstrator was based in simulation, and used accurate physical models of the spacecraft and its components to estimate torque disturbances involved in the deployment, as well as robust controllers to manage them. It included an initial study of the influence of the robot arm motion on the spacecraft attitude control, which is a phenomenon to study in further detail, considering that a continuous attitude adjustment will negatively influence the length of the mission due to excessive energy/fuel consumption, but a too loose attitude adjustment might leave the spacecraft with no communication link to ground for significant periods of time. This is a topic of ongoing research [22].

PULSAR tested the feasibility of key technologies for performing in-space assembly of a large structure, but the same technologies are generally applicable to ISAM (In-space Servicing, Assembly and Manufacturing). Several of the technologies, e.g. torque-controlled robots, visual perception, assembly and task planning, originated from ground applications, where they have proven to be effective to solve challenging problems especially in manufacturing and logistic contexts. Now these automation technologies are proving their feasibility for ISAM, both at the hardware and software levels. Current developments in this direction are considered in the EU project EROSS (European Robotic Orbital Support Services[6]) and successive projects, which will strengthen the European competitiveness in space robotics within the global landscape.

Acknowledgements This work was partially supported by the European Union's Horizon 2020 research and innovation programme under grant agreement No. 821858, project PULSAR.

Appendix

The following links contain videos of the results in the PULSAR project.

- General overview of the PULSAR project: https://www.youtube.com/watch?v=X5PLVNdIm3o
- dPAMT demonstrator: https://www.youtube.com/watch?v=1kyrO8k5rj0
- dLSAFFE demonstrator: https://www.youtube.com/watch?v=o0R-1gai6Ps
- dISAS demonstrator: https://www.youtube.com/watch?v=SaoC67rYp0g

References

1. Space Foundation. (2017). The Space Report.
2. Federal Aviation Administration. (2018). The annual compendium of commercial space transportation.
3. Rognant, M., Cumer, C., Biannic, J. M., Roa, M. A., Verhaeghe, A., & Bissonnette, V. (2019). Autonomous assembly of large structures in space: a technology review. In: Proceeding European conference for aeronautics and aerospace sciences (EUCASS).
4. Roa, M. A., Nottensteiner, K., Wedler, A., & Grunwald, G. (2017). Robotic technologies for in-space assembly operations. In: *Proceeding on advanced space technologies in robotics and automation (ASTRA)*.
5. Lightsey, P., Atkinson, C., Clampin, M., & Feinberg, L. (2012). James Webb Space Telescope: Large deployable cryogenic telescope in space. *Optical Engineering, 51*(1), 011003.
6. Piskorz, D., & Jones, K. (2018). On-orbit assembly of space assets: a path to affordable and adaptable space infrastructure. In *The aerospace corporation*.
7. Rouvinet, J., Ummel, A., Cosandier, F., Nguyen, D., & Schaffter, V. (2020). PULSAR: Development of a mirror tile prototype for future large telescopes robotically assembled in space. In: Proceeding SPIE conference on astronomical telescopes + instrumentation.
8. Letier, P., Siedel, T., Deremetz, M., Pavlovskis, E., Lietaer, B., Nottensteiner, K., et al. (2020). HOTDOCK: design and validation of a new generation of standard robotic interface for on-orbit servicing. In *Proceeding international astronautical congress (IAC)*.
9. Martinez, J., Rodriguez, I., Nottensteiner, K., Lutze, J., Lehner, P., Roa, M. A. (2021). Hybrid planning system for in-space robotic assembly of telescopes using segmented mirror tiles. In *Proceeding IEEE aerospace conference* (pp. 1–16).

[6] https://eross-h2020.eu/

10. Kalakrishnan, M., Chitta, S., Theodorou, E., Pastor, P., & Schaal, S. (2011). STOMP: Stochastic trajectory optimization for motion planning. In Proc. IEEE international conference robotics and automation (ICRA) (pp. 4569–4574).
11. Koch, C. E., Jankovic, M., Natarajan, S., Vyas, S., Brinkmann, W., Bissonnette, V. et al. (2020). Underwater demonstrator for autonomous in-orbit assembly of large structures. In Proceeding international symposium on artificial intelligence, robotics and automation in space (i-SAIRAS).
12. Kalakrishnan, M., Chitta, S., Theodorou, E., Pastor, P., & Schaal, S. (2011). STOMP: Stochastic trajectory optimization for motion planning. In: Proceeding IEEE international conference robotics and automation (ICRA) (pp. 4569–4574).
13. Schulman, J., Duan, Y., Ho, J., Lee, A., Awwal, I., Bradlow, H., Pan, J., Patil, S., Goldberg, K., & Abbeel, P. (2014). Motion planning with sequential convex optimization and convex collision checking. *International Journal Robotics Research, 33*(9), 1251–1270.
14. Sucan, I. A., Moll, M., & Kavraki, L. E. (2012). The Open Motion Planning Library. *IEEE Robotics and Automation Magazine, 19*(4), 72–82.
15. Beeson, P., & Ames, B. (2015). TRAC-IK: An open-source library for improved solving of generic inverse kinematics. In *Proceeding IEEE-RAS international conference on humanoid robots* (pp. 928–935).
16. Smits, R. KDL: Kinematics and Dynamics Library. http://www.orocos.org/kdl
17. Pan, J., Chitta, S., & Manocha, D. (2012). FCL: A general purpose library for collision and proximity queries. In *Proc. IEEE Int. Conf. Robotics and Automation (ICRA)* (pp. 3859–3866).
18. Munoz, M., Montano, G., Wirkus, M., Hoeflinger, K., Silveira, D., Tsiogkas, N., et al. (2017). ESROCOS: a robotic operating system for space and terrestrial applications. In *Proceeding on advanced space technologies in robotics and automation (ASTRA).*
19. Paxton, C., Ratliff, N., Eppner, C., & Fox, D. (2019). Representing robot task plans as robust logical dynamical systems. In: Procceding IEEE/RSJ international conference on intelligent robots and systems (IROS) (pp. 5588–5595).
20. Bissonnette, V., Bazerque, C., Trinh, S., Porte, C., Arcin, G., Rognant, M., Biannic, J.M., et al. (2020), A simulation tool for in-orbit assembly of large structures. In *Proceeding international symposium on artificial intelligence, robotics and automation in space (i-SAIRAS).*
21. Rodriguez, I., Nottensteiner, K., Leidner, D., Durner, M., Stulp, F., & Albu-Schaeffer, A. (2020). Pattern recognition for knowledge transfer in robotic assembly sequence planning. *IEEE Robotics and Automation Letters, 5*(2), 3666–3673.
22. Rodriguez, I., Lutze, J., Mishra, H., Lehner, P., & Roa, M. A. (2022). Hybrid planning to minimize platform disturbances during in-orbit assembly tasks. In: *Proceeding IEEE aerospace conference.*
23. Krogius, M., Haggenmiller, A., Olson, E. (2019). Flexible layouts for fiducial tags. In: *Proc. IEEE/RSJ Int. Conf. on Intelligent Robots and Systems (IROS).*
24. Chaumette, F., Hutchinson, S. (2007). Visual servo control, Part II: Advanced approaches. *IEEE Robotics and Automation Magazine, 14*(1), 109–118.

Chapter 17
An Overview of the Space Servicing Requirements in a Sustainable Space Age

Pablo Lopez Negro, Daniele Filippetto, Claudio Tanzariello, Pierre Letier, and Xiu Tian Yan

Abstract Space-based services are flourishing motivated by new needs in fields like communication, navigation and Earth observation. The increasing demand for space infrastructure has stimulated the deployment of innovative technological solutions; however, among the latest developments, on-orbit servicing of spacecraft is not yet implemented in most commercial programs. This chapter explores the opportunities for developing a sustainable and economically viable solution based on on-orbit servicing to extend spacecraft life-time and to reconfigure its main payloads to adapt to new needs.

- The investigation starts by an analysis of the demand of space based solutions and the prospects for the following years.
- Then, a discussion about the new developments in space infrastructure is proposed, presenting two alternative concepts: large constellations and on-orbit servicing.
- Attention is drawn to the latter, discussing the main technological evolutions needed to sustain its implementation.

The last section proposes a roadmap for the development and progressive application of the new technologies, aiming at implementing on-orbit servicing on a regular basis for commercial spacecraft.

P. L. Negro (✉)
Thales Alenia Space, Cannes, France
e-mail: pablo.lopez-negro@thalesaleniaspace.com

D. Filippetto
Sitael, Italy

P. Letier
Space Applications Services, Leuvensesteenweg 325, 1932 Sint-Stevens-Woluwe, Belgium

X. T. Yan
RAS Group, DMEM, University of Strathclyde, Glasgow, Scotland
e-mail: x.yan@strath.ac.uk

C. Tanzariello
German Research Center for Artificial Intelligence (DFKI), 28359 Bremen, Germany

X. T. Yan and G. Visentin (eds.), *Space Robotics*,
https://doi.org/10.1007/978-3-031-39214-6_17

17.1 Introduction

In recent years we have witnessed a proliferation of newcomers to the space industry. Stimulated by the advancement of technology, the reduction of launch costs and an ever increasing demand for new services on Earth; private initiatives have flourished in a market typically driven by international institutions and national agencies. This explosion of new activities has disrupted a conventionally slow-evolving industry, breaking the typical market and technology barriers that in past decades had limited the space industry to a few players worldwide.

The democratization of the access to space is leading up to two different approaches that will compete in the coming years. Cheaper launch services could be seen as an opportunity to deploy more satellites with production in large series forming constellations of potentially disposable satellites. Conversely, easier and more economic access to space could provide an incentive to finally implement on-orbit servicing, extending satellites' life-time and performances; beginning a new era of sustainable exploitation of resources in orbit.

Although discussed through decades, on-orbit servicing is not yet a viable alternative that is requested on commercial missions. Excluding the missions to the International Space Station, servicing has only been a matter of a few sporadic highlights (example: replacement of Hubble optics by the Space Shuttle). Among other difficulties, orbit servicing has traditionally suffered from a chicken-egg problem: most satellites on orbit are not designed to support orbit servicing, making operations complex and with low added value to the spacecraft owner; as a result, not enough effort is put into the development of the servicer vehicles, which remain exotic and expensive adventures.

However, the evolution of technologies combined to cheaper and more flexible access to space open interesting opportunities to finally see on-orbit servicing deployed on a usual basis on commercial and institutional missions. In this context, MOSAR (MOdular Spacecraft Assembly and Reconfiguration) project pursues the development of key technologies to sustain future space missions by designing:

- Common robotic building blocks, improving autonomy during critical phases such as on-orbit servicing.
- Robotic manipulator to support on-orbit operations such as module assembly or replacement.
- Elaboration of a concept for modular spacecraft, proposing design standards in terms of modules shapes, functions and interfaces.

17.2 Prospects for Space Exploitation in the Near and Mid-Term Future

The space activity has experienced a continuous growth over the past decades mainly supported by institutional programs with a rising contribution from private organizations. Nowadays, commercial activities are flourishing, with promising prospects for the near future [1], which explains the increasing interest on space related initiatives seen in the last few years.

Indeed, space applications have spread beyond its traditional realms (science, exploration, TV broadcasting) and today new markets are emerging supported by private investors (Earth Observation, connectivity, data services, navigation…). The apparition of new segments demanding space infrastructure has caused a proliferation of new actors in the space community all along the value chain: from equipment supply to satellite services, and specially and decisively on launcher manufacturers.

The apparition of private companies on the launcher segment is arguably the most important landmark of the current decade for the space community and is the most decisive factor enabling the developments to be seen in the next decade. Compared to the traditional space exploitation of the twentieth century, the new launchers allow a more affordable (Fig. 17.1) and therefore less restrictive access to space, which is accelerating the change of paradigm of the space community: from a government driven industry to a commercial oriented economy (Fig. 17.2).

In the following years a consolidation of commercial activities in emerging sectors such as Earth observation, navigation and data connectivity is to be expected, yielding more companies involved in space activities improving the overall competitiveness and attractiveness of the space industry.

The apparition of new markets is generating a snow-ball effect propagating upstream to the different players of the space community:

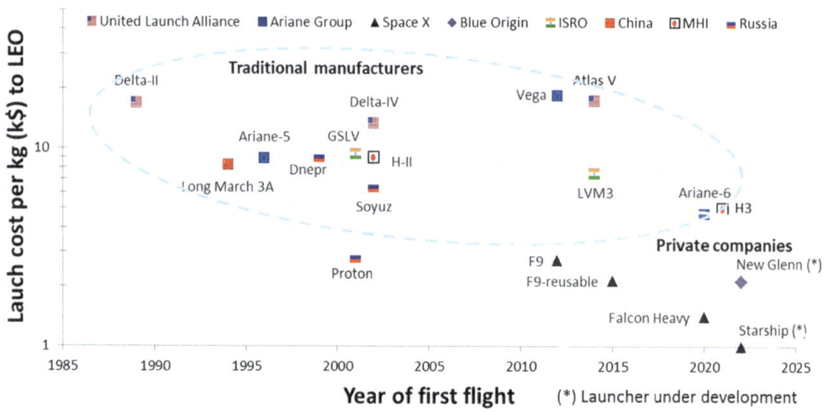

Fig. 17.1 Evolution of launch costs from 1985 to today [2]. The apparition of private companies has lowered down the prices by a factor of two in the last decade

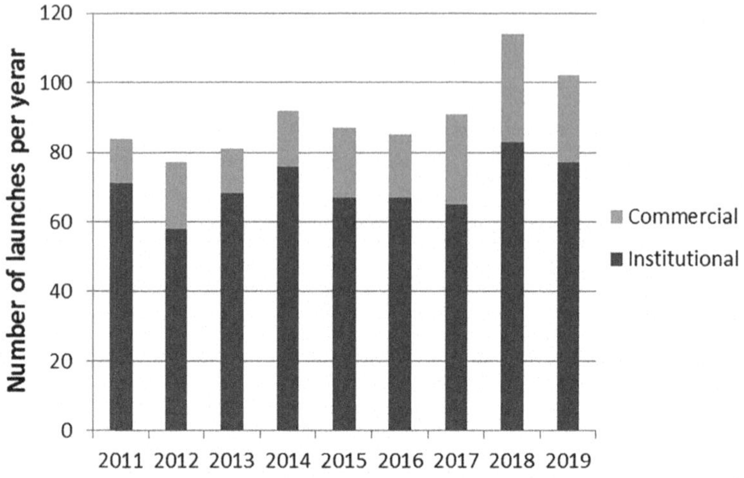

Fig. 17.2 Number of launches in the last ten years [3]. The increasing contribution of commercial initiatives is observed, representing around 15% of the total launches per year in the beginning of the decade and reaching 25% by the end of it

New spacecraft operators appear, motivated by the market demand of services from space. The traditional operators are reacting adapting their activities and business models to the new context and the more competitive market [4, 5]. The main trend that is observed is the diminution of orders for large telecommunication satellites in Geostationary orbit [6] and a progressive substitution by smaller satellites or constellations in Low and Medium Earth Orbit (LEO/MEO). The diminution of order intake for large satellites is partially explained by the wait and see policies adopted by some of the main operators, holding on to the consolidation of key technologies (like digital payloads and electric propulsion) before taking decisions on whether to continue investing in large GEO satellites or look for new operation models based on constellations.

The proliferation of spacecraft programs and business initiatives based on large constellations of small and relatively low performance satellites is a topic of increasing interest to the space community. Success stories like Planet Labs [7] illustrate a promising path for inventive entrepreneurs to start a business with low cost small satellites and end up with a large constellation providing high added value services to final customers.

The spacecraft manufacture industry is also adapting to the market dominated by market forces rather than by institutional and governmental funding. In particular the market demands shorter development cycles and lower costs [8] while implementing faster the latest technology advancements in order to remain competitive. This is motivating a restructuration of the traditional manufacturers towards a cost-driven production, stimulated also by the upcoming threat of new players evolving from the small sat industry. The latter, although lacking the experience and know-how of traditional manufacturers, are inherently better adapted to the new market needs

[9], having experience in modular systems, production in series, introduction of continuous evolution etc.

A traditional space program implied a long development cycle, with careful analysis for the design and selection of components and extensive testing at equipment, subsystem and spacecraft level resulting in a very reliable product but at a high cost both economical and programmatic (long lead time). In this context, the new space actors adopt disruptive techniques based on principles like test as you fly and continuous integration which turns out to be better shaped to respond to the new commercial context and specially in the case of constellations [10] (Figs. 17.3 and 17.4).

- Hardware suppliers are also experiencing a revolution with the progressive introduction of standard components (Commercial-Off-The-Shelf, COTS) to substitute the very expensive parts designed specifically for space applications. This is making the space industry accessible to more companies, some of them

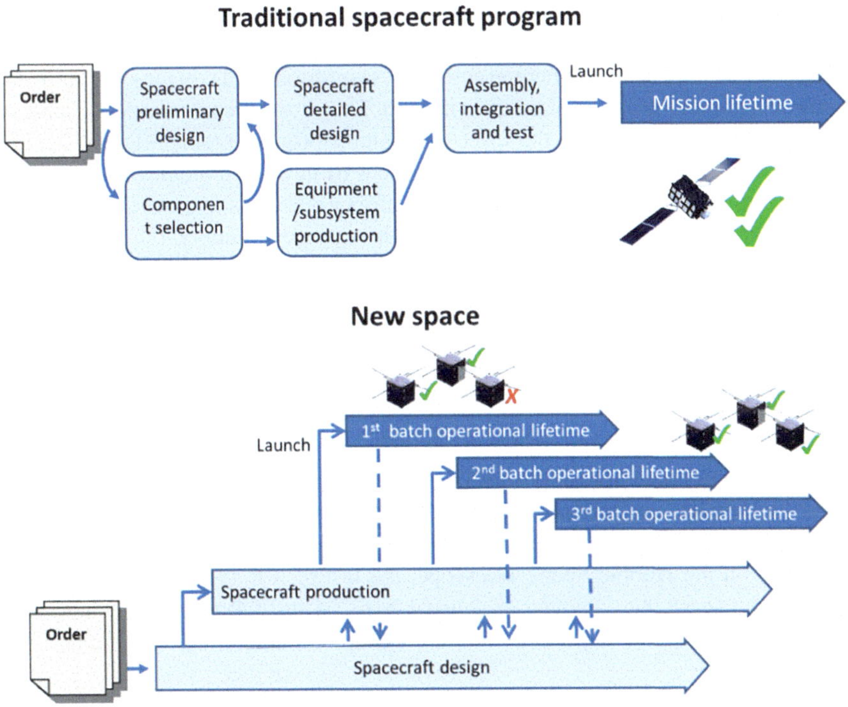

Fig. 17.3 Schematic comparison of the development cycles of traditional spacecraft versus new space paradigm oriented to constellations

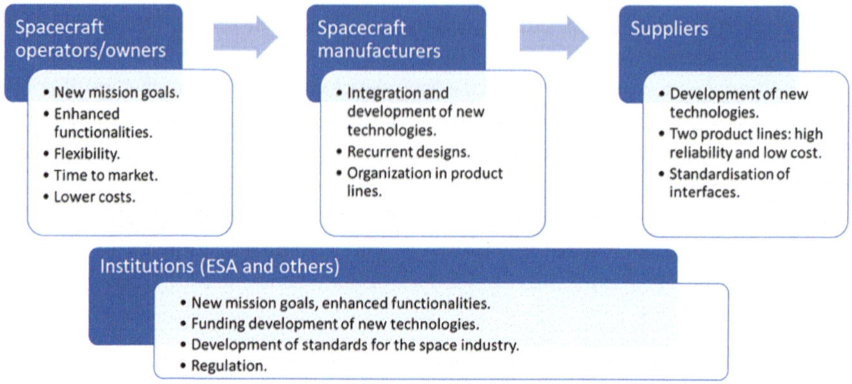

Fig. 17.4 Identified trends for the next decade across different actors of the space community

newcomers to the space community [11]. Some of these new actors are introducing innovative approaches for the development and manufacturing of their products, disrupting the status-quo of a traditionally slowly evolving industry.

- Despite the growth in commercial activities, institutions will still play a prominent role in the space industry of the next decade. They continue to provide the largest economical contribution to the space community, promoting scientific and exploration missions and funding the development of new technologies. Concerning the rising weight of the private sector, institutions develop standards and common procedures to facilitate exchanges among the space community. Furthermore, in the last few years, the rapid increase of the number of objects in orbit around the Earth has motivated an effort in regulation [12–14], and today Space Situational Awareness is one the main fields of activity of governmental organizations and its importance will continue to grow in the following years (together with the number of satellites in orbit).

Among the different innovations that could emerge in the following years, on-orbit servicing is one of the most awaited, as it could be deduced from the number of institutional programs pushing it [15]. Although on-orbit servicing is not properly a new innovation (the Space Shuttle conducted various servicing missions replacing equipment on the Hubble Space Telescope, the last of it in 2009 [16]) it is yet to be seen orbit servicing deployed on a usual basis on institutional and commercial spacecraft. In the last few years efforts in this segment have been strengthened for various reasons:

- In a context of an exponential increase of the number of satellites around the Earth, orbit servicing is seen as an opportunity to repair non-functioning spacecraft, decommission old satellite and give a second use to obsolete but still functioning platforms. These efforts aim towards the mitigation of the risk of collisions between satellites and orbital debris.

- Some commercial missions could benefit from life-extension and spacecraft reconfiguration services [17]. In 2020 the Mission Extension Vehicle (MEV-1) of SpaceLogistics (a subsidiary of Northrop Grumman) performed successfully a life extension operation on a commercial satellite [18], demonstrating the technical feasibility of the concept and leading the way to similar missions in the future.
- The next generation of scientific and exploration missions would require new functionalities. In particular, the possibility of on-orbit assembly allow to envision the implementation of space vehicles with virtually no size limitation, enabling new missions that are not possible today taking into account the size limitations imposed by the launcher vehicles [19, 20].
- Ultimately, on-orbit servicing and on-orbit assembly will be needed for the industrial exploitation of space: asteroid mining, power plants, on-orbit manufacturing etc. These applications, although futuristic, are catching an increasing relevance among the research community [21–23].

The development of technologies enabling orbit servicing is followed in parallel by a development of common industrial standards, operational concepts and regulation framework, in order to ease the large-scale implementation of these techniques [24]. The following years (2020–2030) are decisive for its consolidation, as two opposing solutions are emerging for the future exploitation of space:

- On the one hand the development of large constellations of small (and often disposable) satellites; where services are maintained with successive batches of satellites.
- On the other hand, the development of on-orbit servicing, enabling life-extension of existing spacecraft and potentially reparation and reconfiguration.

As presented in previous pages of this paper, the first approach seems to better respond to the current trends of the commercial segment, but raises questions about the long term sustainability of the model, with a risk of dramatic proliferation of objects in orbit that would yield the space unusable. The second model could help limiting the number of objects in orbit, and furthermore, it could produce solutions for actually recycling and removing obsolete spacecraft and debris. However its economic viability is yet to be proved, as it demands significant development where the return of investment is difficult to establish.

Constellations and orbit servicing are presented as *opposing solutions* although they respond to different technical needs. Some missions are better shaped for one or the other depending on the specific objectives, and only a few kind of missions would withstand a fair comparison between the two concepts based on technical and economic criteria. However, in the author's opinion it is worth comparing them as emerging technologies, presenting distinctive features, but both facing a decisive epoch where its implementation could become a main trend of the industry or a niche segment depending on which solution overcome first its main drawbacks: long-term sustainability for constellations and commercial exploitation for orbit servicing.

For all of the above the 2020–2030 decade presents an interesting landscape for the space community, with the industry involved in an exciting transformation motivated by a conjunction of factors like the increase of commercial activity, the apparition of new companies, the consolidation of key technologies and a renewed fascination for space-oriented initiatives pushed by prominent entrepreneurs.

17.3 Main Challenges and Key Enabling Technologies for On-Orbit Servicing

2020–2030 needs to be the decade where on-orbit servicing is implemented in large-scale on commercial and institutional missions. As presented in the previous section, orbit servicing can count on institutions as a strong ally, pushing these solutions and making significant investments in the development and consolidation of the required technologies. Sadly, from the commercial perspective, on-orbit servicing has been trapped in a vicious circle for some years (Fig. 17.5). The operation is complex, risky and very expensive, and needs the development of some key technologies that are prolific matters of research and technology demonstration projects, but which need additional effort to become industrial solutions.

The common position of commercial operators and large spacecraft manufacturers about these technologies has been wait and see policies; eventually participating in some demonstration projects but not yet adopting orbit servicing as a core product in their portfolios. Eventually, essential technology has become ready for its use, and a servicer satellite is now possible both technically and industrially. However, in the meantime spacecraft are manufactured in a conventional way and the spacecraft that are found in orbit are not prepared to receive orbit servicing. In this context most operators find that the cost and risk of the servicing is not justified by the benefits

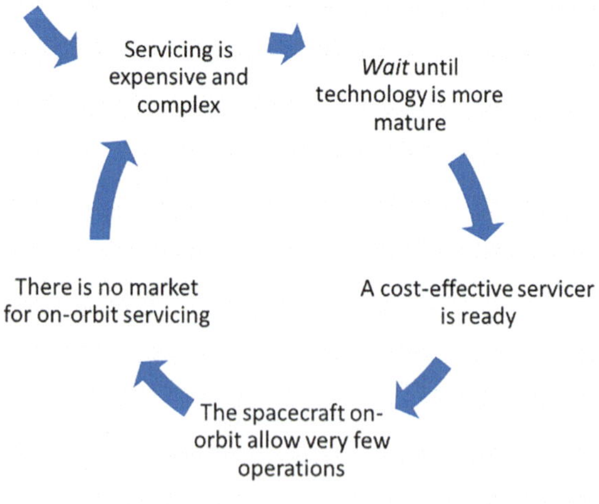

Fig. 17.5 The vicious circle about on-orbit servicing

introduced to the mission and as a result, it is concluded that there is not a current market for on-orbit servicing, and the large investments needed for its development are postponed.

A possible solution to this vicious circle could be the parallel development of technologies in two fields: on-orbit rendez-vous and docking; and technologies and design standards to support operations on spacecraft in orbit. The next generation of spacecraft would need to be manufactured with servicing capabilities, in order to ease on-orbit operations later in the future; offering possibilities to bring added value to client satellites. Indeed, orbit servicing prospects would significantly benefit from target vehicles equipped with docking interfaces, replaceable equipment and software architectures able to handle reconfiguration of spacecraft components.

The current efforts in the orbit servicing field are mainly oriented towards the development of life extension through refuelling or through tugging (the success of MEV-1 is a brilliant milestone, [25]). The spacecraft currently in orbit have very little capabilities to be serviced because its monolithic and completely integrated design. Most of the satellites do not even have a docking interface which difficult the proximity operations and there are not replaceable items that could be easily substituted. For all of the above, the most accessible solution nowadays consists in capturing the client spacecraft with a more manoeuvrable vehicle, and use the servicer to relocate the client or to provide station keeping autonomy using the servicer thrusters and propellant.

Although life extension services can provide significant revenues and offer a solution to mitigate the orbital debris; it is hard to conceive that it will be an economically sustainable solution. In some specific cases, involving very costly satellites or particularly strategic missions, life extension could effectively provide a second life to an operational spacecraft pushing forward the end of its supposed lifetime. However, the cost of the servicing mission including the manufacturing of the servicing vehicle, its launch and the on-orbit operations will be hardly justified by an extension of some few years of the client's lifetime (and that without even considering the cost of the development of the servicing vehicle and the required technologies which is today mainly supported by institutions).

The generalization of electric propulsion, especially applied in the case of geostationary telecommunication satellites, is also a drawback for life extension services. With this technology fuel is not necessarily the limiting factor for the operational lifetime: indeed propellant consumption is drastically reduced and electric spacecraft can add several years of operation at a cost of some ten kilograms of propellant. Therefore, most satellites could afford to include provision for several years of operations without major impacts on its design; making refuelling unnecessary (tugging could still be an option in case of major failures on the propulsion subsystem).

In the near future, it could be foreseen that lifetime of spacecraft will be driven by obsolescence and ageing of hardware elements rather than propellant exhaustion. In this context, it will be essential for orbit servicing to offer a solution to replace equipment on operational spacecraft. This requires that the client spacecraft is specially conceived to support these operations, ideally including hardware interfaces and software architectures to facilitate on-orbit operations.

The substitution of on-board equipment constitutes itself an opportunity to create new commercial initiatives. Although it is not generalized, some commercial and institutional spacecraft embark secondary payloads (generally called hosted payloads) that normally require only a small portion of the platform available resources (mass, power, thermal…) but that are an additional source of funding for the primary operator. In some cases the hosted payloads could be made replaceable, so servicing missions could replace these payloads allowing to sell a part of the spacecraft capacity potentially several times (or even lease the slots for some period of time). In this way orbit servicing could be used to create a new market similarly to what is done in the International Space Station where it is possible to acquire a slot and fly an experiment or a small payload [26]. To some extent, this activity could help fighting against the proliferation of small satellites, offering the possibility of embarking its payloads in more performing satellites rather than having dedicated small satellites [27].

Ultimately, although it is more difficult to imagine because of its inherent complexity, orbit servicing should target the possibility of replacing or reconfiguring elements of the main payload of client spacecraft. That would allow to offer high added value services to client spacecraft, with options to respond to failures on key hardware components, obsolescence and replacement by state-of-the-art components. Ultimately, it could be possible to reconfigure the main payload of the spacecraft to abandon an exhausted market and address a new more promising one.

17.3.1 Key Enabling Technologies

From the technical perspective, the current context offers significant opportunities to deploy orbit servicing, singularly the enhanced performances of launchers vehicles, but also the development of key enabling technologies that could make (finally) on-orbit servicing sustainable both technically and economically.

- Launchers and access to space: the evolution of launcher performances are highly beneficial to the deployment of orbit servicing. New launchers offer more capacity at a lower cost, and with lower costs per launch it is easier to imagine the possibility of launching spacecraft to provide services to other spacecraft (in the past, with higher costs of launch, instead of servicing an existing spacecraft a new one would be launched to substitute the old one). Furthermore, the generalization of co-passengers and combined launches for multiple spacecraft allow to obtain more frequent access to space.
- Orbit rendezvous: stimulated by institutional programs, orbit-rendez-vous technologies are coming of age. Among others, it is worth mentioning the advancements in position and attitude determination (often aided by visual sensors), trajectory optimization and the development of operational standards for the different phases of the approximation. From a technical perspective the technologies are

available, the challenge now is its large scale implementation and a regulation framework and common standards to make possible its generalization.

- Space robotics: the current spacecraft in orbit lack of a specific docking interface which make necessary the use of some kind of sophisticated mechanism or robot to grasp somehow the client spacecraft and physically connect client and servicer. In some cases the rendezvous trajectory could end with a final phase of direct docking but in most cases (especially if the client spacecraft does not implement a docking interface) the servicer would just approach the client and use a robotic manipulator to grasp the client and force the physical connection (similar techniques have been tested in the International Space Station). Furthermore, robotic manipulators are needed to perform operations on the spacecraft like substitution of equipment. Extensive research has been done on the subject of designing and operating robotic manipulators in space, with the main challenge of prove its autonomy without human intervention (most of the experiences in the ISS have constant surveillance while not directly an astronaut driving the mechanisms).
- Robotics are arguably essential for deploying orbit servicing and orbit assembly on applications that go beyond refuelling and space tugs. The main servicing operations carried out in the past that aimed at repairing or replacing equipment (ISS, Hubble telescope) always involved human intervention. However, in order to see orbit servicing deployed on a usual basis it is essential that it could be done without human intervention both in space (without astronauts) and on-ground (without constant surveillance from a ground station). In this context, the application of robotics in space provides the technological building blocks for unmanned operations.

17.3.2 MOSAR: Modular Spacecraft Assembly and Reconfiguration

MOSAR: Modular Spacecraft Assembly and Reconfiguration: an European project developing key technologies for on-orbit servicing and assembly.

MOSAR is a project involving a consortium of European companies aiming at the development of key enabling technologies for on-orbit servicing and on-orbit assembly. MOSAR targets specifically the field of space robotics (MOSAR is part of the Strategic Research Cluster of the European Union dedicated to Space Robotics [28]). Compared to other research initiatives dedicated to orbit servicing, MOSAR focuses on the operations after successful docking of client and servicer spacecraft, and develops technological building blocks enabling on-orbit operations like equipment substitution, spacecraft reconfiguration and module assembly.

Four are the main innovations that MOSAR introduces: the development of a walking robotic manipulator, a multi-functional standard interface, introduction of modular approach for the design of spacecraft and on-board and ground software to support reconfiguration of the spacecraft (Fig. 17.6).

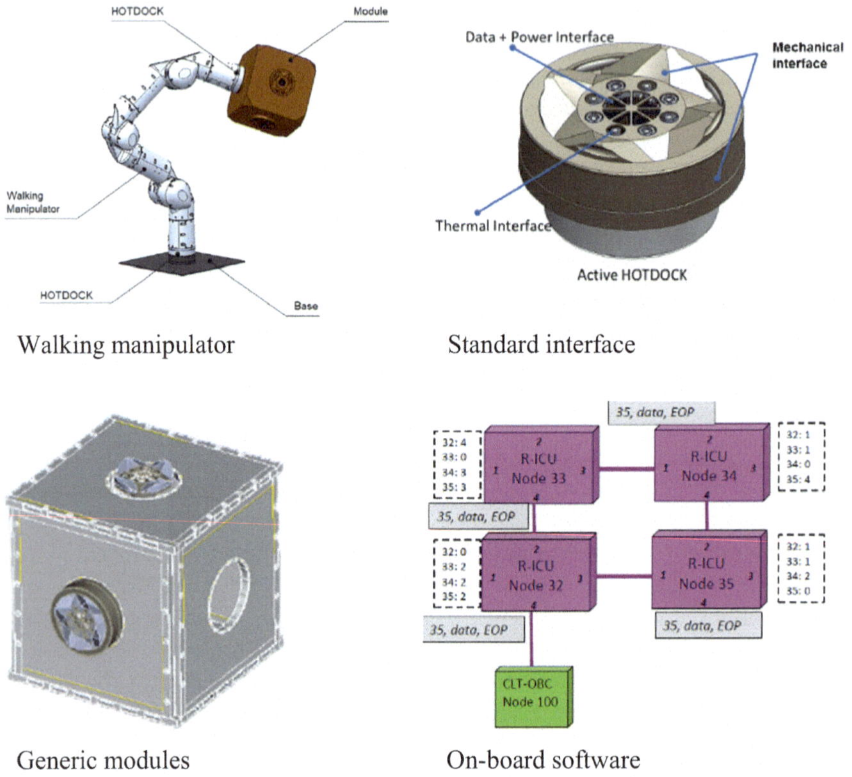

Fig. 17.6 MOSAR technologies: walking manipulator, standard interface, spacecraft modules and dedicated software

- **Walking robotic manipulator**: a walking robotic manipulator is an essential building block to dissociate the operations to be carried out on a spacecraft from its specific shape and size. This way the solution (robotic manipulator) could be generic and still address a wide range of potential end users. Also, the robotic manipulator is better shaped to assemble large structures, being able to move through it and not imposing a limiting constraint on the size of the assembly. The walking manipulator is a symmetric robot, implementing end-effectors in both extremities. These end-effectors allow to grasp objects and also serve as feet to the walking manipulator when they are empty.
- **Standard interface**: also aiming at standardization and generalization of operations, MOSAR implements a multi-functional interface with mechanical, power, thermal and data functionalities, able to serve as end-effector of the robotic manipulator and also to interconnect spacecraft modules.
- **Modular approach**: in order to support replacement of equipment and subsystem in orbit, a modular concept is proposed. The modules have common interfaces and follow specific design standards. Variability is managed similarly to the cubesat

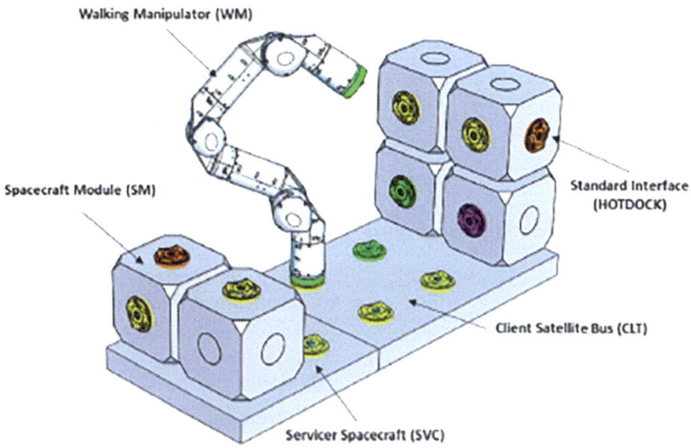

Fig. 17.7 MOSAR demonstrator setup

concept, with the definition of a standardized elementary unit (1U) that could be replicated in multiples to host equipment of different sizes (2U, 3U, 2 × 2U etc.). The modules implement a generic avionic architecture that allow to control the different interconnectors and provide power, data and thermal interface to the payload. One or more faces of the module could be dedicated to a specific element requiring a field of view (like a camera or antenna) or a free surface (solar panel, radiator…).

- **On-board and ground software**: dedicated on-board software is developed to manage the reconfiguration features needed to control a spacecraft made up of several modules (each module acts as a node in a network). Additionally, other software tools are developed for system analysis and for simulation of operations from ground. These tools allow to generate trajectories for the walking manipulator and simulate the behaviour of the system when connecting a new module.

The objective of MOSAR is to develop these technologies and to demonstrate them on a ground facility with a set of scenarios covering key functionalities that are needed for on-orbit servicing (Fig. 17.7, Table 17.1).

17.4 MOSAR Exploitation Roadmap

Based on the specific objectives of MOSAR, and the research carried out on state-of-the-art technologies and current market trends, three are the main postulates (or high level design objectives) that drive the exploitation plan of MOSAR technologies:

(1) Orbit servicing is better applied on *servicing-ready* spacecraft.

Table 17.1 MOSAR demonstration scenarios

Scenario	Main functionalities being tested
Autonomous assembly of spacecraft modules	• The walking manipulator is able to place the modules on the client spacecraft • The modules are powered on and operational • The modules can receive and execute telecommands and deliver telemetries
Replacement of failed modules	• The faulty module is taken from the client and brought to the servicer • A new module replaces the faulty one on the client • The new module is powered on and set operational
Thermal transfer between modules	• Demonstration of active thermal transfer between a module producing heat (payload) and a module dedicated to thermal management
Automatic client network reconfiguration	• Demonstration of the capability of the network to autonomously detect faulty interfaces and reconfigure the system to remain operational

On-orbit rendez-vous and servicing is a complex and risky operation. However its prospects are notably enhanced when the client satellite embark hardware elements and count with operational modes that facilitates the tasks of the servicer. MOSAR targets specifically a new generation of spacecraft that implement a set of elements that allow to dock on the platform and that cooperate during the approximation, final connection and post-mating operations.

(2) Orbit servicing needs to address high added value operations.

Following the reasoning of the previous postulate, the client spacecraft of MOSAR are equipped with interfaces that allow assembly of new modules or replacement of the existing ones. Based on the market analysis, life-extension is usually not enough to justify the costs of servicing missions and therefore additional operation that brings revenues to the mission needs to be done in order to see orbit servicing applied on commercial missions.

(3) A servicer spacecraft shall be able to provide services to more than one satellite.

Standardization is the key for a successful implementation of orbit servicing. The operations that could be done on a spacecraft will hardly justify the cost of sending a dedicated mission to repair it or extend its lifetime. However, if the servicer satellite were to perform operations on a fleet of spacecraft, its commercial prospects would be significantly better.

Based on these main postulates, an exploitation roadmap is created aiming at introducing the technologies on a step-wise approach (Fig. 17.8).

- Indeed, as discussed in previous sections, one of the main drawbacks of orbit servicing is that the spacecraft in orbit have very little capability to introduce

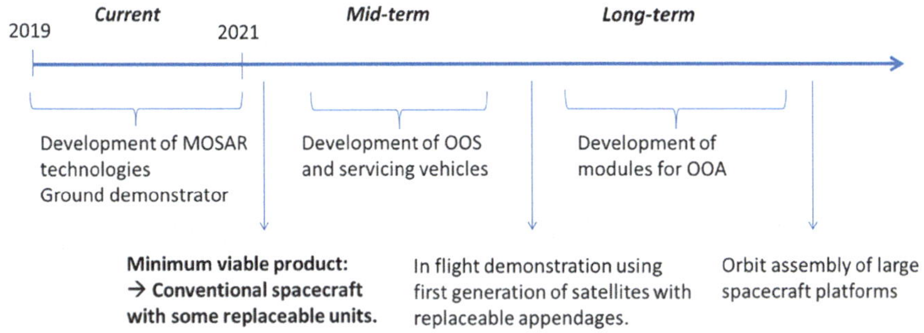

Fig. 17.8 MOSAR exploitation roadmap

modifications to its hardware. MOSAR proposes to start by selecting a conventional spacecraft and introduce a few modular items making use of standard interfaces. This satellite, later in the future, would have the capability to have some of its equipment replaced.

- In the mid-term MOSAR would consolidate the technologies needed for the servicer vehicle and the robotic manipulator. In the meanwhile, more spacecraft are launched with replaceable items, in order to have a fleet of servicing ready spacecraft available in orbit the day that the servicer satellite is available. A first demonstration mission will prove the capability to substitute equipment in orbit or to add new modules to already flying spacecraft.
- In the long term, technologies will be mature enough to support more complex operations, with extended use of on-orbit assembly.

17.4.1 Best Candidate Mission

The best candidate mission to support orbit servicing is selected attending to two criteria:

- Spacecraft value: the value of the spacecraft to be serviced needs to be significant enough to justify the effort (investment and risk) consecrated to it. The value could be understood as the investment made to develop and manufacture the spacecraft (for scientific missions) or as the revenues that it produces (commercial).
- Standardization: allowing to develop a common solution for the servicer vehicle and its operations that could be deployed on multiple satellites.

Attending to these criteria, the main types of spacecraft are evaluated (Fig. 17.9).

- Constellations presents the advantage of having very high standardization. However, the value of each spacecraft (individually) is low, and in case of need it would be probably more interesting to launch additional batches of spacecraft rather than investing in servicing the ones already in orbit.

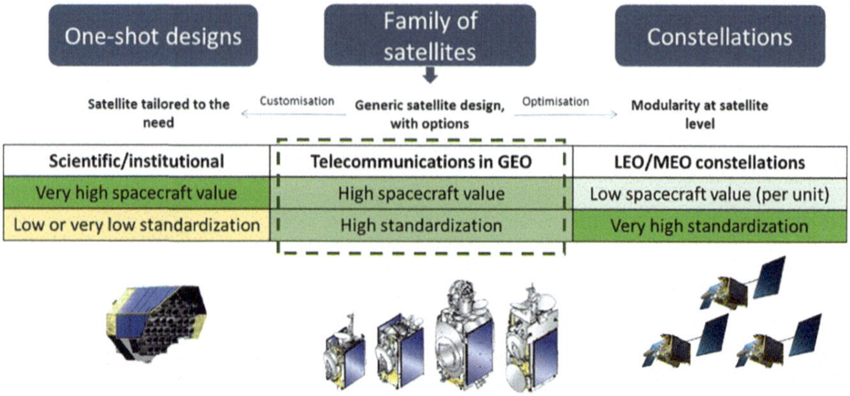

Fig. 17.9 Segmentation analysis of the main products of a large spacecraft manufacturer

- On the other extremity of the chart, scientific spacecraft have often very high value, as they have demanded significant investment and a long development cycle. In this case, maintaining the spacecraft operational for some additional years could justify the investment in a servicing mission. However, these satellites are usually highly customized depending on the mission objective, and probably request also a customized servicing vehicle.
- In the happy middle between these two alternatives lies the concept of product lines (or family of spacecraft). These satellites are usually the main products of large spacecraft operators. The satellites have a great part of common elements, with a few customizable elements that depend on the specific mission and client. Thanks to its common elements, the deployment of generic solutions for orbit servicing will be easier. Also, given the nature of the typical missions that they address, the spacecraft has enough value (or generate enough revenues) to implement servicing.

Among the different product lines, the geostationary telecommunication segment is specifically suited to the MOSAR approach.

Thanks to the specific orbital conditions of the geostationary orbit, a servicer satellite will be able to serve multiple spacecraft without requiring very large quantities of propellant (compared to the case of Low Earth Orbits, GEO satellites lie on the same orbital plane).

There are a large number of GEO spacecraft in orbit, and the trend of new launches is quite stable over time. This means that with some ten satellites launched per year it could be envisioned to have at least one servicing mission on a yearly basis. Also it is common practice to launch GEO satellites by pairs on the same spacecraft, so the servicer satellite could eventually be launched as a co-passenger of a conventional spacecraft.

A handful of operators own most GEO satellites, and these are purchased to only a few manufacturers all over the world. In this context of limited actors it is easier to think about the implementation of common practices and design standards to support on-orbit maintenance of spacecraft. Furthermore, because of the specific features of the GEO orbit there is a significant interest in maintaining the orbit clean and free of obstacles, obsolete and faulty spacecraft are driven away from this orbit to leave room for the operational satellites.

The design of product line spacecraft usually respond to a common platform and a set of options for the primary payload. The platform is sized to support the most demanding case of primary payload (Fig. 17.10) and in some cases it could present some variability (different size of solar panels, external appendages etc.). From a system point of view, the biggest payload of the family drives the design of the different subsystems (avionics, thermal, power…). In some cases an analysis is done to identify the sizing case for each subsystem taking into account the variability of payloads and options. As a result, the platform is usually slightly oversized for the final need.

MOSAR envisions the introduction of standard interfaces on the platform to add external modules to manage the variability of options of the spacecraft and introduce

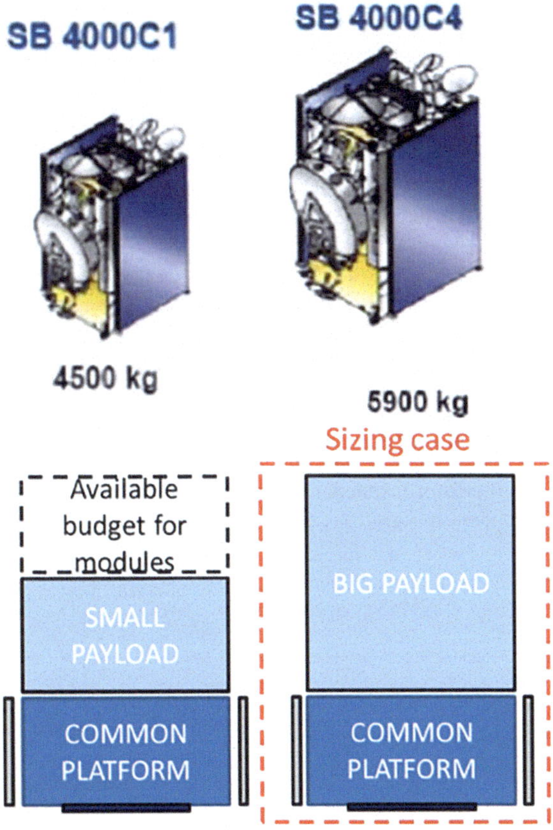

Fig. 17.10 Schematic representation of the implementation of MOSAR modules in a product line of commercial spacecraft

secondary payloads to make use of the additional capacity that the platform could offer when the primary payload is not the sizing case of the catalogue. The use standard interfaces respond to a double purpose in this case: first serve as generic interconnectors to ease assembly and integration on ground; and secondly constitute the first generation of replaceable elements on operational spacecraft a few years later.

17.4.1.1 Commercial Exploitation

In the short term, the secondary payloads constitute an additional business opportunity if they could be replaced in orbit. In some cases, selling the available slots or replacing the hosted payloads on already flying spacecraft could constitute an interesting alternative source of revenues for the satellite owner, that could find interesting to add this option to its life-extension or refuelling mission.

The long-term exploitation of orbit servicing oriented to spacecraft reconfiguration and on-orbit assembly depends on the capability to implement a modular approach for the design of spacecraft in order to being able to introduce progressively higher complexity on the orbit operations, and potentially high added value operations on the client spacecraft.

As presented before, life-extension services are not enough and on the contrary it could become almost a drawback in a context of fast evolving technology and emerging markets. In this context, the main benefits of orbit servicing shall be linked to the capability to introduce changes on existing satellites, introducing new technology that makes the satellites more competitive or changing its payloads if they become obsolete.

In this context MOSAR proposes a shift of paradigm of the life cycle of commercial spacecraft (Fig. 17.11):

- Today, conventional spacecraft have a significant lead-time (around two years) and are designed for 15 years of operation. This means that a spacecraft operator needs to identify a market that will last around two decades, with the uncertainty and risk associated to it.
- In the future, spacecraft operators could be interested by the possibility of reconfiguring its satellites at some point of its operational life-time in case their service

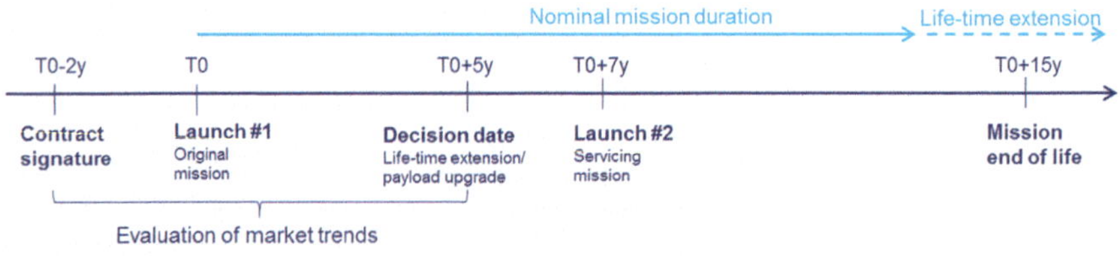

Fig. 17.11 Overview of the life cycle of a commercial spacecraft implementing on-orbit servicing for payload reconfiguration purposes

become less profitable than expected. This way, operators would find opportunities to address emerging markets and to respond to business opportunities that have high initial retribution but unclear long-term prospects.

17.5 Conclusions

This chapter summarizes the results of the literature research, the market analysis and the technology review that has served as foundation for the exploitation roadmap of MOSAR technologies.

MOSAR ambition is to lead the way into a new paradigm of space missions based on mission reconfiguration, on-orbit maintenance and life-extension services; with two main objectives:

- Offer new commercial opportunities for space exploitation.
- Ensure the long term sustainability of space activities.

Acknowledgements The authors would like to acknowledge and express their appreciation. The authors would like to thank our H2020 project officer, the European PERASPERA members reviewing the technical progresses of the project, and all the partners involved in this project: Space Applications Services (Belgium), DLR (Germany), Thales Alenia Space (France and United Kingdom), GMV (Spain), Sitael (Italy) University of Strathclyde (United Kingdom), Magsoar (Spain) and Ellidiss (France). MOSAR project has received funding from the European Union's Horizon 2020 research and innovation programme under grant agreement No. 821966.

References

1. Space Foundation. (2017). *The Space Report 2017.*
2. Federal Aviation Administration. (2018). *The annual compendium of commercial space transportation 2018.*
3. www.spacelaunchreport.com
4. Euroconsult. (2018). *FSS Operators—benchmarks & performance review.*
5. Nothern Sry Research. (2018). *Turning the satellite business model upside-down.*
6. Space News. (2019). *GEO satellite orders continued to underwhelm in 2018.*
7. Vance, A. (2017). *The tiny satellites ushering in the new space revolution.*
8. Via Satellite. (2019). *Satellite manufacturing in a state of transition.*
9. dos Santos Paulino, V., & Le Hir, G. (2016). Industry structure and disruptive innovations: The satellite industry. *Journal of Innovation Economics Management, 2*, 37–60.
10. Sweeting, M. (2018). Modern small satellites-changing the economics of space. *Proceedings of the IEEE, 106*(3), 343–361.
11. European Space Policy Institute. (2017). *The rise of private actors in the space sector.*
12. Oltrogge, D. L., & Christensen, I. A. (2020). Space governance in the new space era. *Journal of Space Safety Engineering, 7*(3), 432–438.
13. Muelhaupt, T. J., Sorge, M. E., Morin, J., & Wilson, R. S. (2019). Space traffic management in the new space era. *Journal of Space Safety Engineering, 6*(2), 80–87.

14. Jakhu, R. S., Nyampong, Y. O. M., & Sgobba, T. (2017). Regulatory framework and organization for space debris removal and on orbit servicing of satellites. *Journal of Space Safety Engineering, 4*(3–4), 129–137.
15. Wilde, M., Harder, J., & Stoll, E. (2019). Editorial: On-orbit servicing and active debris removal: Enabling a paradigm shift in spaceflight. *Frontiers in Robotics and A, 1*, 136.
16. NASA. (2019). *About—Hubble servicing missions | Servicing mission* 4.
17. New Space International Magazine. (2018). *Satellite life extension: servicing satellites in space.*
18. C4ISRNET. (2020). *Intelsat satellite returns to service after first commercial life extension mission.*
19. Piskorz, D., & Jones, K. L. (2018). *On-orbit assembly of space assets: A path to affordable and adaptable space infrastructure.* The Aerospace Corporation.
20. Boyd, I. D., Buenconsejo, R. S., Piskorz, D., Lal, B., Crane, K. W., & De La Rosa Blanco, E. (2017). *On-orbit manufacturing and assembly of spacecraft. Institute for Defense Analyses.*
21. Hein, A. M., Matheson, R., & Fries, D. (2020). A techno-economic analysis of asteroid mining. *Acta Astronautica, 168*, 104–115.
22. Badescu, V. (2013). *Asteroids: Prospective energy and material resources.* Springer Science & Business Media.
23. Jakhu, R. S., & Pelton, J. N. (2017). Space-based solar power. In *Global space governance: An international study* (pp. 205–229). Springer, Cham.
24. Barnhart, D. A., & Rughani, R. (2020). On-orbit servicing ontology applied to recommended standards for satellites in earth orbit. *Journal of Space Safety Engineering, 7*(1), 83–98.
25. SpaceNews. (2020). *Northrop Grumman's MEV-1 servicer docks with Intelsat satellite.*
26. NASA. (2019). *The emerging commercial market place in Low-Earth Orbit.*
27. Behrens, J. R., & Lal, B. (2019). Exploring trends in the global small satellite ecosystem. *New Space, 7*(3), 126–136.
28. www.h2020-peraspera.eu

Chapter 18
Building a Lunar Infrastructure with the Help of a Heterogeneous (Semi)Autonomous Multi-robot-Team

Shashank Govindaraj, Wiebke Brinkmann, Francisco Javier Colmenero, Irene Sanz Nieto, Alexandru But, Matteo De Benedetti, Leon C. Danter, Mercedes Alonso, Enrique Heredia, Simon Lacroix, Dominik Kleszczynski, Joseph Purnell, Kevin Picton, Nabil Aouf, and Luis Lopes

Abstract Exploring planets requires cooperative robotics technologies that make it possible to act independently of human influence. So-called multi-robot teams, consisting of different and synchronized robots, can solve problems that cannot be handled by a single robot. The PRO-ACT (Planetary RObots deployed for Assembly and Construction Tasks) project aimed to develop and demonstrate key technologies for robot collaboration in the construction of future ISRU (In-Situ Resource

S. Govindaraj (✉) · I. S. Nieto · A. But · M. De Benedetti
Space Applications Services, Leuvensesteenweg 325, 1932 Sint-Stevens-Woluwe, Belgium
e-mail: shashank.govindaraj@spaceapplications.com

W. Brinkmann · L. C. Danter
Robotics Innovation Center DFKI GmbH, Robert-Hooke-Str. 1, 28359 Bremen, Germany
e-mail: wiebke.brinkmann@dfki.de

F. J. Colmenero · M. Alonso · E. Heredia
GMV Aerospace and Defence, Calle Isaac Newton 11, P.T.M. Tres Cantos, 28760 Madrid, Spain
e-mail: fjcolmenero@gmv.com

S. Lacroix
LAAS-CNRS, Université de Toulouse, CNRS, Toulouse, France
e-mail: simon.lacroix@laas.fr

D. Kleszczynski
PIAP-Space Sp. Z o.o., Al. Jerozolimskie 202, 02-486 Warszaw, Poland
e-mail: dominik.kleszczynski@piap.space

J. Purnell
Thales Alenia Space, Avenue, Bristol Business Park, Bristol B660, UK
e-mail: joseph.purnell@thalesaleniaspace.com

K. Picton
AVS SLU, Calle Xixilion 2 Pabellon A Y, Elgoibar, Spain
e-mail: kpicton@a-v-s.es

N. Aouf
City University of London, Northampton Square, London EC1V 0HB, UK
e-mail: Nabil.Aouf@city.ac.uk

X. T. Yan and G. Visentin (eds.), *Space Robotics*,
https://doi.org/10.1007/978-3-031-39214-6_18

Utilization) facilities on the Moon. To this end, the following robots were used: Veles—a rover with six wheels and a 7-DoF (Degree of Freedom) arm, Mantis—a six-legged walking system, and a mobile gantry that can be used for payload manipulation or 3D printing. The project further developed existing software and hardware developed in previous space robotics projects and integrated them into the robotic systems involved. The software enables collaborative tasks such as transportation, mapping and navigation. Due to the Covid-19 situation, the final demonstration was performed remotely for defined mission scenarios. The intensive remote test campaigns provided valuable lessons learned that are directly applicable to future space missions. In addition, PRO-ACT opens a new way for multi-robot collaboration. The paper describes the developed robotic software and hardware as well as the final mission scenarios performed in lunar analogues with Mantis tested in the test field with granules in the DFKI Space Hall in Bremen, Germany, with Veles tested in Warsaw, Poland and with the mobile gantry tested in Elgoibar, Spain. In addition, one mission scenario, manipulation tasks with two robotic systems, was performed with two Panda robotic arms in Toulouse, France. The paper concludes with the results of the final demonstration of the multi-robotics team.

18.1 Introduction

For exploration of planets far from Earth, the Moon is now being considered as a future exploration outpost. The idea is to use existing resources on the Moon, rather than resources from Earth, which would, among others, save weight on launches from Earth. To be able to operate In-Situ Resource Utilization (ISRU) on the Moon without permanently exposing humans to the damaging environment, robotic systems should not only serve as support, but also as independently acting systems to build up an infrastructure, for example. Multi-robot-teams are well suited for such tasks since different robotic systems can be used to cooperate with each other and perform different tasks together. This was one of the main goals of PRO-ACT (Planetary RObots deployed for Assembly and Construction Tasks) [1]. PRO-ACT is part of the European Commission's H2020 strategic research cluster (SRC) in space robotics technologies managed by PERASPERA [2]. To demonstrate these capabilities, the following robots were used: Mantis—a six-legged walking system, the Veles—a six-wheeled rover and a mobile gantry for payload manipulation and 3D printing.

However, independent of the field of application and the specific sensor types and sensor versions used, the development of such sensors and algorithms is facing various challenges (in terms of verification, validation and optimization of all system components), which increasingly require the complementary use of advanced 3D simulations to be efficiently solved. The creation of Digital Twins and their execution

L. Lopes
LPRC, Calle el Castillo – El Fronton 37, Santa Cruz de La Palma, Spain
e-mail: luislopes@lapalmacentre.eu

by means of 3D simulation allow for an efficient and systematic investigation of component and overall system functionality with a sufficient test coverage—thanks to safe and reproducible virtual 3D environments.

In the following sections, we firstly take up the most relevant challenges that arise in the development of sensor-based systems. Here we also include current approaches, which target to solve these challenges from the simulation perspective but still do not meet the needs for flexible, continuous and realistic testing during development. Consequently, we then introduce our solution and overall concept of Hybrid Testbeds (HTBs) addressing their multidimensional scalability. Afterwards, we highlight underlying essentials of physical sensor simulation and decomposition of simulation tasks that ensure the reliability as well as real time capability of our HTBs. Finally, we present the benefits of our HTBs achieved for their application in open-loop and closed-loop testing. In this context, we present the adaptability of HTBs to support the development of a lidar sensor used for automation of Rendezvous- and Docking (RvD) in various space missions. This is also the case for the currently planned Mars Sample Return mission, which we address at the end of this article by presenting the benefits of our HTBs in supporting the verification, validation and optimization of sensor and system configuration for this application.

18.2 Project Overview

The target was to have heterogeneous autonomous robots (Mantis, Veles and the mobile gantry) cooperate with each other to plan high-level goals; jointly explore an unknown area, cooperatively unload and transport large components, and assemble an ISRU facility on the Moon. This was to incorporate and further develop outcomes from previous operational grants (OGs) of PERASPERA, including among other things developed framework software and robot guidance software. The focus was primarily on (1) multi-robot system architecture, (2) task assignment through mutual negotiation, (3) cooperative task execution, (4) single and multi-agent perception, mapping, and localization, (5) cooperative manipulation planning and control, and (6) robot hardware adaptation.

Mission cooperative scenarios (Fig. 18.1), like (i) mission planning, (ii) manipulation, (iii) transportation and (iv) mapping, were defined to demonstrate the required capabilities. Due to the Covid-19 pandemic, these defined scenarios had to be split for the final demonstration, due to hygienic restrictions not allowing all robotic systems to be operated together in the same place. The scenarios will be described later in this paper.

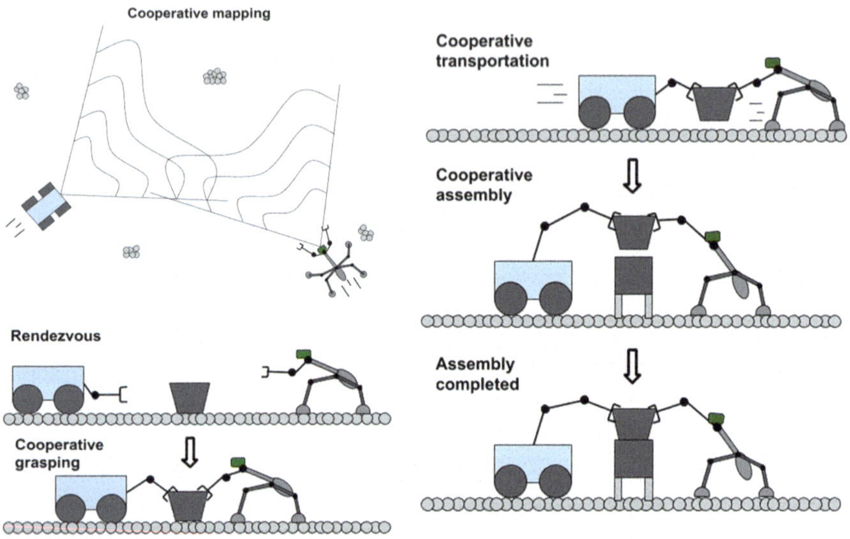

Fig. 18.1 Cooperative scenarios description

18.3 Robotic Systems

This section describes the three involved robotic systems Mantis, Veles and the mobile gantry.

18.3.1 Mantis

The Mantis (Fig. 18.2) was developed with the aim to provide high mobility and manipulation capabilities in uneven and unstructured terrain [3].

The robot possesses six extremities for locomotion. The four hind legs have six degrees of freedom (DOF) each. The two foremost limbs have only five DOFs. However, they are each equipped with a simple gripper and can be used for dual arm manipulation as Mantis is able to erect its body and free the two fore-most extremities to use them as arms. The main electronic, compartment (power management, high-level processing and overall robot control) is located in the rearmost body segment, the abdomen. The abdomen further serves as a counterweight to the upper body and thus shifts the centre of mass towards the centre of the four rear legs. This feature facilitates switching between locomotion and manipulation postures. The dimensions of Mantis are 1.96 m × 1.84 m × 0.32 m (locomotion posture) with a mass of 107 kg.

During the PRO-ACT project Mantis was adapted and improved on the hardware and software side [4]. Mantis is able to walk in six-legged, five-legged and four-legged poses. The five-legged locomotion mode allows Mantis to walk on its four

Fig. 18.2 Mantis (*Credit* DFKI GmbH)

legs and one forearm whereas the second forearm with gripper can carry a load or can be used for cooperative transportation. Mantis is extended by a Robot Client API which is independent and based on the open source robot_remote_library [5], which is under current development at DFKI. The communication layer is based on ZeroMQ [6] and de-/serialization is handled with Google's protocol buffers (proto3). The core library was written in a way that facilitates extensibility through inheritance [7].

18.3.2 Veles

Veles is based on the IBIS [8] mobile manipulator (Fig. 18.3), a robot designed by the PIAP institute for pyrotechnic operations and reconnaissance. Base IBIS device was teleoperated with limited autonomy, which was modified to host multi-robot autonomy functions. Veles was designed as a dedicated Robotic Working Agent (RWA) for the PRO-ACT project.

It is dedicated to perform cooperative tasks with Mantis: cooperative manipulation, mapping, transportation and deployment of the mobile gantry. Veles is a highly off-road capable mobile platform with high towing capabilities and the ability to carry a heavy payload (Fig. 18.4). Following modifications to the IBIS platform, on which Veles is based were carried out: modifications of the communication suite and monitoring and control station (MCS), a redesign of the robotic arm, including

Fig. 18.3 IBIS mobile manipulation robot (*Credit* DFKI GmbH)

motors and controllers, modifications to the platform gearing and packaging and finally major changes of the safety systems of the platform.

The Veles Robotic Arm is a 7 DoF manipulator with relatively high dexterity (Fig. 18.5). Manipulator joints are divided into two main types called "Horizontal" and "Vertical" according to their orientation with respect to arm home position. The arm consists of seven revolute joints: three Horizontal and four Vertical. Joints have equal position angle limits. The arm is equipped with a force and torque sensor and HOTDOCK [9] interface enabling connection with end-effectors (grippers, shovels, etc.).

The arm is capable of manipulating the payload of maximum 20 kg. The payload is manipulated from 25 cm below ground level to minimum 150 cm (maximum 200 cm) above ground level. The Manipulator is equipped with brakes on each joint that prevents motion without communication with the MCS. Brakes are normally engaged (have to be actively unblocked by the system). Veles is equipped with a safety emergency stop. The manipulator power bus is controlled by redundant relays.

18.3.3 Mobile Gantry

The Mobile Gantry has been developed within PRO-ACT with the aim of providing a flexible, high load manipulation and large work volume robotic system. The Mobile

Fig. 18.4 Veles Mobile robot with a manipulator and gripper, grading blade and shovel

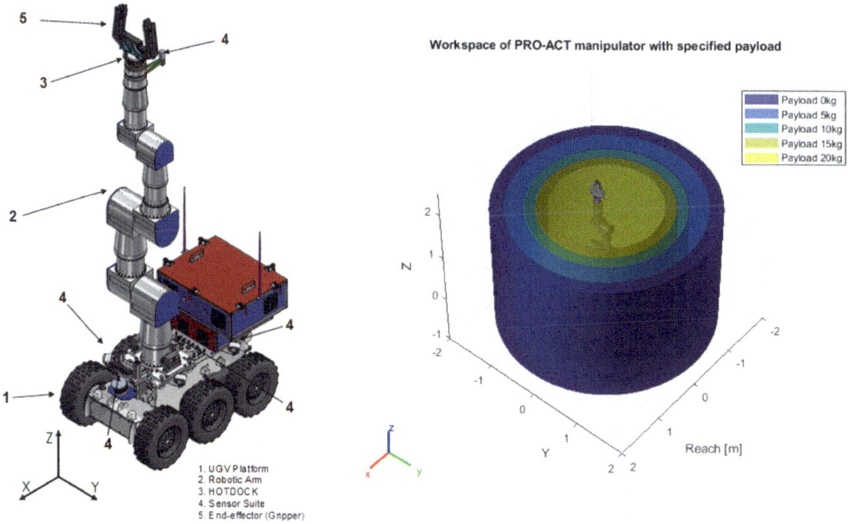

Fig. 18.5 Veles manipulator and the manipulator workspace volume for different weight payloads

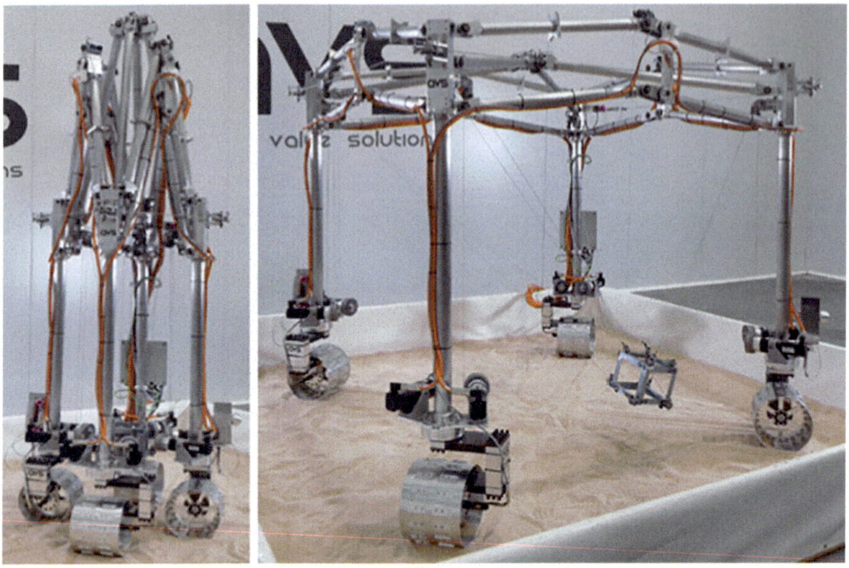

Fig. 18.6 Mobile gantry deployment

Gantry is a cable robot having two well differentiated systems, the deployable structure and the end-effector cable system itself.

The end-effector cable system [10, 11] has industrial heritage but has been scaled in order to meet the required dimensions. The deployable structure (Fig. 18.6) is a newly developed high-precision mechanism that provides support for the cable actuators and adds stowing and deploying capacities to the system. It has self-deploying capabilities by using wheel actuation plus an actuated cable that has been added to the structure. Mantis and Veles will also provide auxiliary deployment force in order to be able to achieve deployment in lunar regolith simulant terrain conditions.

When deployed it has a work envelope of 1.7 m × 1.7 m × 1.7 m with a total mass of 230 kg and a lifting capacity of 500 N. The mobile platform can be fitted with different end-effectors such as a 3D printer or a gripper for payload manipulation and it has all 6 DOFs within the work envelope. The actuated wheels [12] on the Gantry provide it with mobility capacities. In this context, the steering of the wheels is performed manually for system simplicity but actuators could be added to provide autonomous mobility.

The control hardware is composed of two industrial Programmable Logic Controllers (PLC) which are separate units from the Gantry itself with power and communication connections to it. From a software perspective deployment control algorithms have been implemented. During deployment, all systems (wheels, deployment cable, end-effector cables, Mantis and Veles) must be actuated simultaneously in order for the deployment to take place.

18.4 Multi-robot System Design

18.4.1 Overall System Design

From a deployment perspective, the systems need adapted processing power and communication facilities to run. There are two levels of distribution of processing power: one between robotic platforms and one within each of the platforms. The system is composed of four main systems: Ground Monitoring Station, Gantry, Mantis and Veles, each one of them having multiple processing units. In Fig. 18.7 one can observe the four systems with all their computers and communication links that tie them together.

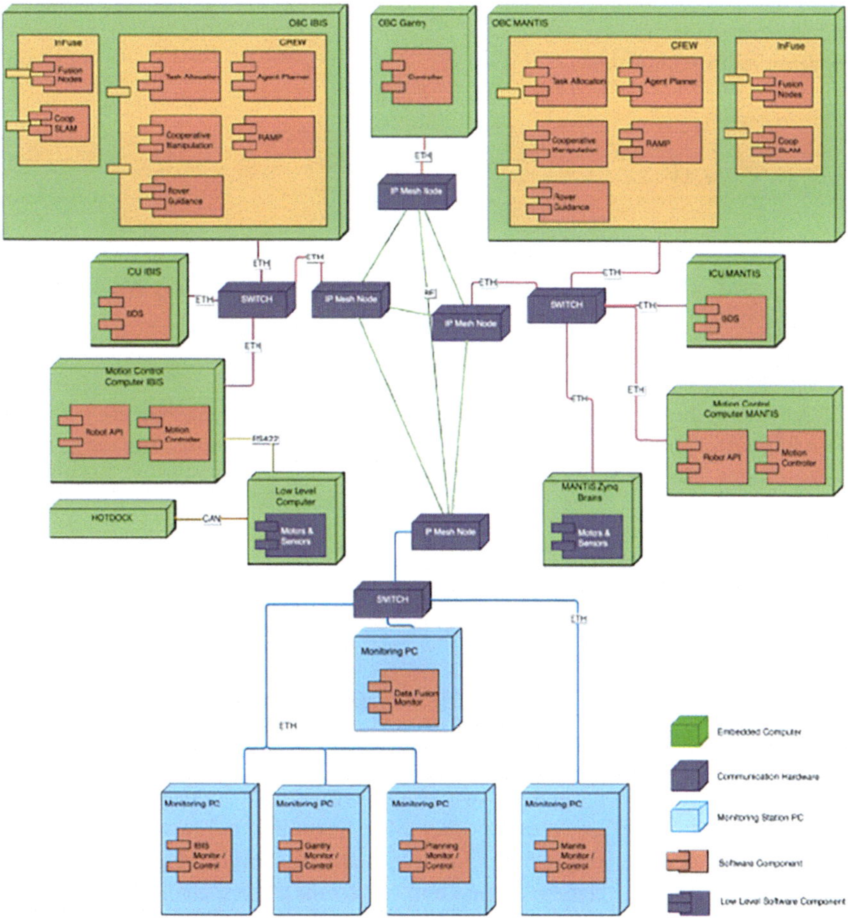

Fig. 18.7 PRO-ACT system design

The Veles and the Mantis are meant to demonstrate multi-agent cooperation and collaboration capabilities within PRO-ACT, while the mobile gantry is to demonstrate the cooperation only. The monitoring stations are standalone systems for defining the cooperative goals and monitoring the execution of a mission. One of the properties of the multi-agent system is scalability to the number of robots which can be added. Even though the two robots are very different in terms of their mechanical design, the PRO-ACT architecture is created to keep the high level software design of the two applications to be the same and are built around the results of OG1–OG4 [13–16].

From the top-level view of the deployment diagram, the Veles and Mantis have an on-board computer (OBC) that hosts both the Cooperative Robotics for Enhanced Workforce (CREW) and InFuse data fusion processes. A sensor-processing unit—the adapted OG4/I3DS ICU (Inertial Components Unit), is also present on both robots to interface with sensors and pre-process the raw sensor data for the high level of sensor data fusion processes. The final computational platform, that is close to the electronics and electrical controllers, is the robot control OBC that interacts with the low-level motor controllers.

18.4.2 CREW

Cooperative Robotics for Enhanced Workforce (CREW) is the result of the design process based on the reuse and adaptations of the previous building blocks (OG1-OG5) extended with new functionalities focused on maximizing the collaboration and cooperation activities for a multi-robot system. Figure 18.8 shows an integrated overview of the CREW Agent Autonomy component design.

It represents the major software components and interfaces deployed on each RWA of the PRO-ACT multi-robot system. This section has been split in the following three subsections focused on the cooperation design drive required by all multi-agent robot system:

- Cooperative Mission Task Allocation
- Cooperative Mission Planning Execution
- Cooperative Manipulation and Transport

18.4.2.1 Cooperative Multi-Robot Mission Task Allocation

The task allocation algorithm is a decision-making algorithm for decentralized task distribution between agents that can perform a task cooperatively. Advantages over a centralised one are, especially, the flexibility and task scalability are observed while dealing with dynamic environments. However, the potential excess of communications between each agent unit represents the big challenge of using a fully decentralised approach as this can increase power consumption to manage the communication between the agents which leads to the loss of data.

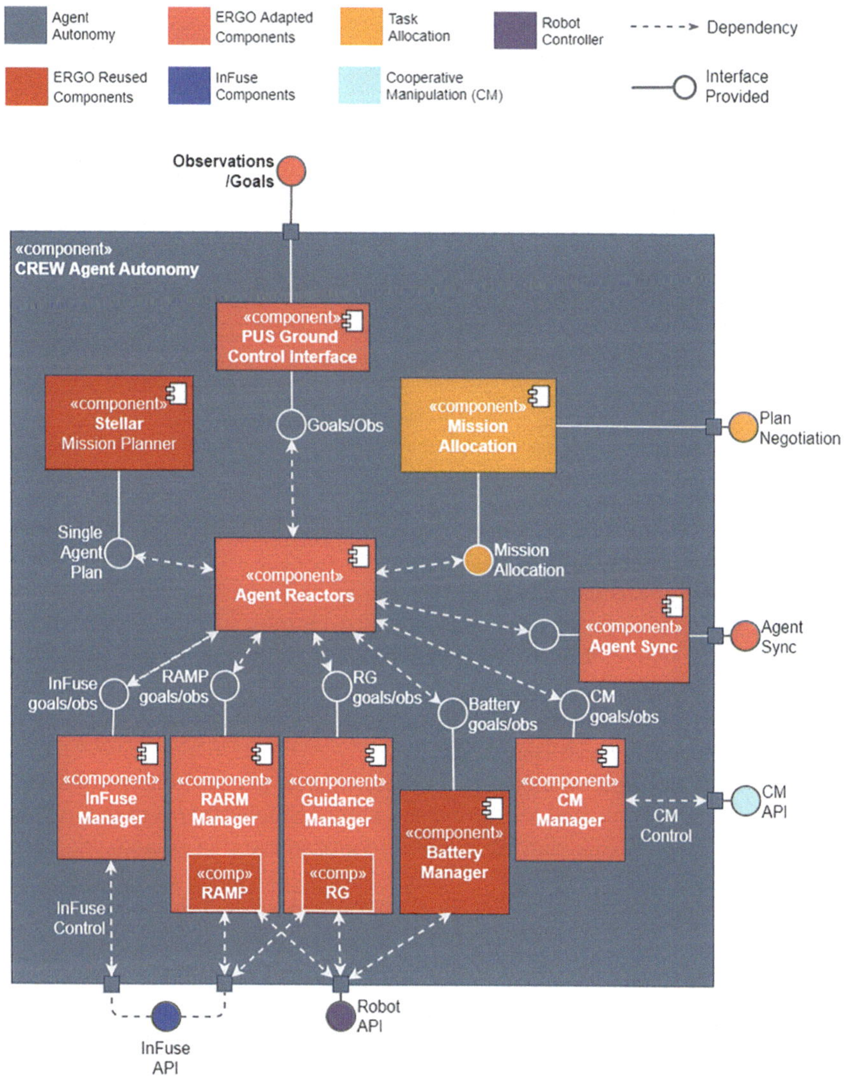

Fig. 18.8 Multi-Agent Autonomy Component Diagram

To tackle this task allocation problem in the PRO-ACT project, new optimisation techniques have been explored and used based on integer or mixed-integer linear programming (ILP or MILP). The latter is NP-hard in the way it is formulated for distributed task allocation to avoid the Inter-RWA path conflicts (between different agents) and Intra-RWA path conflicts (same agent). Based on distributed techniques, the optimal or near-optimal solutions have been investigated for such multi-robot goal assignment and present the best solution for task allocation issues also to reach real-time computation.

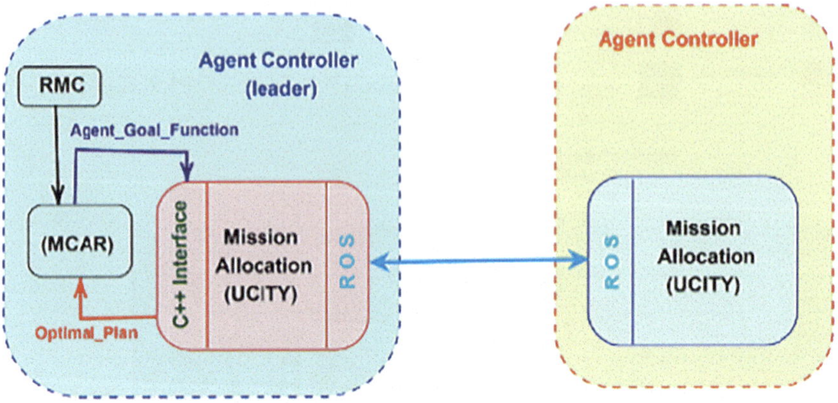

Fig. 18.9 Task allocation with mission coordinator

The share of the mission parameters between the PRO-ACT platform agents while resolving conflicts in task allocations permitted to manage and produce an efficient decentralised task decision-making process among cooperative agents. The ground operator creates a mission and uploads it to all available robots, being the robot Agent granted with the role of "leader", which passes the mission tasks to the Auctioneer as an AgentGoal function. After receiving the AgentGoal function, each agent evaluates the mission tasks based on its: skills, availability, capability, and suitability to contribute to the auctioned Mission Primitive.

At the end of the task allocation process, the auctioneer evaluates bids and declares the winner of the bid (auction) and attributes it to the identified robot or set of robots. The optimal plan is shared with the mission coordinator to start planning the task, seen in Fig. 18.9.

18.4.2.2 Cooperative Multi-robot Mission Planning Execution

This subsystem deals with the autonomous multi robots planning, coordination and execution of high level goals using the available resources of the fleet (robots and instruments) in an optimal way. Design of the Cooperative Multi-Robot Planning and Execution component is based on the OG2/ERGO autonomous Agent concept. An overview of the CREW architectural design (showing only two robot instances) is depicted in the following Fig. 18.10.

Each CREW Multi-Agent instance is composed of the following two main components:

1. *Agent Taste Function*; Composed by an Agent main controller and a set of

   Reactors responsible of managing and handling of the topics below:

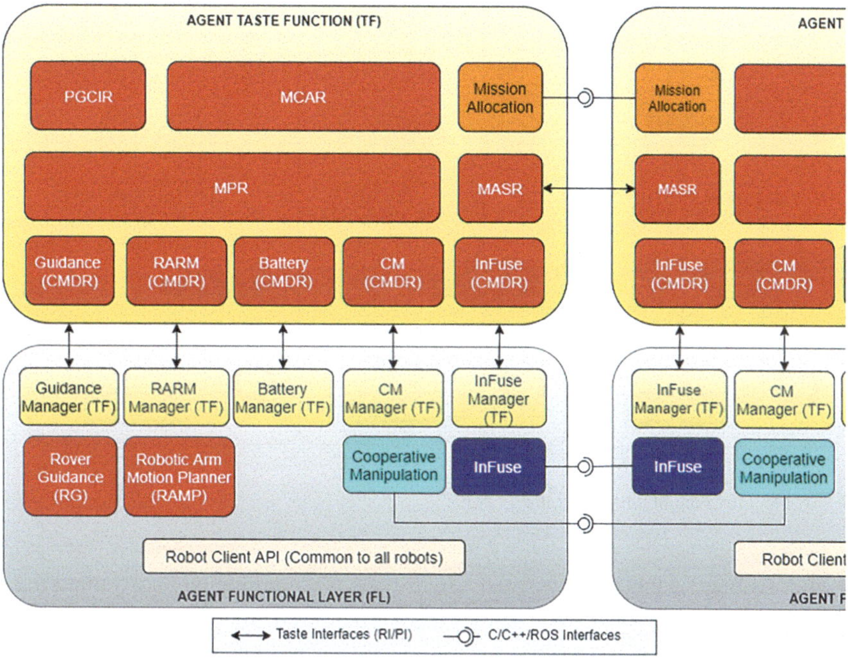

Fig. 18.10 CREW multi-agent design, showing only two robot instances

1.1 PGCIR (PUS Ground Control Interface Reactor) responsible for the interactions with ground telecomanding (TC) and telemetry (TM) using PUS (Packet Utilisation Standard) services.

1.2 MCAR (Mission Coordination and Allocation Reactor) responsible for the multi-Agent mission Coordination and synchronisation, including the interaction with the Cooperative Multi-robot Tasks Allocation.

1.3 MASR (multi-agent Synchronization Reactor); responsible for the handling of exchanged messages between different agent instances.

1.4 MPR (Mission Planning Reactor) for the planning of the goals allocated to each individual robot. It also includes the definition of the PDDL (Planning Domain Definition Language) domain and problems files in line with timelines and predicates defined based on functionalities and features statuses and the final demonstration scenarios selected.

1.5 A set of Command Dispatcher (CMD) Reactors for the final interaction with the Agent Functional Layer.

2. *Agent Functional Layer*; it is the executive layer of a CREW agent instance, that is

   the part of the autonomous Agent implementing the commanding generated and sent (requests) from the Agent Controller to the low level robot control drivers.

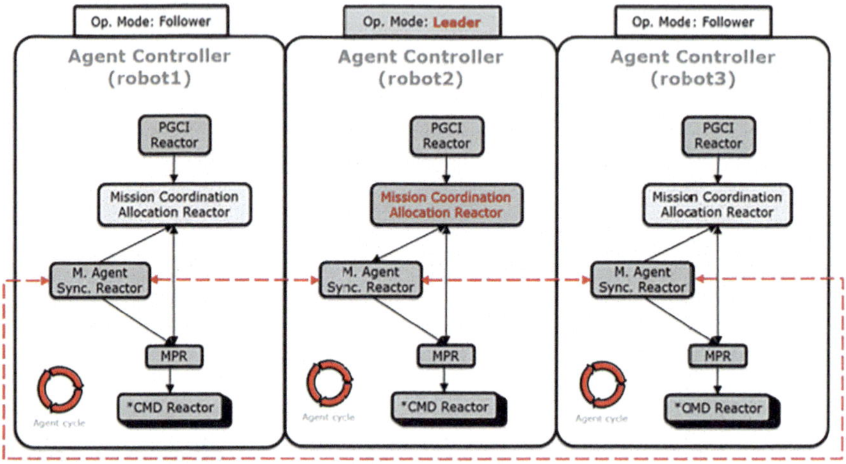

Fig. 18.11 Multi-agent coordination

Multi-robots/agents systems need a high level of coordination and communications at very low rates (close to real-time). Leader Agent's controller and reactors are responsible for the coordination of the Cooperative multi-Agent mission planning, execution and monitoring of the actions to be developed by each robot in an autonomous way. Two different roles, "Leader" and "Follower", were implemented as part of the Multi-Agent system for supporting the coordination of the robot fleet during the mission execution (Fig. 18.11). A "Follower" could become "Leader" in case the latter suddenly disconnects from the network fleet.

For this purpose, each relevant "functionality", from the logical mission plan point of view, is modelled as a timeline that embeds a state machine representing all the statuses in which this functionality can be. This also manages coordination, synchronisation and fault recovery features based on wireless (RF) communications for reacting to unexpected events or problems during the mission execution.

18.4.2.3 Cooperative Multi-Robot Mission Planning Execution

Cooperative manipulation (CM) consists in having the two robots Mantis and Veles jointly use their arms to transport and/or manipulate large objects, thus achieving a task no single robot can achieve. A CM task always encompass three different phases:

1. CM mission definition: specifies the object to manipulate, the picking configurations, and the place at which it must be set. Within the whole PRO-ACT system, this phase is a result of cooperative mission planning. It could also be defined by a remote operator.

2. CM mission planning (CM2P): on the basis of the mission definition, the CM2P produces the trajectories that the robots must follow to achieve the mission. The trajectory defines the motion of the robots' base and arm. This phase is a non-real time that may take up to several minutes.
3. CM mission control (CM2C): CM2C ensures the proper execution of the planned trajectories, by sending control inputs to both the base and arm of the robots. A high rate real-time controller runs on each robot, the controllers being coordinated implicitly through force exchanges, and by exchanging status reports at a low frequency.

The execution of motion control is performed with an admittance controller in CM2C. The admittance controllers regulate position or velocity of joints based on the force errors on the end effector, using a relation describing the impedance of the mechanism (e.g. springs and dampers). To coordinate themselves, the robots exchange at low frequency status messages, e.g. to synchronize the start of the motion controllers and to monitor the motion execution. The control coordination is implicitly handled at the high controller frequency through the force interactions between the manipulator induced by the manipulated object.

18.4.3 Perception, Mapping and Localisation

18.4.3.1 Object Detection and Tracking

For object detection and tracking, two different approaches were used: model-based and marker-based algorithms.

Model-based detection and tracking.

The Mid-Range 3D Model Detection and Tracking Data Fusion Compression Compounds (DFPCs) from the Common Data Fusion Framework (CDFF) [17] were reused and extended.

The model-based detection and tracking was used in PRO-ACT to detect the lander used in the work scenarios (Fig. 18.12). The relative pose was then passed to the CREW in order to guide the two RWAs towards the lander. The object detection and tracking algorithms were combined in order to obtain a consistent object pose throughout the scenarios.

Marker-based detection and tracking.

The AruCo marker detector was used to obtain a more robust object pose estimation primarily for grasping purposes.

The algorithm was benchmarked with a dataset recorded using Mantis in DFKI's Lunar Hall. In said test, different-sized AruCo markers were placed in the environment and Mantis was recorded moving towards them (Fig. 18.13).

This allowed the selection of the final AruCo marker size that was later used in the posterior tests and the final demonstration. Figure 18.14 shows different sizes of

Fig. 18.12 ISRU mockup on lander detection and tracking

Fig. 18.13 Initial tests AruCo marker detect

AruCo markers on the ISRU mock ups used in the preliminary tests and within the final demonstration.

18.4.3.2 Self-localization

The algorithm that was used to obtain the pose of each of the RWAs is the ROS viso2 implementation.

Both algorithms were selected by testing their performances using datasets gathered from OG3/InFuse (Fig. 18.15), obtaining good preliminary performances.

Further tests were performed using outdoor datasets from both RWAs collected remotely, in which the CPU and memory consumption of the algorithms were measured, as well as the accuracy of each algorithm in the different scenarios. During

Fig. 18.14 Different-sized AruCo markers on the demonstration ISRU mockups

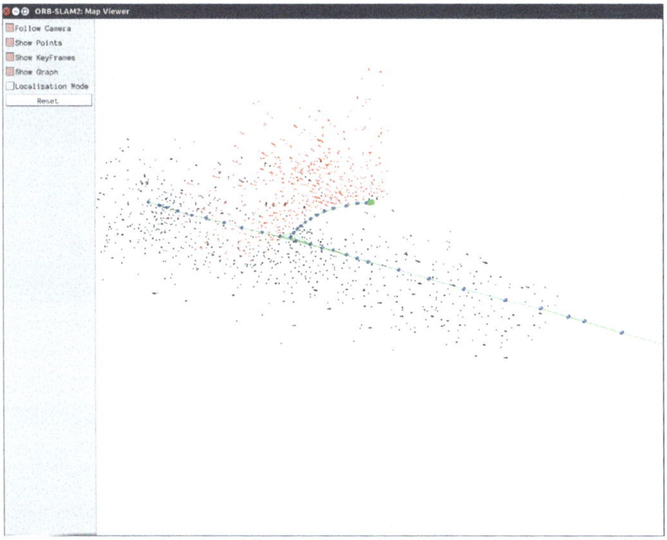

Fig. 18.15 Visual representation of the estimated poses (blue dots) and map (point cloud) created by the CDFF's Visual SLAM DPFC

the final demonstration the Visual Odometry (VO) algorithm was used, fused with the wheel or leg odometry and the Inertial Measurement Unit (IMU) information of the RWAs using an Extended Kalman Filter (EKF) to obtain a more robust estimation.

18.4.3.3 DEM Building

The selected algorithm to be used in PRO-ACT is the ROS elevation mapping node. Initial testing of the algorithm was performed to test the accuracy of the Digital Elevation Model (DEM) building algorithm, using the datasets gathered during OG3/ InFuse in the PEL (Planetary Spectroscopy Laboratory) in DLR (datasets under DLR ownership), as seen in Figs. 18.16 and 18.17. These datasets contain lidar and stereo camera data, the pose of the robot using a precise indoor localisation system, transforms between the tracking device for the indoor localisation system on the robot to the sensors and an elevation map ground truth.

The DEM building algorithm was used during the different tests as an input to CREW in order to help with path planning and robot motion.

Fig. 18.16 Ground truth from the DLR dataset

Fig. 18.17 Initial DEM reconstructed from the DLR dataset

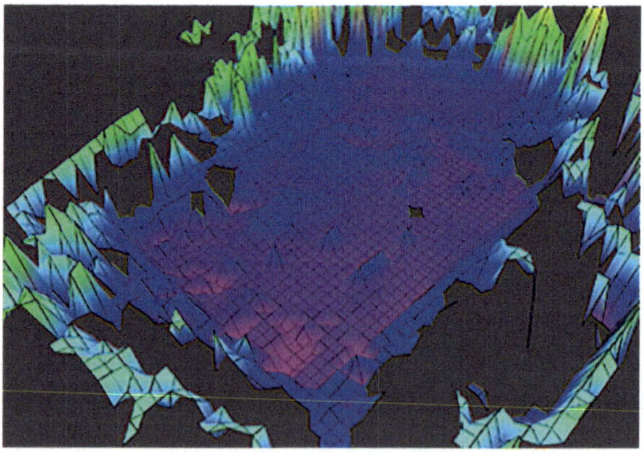

18.4.3.4 Collaborative SLAM

The collaborative Simultaneous Localization and Mapping (SLAM) proposed here fuses inputs from multiple agents to perform the localization and mapping. This cooperative SLAM/VO (CSLAM/VO) algorithm, is composed of two main blocks that are triggered concurrently on each robot, including VO and Loop Closure Detection (LCD) algorithms (Fig. 18.18). The former obtains the pose of the robot using a robust motion estimation algorithm. The latter compares the current view of the robot obtained from the stereo camera with the saved key-images in its database to detect the loop closure. The loop closure can be either intra loop closure when the robot detects a loop from its previously observed area or inter loop closure when the robot detects a loop from the shared images from the other robots. The CSLAM is using ROS messages to send and receive the images from the database/agents. Although the CSLAM uses only stereo images for VO, other useful information such as IMU data can be used by subscribing to Infuse. The CSLAM/VO publishes the pose and key-images and updates the map which can be used by other agents or even other purposes like cooperative manipulation. The proposed algorithm is a distributed CSLAM/VO, then each agent is self-localised while it is trying to detect the shared area from the other agents. The proposed LCD is based on binary descriptors to retrieve previously seen similar images. It should be noted that the shared key-images are used to exchange between RWAs instead of Maps for VO (Fig. 18.18). The local DEMs are being registered to the outcome of pose estimates obtained from cooperative VO. A simple but effective mechanism is used to group images close to the pose of the robot in a gating area and search within them, which reduces the computational times.

18.4.4 Sensors

The I3DS software framework was designed during the OG4 to support a modular suite of sensors which can be tailored or extended over time.

The PRO-ACT sensors implemented within I3DS are shown in Fig. 18.19. The core of the framework defines a number of ASN.1 types on which all communications between sensor nodes and subscribing clients are based. These ASN.1 types are also the key enabler in allowing integration between different OGs. The I3DS software thus consists of the 'i3ds-interfaces' library, the 'i3ds-framework-cpp' library and the specific sensor implementation e.g. 'i3ds-basler' for the Basler Ace series of cameras. Each of these are provided as separate repositories, publicly available on GitLab [16]. These dependencies are shown in Fig. 18.20 by dashed arrow-lines.

The Zynq US + MPSoC combines multiple ARM processors (the "Processing System") and FPGA (the "Programmable Logic") on a single chip, connected by a high speed AXI4 bus. This allows data to be passed to processing kernels in the Programmable Logic (which may access DDR directly) with minimal overheads. In turn, this allows the Processing System to handle the complex general tasks that it is suited to—networking, TCP/IP, zmq, USB camera interfacing etc.—while employing

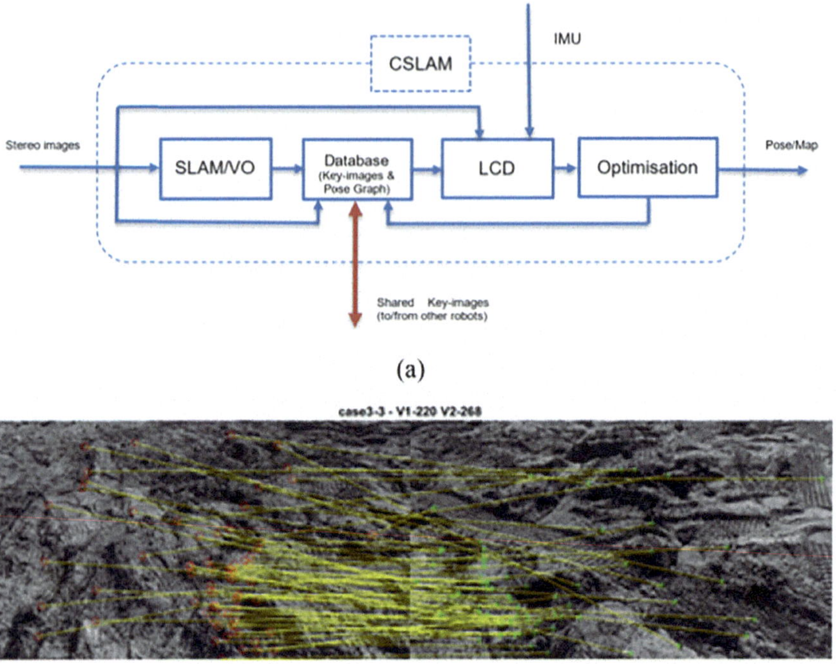

(a)

(b)

Fig. 18.18 CSLAM algorithm overview (**a**); Loop closure detection across different view points (**b**)

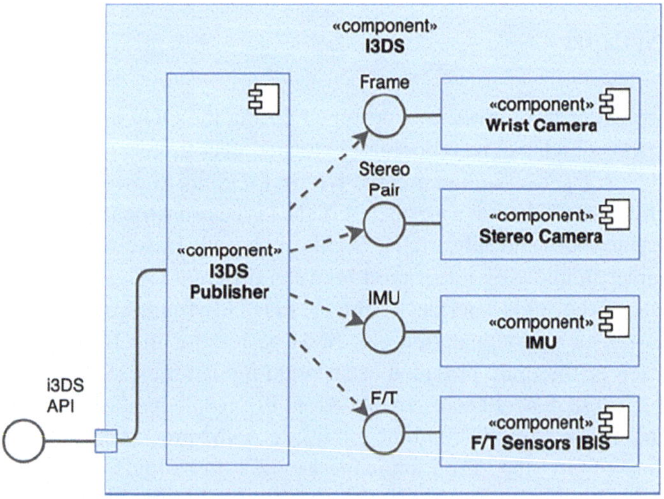

Fig. 18.19 I3DS architecture overview in PRO-ACT

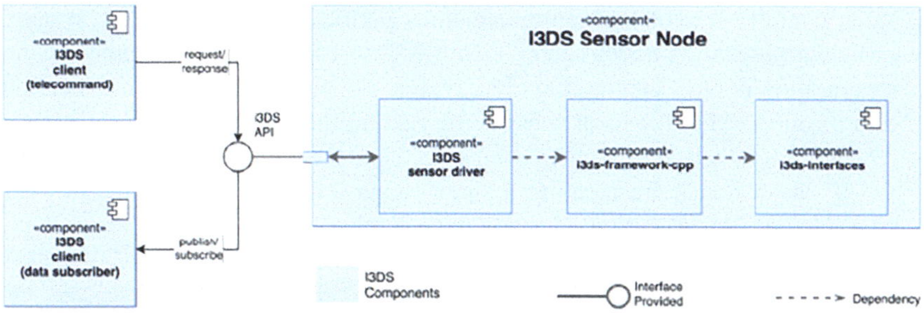

Fig. 18.20 I3DS sensor architecture

the Programmable Logic for specialised, highly parallelizable tasks to reduce latency, increase performance and reduce the load on the processors. For specialised, parallel tasks FPGAs typically provide the best performance-per-watt (bested only by ASIC technology), along with low, deterministic latency. These are all key performance metrics for battery-powered, autonomous robotics that are moving within a complex environment.

18.4.5 Communication Systems

All the RWAs are within the same mesh network, to enable intra-system and inter-system communications. A mesh network has been chosen for this application as a baseline to allow scalability of RWAs that can be added progressively, providing intermediate hops to nodes that are not in line of sight between 2 communicating nodes. Figure 18.21 shows the selected IP Mesh nodes. The availability of robust, long range and reasonable bandwidth solutions such as the IP mesh or short-medium range Wi-Fi mesh solutions with higher bandwidths make this type of communication interesting for PRO-ACT applications.

Fig. 18.21 IP mesh Node—DTC SOLO8 Software Defined Radio (Concealment—left, Robust—middle), MIMO antenna right

As an overall view of the communications performed throughout the system, four RF communication nodes can be identified, corresponding to the three robotic devices and the control centre (Fig. 18.22). Each of the IP Mesh communication nodes are connected to the local subsystems via an Ethernet switch, which forms a local network for each of the robots (Fig. 18.23) and the control centre.

The IP Mesh nodes are creating a single subnet, thus making all the devices discoverable and able to communicate with each other.

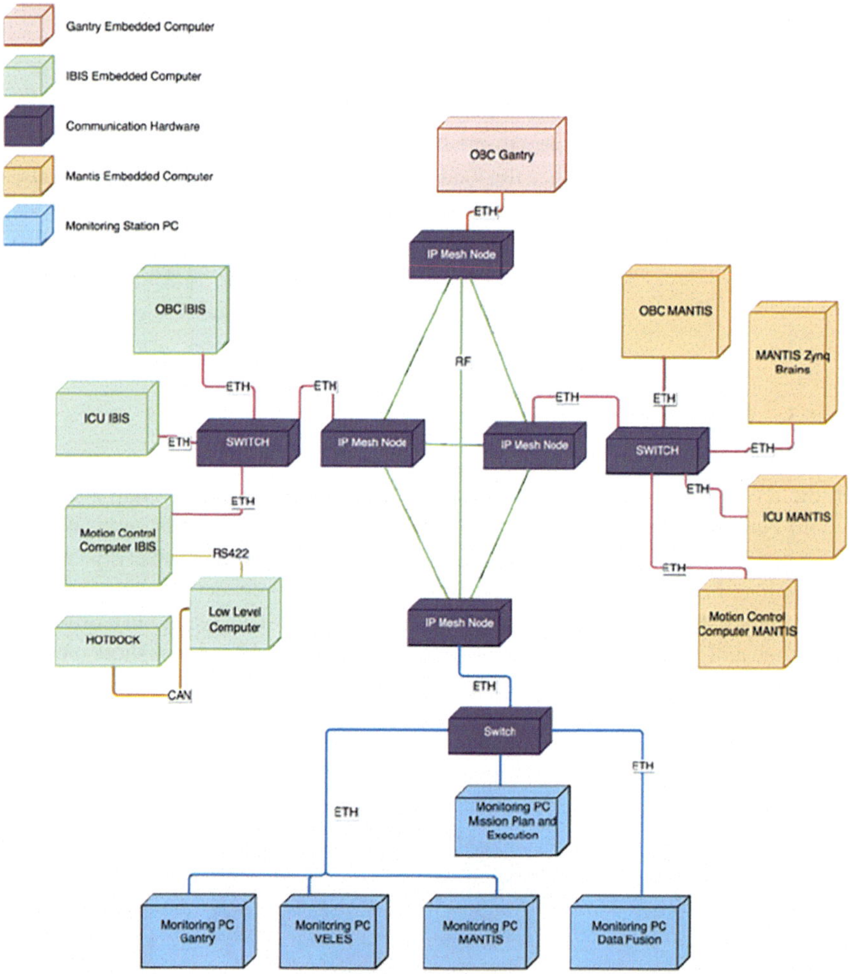

Fig. 18.22 IP mesh communication architecture

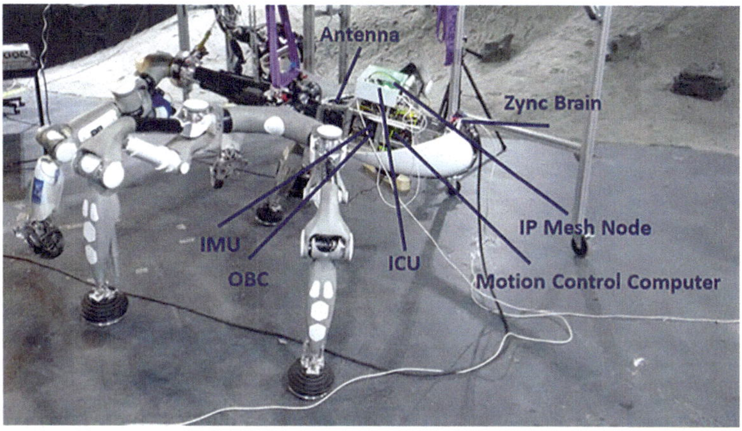

Fig. 18.23 IPMesh node on mantis

18.4.6 MCS System Architecture

The Monitoring Station for the PRO-ACT project needs to perform different functions in the system, such as monitoring the status and health of the RWAs, planning, commanding and monitoring the paths each of the RWAs follows and monitoring the data fusion algorithms.

The RWAs'providers already have Monitoring and Control Systems for the low level functionalities and control of the robots. On the same line, during the previous OGs Monitoring and Control Stations were developed.

Due to the heterogeneity of the system with RWAs with different design, sensors and actuators and the different software components as well as the legacy of the components developed in the previous OGs, it was decided to create separate monitoring and control stations for the functionalities previously presented. The Monitoring and Control System Deployment diagram can be seen in Fig. 18.24.

18.4.7 Common Electromechanical Interface

The Active HOTDOCK [9], a multi-functional standard interconnect, integrated on the tip of the end effector of the Veles manipulator is able to connect to:

- the Passive HOTDOCK integrated with the Gripper, to enable power and control of it through the data and power interface
- the mechanical HOTDOCK fixed with the shovel and on the ISRU elements, to provide a mechanical connection for manipulation

The HOTDOCK firmware is responsible for the management of the HOTDOCK controller PCB modules (motor, sensors, CAN transceivers, relays, LVDS switches)

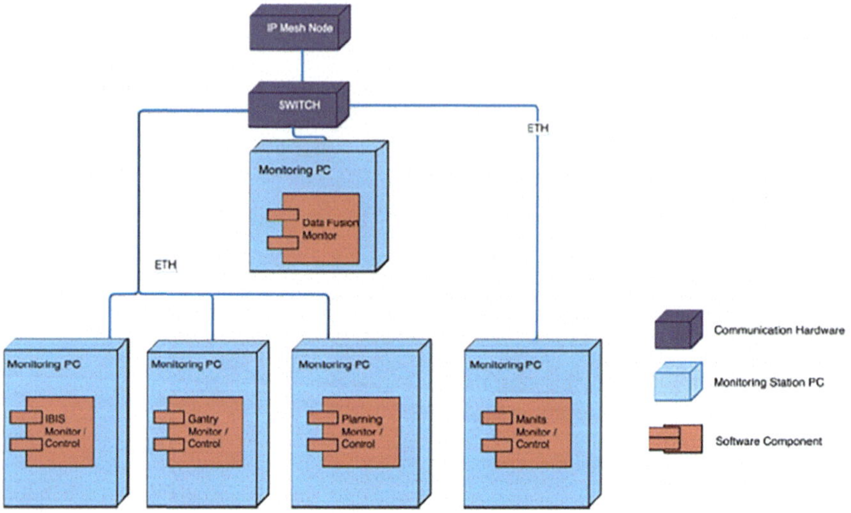

Fig. 18.24 Monitoring and control system deployment

as well as the TM/TC with external components through CAN communication. It is composed of a single software running on the PCB microcontroller.

18.4.8 Software Interfaces Definition

The Matrix of Software Interactions (Fig. 18.25) represents all the data and commanding exchanges between components. We have identified with blue the exchange of commands and with purple, the exchange of data payloads. The antidiagonal of the interface matrix is expected to be blank since the commanding and data exchanges of the same component are unlikely. In our case we represented these exchange points in the matrix to show the communication between the same components running on different RWAs.

18.5 Test Areas and Test Setup

Due to force majeure COVID-19 conditions, the multi-robot cooperation scenarios were performed with the robots deployed in the local analogs, but connected virtually via a common network (VPN).

FROM \ TO		A I3DS Client	B InFuse Orchestrator	C InFuse Data Fusion	D Cooperative Manipulation - Motion Planning	E Cooperative Manipulation - Motion Control	F Cooperative SLAM / Map	G MissionAllocation	H Agent	I InFuse Manager	J Guidance Manager	K RARM Manager	L CM Manager	M Rover Guidance	N Robotic Arm Motion Planner	O Robot Client API	P Mission Planning MCS	Q Sensor Fusion MCS	R Gauntry Controller	S Motion Controller IBIS	T Motion Controller MANTIS	U IBIS MCS	V MANTIS MCS	W Gauntry MCS	X Mars Simulation Tool	Y HOTDOCK	Z PGCI
I3DS Client	A		C	D,			D,																				
InFuse Orchestrator	B	C,		C,	C,		C,			C,								C,	C,								
InFuse Data Fusion	C		C		D,	D,								D,	D,				D,								
Cooperative Manipulation - Motion Planning	D		C,	D,		C, D,																					
Cooperative Manipulation - Motion Control	E			D,	C,								C,			C,											
Cooperative SLAM / Map	F		C,	C,			D,							D,	D,				D,								
MissionAllocation	G								C,																		
Agent	H							C,	C,	C,	C,	C,	C,				C,										C,
InFuse Manager	I		C,						C,		C,	C,	C,														
Guidance Manager	J								C,	C,				C,													
RARM Manager	K								C,	C,					C,												
CM Manager	L					C,			C,	C,																	
Rover Guidance	M			D,							C,					C,											
Robotic Arm Motion Planner	N			D,								C,				C,											
Robot Client API	O			D,		D,	D,			C,				C,	C,					C,D	C,D,				C D		
Mission Planning MCS	P								C,																		D,
Sensor Fusion MCS	Q																										
Gauntry Controller	R																										
Motion Controller IBIS	S															D						D				C, D	
Motion Controller MANTIS	T					C,																	C, D				
IBIS MCS	U																			C,,D							
MANTIS MCS	V																										
Gauntry MCS	W																		C								
Mars Simulation Tool	X															C,D											
HOTDOCK	Y																			D							
PGCI	Z								C,								D,										

Fig. 18.25 Mapping Matrix of software interfaces

18.5.1 Space Exploration Hall (Mantis)

Experiments with Mantis were conducted in the Space Exploration Hall in Bremen, Germany. The test area was a sandbox in front of the existing crater with an area of 48m2 with Baltic sea sand (grain size 0—1 mm) and a fill level of 25 cm (see Fig. 18.26).

Furthermore, a VICON tracking system consisting of six cameras at a height of 6.5 m covering a volume of 770 m^3 was used to record the 3D movement of Mantis in the test field. Since cooperative manipulation requires very precise localisation an additional localisation source was added, using a COTS UltraSound System from Marvelmind Robotics. This system can be used both indoor and outdoor, with an accuracy around 3–5 cm RMSE. To ensure that the project partners from outside had access to Mantis, a VPN connection was set up. For this purpose, access to the robot was available via LAN as well as WLAN.

Fig. 18.26 Mantis in the space exploration hall

18.5.2 Test Area in Poland (Veles)

PIAP Space rented a former thermal tests room in PIAP Institute located in Warsaw and arranged it for the final tests/demonstration campaign.

The test room's (Fig. 18.27) dimensions were 6 × 6 m and about 6 m high. The floor was divided in two sections: work area for the test team, computers and supplementary hardware, and a test area covered with regolith analogue. The work area was 3.5 × 1.5 m. To simulate regolith, 3 tons of clear and dry silicon sand was used (1.5 tons of 0.1 to 0.5 mm grains and 1.5 tons of 0.2 to 0.8 mm grains).

The test room was equipped with a network infrastructure, camera and ultra sound system for robot localisation.

18.5.3 Test Area in Spain (Mobile Gantry)

A dedicated test area was installed in AVS' premises in order to perform tests for deployment algorithm calibration and system validation. It consists of a sandbox

Fig. 18.27 Veles at PIAP SPACE lunar analogue

of 3.5 × 3.5m containing dry sand with an average grain size of 0.2 mm. Load cells and encoders were used for deployment algorithm calibration. Gantry software interface with internet connection was used for gantry remote deployment by other consortium members. Shovels were used for introducing sand features (obstacles) in order to test deployment and mobility capabilities. Adapted payload boxes were assembled in order to demonstrate end effector manipulation capabilities.

18.5.4 Testbench at LAAS (Coordinated Manipulation)

The cooperative manipulation developments have been validated with two Franka Emica Panda robotics arms, mounted on a fixed base. Sequences of pick/move/place actions are planned offline, and the resulting motions are executed in real time on two separate PCs that exchange low frequency coordination messages.

18.6 Results

18.6.1 Cooperative Mapping

The initial phases involved demonstrating the capability of the Veles and Mantis receiving goals, planning traverses independently with maps and localisation as inputs. These initial pre-demos were important in preparing for the cooperative demonstration by debugging the integration problems and functional issues related to RG, Infuse, frames, robot control, hardware failures, etc.

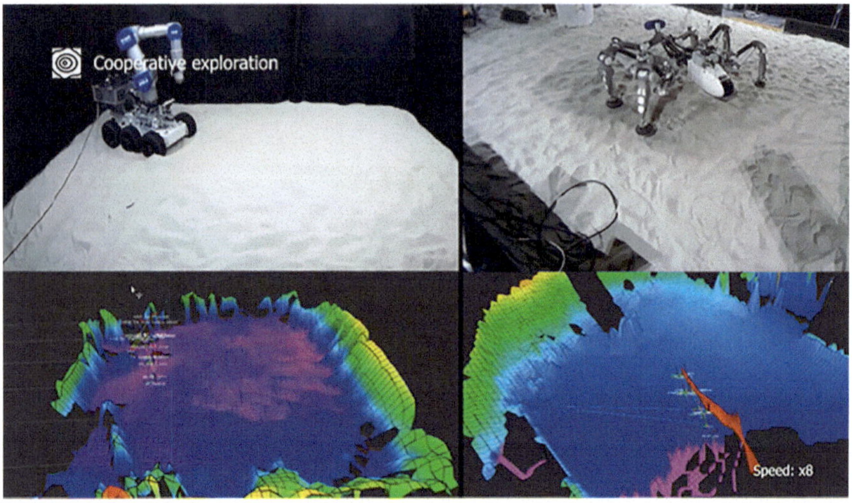

Fig. 18.28 Cooperative mapping performed remotely using Veles (left) and Mantis (right). Reconstructed maps from each robot shown below

This cooperative mapping demo involved both robots at different locations connected by a common network for the multi-agent task planned to schedule and issue goals to each other for cooperative exploration and mapping (Fig. 18.28). The agents were able to receive and plan traverses for their respective mapping goals correctly and plan traverses in the allocated area using the Rover Guidance (RG).

The local/rover DEMs and global DEMs (merged) for each robot were generated and provided representative maps of the test areas. The local DEMs together with a relatively reliable localisation estimation was sufficient for the robots to navigate along their reference trajectories to generate global maps of their respective environments.

A cooperative goal is sent to both robots to execute a cooperative mapping, moving around in each of the test environments autonomously. Waypoints are sent to both robots and executed at the same time. RG uses feedback from InFuse (Pose Estimation and DEM) and RobotClientAPI to plan and then execute the motion through the waypoints.

The reconstructed DEMs occasionally had issues on the Mantis robot, showing some noise in the map requiring a reset in some instances to allow the continuation of the test. With the DEMs, path and motion planning by RG encountered issues in some instances while moving the robot autonomously, requiring multiple restarts of the demo and for some sequences of the test, a flat DEM was provided to allow RG to plan the path and continue with the tests. The agents were able to receive, plan and execute their respective goals correctly. The generation of DEMs was stable and provided representative maps of the test areas which can be used both as the orbital or robot's local maps.

18.6.2 Grasping and Tool Exchange

The tool exchange and grasping scenario was performed with Veles in the PIAPs' Space tests area in Warsaw. The goal of the activity was to present the ability of the robot to use multiple, dedicated end effectors on a single robotic arm as well as effectively prepare a flat working area in a lunar environment.

For the tool exchange task, initial tests consisted of commanding the robot arm to define pre-docking poses of the HOTDOCK interface of the gripper and shovel. Following this, the Robotic Arm Motion Planner (RAMP) was used to command the arm to the tool interfaces pre-docking poses and perform a straight line trajectory for the docking phase. The HOTDOCK interface was commanded to lock/unlock via the Veles MCS automatically using a state machine which matched the goal when within range of the alignment sensors. The gripper was dropped into its holster (Fig. 18.30) and later connected to the shovel (Fig. 18.29) using the RAMP for commanding the arm and marker poses from InFuse as a predefined input.

The grading operation was validated first via teleoperation to adjust the height of the grader blade and check its interaction with the sandy surface. The next step was to perform the grading test using RG with waypoints, perception and localisation which was verified successfully. Localisation of the robot and the mapping functioned correctly and consistently with the surroundings of the robot.

A DEM was constructed before the grading operation to identify areas to be graded by the operator (Fig. 18.31). The waypoints were issued to the RG for the actual autonomous grading motion and the final DEM was used to visualize and validate the graded area (Fig. 18.32). In one case, a marker was used as the waypoint for RG to send the Veles on a straight line motion to the marker while grading the surface.

Fig. 18.29 Veles arm connected to shovel after swapping the gripper via HOTDOCK

Fig. 18.30 Tool exchange procedure using HOTDOCKs

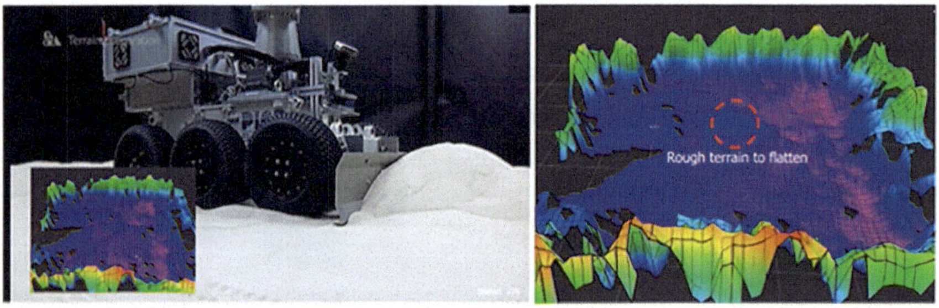

Fig. 18.31 Veles before flattening rough area in DEM

Fig. 18.32 Veles and its DEM after the grading task

18.6.3 Cooperative Unloading and Assembling Modules

Coordinated manipulation tests with the two fixed Panda robotics arms manipulating a rigid bar have shown the benefit of the admittance controller, which is able to cope with both internal and external perturbations, i.e. small errors in arm trajectory tracking and significant forces manually applied to the manipulated object.

These tests foresaw that the arms controller is able to cope with slight deviations of the trajectory of mobile bases when executing a cooperative transportation task.

This scenario demonstrated cooperative manipulation using two Panda arms with their base fixed at LAAS-CNRS' premises. A cooperative trajectory to move a bar from its initial position to a defined final position was planned and executed.

The planner generated a trajectory without any collision between the robots, with the object and the environment. The planned trajectory was properly executed with the closed kinematic chain. The sequence of pick, move and drop was confirmed without dropping or damaging the robotic arms (Fig. 18.33).

18.6.4 Cooperative Manipulation (Virtual) and Transport

This scenario was initially planned to be performed with both the Mantis and the Veles grasping an object together via handles and moving from one location to another. Due to force majeure COVID-19 conditions, the robots were now deployed within their local analogs and with separate objects to demonstrate the manipulation and transport actions, cooperating logically by synchronizing their tasks. Hence the cooperative transport demonstration with Mantis and Veles, each at a different location and connected through a VPN. The Mantis robot was located at DFKI's premises in the Lunar Hall while Veles was located in PIAP-SPACE's premises test area.

This virtual cooperative demo focused on the synchronous planning and motion control for executing each robot's task, to detect the marker on an object, approach the object and grasp it virtually (grasping was demonstrated independently) and finally move it to another location. During this demo, the Mantis and Veles were commanded by the multi-agent ERGO planner to move towards a target object using the detected object pose from markers.

A cooperative goal is sent to both robots to execute a cooperative object manipulation goal. This goal is received by both robots, which move autonomously and at the same time. RG uses feedback from InFuse (Pose Estimation and DEM) and RobotClientAPI to plan and then execute the motion through the waypoints.

Once both robots are at the grasping pose, they request the pre-grasping pose from InFuse (obtained using ArUco markers). In Figs. 18.34, 18.35 and 18.36 both robots perform virtual graspings (opening and closing the grippers). When both agents have finished the grasping, the virtual cooperative transportation starts, executing

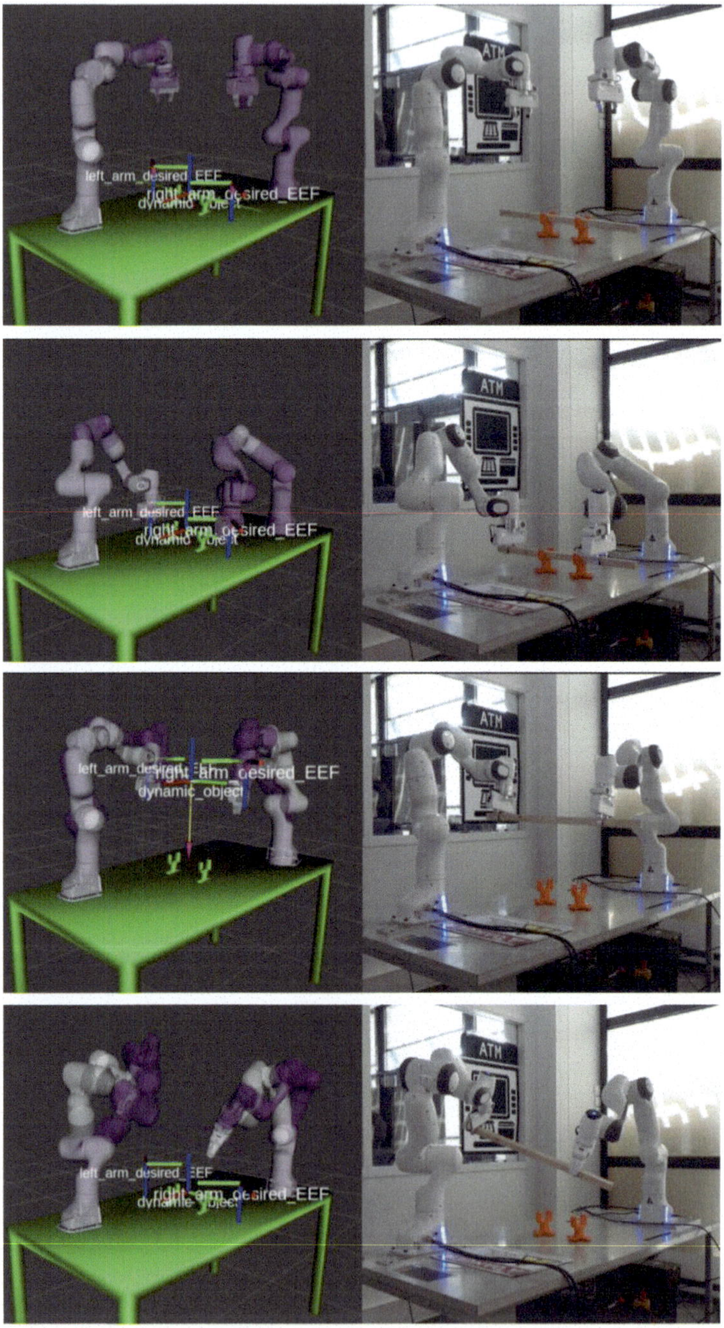

Fig. 18.33 Two Panda robotics arms manipulating a wooden bar

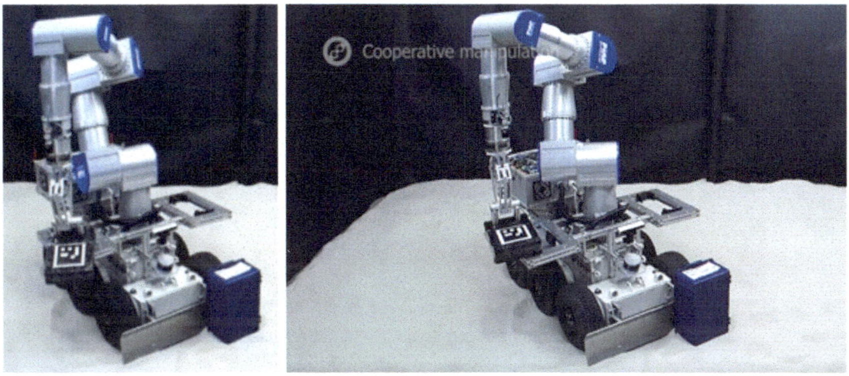

Fig. 18.34 Grasping tests with veles

Fig. 18.35 Grasping approach to ISRU mockup

a traverse to the target object pose (Fig. 18.37). Once the robots reach the target location, they perform the object drop motion by opening and closing the grippers.

18.6.5 Gantry Deployment and Control

The initial scenario of the gantry deployment over uneven terrain was planned using the Veles assisting the deployment by pulling one of the legs with a constant force and velocity, while the Mantis would be anchoring one of the legs. The CM2C would coordinate this by commanding the Gantry's motion for this deployment together with the RWAs. Due to the updated scope of the demo, validation of this setup was performed remotely to check the commanding of the deployment and reception of the telemetry from the client interface over a virtual network.

Fig. 18.36 Grasping of Mockup

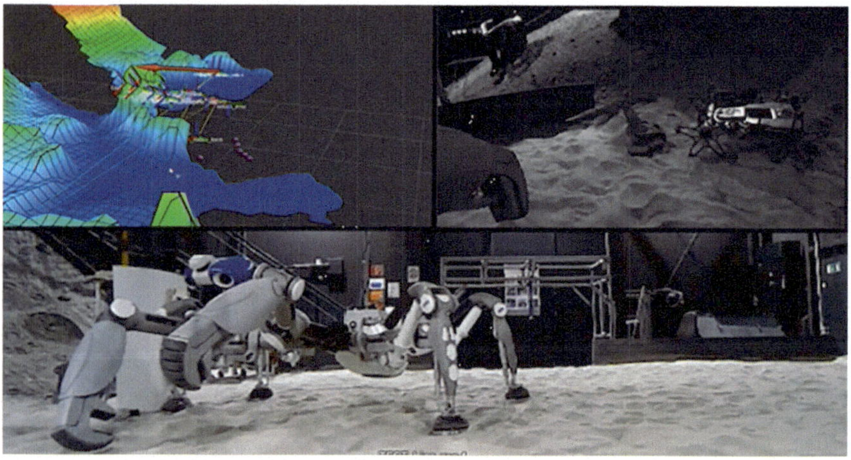

Fig. 18.37 Mantis' perception and localization during the transportation scenario

The gantry was successfully deployed and reached the deployed-secured state with the desired velocity, and then retracted back to the retracted-secured state on the flat terrain (Fig. 18.38). The sequence of photos below shows the transition from retracted to secured, as seen from SPACEAPPS viewpoint during the remote demonstration, while the Gantry was actually operating on AVS premises.

The cable driven end effector was commanded to desired coordinates just above the object to pick up the mock-up (via a set of magnets) and to then move it to

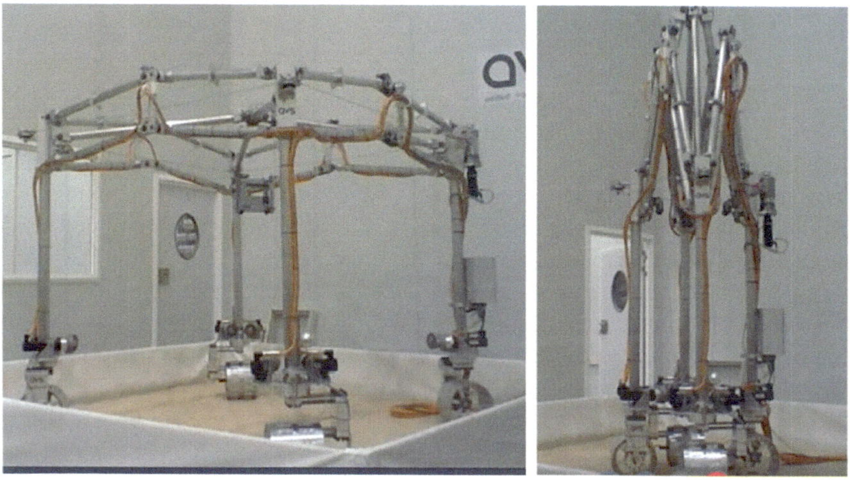

Fig. 18.38 Gantry deployment sequence. Deployed (left) and stowed (right) configurations

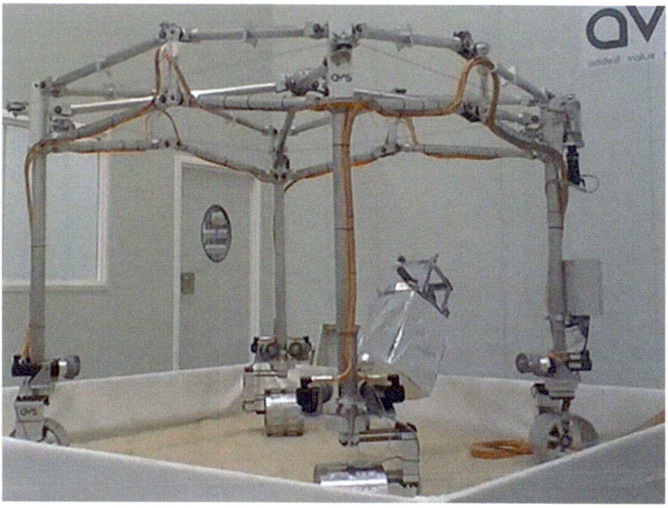

Fig. 18.39 Gantry's end effector manipulating a Mockup

other coordinates, demonstrating also the capability of changing the end effector's orientation. Figure 18.39 shows.

18.7 Conclusions

Remote demonstration of the cooperative mapping, grading, end effector tool exchange and gantry deployment were achieved with significant success. The remote virtual and physical cooperative manipulation scenarios were very challenging due to the remote nature of understanding issues and solving them with minimum information. The virtual manipulation demo was achieved with certain simplified steps due to limited time and feasibility due to the remote nature of the demo. The physical cooperative manipulation demo experienced further challenges due to a change of manipulators to industrial manipulators at the last moment in the project which required extensive re-work of the cooperative manipulation planner and controller. This was also demonstrated successfully towards the end of the project.

Overall, the project managed to achieve the objectives for each of the scenarios, even if they were logically cooperating, instead of physically collaborating on the same location or environment. This project demonstrates that multi-robot cooperation and collaboration is feasible to develop, integrate and deploy in a future lunar exploration and construction context. The challenges with multiple robots working together are significant and require novel approaches as demonstrated in PRO-ACT to make the team of heterogeneous robots cooperative seamlessly.

Acknowledgements PRO-ACT is funded under the European Commission Horizon 2020 Space Strategic Research Clusters—Operational Grants number 821903.

References

1. Govindaraj, S., Gancet, J., Urbina, D., Brinkmann, W., Aouf, N., Lacroix, S., Wolski, M., Colmenero, F., Walshe, M., Ortega, C. & Balazs, B. (2019). PRO-ACT: Planetary Robots Deployed for Assembly and Construction of Future Lunar ISRU and Supporting Infrastructures. In *Proceedings of the 15th Symposium on Advanced Space Technologies in Robotics and Automation (ASTRA-2019)*.
2. EU Project PERASPERA: Plan European Roadmap and Activities for Space Exploitation of Robotics and Autonomy, https://www.h2020-peraspera.eu. (Accessed 21 September 21)
3. Bartsch, S., Manz, M., Kampmann, P., Dettmann, A., Hanff, H., Langosz, M., von Szadkowski, K., Hilljegerdes, J., Simnofske, M., Kloss, P. & Meder, M. (2016). Development and Control of the Multi-legged Robot Mantis. In *Proceedings of ISR 2016: 47st International Symposium on Robotics. VDE, 2016*. pp. 1–8
4. Brinkmann, W., Dettmann, A., Danter, L., Schulz, C., Stark, T., & Brandt, A. (2020). Enhancement of the six-legged robot mantis for assembly and construction tasks in lunar mission scenarios within a multi-robot team. In *Proceedings of International Symposium on Artificial Intelligence, Robotics and Automation in Space 2020 (i-SAIRAS 2020)*.
5. Code: github.com/dfki-ric/robot_remote_control
6. ZeroMQ, https://zeromq.org/. (Accessed 27 September 21)
7. Danter, L., Planthaber, S., Dettmann, A., Brinkmann, W., & Kirchner, F. (2020). Lightweight and Framework-Independent Communication Library to Support Cross-Platform Robotic Applications and High-Latency Connections. In *Proceedings of International Symposium on Artificial Intelligence, Robotics and Automation in Space 2020 (i-SAIRAS 2020)*.

8. An IBIS robot on watch for safety of Polish airports, https://piap.pl/en/2018/01/26/an-ibis-robot-on-watch-for-safety-of-polish-airports/. (Accessed 21 September 21)
9. Letier, P., Siedel, T., Deremetz, M., Pavlovskis, E., Lietaer, B., Nottensteiner, K., Roa Garzon, M. A., Sánchez Garcia, J., Corella, J. L., & Gancet, J. (2020). HOTDOCK: Design and validation of a new generation of standard robotic interface for on-orbit servicing. In *71st International Astronautical Congress, IAC 2020.*
10. Nguyen, D. Q., Gouttefarde, M., Company, O., & Pierrot, F. (2013). On the simplifications of cable model in static analysis of large-dimension cable-driven parallel robots. In *IEEE/RSJ International Conference on Intelligent Robots and Systems,* pp. 928–934.
11. Jean-Baptiste, I., Dubor, A., Herve, P.-E., Cabay, E., Culla, D., Rodríguez, M., & Barrado, M. (2017). Large-scale 3D printing with cable-driven parallel robots. *Construction Robotics, 1,* 69–76.
12. Ishigami, G. (2008) Terramechanics-based analysis and control for lunar/planetary exploration robots. *Doctoral Thesis, Tohoku University.*
13. European Space Robotics Control and Operating System (ESROCOS), https://www.h2020-esrocos.eu/ (Accessed 21 September 21)
14. European Robotic Goal-Oriented Autonomous Controller (ERGO), https://www.h2020-ergo.eu/ (Accessed 21 September 21)
15. Infuse, Data Fusion for Space Robotics, https://www.h2020-infuse.eu/, (Accessed 21.09.21)
16. I3DS, https://github.com/I3DS. (Accessed 21.09.21)
17. OG3-InFuse, https://gitlab.com/h2020src/og3. (Accessed 21 September 21)

Chapter 19
The AgriRover: A Space Robotics Inspired Mechatronic Platform Technology for Precision Farming

Xiu Tian Yan, Mark A. Post, Alessandro Bianco, Roberto Palazzetti, Ying Lu, Youhua Li, Aron Kisdi, and Wayne Tubby

Abstract This paper reports on the findings of AgriRover project, which is the first of its kind in exploiting and applying space robotic technologies in precision farming. A novel mobile platform inspired by space robotics is presented and its unique features for farming applications are highlighted. To measure energy performance of mobile platform, a new dynamic total cost of transport is proposed and validated. An autonomous navigation system based on a rover control architecture has been developed to enable the AgriRover to traverse in unstructured farming environments. A novel and agriculture- specific object recognition algorithm has been investigated and implemented to enable higher degree of intelligence. A novel soil sample collecting mechanism has been designed and prototyped for on-board and in-situ soil quality measurement. A three-level task planning module has been developed, providing a user friendly interface to simultaneously define high level soil sampling tasks and to perform sophisticated autonomous and dynamic task planning based on the field terrain information. The design of the whole system has benefited from the use of a mechatronic design process known as the Tiv model. The AgriRover system has gone through three field trials in the UK and their results are reported.

19.1 Introduction

Agriculture is a vital industry for 7.3 billion citizens in the world in 2015 and it is projected the world population will rise to 8.5 billion by 2030. With continuing increased demands for more food due to increased population and better food from

X. T. Yan (✉) · M. A. Post · A. Bianco · R. Palazzetti · Y. Lu · Y. Li
RAS Group, DMEM, University of Strathclyde, Glasgow, Scotland
e-mail: x.yan@strath.ac.uk

A. Kisdi
GMV UK, Madrid, Spain

W. Tubby
RAL Space, Didcot, UK

X. T. Yan and G. Visentin (eds.), *Space Robotics*,
https://doi.org/10.1007/978-3-031-39214-6_19

a growing middle-class population, the world agriculture has entered a new era characterized by the following trends: clearer needs for sustainable farming, closer monitoring for better crops production, and urgent needs for precision farming in terms of precision delivery of fertilisers for minimal environmental impacts. These requirements can only be met and enabled by new technology development such as earth observation, precision farming equipment and agricultural robotics.

Intensive farming through often over-fertilisation can cause potential environmental issues and it starts to show potential challenges for sustainable farming and environment impacts. Specifically, the following problem and challenges exist, putting pressure on environment and crop growth:

A. Soil quality measurement challenge

Soil quality measurement is a key challenge to precision farming in order to be able to apply the right amount of fertiliser for a targeted farming land with the right type of fertiliser at the right time and at the right intervals. Due to the vast amount of fields, the current practice in countries such as the UK is either to measure the nutrients, such as nitrogen, potassium and phosphate at much less frequent intervals e.g. once every two years, or at much lower spatial resolution conducted by human operators, typically approximately 100 m resolution.

Figure 19.1 shows the average fertilizer usage by farmers in the UK and World from 2006 to 2013 and the data are from the World Bank. These analyses show that farmers in the UK and around the world uses significantly varying amount of fertilisers. It is clear that there is a need to optimize the usage of fertilisers.

Satellite data is already being used for soil quality monitoring with model based calibration of data products. In-situ point measurements are still the most commonly used technique taken by people in agricultural lands, though some ground vehicles are also used on an ad-hoc basis. Satellite data are mostly still used for research investigation. The feedback from end users is that both the spatial and temporal resolutions in satellite based monitoring is still not adequately sufficient for precision farming. Specifically, the data acquisition rate cannot meet agricultural production demands. Manual measurement precision is not reliable and objective and the coverage of this approach is typically limited for real use.

To make remote sensing techniques generally available and more useful for soil quality measurement, it is essential to have frequent, reliable, larger areas covering calibration data to support the satellite based modelling and predictions. The current

Fig. 19.1 Fertilizer usage in kilograms per hectare of arable land by farmers in World and UK

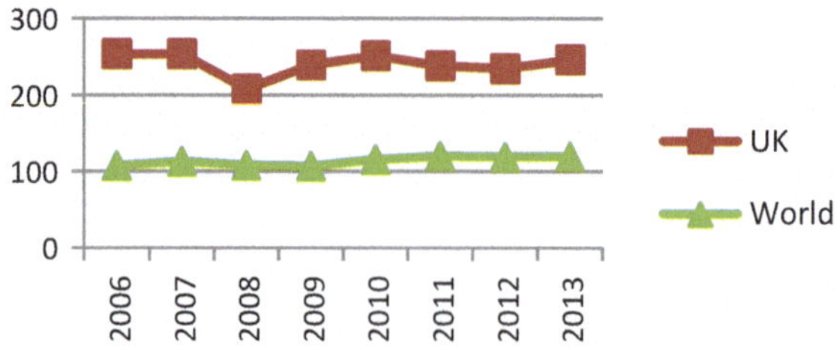

challenges for earth observation modelling are related to a lack of this calibration data.

To address these challenges, it is believed that space based robotics such as ExoMars Rover technologies could offer a novel solution to revolutionise the soil monitoring practice. These systems could potentially be used for the automation of the ground soil quality measurement and subsequent calibration data acquisition so that high resolution data are captured to support earth observation technology. This complimentary approach to generate real ground data and model derived data to cover large data areas is believed in this research to achieve the required spatial and temporal resolution for precision farming soli quality measurement.

Robotic technologies provide a novel means to address these challenges. In the following we introduce the state of the art in robot technologies upon which our platform is built.

B. Rover and legged mobile platform technologies

Legged robots have been popular research solutions and have been developed as a mobile platform over last decade. These have been demonstrated by MIT's Cheetah robot [1] and ASIMO and Boston Dynamics BigDog. Researchers have developed a criterion to measure the energy efficiency of these legged locomotion method. It is suggested that the ratio of the total power consumption to the product of the weight of the legged robot and velocity that the robot can achieve can be used to measure the total cost of transport (TCoT) [2]. Analyse show that the challenge for these mobile platforms are that they have significantly higher TCoT compared with the humans or animals [3].

For farming applications, in order to cover vast areas of many fields, the energy efficiency is a critical design parameter to be optimised. Based on this design criterion, it is sensible to find an alternative design solution of higher energy efficiency or lower TCoT to address the farming soil sensing requirements.

C. Current soil monitoring technologies

The most basic tasks that can be automated on a farm involve ground coverage such as ploughing & planting, monitoring, and harvesting [4]. Ploughing & planting involves heavy equipment and many farmers have indicated a preference to do it personally because of the need for careful planting of a field. Many robots have been developed over the last 50 years specifically for harvesting tasks, but harvesting success remains in the range of 66% on average with fruit damage occurring on average in 5% of cases and an average cycle time of 33 s [5]. The most successful robots have been designed to enable monitoring and treatment of crops for precision agriculture applications. Additionally, many of these robots never left the research and development stage. Successful robots will need to perform successful fusion of global and local references to navigate real fields autonomously as well as being cost-effective enough for farm use [6]. This technical need can now be filled using available mapping and navigation technologies available for planetary rovers.

Aerial visual monitoring of farms is already performed using unmanned multi-rotor or fixed-wing aircraft at long range, but there are very few autonomous systems in the literature for close-range monitoring. Compared to intervention or harvesting systems there is a notable lack of autonomous ground-level monitoring robots, which are essential for lowering the costs of precision agriculture. Modular robotic arms and other manipulators can be used for monitoring, intervention, or harvesting in small greenhouses [7, 8], but fields require mobile robots with significant range and on-site charging capability. Several studies of autonomy methods for agricultural robots have been performed such as simulation of multi- agent harvesting [9] and the use of neural networks and genetic algorithms [10], but very few have focused on the design itself of an agricultural rover for monitoring.

One of the first agricultural mobile robots was AU- RORA, but its sensor system restricted it to the interior of greenhouses [11]. A small skid-steered inspection robot that integrated a monocular camera and GPS receiver for simple automated inspection duties across an entire field is described in [12]. The use of monocular vision simplifies the image processing pipeline, but does not consider making detailed maps of tall crops or navigation of overgrown areas of farm. Another mobile robot with monocular vision for weeding duties was developed on a similar platform [13]. Also, precise vehicular control requires wheel rotation steering and path tracking control such as that in [14].

D. Object Recognition

Object recognition has been investigated for several decades and still remains a main challenge in computer vision. It has been addressed by various techniques ranging from image matching of characteristic points [15], to bags of words of features descriptors [16], from 2d image analysis to 3d point cloud processing. Among these techniques, pixel labelling allows simultaneous detection, classification and localization of objects within a 2D image and, as such, it is the ideal way to completely parse an outdoor scene and allow a mobile robot to understand its surroundings. When the pixel labels are coupled with depth information, the recognized objects can also be located on a 2D or 3D map. Solutions to the pixel labelling problem are based on machine learning algorithms and are still under active investigation. Earlier promising techniques [17, 18] use conditional random fields to represent the statistical distribution of texture, spatial layout and context saliency of pixels. The original work reported in [19–22] have been expanded to incorporate additional pixel features, and source code for developed frameworks [23] have been published. More recent pixel labelling techniques [24] are based on convolutional neural networks and show promising results. Scene understanding in mobile robotics has focused on obstacle avoidance, path tracking, or object detection of particular object instances. A complete classification of a robot surroundings has been only recently addressed [25, 26]. Since robots uses a stream of camera images, relevant work to robot autonomous navigation has been done in the context of image parsing of video sequences [27].

Classification of agricultural features such as crop and plants have also been a key challenge toward automatic control of cultures, and machine learning techniques based on recognition of colours and shape have been developed in [28] and [24].

19.2 Mechatronic/Robotic System Design Methodology

The theoretical and novel aspect of this system AgriRover will be reported and derived from this methodology and its use. This will show novelty of the AgriRover project and its findings and illustrate its applicability.

A. Tiv-Model

The Tiv-Model detailed in [29] is a complex systems engineering (CSE) design methodology that possesses multiple traits that make it highly beneficial for use in the complex systems design. The methodology was designed with a focus on space robotics development and their terrestrial applications. The development focused on taking an existing model platform and adjusting certain characteristics. It is intended to provide a smooth process for a complex system designed to be realised at manufacturing, assembly, logistics and maintenance phases of a system life-cycle. These are achieved through multiple manufacturing model, assembly mode as terms MA models, Logistic model and maintenance model.

1. Categorised Sequential Task Process –

By categorising tasks into stages as a sequential process, much like traditional design processes, the process can be simplified into a step-by-step program of tasks. At a glance one would think that this hinders the application of concurrent engineering, however by using knowledge databases, the general stage tasks can be partitioned into discipline specific tasks and goals. This will allow either a traditional approach to the design process, or the more modern concurrent approach depending on the preference of the organization or team. Allowing for both of these approaches ensures the general flexibility of the model.

2. Goal Oriented Process –

Traditional design methodologies will sometimes incorporate specific methods as part of the design process; having methods that are well proven to work for that specific application can be beneficial, but ultimately means that the overall flexibility of the methodology is compromised. Additionally, by focusing on the task rather than the short-term goal, new designers may not understand the purpose of doing such a task, leading to a possible chance of failure. The Tiv Model instead states which deliverables are required at that stage, while providing possible, but not definite, methods to accomplish the task.

This allows organisations to adopt the model without changing their pre-implemented methods or tools. This also eliminates the risk of the design team performing a task simply for the sake of performing a task, which can occur if leaders do not specify the "Why?" behind it. By focusing on what is needed of them as an end result, designers can understand the process, focus on the output and are still free to use whichever method preferable to obtain that output. Further details of the Tiv model description can be found in [29].

B. AgriRover System architecture

Using Tiv model as the design methodology, the research team proposed a system architecture which consists of four key blocks as shown in Fig. 19.2: namely sensing and sensing information processing, AgriRover three level task planning, user interaction with the system and finally mobile platform supported by two additional task specific actuation modules. Soil sample collection mechanism module is designed to help to obtain right amount of soil with correct pre-test conditions for a key functionality of the AgriRover soil sensing. Haptic robotic arm system module is designed to enable a farmer to remotely control the arm so that the system can help farmers to manipulate an end-effector in a harsh environment, such as humid and hot greenhouses.

Another consideration of the design is to ensure the final design looks cool and this is intended to attract younger generations to pursue the farming industry as an attractive and interesting profession. This is the outcome of conceptual model and mechatronic system model applying qualitative and evaluative design approaches and the final conceptual model resembles the prototype shown in Fig. 19.3.

Moving from the conceptual design stage of the Tiv model to the detailed design stage through multi-perspective modelling and simulation, a prototype AgriRover system has been constructed and its major components are shown in Fig. 19.3.

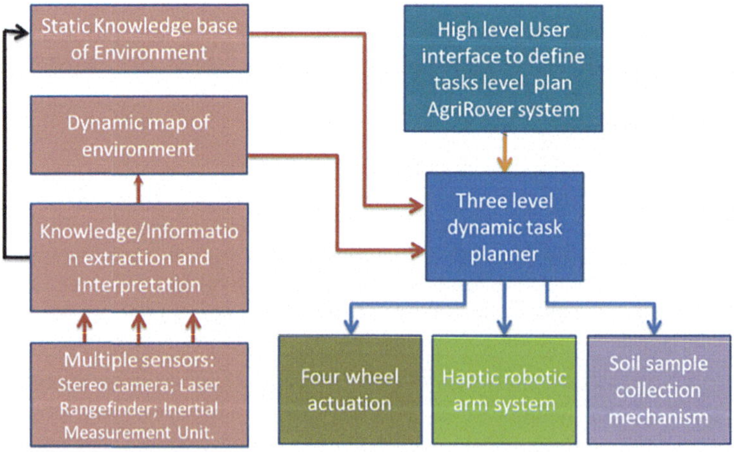

Fig. 19.2 AgriRover system Architecture

Fig. 19.3 Final AgriRover Platform system prototype and main control and sensing design feature components of the AgriRover

19.3 The AgriRover Mobile Platform Technology Description

The AgriRover platform technology is inspired by space robotic research in planetary exploration, represented by the ExoMars rover. It also addresses the specific needs of soil quality monitoring. The platform technologies consists of a four wheel mobility structure, rover control architecture, soil sample collection mechanism and finally a user interface for high level rover control and monitoring.

A. Four wheel mobility design

Commercial off-the-shelf robots were assessed but it was necessary to design a new system to comply with the following key drivers:

- Off the Shelf Parts/Low Cost: The platform needs to be easy to fix or to replace parts. This usually requires off the shelf or easily available parts.
- Easy to Assemble or Maintain: Multiple platforms are required so it should be easy to build multiple of them easily and relatively quickly
- Low Weight: The platform should be easy to ship to various fields as needed. It should also be easy for a person to lift and position once in the field.
- Small footprint: This is to allow the robot to fit in tramlines and between the rows of crop (designed for Maze or Wheat) without damaging the crops.
- Long battery life: The robot is required to run for a full day throughout the field trials.

- Payload: The platform is required to carry a variety of instruments. It should also be able to interface with them where needed.
- Rugged Design: For outdoors all-weather agricultural use.
- Platform Design

The platform has been designed to cope with driving off-road. It has four wheel drive and four wheel steering with enough rotational range to allow it to drive sideways, giving it higher mobility within potentially cramped spaces. It uses four in-wheel motors to reduce the weight and the size, while also allowing it to carry relatively heavy payloads. The Platform is roughly 400 mm long, 400 mm wide and 254 mm tall and weighs only ~ 15.5 kg when fully loaded with hardware and batteries.

Assembly only uses commonly used hand tools. The platform has been designed so that it is possible to access all sections easily without needing to be fully dismantled. It is also possible to swap in sections such as the wheels or steering blocks with minimal disassembly.

Building on the same concept of measuring the energy efficiency of a mobile platform, regardless it being a legged one or a rover based one, it is proposed that in this research a dynamic energy efficiency measurement should be derived in order to precisely measure the dynamic nature of the energy efficiency of a mobile platform. This measurement will allow one to evaluate the instant and peak performance of the mobile platform in a typical soil sensing operations, where it is necessary to stop the rover, restart the rover to transverse in a sloped framing field. For these types of applications, it is believed that static energy efficiency measurement proposed is not sufficient.

The new proposed measurement therefore has a new dimension of time included in the expression as show below:

$$\text{Total cost of transport (TCoT)} := \text{P(t)}/(\text{W(t)} * \text{V(t)} \qquad (19.1)$$

This is the ratio of the power consumption to the product of weight and velocity in real time enhancing the static measurement introduced in [2]. It is important for agricultural applications to have this measure as it gives a more precise and continuous measurement of cost of moving the AgriRover in the filed for soil sampling. This can be used as a critical criterion to guide the motion planning discussed in path planning in section IV. Using this expression, a dynamic energy efficiency plot can be obtained and shown in Fig. 19.4. For two patters of operation, it is clear from these two graphs in Fig. 19.4 that the AgriRover starts with an increasing with a typical higher TCoT value of greater than 10. As it travels further, TCoT decreases to a value less than 10.

a. Chassis and Suspension

The chassis is split into two identical pieces with an unactuated pivot in between them. This allows the chassis to twist in the centre keeping all four wheels on the floor at all times for stability and traction.

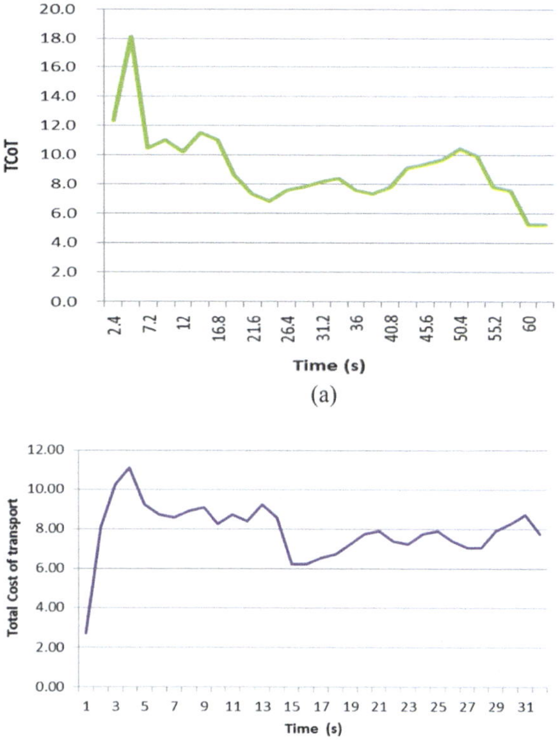

(a)

(b)

Fig. 19.4 Dynamic energy efficiency plot examples **a** at 30% of the maximum speed set running for 18.3 m, 63 s and with an average speed of 0.3 m/s; **b** at 50% of the maximum speed set running for 20 m, 43 s and with an average speed of 0.46 m/s

Each half has a large hollow space for either payload or batteries. This can be used for potentially sensitive devices requiring protection from dirt or water, such as a small computer, and is used to house the Jetson TK1 mainboard and other control electronics.

b. In-Wheel Motor

This design has been used to minimise weight and space while still keeping high torque. Each wheel has an inbuilt motor controller and an encoder making it easy to control and simple to swap if needed.

As there is no drive train to the wheel it is possible to steer the wheel through the wide angle required without loss of efficiency.

c. Battery/Control Block

The battery packs and control board are built into one unit that controls the platform and can be easily removed. The block has multiple IO allowing the on-board controls to be connected to the cameras or devices on top of the platform.

B. Rover control architecture

The AgriRover uses Ubuntu Linux 14.04 and the Robot Operating System (ROS) "Indigo" release as system software on the Jetson TK1 motherboard. The information flow in the AgriRover is shown in Fig. 19.5 and follows the conventions used in ROS to maximize re-use of available functionality within ROS. A local navigational costmap that provides local spatial probabilities of object presence is created from three sources: a Stereolabs ZED camera, a Hokuyo UTM-30LX-EW scanning laser rangefinder, and a pair of MaxBotix WR1 ultrasonic rangefinders both front and back. The ZED is used as the primary sensing method for obtaining three-dimensional maps of the environment from depth images, and also to recognize three-dimensional objects. The laser is used to add high-speed and high-resolution obstacle detection, while the ultrasonic rangefinders provide an additional layer of collision detection in the event that obstacles are not fully recognized by other sensors. All three sensors are used to provide a local navigational costmap by fusing their occupancy estimates by superimposing them in the map frame of reference.

A persistent global costmap is built from the local costmap as the AgriRover travels and kept up to date by storing it in a map server ROS node. The positions of recognized objects are also stored for display to the farmer. Global inertial sensing is provided by a Sparkfun 9 degree of freedom (DOF) Razor board using the razor imu 9 DOF of ROS node for orientation measurement, and a U-Blox EVK-6 T is used with RTKLib to provide Real-Time Kinematic (RTK) GNSS localization within the global map, which is created to be impassably bounded at the extents of the field.

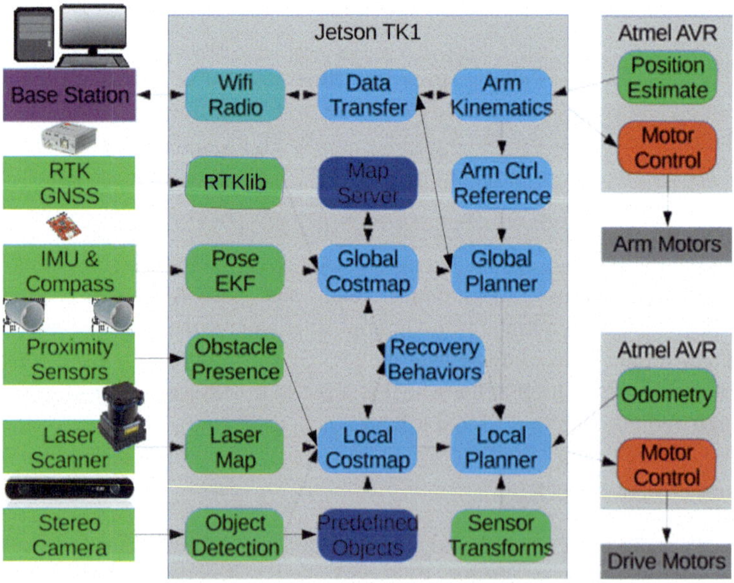

Fig. 19.5 Diagram of AgriRover Systems (implemented in ROS)

A move base ROS node is then used to perform local planning to produce a ROS twist message that is sent via SPI bus to an Atmel AVR microcontroller on an Arduino Mega prototyping board for distribution to the motor controllers via serial link. The twist message of velocity v and rotational rate $\theta^{\cdot}$ is converted to a forward velocity v and wheel angle φ for each of the four wheels by rearranging the classic Ackermann steering equation for four-wheel steered vehicles [30] with wheelbase length l as

$$\varphi = \arctan(\theta l/(2v)) \tag{19.2}$$

If forward velocity is zero, wheels are simply set to their maximum angle of 45° outwards, and velocity for all wheels set to $r\theta^{\cdot}$ with r as the radial distance to the wheel from the body centre and sign as to produce clock- wise or counter-clockwise rotation. Control messages for the arm kinematics are also sent as position vectors to a separate AVR board that computes arm kinematics for the inspection arm and distributes motor messages. The distribution of tasks across separate devices increases reliability and reduces the processor load and real-time requirements for the Jetson TK1 motherboard, and the use of SPI increases the speed and reliability of serial communications, and is also derived from space hardware technology.

C. Soil sample collection mechanism

The rover is equipped with a laser induced breakdown spectrometer (LIBS) system for soil sensing and detection of nitrogen. The design and operation of the LIBS system in analysing a soil sample will be the subject of another paper to be published later. This paper focuses on the mechatronic aspect of the AgriRover system, i.e. how the same will be collected and prepared ready for LIBS analysis.

It is therefore necessary to extract a soil sample from the ground. Depth of soil sampling is the first issue to be addressed; few works on the topic of soil sampling for nitrogen measurement [31], Hinckley et al. [32] suggested a depth between 30 and 100 cm. Due to the limited payload weight on the rover, it is important to keep the driller's weight to its minimum, and deep penetration will lead to longer and heavier sampling mechanisms. It has been therefore decided to set a depth of 50 cm based on the consultation with soil experts. Renderings of the overall design are shown in Fig. 19.6a.

The system is made of 4 tube steel columns, 600 mm long, a vertically sliding body guided by them, and a plate on the bottom and one on the top that keep the tubes together and give torsional stiffness. The plate is moved up and down by a threaded bar powered with a 12 V DC gear-motor that guarantees the needed force for penetration. The plate itself mounts another 12 V DC gear-motor, faster than the first one, that spins the drill-head (Fig. 19.6b) and guarantees the sampling of the soil.

In order to reduce the total height of the driller and minimize the risk of tipping, a 12 V and 300 mm stroke linear actuator is deployed to make the whole drilling system move up and down (see Fig. 19.6a). In this way, the driller has just 500 mm stroke,

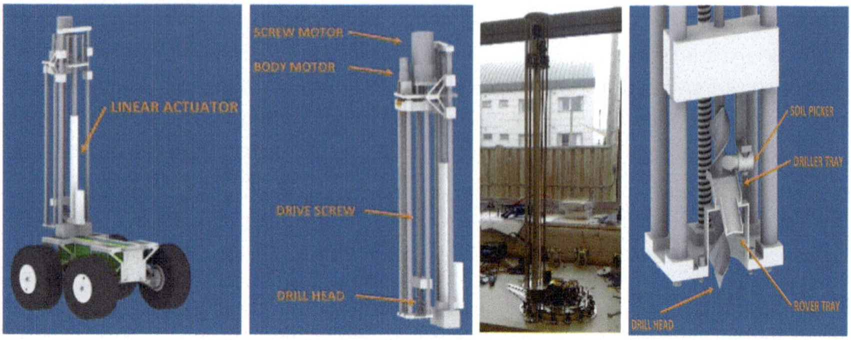

(a) (b) (c) (d)

Fig. 19.6 Driller design: **a** Rendered views of the driller, **b** Detailed views of the driller, **c** Prototype of the driller consisting of a linear actuator to lift and lower the sample collector

and starts to work from the ground level. The above design has been prototyped and the final driller can be seen in Fig. 19.6c.

Once the drill-head has descended deep enough in the soil, it is brought back on the initial position. On the way, a soil picker will remove the soil from the cochlea (see Fig. 19.6d), causing it to fall over the trays and from there be deposited into the LIBS soil sample holder.

19.4 Autonomous Navigation and Task Planning

A key capability of the designed AgriRover system is its capability to navigate intelligently and safely in a farming environment, where there exist typical agricultural objects such as cows, agricultural equipment e.g. tractors and so forth, fence posts and fence wires. The AgriRover realizes this capability and navigates through an environment chosen for the project by simultaneous localization and mapping using RTAB-Map for depth maps and Hector SLAM for planar obstacle maps. Initially, the map is built as it is traversed when the AgriRover is directed to visit an array of given locations for soil sampling and monitoring in a whole field within pre-defined boundaries. A user interface has been designed and implemented to allow a user to select a field for soil sampling. This high level planner then generates a soil sampling plan for the AgriRover to implement and execute.

The map is also stored to improve global planning and also to aid in visual monitoring of a field by the farmer. The AgriRover is localized within the global map using a GNSS receiver, but local sensing is used for odometry and mapping to ensure precise movement between crop rows.

A. Autonomous Localization and Mapping

The primary mapping method for the AgriRover is implemented using a ZED stereo camera providing depth maps and full colour images to a ROS rtabmap node. RTAB-Map (Real-Time Appearance-Based Mapping) makes use of an RGB-D SLAM algorithm for visual odometry with a global loop closure detector that compares visual similarity of visited scenes [33]. This is supplemented by the Hokuyo scanning laser rangefinder that provides high-rate obstacle detection and planar odometry using a ROS hector slam node [21]. Both nodes project a local costmap that are combined with simple obstacle detection front and back from the ultrasonic rangefinders, and stored in a ROS map server node to form a global navigation costmap.

While a navigational global costmap of a field can be stored efficiently in ROS using only two bits for each grid cell, three-dimensional point cloud maps of fields produced by RTAB-Map are too large for the limited space in on-board random-access memory, and are not kept persistent in memory but are swapped out to the local solid-state hard disk in the form of a sub-map so that when the AgriRover leaves a given map grid region of defined size, currently set to 10 m square. The ability of RTAB-Map to identify and localize separate map sections is utilised by simply loading in new map areas adjacent to the current estimated global position, though for this to work the entire visual map created must be stored in its original form to allow loop closure to operate, which is space intensive on the order of 100 GB for a moderate-sized field.

B. Agricultural Objects Recognition

Our module is a novel attempt to apply pixel labelling to the recognition of objects in a farmland environment. The purpose of the module is twofold: (i) to enhance the autonomous navigation capabilities by recognizing agricultural landmarks such as fences, trees and buildings in order to aid its autonomous localization, (ii) to provide an informative map to an end-user by detecting and localizing patches of cultures and moving animals in order to provide an informative map to the user. In our work we made use of the Darwin framework: it showed promising results on a set of pictures taken during a field trials and its source code is publicly available with clear documentation. Figure 19.7 shows the class architecture of our module. The entry point is a ROS node that takes as input the RGB and the depth images provided by the Zed Camera, and gives as output a list of classified objects described by their class labels and the list of boundary points on a 2D map.

The classifier outputs an image representation of the decomposition of the input scenes into classified regions. The core pixel labelling work is executed by the Darwin framework. A custom segmentation algorithm decomposes the scene into relevant regions by decomposing RGB and depth information. Finally the regions are classified according to a majority vote of the contained pixels. The cloud locator has the responsibility of projecting the image on a 2D map. At first a point cloud is reconstructed from the depth information, then all objects whose distance from the movement plane is contained within the robots height are represented as classified obstacles on the map.

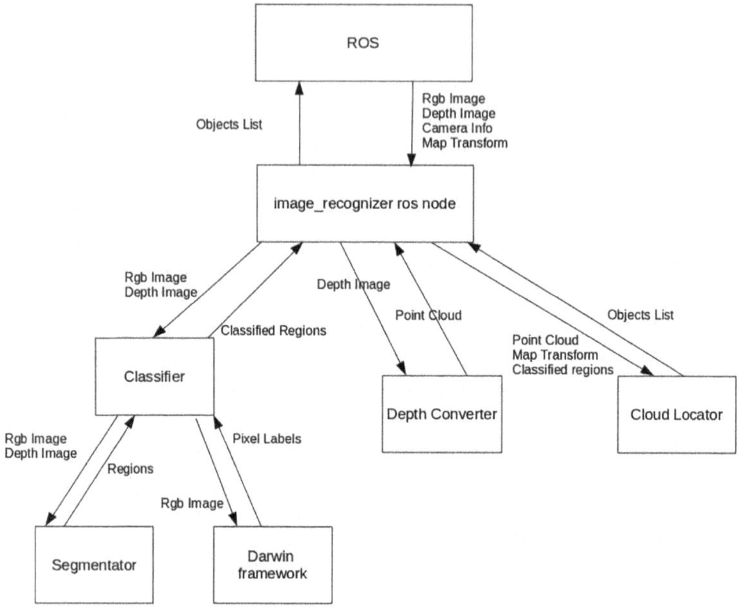

Fig. 19.7 Image recognition class diagram

The first main component class is the Classifier, its purpose is to take RGB and depth information and determine what object type each image pixel represents. The main output is a new RGB image composed by a set of standard colours each one of which represents a given object type. It works in three steps. During the first step the Darwin module is used to perform pixel labelling. The Darwin module is based on a supervised boosted classifier which has been pre-trained with a set of manually labelled images taken from our field tests. At execution time, the whole RGB image is provided to the Darwin classifier in order to obtain a pixel labelling. During the second step the original image is decomposed in meaningful regions according to the depth information and colour distribution. In the third step, a class is attributed to each region according to majority rule among the object types detected by Darwin within the regions. The second and third step has been added under the rationale that Darwin may attribute a label to pixel in case of high uncertainty, hence the uncertainty is reduced if pixel within the same region are first grouped together.

The depth converter takes the depth information and converts it into a point cloud. By adding the information provided by the classifier we obtain a labelled cloud and each object can be located in space.

The cloud locator is to localise the point cloud into the coordinate system of the reference map and project it onto a 2D plan in order to get a 2D representations of maps and object. It works in three steps. During the first step, a transform is applied to the point cloud in order to locate it within the reference system of the local map under construction. During the second step, the point cloud is projected on the horizontal plane in order to generate a 2D object map. However, only some selected significant objects are projected on the map. A height limit also defines the maximum and minimum height for the object representation. During the third step, the list of detected objects is extracted and their boundary points are computed.

Figure 19.8 shows a classification example. The top left image is the original image. The top right image is Darwin output. The bottom left image is the result of the decomposition into regions. The bottom right image is final classification. In the classification, a red colour refers to cows, the lighter blue colour refers to the sky, the darker blue colour refers to the ground, the lighter green colour refers to grass and the darker green colour refers to bushes, the black colour represents unclassified area.

C. Path planning

The main purpose of the AgriRover is to travel through a targeted field of an end-user's choice freely and safely to undertake soil sampling operations. Farming fields are typically unstructured and can be dynamically changing in terms of obstacles and its "actors" in the field. Path planning involves using the information gathered by the AgriRover's multiple sensors detailed in Fig. 19.3 and fuse them into a correct understanding of the AgriRover's environment so it can plan to complete soil sensing safely and with minimum cost of energy. This involves three levels of planning, with each of which focus on different tasks of overall path planning.

1. Task planning

At its highest level planning, AgriRover planner receives user's instruction on a targeted field, soil sampling frequency. A user can select a soil sampling field using

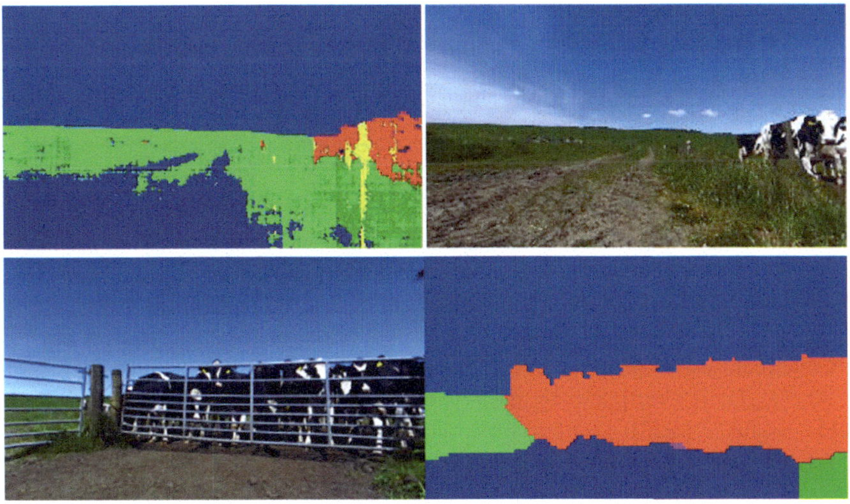

Fig. 19.8 An object recognition classification example

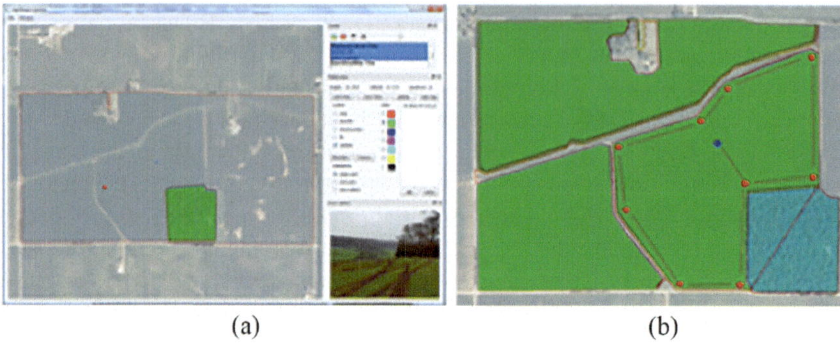

Fig. 19.9 a GUI design of the user interface; b Field Boundary Extraction

a map based interface where a user can zoom in or zoom out to find a target field. A user can simply select any point within the field shown as a red point in Fig. 19.9a.

2. Global planning

For farm operational planning, a good understanding of the field's three dimensional terrain information is critical for energy efficient planning, especially for the AgriRover to complete its soil sampling tasks with the shortest travel distance and minimum energy consumption. In addition, the total cost of transport measurement from tests is also used to guide the energy efficient planning. Preliminary results from field trial show the total cost of transport in the following table.

Set speed (%)	Distance Travelled (m)	Time (s)	Average Speed (m/s)	TCoT
15.00	37.1	58.0	0.64	4.67
30.00	18.3	62.0	0.25	9.1
50.00	20.0	43.0	0.46	7.91
100.00	18.3	22.0	0.83	6.70

It is clear that the set speed, duration, distance travelled and average speed has impact on the rover's energy consumption. These variations show that the AgriRover has lower cost at either full speed or lower speed. This TCoT knowledge is embedded into the planner to achieve maximum energy efficiency and at the same time to achieve required speed for soil sampling in a target field by selecting the most suitable speed. This body of knowledge and other field terrain information forms the basis of energy efficient planning at the global level.

At the global level planning, the two dimensional boundary information of the field is the first critical piece of information to be extracted before the global planning. This boundary information is intended to constrain the rover to the chosen target field. Altitude information is also critical for the global planning as this terrain information is needed to plot the most energy efficient path plan. In this work both types of information have been used to generate a safe, energy-efficient global plan for the AgriRover to execute. The boundary is presented by a polygon with vertices in order.

An automatic boundary extraction algorithm is implemented and it extracts the boundary of the field by using the GIS map tile or satellite map tile. The algorism is based on a seed filling algorithm to extract the field area and the boundary pixels in the field image and it works based on the assumption that pixels of the field area have the same or similar colour in a map tile. Recognized boundary pixels are then transformed into approximate straight lines of the boundary which has several intersection points at the vertices of the boundary by using Hough transformation algorithm (Fig. 19.9b).

Based on the extracted boundary polygon of a targeted field and sampling resolution information defined by a user, the global planner generates a set of ordered vertices as the sampling command target locations and the rover is expected to follow and visit these ordered target points shown as vertices on the planner map. This plan is sent to the AgriRover to travel the field as a global plan for soil sampling exploration (shown in Fig. 19.9b).

3. Local planning and testing:

Local planning is performed using the ROS move base node, which maintains both local and global costmaps derived from the sensors [25]. The local planner functionality is mainly for obstacle avoidance, and makes use of the high- resolution map of 10 cm resolution provided by the laser scanner and ZED camera. By discretely testing values in the two-dimensional control space (dx, dy, dθ) over a short period of time, the planner predictively scores a set of possible movements and determines the best control combination for the next time step in movement [26].

19.5 Field Trials

On the first day of trial, the autonomous navigation did not perform well. The robot could not locate itself within the field, and lost track of its position soon after initiating movement. The visual odometry technique was not sufficient, and the robot could not compute its own translation and rotation. The visual odometry algorithm has been subsequently re-configured for the tests on the following days. The robot could be easily manoeuvred with a portable joystick. It mechanically performed properly on rough terrain and grassy areas. It showed its ability to overcome small obstacles, such as logs and molehills. However, its power was not enough to climb uphill, especially after a long exposure to cold winds interfered with the battery output capacity.

On the following days, the navigation system parameters were changed. The visual odometry algorithm was allowed to be less demanding on the number of features require to localize the robot, and less precise. A camera supporting pole was added to improve the field of view and allow the system to gather with more significant visual features. The robot could move autonomously and reach a new position without halting. A displacement error in the final pose was detected.

The mapping system produced good maps on approximately flat surfaces. However, it could not reconstruct a correct map of the environment in non- flat wide areas. As an example, the reconstructed map would show a path leading uphill, but in reality, the path lead downhill.

The actual maps generated from the field trials by RTAB-Map are very high resolution, and accurate enough to be used for navigation when correlated with accurate global localization data. The best results were found when mapping was done in a forest grove where a lot of recognizable features were present. Figure 19.10a shows maps that were made manoeuvrings through a forest grove. Note that loop closure in mapping worked well here due to abundance of features and reliable odometry.

In open fields, RTAB-Map did not perform as well. The absence of distinct features made loop closure much less likely and the addition of better inertial estimation

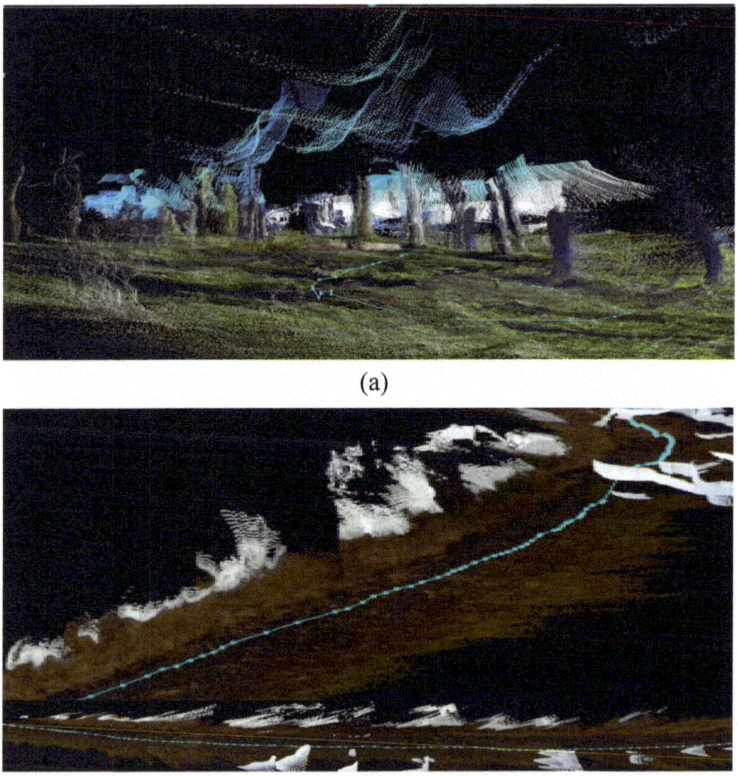

(a)

(b)

Fig. 19.10 Mapping and Loop Closure in Forest Grove (**a**) and field test navigating maps generated (**b**) using GNSS, IMU data, and EKF filtering is needed, which will be done as the next step in mapping development. Figure 19.10b shows field test maps generated while navigating the boundary of a field where adequate local features are available for good visual odometry.

19.6 Discussions and Future Work

19.6.1 Discussion

Many useful observations were gained during the field trials. This includes the following points which will be addressed in upcoming work for the AgriRover:

- The AgriRover can get lost if odometry fails, but if the drive control accurately follows the planned path then it successfully recognizes this and is able to set new goals.
- It copes well with uneven terrain, despite the camera wobbling. Orientation due to terrain could have implications for the laser scan object detection but in the field tests the gradient of tilt was never this steep.
- The field tests were particularly useful as testing in the lab was the wrong environment to get a good idea of how the system operates in its intended surroundings.
- The local planners most commonly used in ROS do not cope well with uneven terrain and four-wheel or Ackermann steering as they attempt to execute small turning and correction movements that are ultimately lost due to the long feedback loop through RTAB-Map and Hector SLAM localization and the long delays in wheel rotation to execute turning manoeuvres.
- There is difficulty with odometry when turning on the spot. Due to motor torque limitations, there is a minimum speed at which the wheels must move in order to move the rover; this affects the overall speed, particularly when turning on the spot. Visual features change rapidly when the rover rotates or travels at high speed, causing odometry to fail. Tuning of the motor controllers for better low-speed performance has largely solved this problem.
- If the images that the odometry & mapping are based on are not good - for example, if they are subject to motion blur - the resulting odometry/maps will be of poor quality. This follows the law of'garbage-in-garbage-out'. Ideally, the rover showed move in a slow and controlled manner, but of course this has a trade-off with field size and maximum allowed runtime. Again, the minimum speed required by the motors to move the rover factors into the overall rover speed.
- Visual mapping worked well when there were features like shadows/trees or nearby obstacles that can be sensed by the Hokuyo scanner. Changing or absent features (such as when the sun goes down and shadows change or disappear) can cause problems with odometry over a long period of time.

19.6.2 Future Work

During the operation of the robot platforms the project team identified some additional opportunities for future hardware development. To improve the system in future, wheel odometry, gyro and IMU information is being more thoroughly integrated to provide correction measures to the visual odometry and reduce the impact of the errors on the final maps. Many ROS nodes are not configured to use each other's sensor information dynamically.

The types of the agricultural objects covered within the project are limited and these will be extended in future to make the AgriRover system more applicable for wider crops and fields.

The rate of processed images by Rtab-map should be adaptive: more images should be processed at higher speeds of visual change, and fewer images should be processed at lower speeds. Some images may not be processed when other sensors detect bumps or other sudden and unexpected quick variations of the robot's pose. A map refinement system should be added, in order to ensure the correctness of the detected obstacles within the map and allow the navigation algorithm to compute correctly the path to the destination. A terrain detection module should be added to the object detection module in order to allow the robot to choose the best image sampling strategy, and the best odometry correction procedure. Finally, the procedure for executing manoeuvres in ROS should be modified to involve sending discrete manoeuvres of limited duration to the mobility system to allow the localization system time to catch up before executing the next manoeuvre, and these manoeuvres should be large enough to move the rover deterministically before re-localization is performed.

19.7 Conclusions

The AgriRover system represents a large, but necessary, leap forward in thinking for agricultural planning and management. For the farms of the future to keep pace with demand in a competitive economy while maintaining and being conscious of the environment, reliable and timely information is vital. Robotic systems that can use cutting edge technology, both from on Earth and for Space will be a key solution.

Acknowledgements The authors would also like to thank the James Hutton Institute and Rushyhill Farm for the use of their Hartwood Home Farm facility for field trials of the AgriRover. The AgriRover project is funded by the UK Space Agency under its International Partnerships in Space Programme and the authors would like to thank the Agency for its financial support. The authors would also like to thank the following students working on a group project to support this work: Kyle Burnett, Giacomo Corvi, Finlay Harris, Jennifer Miller and Jonathan Watson.

References

1. Seok, S., Wang, A., Chuah, M. Y. (Michael)., Hyun, D. J., Lee, J., Otten, D., Lang, J., & Kim, S. (2015). Design Principles for Energy-Efficient Legged Locomotion and Implementation on the MIT Cheetah Robot. *IEEE/ASME Transactions on Mechatronics, 20* (3).
2. Bhounsule, P., Cortell, J. & Ruina, A. (2012). Design and control of ranger: An energy-efficient, dynamic walking robot, in Proc. 15th Int. Conf. Climbing Walking Robots Support Technol. Mobile Mach., Jul. 2012, pp. 441–448.
3. Tucker, V. A. (1975). The energetic cost of moving about: Walking and running are extremely inefficient forms of locomotion. Much greater efficiency is achieved by birds, fish and bicyclists. *Amer. Scientist, 63*(4), 413–419.
4. Bloss, R. (2014). Robot innovation brings to agriculture efficiency, safety, labor savings and accuracy by plowing, milking, harvesting, crop tending/picking and monitoring. *Industrial Robot: An International Journal, 41*(6), 493–499.
5. Bac, C. W., Henten, E. J., Hemming, J., & Edan, Y. (2014). Harvesting robots for high-value crops: State-of-the-art review and challenges ahead. *Journal of Field Robotics, 31*(6), 888–911.
6. Yaghoubi, S., Akbarzadeh, N. A., Bazargani, S. S., Bazargani, S. S., Bamizan, M., & Asl, M. I. (2013). Autonomous robots for agricultural tasks and farm assignment and future trends in agro robots. *International Journal of Mechanical & Mechatronics Engineering IJMME-IJENS, 13*(03), 1–6.
7. Ulbrich, H., Baur, J., Pfaff, J. & Schuetz, C. (2015). Design and realization of a redundant modular multipurpose agricultural robot. *In Proceedings of the XVII International Symposium on Dynamic Problems of Mechanics (DINAME),*
8. Baur, J., Pfaff, J., Ulbrich, H. & Villgrattner, T. (2012). Design and development of a redundant modular multipurpose agricultural manipulator. In *2012 IEEE/ASME International Conference on Advanced Intelligent Mechatronics (AIM).* (pp. 823– 830).
9. Arguenon, V., Bergues-Lagarde, A., Rosenberger, C., Bro, P. & Smari, W. (2006). Multi-agent based prototyping of agriculture robots. *In International Symposium on Collaborative Technologies and Systems (CTS'06)* (pp. 282–288).
10. Noguchi, N., & Terao, H. (1997). Path planning of an agricultural mo- bile robot by neural network and genetic algorithm. *Computers and electronics in agriculture, 18*(2), 187–204.
11. Mandow, J., Gomez-de Gabriel, J. L., Martinez, V. F., Munoz, A., Ollero, A., & Garcia-Cerezo, A. (1996). The autonomous mobile robot aurora for greenhouse operation. *IEEE Robotics & Automation Magazine, 3*(4), 18–28.
12. Bengochea-Guevara, J. M., Conesa-Mun˜oz, J., Andu´jar, D., & A. Ribeiro. (2016). Merge fuzzy visual servoing and gps-based planning to obtain a proper navigation behavior for a small crop-inspection robot. *Sensors, 16*(3), 276.
13. Asstrand, B., & Baerveldt, A.-J. (2002). An agricultural mobile robot with vision-based perception for mechanical weed control. *Autonomous robots, 13*(1), 21–35.
14. Zhu, Z.-X., Chen, J., Yoshida, T., Torisu, R., Song, Z.-H., & Mao, E.-R. (2007). Path tracking control of autonomous agricultural mobile robots. *Journal of Zhejiang University SCIENCE A, 8*(10), 1596–1603.
15. Prince, S. (2012). *Computer vision: Models learning and inference.* Cambridge University Press.
16. Andreopoulos, A., & Tsotsos, J. K. (2013). 50 years of object recognition: Directions forward. *Computer vision and image understanding, 117*(8), 827–891.
17. Follett, R. F. (2001). Soil management concepts and carbon sequestration in cropland soils. *Soil and tillage research, 61*(1–2), 77–92.
18. He, X., Zemel, R. S., & Carreira-Perpinán, M. A. (2004). Multiscale conditional random fields for image labeling. *In Proceedings of the 2004 IEEE Computer Society Conference on Computer Vision and Pattern Recognition, 2004. CVPR 2004.* Vol. 2, pp. 695–702.
19. Shotton, J., Winn, J., Rother, C., & Criminisi, A. (2006). Textonboost: Joint appearance, shape and context modeling for multi-class object recognition and segmentation. *In Computer Vision - ECCV 2006, 9th European Conference on Computer Vision Proceedings.* Part I, pp. 1–15.

20. Shotton, J., Winn, J. M., Rother, C., & Criminisi, A. (2009). Textonboost for image understanding: Multi-class object recognition and segmentation by jointly modeling texture, layout, and context. *International Journal of Computer Vision, 81*(1), 2–23.
21. Kohlbrecher, S., Meyer, J., von Stryk, O., & Klingauf, U. (2011). A flexible and scalable slam system with full 3d motion estimation. In *Proc. IEEE International Symposium on Safety, Security and Rescue Robotics (SSRR)*. IEEE, November 2011.
22. Gould, S., & He, X. (2014). Scene understanding by labeling pixels. *Communications of the ACM, 57*(11), 68–77.
23. Gould, S. (2012). DARWIN: A framework for machine learning and computer vision research and development. *Journal of Machine Learning Research, 13*, 3533–3537.
24. Couprie, C., Farabet, C., Najman, L., & LeCun, Y. (2014). Convolutional nets and watershed cuts for real-time semantic labeling of RGBD videos. *Journal of Machine Learning Research, 15*(1), 3489–3511.
25. Reinaldo, J. O., Maia, R. S., & Souza, A. A. (2015). Adaptive navigation for mobile robots with object recognition and ontologies. *in 2015 Brazilian Conference on Intelligent Systems (BRACIS)*, pp. 210–215.
26. Mousavian, A., Košecká, J., & Lien, J. M. (2015). Semantically guided location recognition for outdoors scenes. *In 2015 IEEE international conference on robotics and automation (ICRA)*. pp. 4882–4889.
27. Liu, B., & He, X. (2015). Multiclass semantic video segmentation with object-level active inference. *IEEE Conference on Computer Vision and Pattern Recognition, CVPR, 2015*, 4286–4294.
28. Guo, W., Rage, U. K., & Ninomiya, S. (2013). Illumination invariant segmentation of vegetation for time series wheat images based on decision tree model. *Computers and Electronics in Agriculture, 96*, 58–66.
29. Melville, C., & Yan, X. T. (2016). Tiv Model - An attempt at breaching the industry adoption barrier for new complex system design methodologies. *Mechatronic Futures* (pp. 41–57). Springer.
30. Mitchell, W. C., Staniforth, A., & Scott, I. (2006). Analysis of Ackermann steering geometry (No. 2006–01–3638). *SAE Technical Paper*.
31. Miransari, M., & Mackenzie, A. (2014). Optimal n fertilization, using total and mineral n, affecting corn (zea mays l.) grain n uptake. *Journal of Plant Nutrition, 37*(2), 232–243.
32. Hinckley, E.L.S., Bonan, G.B., Bowen, G.J., Colman, B.P., Duffy, P.A., Goodale, C.L., Houlton, B.Z., Marín-Spiotta, E., Ogle, K., Ollinger, S.V. and Paul, E.A., Vitousek, P. M., Weathers, K. C. & Williams, D. G. (2016). The soil and plant biogeochemistry sampling design for The National Ecological Observatory Network. *Ecosphere*, 7(3), p.e 01234.
33. Labb, M., & Michaud, F. (2013). Appearance-based loop closure detection for online large-scale and long-term operation. *IEEE Transactions on Robotics, 29*(3), 734–745.

Chapter 20
Space Robotics Standard Interface: SIROM Evolution

Javier Vinals

Abstract Since the launch of SIROM space robotic standard user interface, there has been considerable further research and investigation undertaken to advance the standard to a more capable level. This chapter provides an overview of these new development for this important aspect of the space robotics.

20.1 Introduction

SIROM stands for Standard Interface for Robotic Manipulation. SIROM project initially designed and realized a set of integrated and inherently optimized interfaces for space robots. These interfaces include mechanical, data, electrical and thermal connectivity. These collectively enable reliable, robust and multi-functional coupling of payload to robot manipulators, payload to another payload or client to server. SIROM project demonstrated an "Intelligent" interface combining mechanical, electrical, data, thermal into a unified one. This design concept has greatly advanced the standardization of common building blocks for more connectivity among modules within a robotic system and also among satellite modules.

Back in 2018, SIROM first design was functionally tested and validated during the PERASPERA OG5 test campaign at DLR and Airbus Defence and Space facilities. There, SIROM reached a TRL 4 (Technology Readiness Level). This was the first step on developing a family product of SIROM interfaces developed against different requirements and suitable for different applications (Fig 20.1).

Since then, SIROM has been part of different projects such as EROSS, PERIOD or MIRROR, where it has been optimized and enhanced. In this chapter, we will give an overview of such projects and other internal developments lead by SENER Aeroespacial to further develop SIROM. In order to present a clear perspective of SIROM evolution, the projects and internal developments will be presented in a chronological manner, summarized as follows:

J. Vinals (✉)
Sener Aeiroespacal, Getxo, Spain
e-mail: javier.vinals@aeroespacial.sener

X. T. Yan and G. Visentin (eds.), *Space Robotics*,
https://doi.org/10.1007/978-3-031-39214-6_20

Fig. 20.1 Orbital test set up at ADS for PERASPERA OG5 project

- Electro-mechanical integration
- EROSS project
- TRL 6 qualification test campaign
- MIRROR project
- PERIOD project
- SIROM with Refuelling capabilities
- Additional developments
- SIROM family product.

20.2 Electro-Mechanical Integration

The first version of SIROM that was developed during H2020-OG5 project did not integrate the electronics in a single envelope. Also, these electronics consisted of commercial off-the-shelf (COTS) components not flight proof.

To continue with SIROM development, SENER has developed an integrated product combining the mechanics and electronics suitable for flight, resulting in a simple and compact mechanism. Figure below shows a view of the developed solution (Fig. 20.2).

As an asset, this new SIROM version also features spring-loaded pins allowing the implementation of more transmission lines for power and data transmission, in comparison with previous connectors used. Also, other modifications have been introduced based on the lessons learnt from OG5, such as decreasing the latching time (down to 15 s), smoother guiding petals geometry, removal of latch backlash, reduction of the number of moving parts, optimization of assembly procedure, etc.

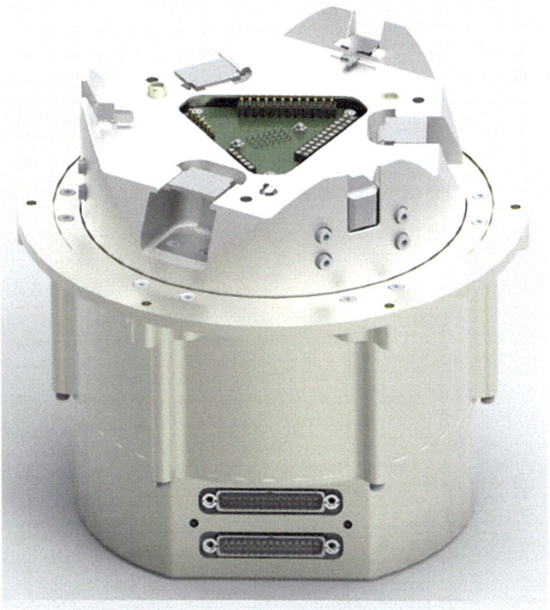

Fig. 20.2 SIROM electro-mechanics upgrade

20.2.1 Flight Electronics Design

The electronics subsystem for the SIROM follows a NewSpace philosophy. The components of the current Engineering Model design have been chosen to function on TRL 6 level, prior to the Flight Model design.

The main functions of the SIROM Electronics are:

- Actuation of the BLDC motor
- Communication over CAN bus of telemetries and telecommands
- Bidirectional power transference
- Bidirectional data transference
- DC/DC power supply and isolation
- Monitoring and protections

The electrical and data rate performances of the SIROM interface are shown in Table 20.1. For further information regarding the mechanical and electrical performances, refer to the SIROM datasheet (SENER Aeroespacial, s.f.).

The SIROM electronics is divided in two PCAs, each with a distinct functionality, following a modular approach. It is divided on a Power PCA and a Motor Driver PCA (Fig. 20.3).

The SIROM electronics is connected to the SIROM mechanism (including BLDC motor and MR sensors) and the Connectors Plate through the Driver PCA. The Driver PCA is connected to the Power PCA through a flex-cable to minimize volume. The

Table 20.1 Electrical and data performances

Electrical	Voltage input	20–34 V
	Power consumption (idle)	3 W
	Power consumption (motor driving)	5 W
	Power transfer (bidirectional)	3.5A @ 28V (98 W)
High-speed data	Protocol	Gigabit Ethernet (802.3). Others also valid
	Data rate	1 Gbit/s
Low-speed data	Protocol	CAN bus, Standard
	Data rate	1 Mbit/s

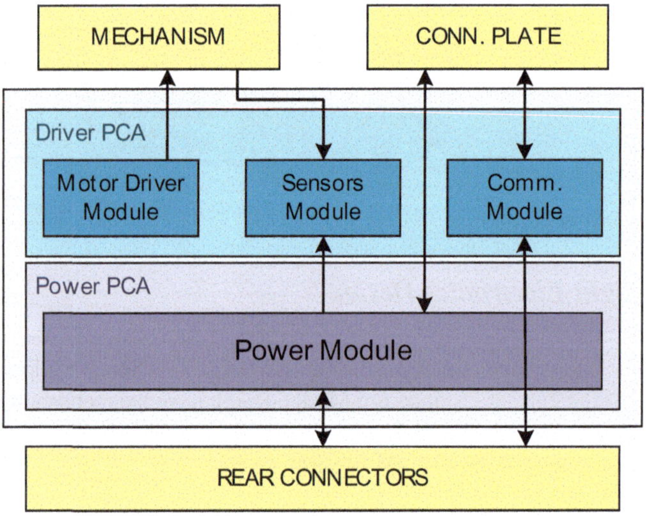

Fig. 20.3 Electronics subsystem block diagram

Rear Connectors, the only access to the power and data buses, are connected to the Power PCA.

Fig. 20.4 Power PCA (left), motor driver PCA (right). (SENER©)

The four functional blocks are distributed on both PCAs:

- *Motor Driver Module*: drives the BLDC motor for the latching operation.
- *Sensors Module*: includes acquisition for MR sensors to control the latching status, NTC sensors to monitor the temperature, motor voltage and speed monitoring circuits, and current sensing circuits for the BLDC phases.
- *Communication Module*: this module is in charge of processing TC and reporting TM on demand through the CAN bus.
- *Power Module*: handles the bidirectional power transfer between SIROMs and the power supply for all the electronic components. Includes isolation an EMI filters (Fig. 20.4).

20.3 EROSS Project

The purpose of the H2020 project entitled "European Robotic Orbital Support Services" (EROSS) is the mission and system design of a servicing spacecraft to perform rendezvous, capture and servicing of a client satellite, while validating this design with end-to-end ground experiments [1]. EROSS is led by Thales Alenia Space, with support from ten space companies and organisations, including SENER Aeroespacial.

The EROSS demonstration has four distinct mission phases: rendezvous with forced motion, robotic capture, berthing, and servicing (Figs. 20.5 and 20.6).

The servicing scenario includes docking and refuelling of the client satellite, and a robotic exchange of an ORU between the Servicer and the Client. For this mission, the ORU is a generic payload that includes a Raspberry Pi board with a small camera, allowing the control station to request a picture from the ORU through Ethernet. The ORU has two SIROM interfaces on both sides to allow latching to the client or servicer side and being manipulated by the robotic arm.

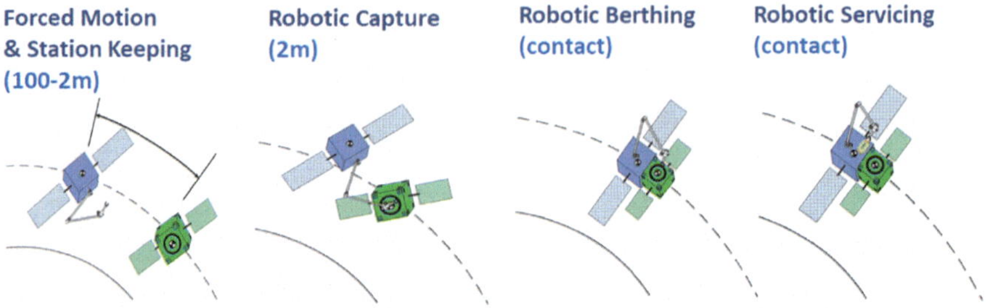

Fig. 20.5 Main steps of EROSS mission scenario

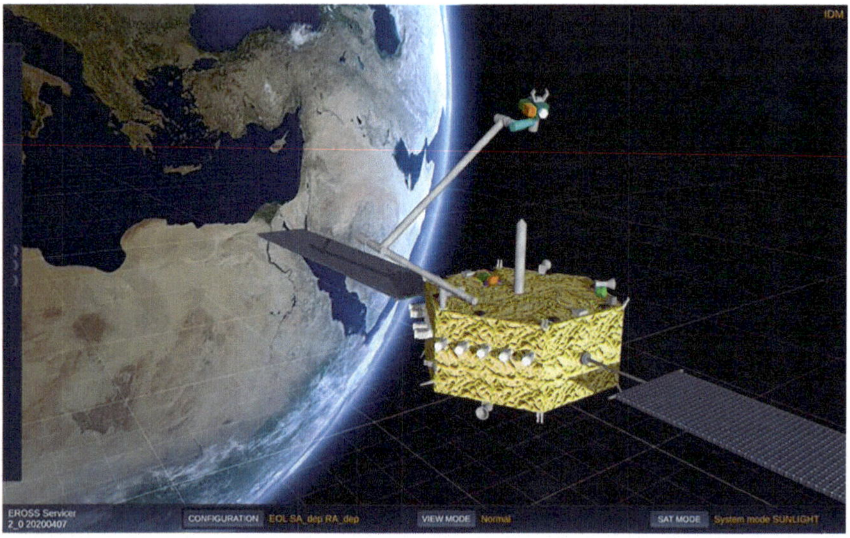

Fig. 20.6 EROSS servicer deployed configuration (Courtesy of Thales Alenia space)

For the proposed scenario, there are five SIROMs involved. One as the end-effector of the robotic arm, two on the ORU, and one on each satellite (client and servicer). The mission consists of the following steps:

- Latching the robotic arm to the ORU on the servicer side.
- Unlatching the ORU from the servicer satellite.
- Transporting the ORU to the client side's SIROM interface using the robotic arm.
- Latching the ORU to the client side.
- Unlatching the robotic arm from the ORU.

Prior to the EROSS demonstration with a full integration with the other building blocks of the project, there was an internal demonstration on an equivalent scenario to test the performance of the SIROM interface independently. A video of the demonstration is shown in [2] (SENER Group, 2021).

Fig. 20.7 SIROM internal validation in SENER (December 2020) (SENER ©)

The demonstration was carried out at the SENER facilities in Bilbao in December 2020 [3] (SENER Aeroespacial, 2021). The robotic arm used for the test is the UR10 from Universal Robots. The positioning of the robotic arm and the motion between SIROM interfaces was done through the programming console. The full set-up is shown in Fig. 20.7.

The performance observed in the internal validation was satisfactory. The SIROM interfaces were able to latch and unlatch without issues. Once latched, they were able to transfer power on the different SIROMs, to send commands over the CAN bus to control de latching operations, and to transfer Ethernet data to get pictures from the ORU.

After validating the on-orbit servicing scenario internally, the SIROMs were tested on the EROSS project along with the rest of technologies developed in past OGs from the SRC on Space Robotics Technologies.

The demonstration was carried out at the facilities of GMV in Madrid in April 2021 and at TAS-F facilities in Cannes in June 2021. It included mock-ups for the servicer and client satellite and two robotic arms to simulate the tasks of rendezvous and servicing.

The replacement of the ORU was very similar to the internal demonstration. In this case, the tasks of placing the SIROM interfaces in front of each other to allow the latching operation were carried out by visual-based GNC software developed by collaborating partners on the EROSS project, as well as the automation of commands to the SIROMs and its integration on the general control software (Figs. 20.8 and 20.9).

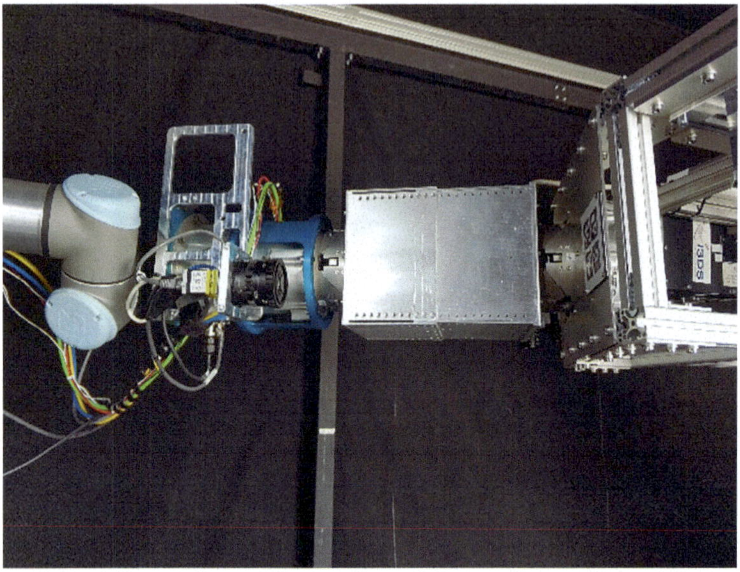

Fig. 20.8 SIROM in EROSS demonstration at GMV premises in Madrid, Spain (April 2021)

The demonstration on the EROSS scenario was successful, testing the performance of the SIROM interface in open loop. The project was concluded in June 2021 with satisfactory results (GMV, 2021).

20.4 TRL 6 Qualification Test Campaign

In 2019, the SIROM mechanism was validated for TRL 4 (Viñals, Gala, & Guerra, 2020). In that phase, the SIROM interface did not include integrated electronics and thus the Electronics subsystem based on COTS was only qualified for TRL 3. One of purposes of designing and integrating the Electronics subsystem is reaching TRL 6 level (Fig. 20.10).

20.4.1 Vibration Test

The first test that was performed on the SIROM was a vibration test, focused on the mechanism and its performance.

Vibration testing was performed under two different configurations:

- Stand-alone configuration: one SIROM is fixed at its interface to the vibration plate.

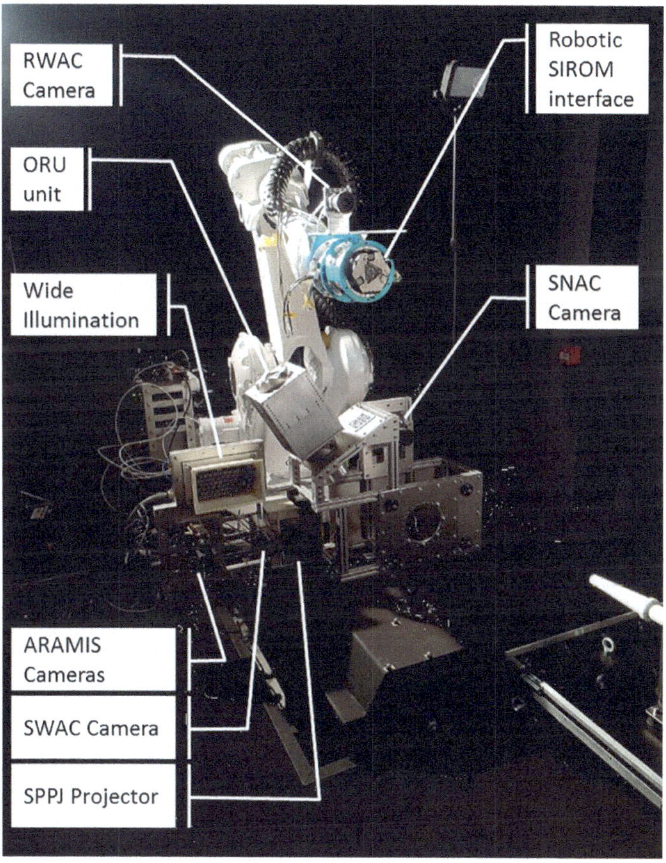

Fig. 20.9 EROSS demonstration setup at Thales Alenia Space premises in Cannes, France (June 2021) (Courtesy of Thales Alenia Space)

Fig. 20.10 Stand-alone vibration test (left). Mated pair vibration test (right)

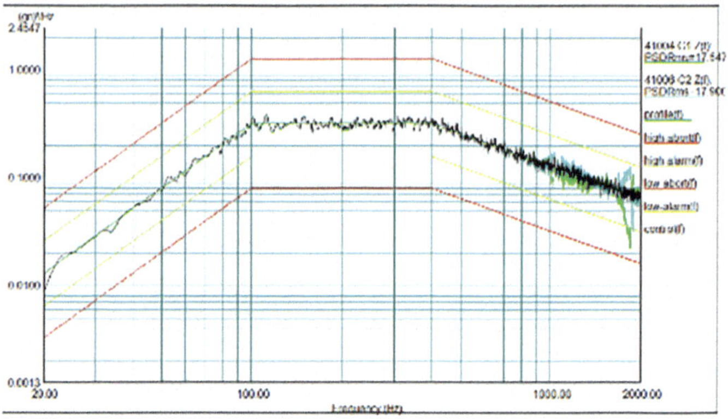

Fig. 20.11 Random vibration profile on Z axis on standalone configuration

- Mated pair configuration: one SIROM is fixed to the vibration plate while latching a passive SIROM. The aim of this test is to validate SIROM latching capability to survive launch loads when mated.

An example of one output graph from the vibration test is shown in Fig. 20.11, which corresponds to the stand-alone configuration and vibration on its Z axis (perpendicular to the vibration table). All tests in both configurations were carried out successfully.

20.4.2 Life Testing

The purpose of this test is evaluating the durability of the device. The test was carried out in a climatic chamber, at temperatures ranging from 0 to 70 °C, where the SIROM was commanded to latch and unlatch multiple times.

The electronic subsystem, while keeping a low-cost NewSpace approach, has been designed on this first iteration with components that are able to withstand up to 80ºC or more. As such, the test was successfully passed.

20.4.3 TVAC

The TVAC test was carried out on the same temperature scale that the Life Testing test: from 0 to 70 °C, with the novelty of being in a vacuum. In this test, the SIROM was commanded to latch and unlatch once in each state.

As has been previously stated, the SIROM components for the Electronics subsystem have been carefully chosen to withstand harsh conditions. Additionally, to prevent outgassing or malfunctioning of any of the components both PCAs were manufactured with a conformal coating layer.

The results of the test were satisfactory. The Electronic subsystem was able to function correctly at both points of operation (0 and 70°C).

20.4.4 Load Testing

Finally, SIROM mechanisms have been submitted to a load test. Here, an Active SIROM latches a Passive SIROM, while a load is progressively applied through the Passive SIROM interface. The loads applied were:

- Traction: 820 N
- Bending: 98 Nm
- Torque: 50 Nm

The loads that could be applied were limited by the test set-up, and not by SIROM capabilities. The tested interfaces were able to withstand the loads without degradation of SIROM performances.

20.5 MIRROR Project

The Multi-arm Installation Robot for Readying project (MIRROR) is an ESA-funded project (ESA contract No. 4000129486/19/NL/RA) focusing on the design and breadboarding of a robotic system for on-orbit operations. The MIRROR activity consists of two parallel contracts, one of which is led by GMV with support from SENER and other space companies [4].

In MIRROR project, there are two use cases: assembly of a tiled reflector and reconfiguration of a modular spacecraft. Here, SIROM is the baseline solution for the assembly of hexagonal mirror tiles to form a larger structure.

For the mission scenario, individual mirror tiles are provided with SIROMs interfaces allowing the assembly of the system. Additionally, three SIROMs are provided at the extremities of a 3-legged robot, providing the robot walking and tile handling capabilities. The scope of SENER's role within this contract includes the delivery of 11 SIROMs.

In the frame of the MIRROR project, the SIROM mechanism was upgraded to simplify the gearing stage and allow smoother motion. Additionally, the electrical interface on the Connectors Plate has been redesigned to include two additional lines for a High-Power bypass, required to charge the robot batteries through the SIROM. The distribution of the Connectors Plate has also changed to provide 120° rotational symmetry. The electronics subsystem is mainly unaltered in this iteration.

The following picture shows a functional test performed at SENER facilities in September 2021, prior to the delivery of the SIROMs to GMV [5]. In this test, it was

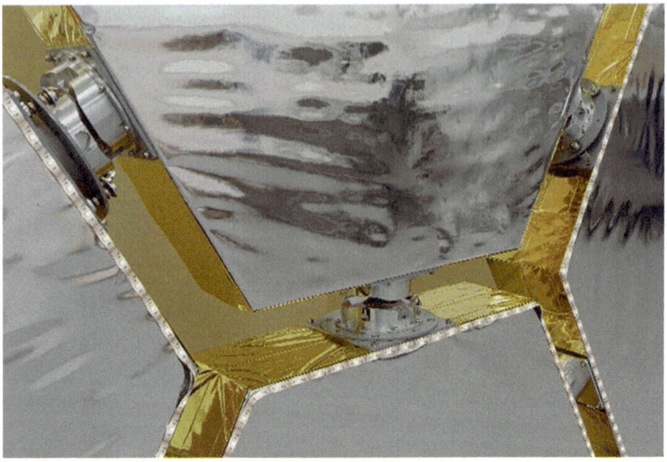

Fig. 20.12 Triple docking functional test prior to SIROM delivery (SENER ©)

tested the handling capability of a 1.2 m hexagonal tile by a UR10 robot provided with SIROM, focusing on the most challenging operation: triple docking. A video of the demonstration is shown in [2] (Fig. 20.12).

20.6 PERIOD Project

Additionally, SIROM is under current evaluation in the frame of PERIOD project [6], which is a Horizon 2020 project in phase A/B1. PERIOD is proposing as baseline solution an ambitious demonstration concept that will allow building a functional satellite in a robotised orbital "Factory", including antenna fabrication, satellite assembly, reconfiguration, and verification (Fig. 20.13).

20.7 SIROM with Refuelling Capabilities

Standard Interfaces (SI) are a central pillar for the On-Orbit Servicing, Assembly and Manufacturing (OSAM) applications. Inside the OSAM market it is expected that life extension and refuelling applications will represent a large share of the market demands, as shown by several studies [1]. Thus, SENER Aeroespacial has developed a SI product with fluid transfer capabilities [7] based on the combination and leveraging of SIROM (Standard Interface for Robotic Manipulation) (Standard Interface for RObotic Manipulation, n.d.) and RIDER (Refuelling Interface Development for ESPRIT).

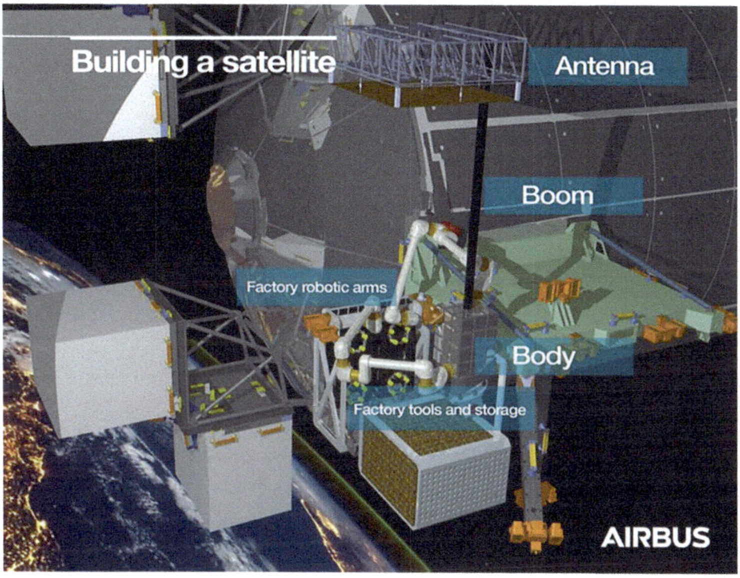

Fig. 20.13 PERIOD mission architecture

20.7.1 RIDER Fluid Transfer Connector

SIROM FTC (Fluid Transfer Connector) is based on RIDER connector. This Connector has been developed for the IBDM (International Berthing Docking Mechanism), in particular for the HCS (Hard Capture System).

RIDER has reached technological maturity of TRL 6 and is composed of the three following main subassemblies:

- Fluid Coupling, with only one movable part, Pin Sleeve
- Alignment System, for mating & docking misalignment compensation
- Traveling Mechanism, for mating / demating the fluid coupling active sleeve

Within SIROM, only the fluid coupling is maintained, while alignment correction is provided by the SIROM mechanical interface (latches and guiding petals) and the traveling mechanism is provided by a leadscrew system inside SIROM, as in RIDER solution (Fig, 20.14).

The probe tip is designed to minimize wear of the seals during the stroke of the external sleeves of the FTC. In addition, both active and passive tips feature a conical design to improve centring and alignment between the sleeves (Fig. 20.15).

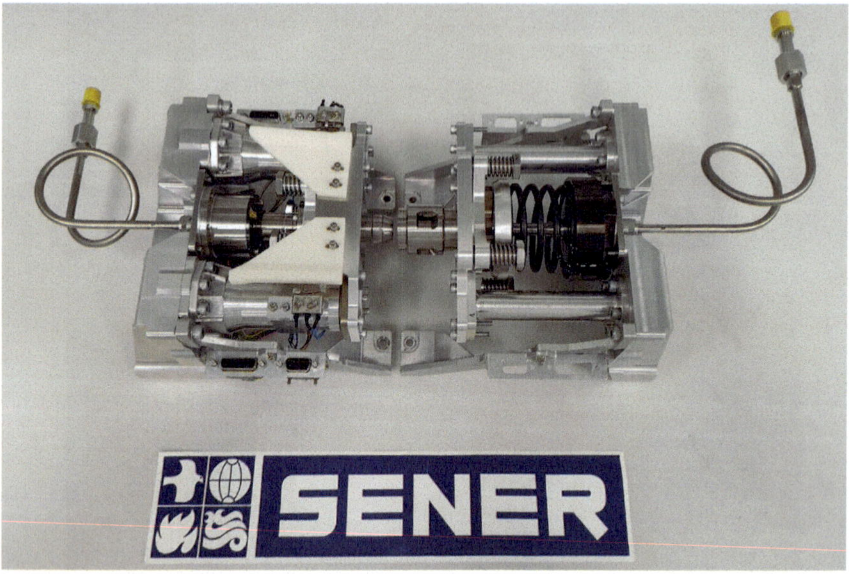

Fig. 20.14 RIDER fluid transfer connector

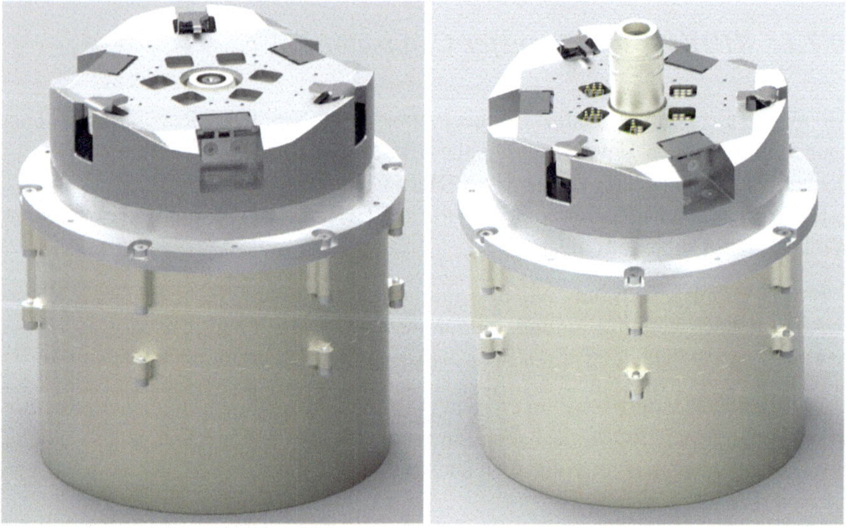

Fig. 20.15 SIROM-A for refuelling: latched position (left), connected position (right)

20.7.2 General Design

SIROM with resupply consists of an active interface (SIROM-A) mating towards a passive interface (SIROM-P). Initially, SIROM-A and SIROM-P have the data and electrical POGO connectors placed inside the housing, being protected to dust and ESD (electrostatic discharge) by an opening/closing cover. As a result, the connectors being hidden under the cover during the latching manoeuvre offers protection in case of docking collisions and misalignment correction between SIROMs.

The mating procedure starts with SIROM-A mechanically capturing, correcting misalignments and hard-docking with its passive counterpart. Refuelling procedure stars after positive latching (Fig. 20.16).

Once the latching is completed, SIROM-A deploys the external sleeve of its FTC via a set of three leadscrew-nuts, while the central needle tubing remains static. This way, it is avoided the need of using flexible hoses inside SIROM. The central needle features four radial drills at its tip allowing exit of the fluid and an output fluid fitting placed at the bottom of SIROM.

In the first phase of its axial stroke (0–20 mm), the external sleeve of SIROM-A enters SIROM-P, displacing its FTC external sleeve and making it slide inwards. Then, in the last phases of its axial stroke (20–30 mm), the external sleeve of SIROM-A pushes the connectors plate (containing the data and electrical POGO connectors) which is mounted on a series of springs that allow to return back to the undeployed position in the demating procedure. Simultaneously while deploying, the connectors plate releases the end-stop of three spring-loaded mechanisms that open the three dust/ESD covers.

On its side, during the second phase (20–30 mm stroke) the external sleeve of SIROM-P drives a passive linkage that pushes SIROM-P connectors plate (that in its turn opens the three dust covers).

Fig. 20.16 SIROM interface close to latch

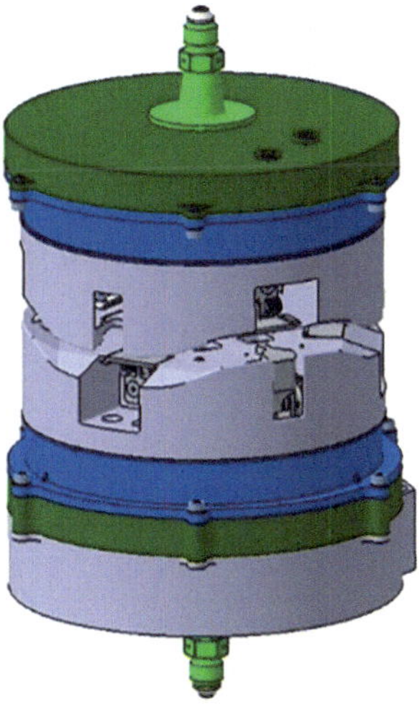

Table 20.2 SIROM FTC performances

Fluid compatibility	Xenon, Hydrazine, MON-3, MMH, gaseous He, gaseous Ni, alcohol, water, coolants, etc
Flow rate	Gas (He): 0.8 g/s @ 1 bar p Liquid (water): 35 g/s @ 1 bar p
MEOP	150 bar (gas), 30 bar (liquid)
Burst pressure	375 bar (gas), 75 bar (liquid)
Pressure cycles	> 260 cycles (tested)
Operational life	15 years GEO (with 5 mm satellite shielding)
Leakage	< 10^{-4} scc/s Helium

At the end of the full stroke, data, electrical and FTC mating is completely achieved.

During pressurization and fluid transfer, mechanical forces are transmitted directly via the fixed needles and supported by SIROM latches. FTC position is maintained via non-back driveable leadscrews allowing to maintain the insertion force in the event of loss of power or pressure.

After fluid interchange is finished, the active sleeve performs the reverse movement, returning both SIROMs to their initial position (Table 20.2).

20.7.3 Mechanical Interface

The mechanical interface is responsible for providing SIROM capturing capability against another SIROM and also compliance to loads arising during operation, once the attachment is done. The mechanical interface presents an external housing with three guiding petals and three capture hooks (or latches) evenly distributed with 120º apart. Each latch consists of a titanium four-bar-linkage moved by its own pinion and synchronized by an internal gear. The mechanical interface is complemented with a capture switch based on hall sensors and magnets that automatically triggers a telemetry to the controller once two SIROMs are within the latches range of capture. This switch is not affected by illumination conditions in the surroundings (Table 20.3).

Table 20.3 Mechanical performances

Mass resupply SIROM-A	1.5 kg
Capture range	Axial: + 15 mm Radial: ± 5 mm
Docking time	7 s
Traction load	2.4 kN
Compression load	5 kN
Radial load	5 kN
Torque	770 Nm
Bending moment	270 Nm

20.7.4 Electrical Architecture

As previous SIROM versions, the resupply family features a connectors plate consisting of spring-loaded POGO pins/pads to establish a physical plug for electrical, high-speed and low-speed data transmission. In this version, the plate implements a total of 24 pins and 24 pads distributed as per Figs. 20.17, 20.18 and 20.19.

- Figure 20.17 diagram represents the routing and internal connections between the connectors plate (POGO lines), SIROM electronics, and SIROM lateral connectors (Table 20.4).

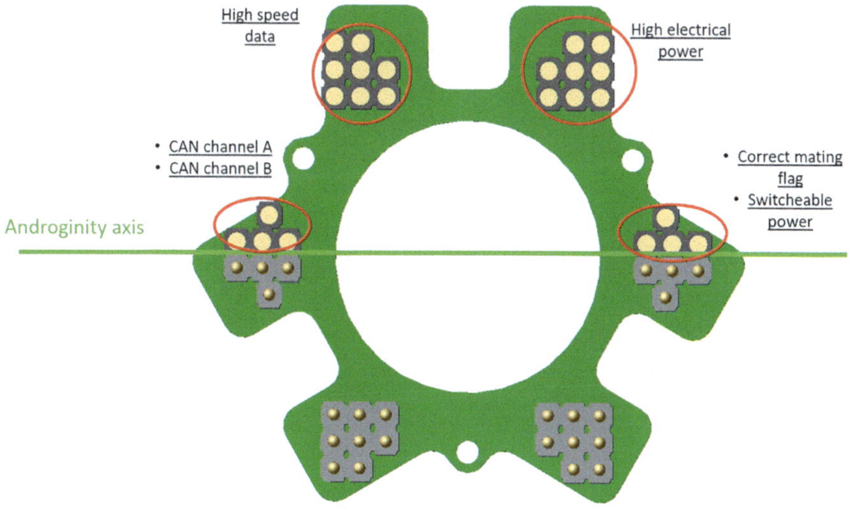

Fig. 20.17 Connectors plate distribution

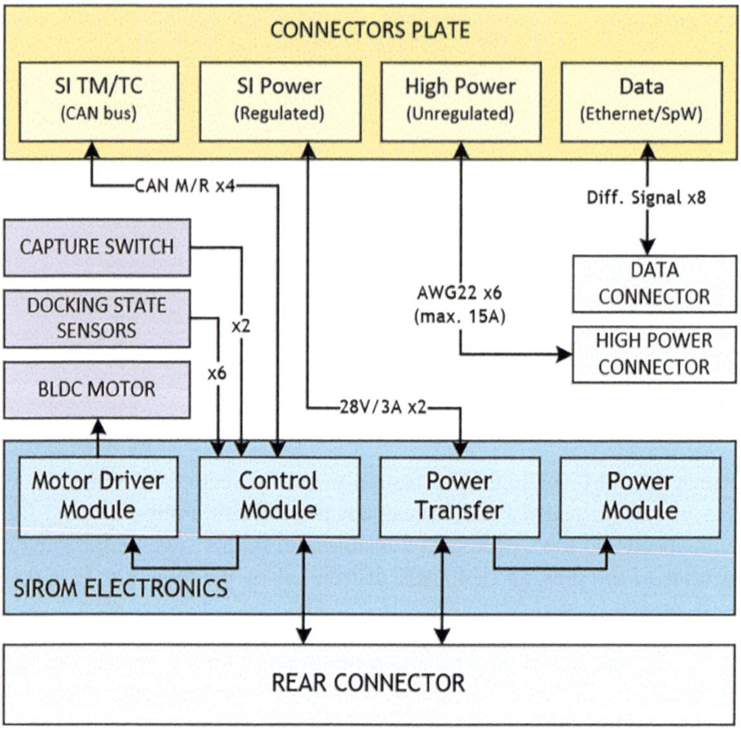

Fig. 20.18 SIROM electrical and data diagram

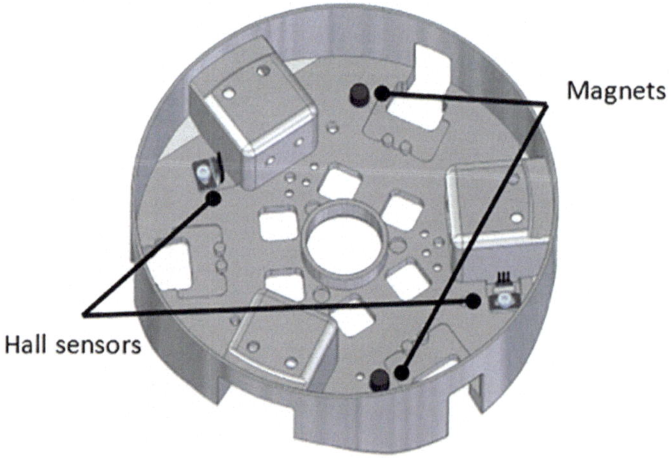

Fig. 20.19 Capture switch

Table 20.4 Electrical, high-speed and low-speed data performances

Electrical	SIROM power input	20-34 V
	Regulated/ switchable	3.5 A @28 V (98 W)
	Unregulated	16 A @150 V (2.4 kW)
High-speed data	Data transfer	Gigabit Ethernet. Other protocols also valid
	Data rate	1 Gbit/s
Low-speed data	SIROM Control	CAN bus

20.7.5 Regulated and Switchable Electrical Interface

SIROM implements lines for regulated/switchable power transfer failure tolerant (with reduced performances) thanks to the POGO connectors symmetric design. These lines are connected to the power module of SIROM electronics, which manages the power transfer between the connectors plate and SIROM lateral connectors. The transference is handled by two MOSFET that allow bi-directional power transfer or disabling power transfer.

20.7.6 Unregulated Electrical Interface

In addition, SIROM includes a set of lines that by-pass SIROM electronics, and that are routed directly to a lateral connector. These lines represent a medium for unregulated power transfer with capabilities to transfer bidirectionally 16 A at 150 V. They are also failure tolerant (with reduced performances) thanks to the POGO connectors symmetric design.

20.7.7 High-Speed Data Interface

Regarding the high-speed data transfer, SIROM includes four differential pairs for Gigabit Ethernet (1 Gbit/s data rate), failure tolerant (with reduced performances) thanks to the POGO connectors symmetric design. The impedance of the lines is adapted to minimize signal distortion.

20.7.8 Low-Speed Data Interface

The main and redundant low-speed data lines consist of two separate CAN channels (channel A & B). The signals from the connectors plate are routed to SIROM electronics control module and they are used to command and control SIROM states, although it can be used to transmit other data via SIROM pairs.

20.7.9 Development Plan

In short, the TRL6/7 development plan for SIROM with resupply will include the following phases and main tasks:

1. Mechanism validation with dummy FTC
    - (i) Mechanism assembly and calibration with COTS electronics
    - (ii) Single SIROM-A functional test
        - (1) Mechanical docking
        - (2) Dust cover opening/closing
        - (3) POGO connectors (electrical, data) stroke
        - (4) Dummy FTC stroke
    - (iii) Single SIROM-P functional test
        - (1) Passive mechanisms test
    - (iv) Paired SIROM functional test
        - (1) Mechanical mating
        - (2) POGO lines continuity test
2. Lessons learnt applied to SIROM with resupply version

(3) SIROM with resupply

- (i) Physical measurements test
- (ii) Functional tests (before/after vibration, shock, TVAC, thermal life, load test)
    - (1) Motorization margin
    - (2) Electrical/data transmission
    - (3) Fluid transfer
    - (4) Leakage test
- (iii) Vibration
- (iv) Shock
- (v) TVAC
- (vi) Thermal life
- (vii) Load test
- (viii) Disassembly
    - (1) Seals wear evaluation
    - (2) Internal mechanisms inspection

20.8 Additional Developments

20.8.1 Capture Switch

Many of today's robotic operations rely on systems providing visual feedback for situational awareness and control of robot motion. Up to now, SENER approach has been to transfer completely this responsibility to robot operators and designing SIROMs with limited AI (Artificial Intelligence) capabilities. However, a deeper understanding of robotic servicing requirements as well as client's concerns about safety in on-orbit servicing operations, has led SIROM team to study aiding features to be introduced in the design. Also, this strategy will help to differentiate SIROM from competitors in the market of the standard interfaces.

As a result, a capture switch has been implemented inside SIROM. The goal is to add a sensor that automatically triggers a telemetry to the controller once two SIROMs are within the latches range of capture. The baseline to this solution is to implement a switch based on hall sensors and magnets which have demonstrated good performances on mock-up models.

20.8.2 Distributed Electronics Design

As an option allowing further customization, SENER has developed a modular stackable electronics module that allows to control several SIROMs inside an ORU (Orbital Replacement Unit) with a single electronics unit. The baseline for the design considers a single electronic unit for all the Standard Interfaces that are attached to the ORU. In this sense, the mechanics and electronics of the SI are separated, being the latter integrated inside the ORU and the former on the laterals of the ORU. The electronics would not only drive the SI mechanics, but also serves as an interface between the SIs and the payload.

The electronic unit is aimed to be modular, cost-effective and easily adapted to other configurations and even other commercial Standard Interfaces found in the market. The general architecture of the Electronic Unit is shown in Fig. 20.20.

The architecture has been designed to maximize flexibility as one of the main goals. The Motor Driver boards and Interface board are connected through a standard PC/104 stackable connector widely used in CubeSats. The PC/104 standard also defines the dimensions of the boards stacked in the connector. All boards must be 90 × 96 mm with pre-defined mechanical interfaces.

The aim of using this standard connector is allowing the system to support from 1 to 6 SIs without major modifications to the electronic unit. The addition of more SIs would only imply stacking another Motor Driver board to the Power/IF board.

This concept is key for the modularity of the system, as each Motor Driver board is identical to the rest. With the proposed architecture, only the Interface board

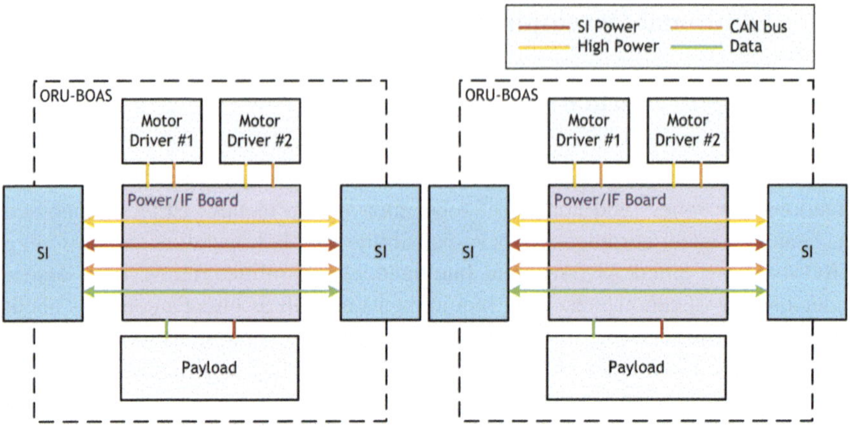

Fig. 20.20 Architecture diagram of the Electronic Unit inside each ORU

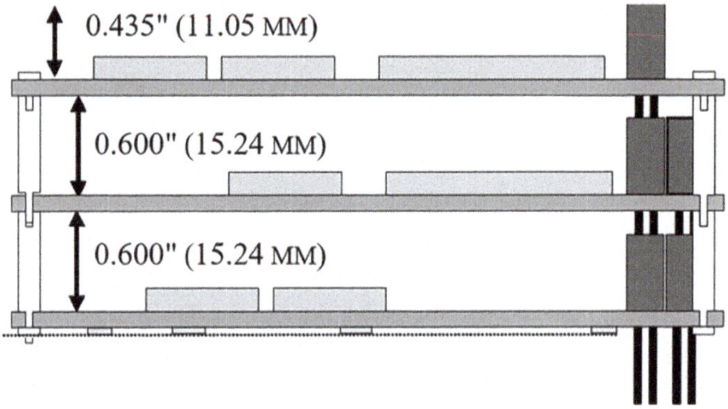

Fig. 20.21 Stackable connector from standard PC/104

would be application-dependent, although it would also be designed to allow as many functionalities as possible (Fig. 20.21).

20.8.3 *Visual-Servoing*

SENER is carrying out activities to add a camera in open-loop as a SIROM add-on that in the future will be used for visual servoing applications, monitoring the pose (relative position and attitude) of one SI relative to the target SI to be mated. The planned approach is to use a COTS electronics board to test the visual servoing algorithms. These electronics will not be included in the qualification test campaign (vibration, vacuum, etc.) but only in functional tests at SENER facilities with a UR10 robot arm (Figs. 20.22 and 20.23).

20.9 SIROM Family Product

20.9.1 SIROM General Design

SIROM is designed as an androgynous interface allowing easy mating/demating with other SIROMs. Its high capture-range latches are based on the docking system for ISS. This, combined with its guiding petals, provide SIROM a self-aligning capability tolerant to very large misalignment conditions. The docking system keeps the locked position without the need of friction brakes or power consumption. Also, SIROM features a capture switch independent of illumination conditions, that gives information once two SIROMs are within the latching capture range. Once mechanically latched, SIROM deploys its connectors board to establish a physical plug for data, electrical power transmission and fluid transmission (optionally). One of the data lines is for CAN (Controller Area Network) protocol and is managed by SIROM electronics. The other data lines are fully customizable and typical uses are high-speed data transmission such as SpaceWire or Gigabit Ethernet. These lines directly bypass SIROM electronics and thus, signal attenuation is minimal. In terms of electrical power each SIROM is provided with at least two lines:

- Regulated/switchable, bi-directional, power transfer at 28V used for SIROM power supply.
- Unregulated power transfer bypassing SIROM electronics allowing high power transfer

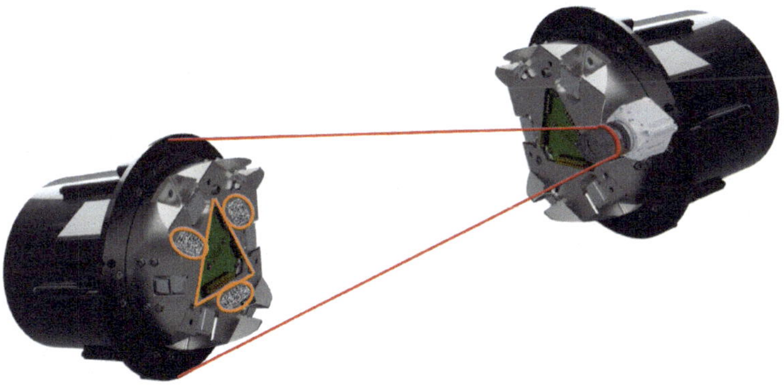

Fig. 20.22 Visual servoing system

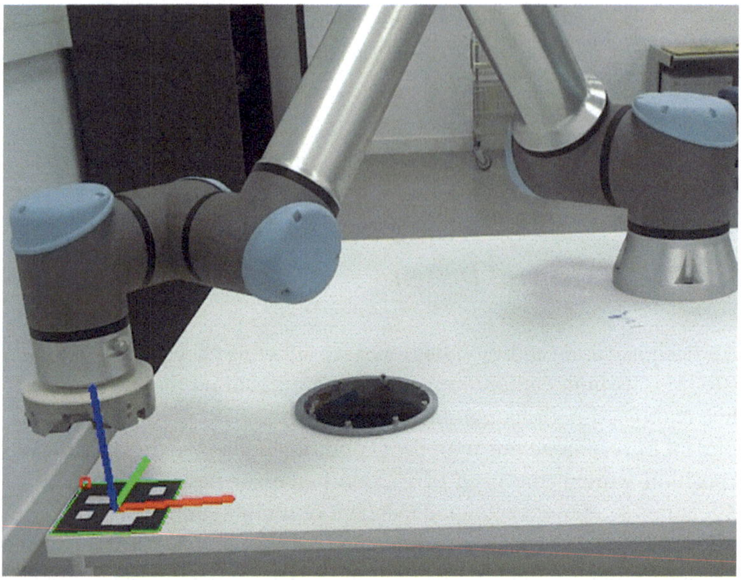

Fig. 20.23 Visual servoing preliminary test

The resupply interface is based on RIDER connector, currently at TRL 6, allowing the transfer of propellant (Xenon, Hydrazine, etc.) or coolant, among others, without the need of flexible hoses.

20.9.2 SIROM Configurations

SIROM is available in three basic configurations:

- Active–Passive (X), which can mate with any other configuration
- Active (A), which can only mate towards a P or X configuration
- Passive (P), which can be mated by an A or X configuration.

Each configuration presents different characteristics in terms of internal mechanisms and electronics and thus results in a different mass, volume, and cost (Fig. 20.24).

20.9.3 Family Product

SIROM is to be understood as a product family where each SI inside family shall be compatible (at least in a mechanical point of view). Most of the parts are common

Fig. 20.24 Active SIROM (left), Passive SIROM (right). Version not for resupply

for all the interface selections with all the standardization and industrialization advantages.

Within each family line, SIROM allows customization of:

- Resupply option
- Data protocols and number of data lines
- Electrical power transfer (number of lines, electrical performances)
- Active–Passive (X), only Active (A) or Passive (P) versions
- Integrated electronics or distributed electronics to allow the control of several SIROMs with a single and modular electronics module
- Visual-servoing system (under development) (Table 20.5; Figs. 20.25 and 20.26).

20.10 Conclusions

This chapter reports the latest development of the space robotic standard interface SIROM and the process of maturing it from initial TRL4 to 6 now. The overall design improvements have been made to address potential improvements identified in the initial SIROM design and testing. A significant improvement has been made in generating a robust and flight ready design demonstrated in this chapter. This has placed the SIROM design to its next level for an on-orbit servicing mission in future.

Table 20.5 SIROM family line

Family version	A-P-X	Functionality
OG5-H2020 (SIROM) (SIROM consortium, n.d.)	X	Mechanical capture Switchable electrical power High-speed data (SpaceWire) Low-speed data (CAN) Thermal fluid
OG7-H2020 (EROSS) [8] (EROSS Partners 2022)	X P	Mechanical Switchable electrical power High-speed data (Ethernet) Low-speed data (CAN)
MIRROR [9] (Estremera et al. 2022)	X P	Mechanical Switchable electrical power High electrical power by-pass High-speed data (Ethernet) Low-speed data (CAN)
Resupply	A P	Mechanical Switchable electrical power High electrical power by-pass High-speed data (Ethernet) Low-speed data (CAN) Resupply (optionally)
	X	Mechanical Switchable electrical power High electrical power by-pass High-speed data (Ethernet) Low-speed data (CAN)

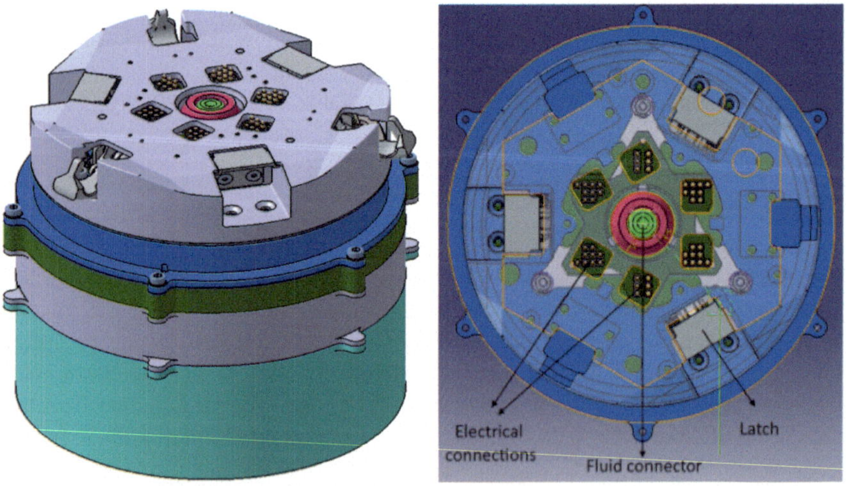

Fig. 20.25 Resupply SIROM-A

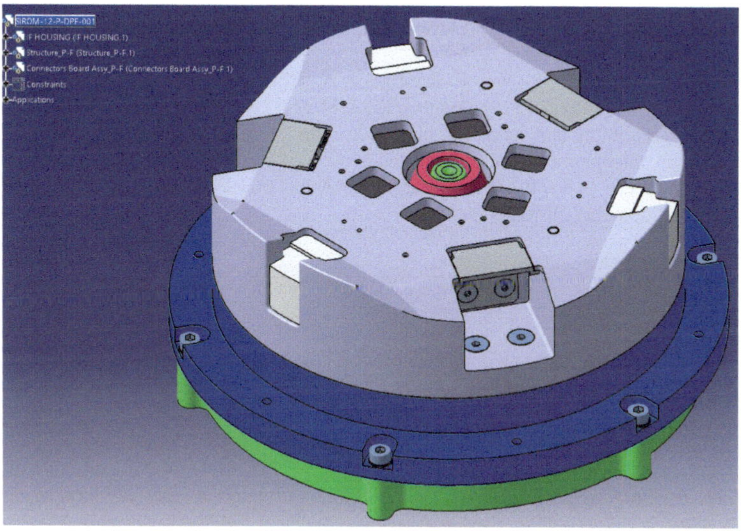

Fig. 20.26 Resupply SIROM-P

Acknowledgements The SIROM project is part of the PERASPERA project on Space Robotics Technologies and has received funding from the European Union's Horizon 2020 research and innovation programme under grant agreement number 730035.

SIROM is also part of EROSS (EROSS Partners, 2022) under grant agreement number 821904, and PERIOD [10] under grant agreement number 101004151.

References

1. Bokhari, S. H. (2021, February 17th). Northern Sky Research. Retrieved April 4th, 2022, from https://www.nsr.com/in-orbit-servicing-stepping-up-to-the-challenge/.
2. SENER Group. (2021, January 20th). Youtube. Retrieved from https://www.youtube.com/watch?v=fO-iVjy4voA
3. SENER Aeroespacial. (2021, January 21st). SENER Aeroespacial finishes the validation of the robotic interface SIROM for the project EROSS H2020. Retrieved July 2021, from https://www.aeroespacial.sener/en/press-releases/sener-aeroespacial-finishes-the-validation-of-the-robotic-interface-sirom-for-the-project-eross-h2020
4. Estremera, J., García, J., Lucia, M., Rusconi, A., Sangiovanni, G., Tsagarakis, N., Guerra, G. (13th–17th September 2021). Multi-arm robotic manipulator for on-orbit assembly and servicing. *Stardust-R - Second Global Virtual Workshop.*
5. GMV. (2021, May 27th). EROSS tests validate the on-orbit servicing technology. Retrieved July 2021, from https://www.gmv.com/en-es/communication/news/eross-tests-validate-orbit-servicing-technology
6. PERIOD Partners. (2021). PERIOD - PERASPERA In-Orbit Demonstration. Retrieved July 2021, from https://period-h2020.eu/
7. Guerra, G., Viñals, J., Soto, I., Díaz-Carrasco, M., & Gala, J. (2022). Development of a robotic Fluid Transfer interface based on RIDER connector. In *16th Symposium on Advanced Space Technologies in Robotics and Automation.* Noordwijk.

8. EROSS Partners. (2022). EROSS (European Robotic Orbital Support Services). Retrieved April 2022, from https://eross-h2020.eu/eross/
9. Estremera, J., García, J., Romeo, P., Rodriguez, A., Colmenarejo, I., Lucia, M., Rusconi, A., Sangiovanni, G., Cordasco, S., Antonucci, D., & Margan, A. (2022). Re-locatable manipulator for on-orbit assembly and servicing. In *16th Symposium on advanced space technologies in robotics and automation (ASTRA)*. Noordwijk.
10. PERIOD partners. (n.d.). PERIOD (PERASPERA In-Orbit Demonstration). Retrieved April 2022, from https://period-h2020.eu/

Chapter 21
Robotics Concepts for Future Planetary Exploration Missions

Thomas Vögele, Roland Sonsalla, Alexander Dettmann, Florian Cordes, Michael Maurus, Raul Dominguez, and Frank Kirchner

Abstract This chapter highlights general requirements for robotic concepts to support long-term exploration missions, for example to Moon or Mars. The chapter is organized following robotics capabilities that are needed for such complex exploration missions. For each capability, a generic description and discussion is provided, always illustrated by an example project or exemplary robot system. The capability to autonomously traverse large distances is discussed using the example of two EU-funded R&D projects. In the first project, core components for autonomous navigation were verified in a large-scale Martian analogue mission, in which more than 1 km of desert terrain could be traversed autonomously. In the second project, the autonomous handling of opportunistic science was added to the long-distance exploration task. To traverse extremely rough terrains, new concepts for mobility are required. This is discussed using the examples of a six-legged walking and climbing robot inspired by ants, and a hybrid rover featuring four wheels with a special legged-wheel design. Combining both long-distance and rough-terrain mobility capabilities in one single rover design is difficult. Nevertheless, for scenarios where a rover

T. Vögele (✉) · R. Sonsalla · A. Dettmann · F. Cordes · M. Maurus · R. Dominguez · F. Kirchner
DFKI Robotics Innovation Center, Robert-Hooke-Str. 1, 28359 Bremen, Germany
e-mail: Thomas.Voegele@dfki.de

R. Sonsalla
e-mail: Roland.Sonsalla@dfki.de

A. Dettmann
e-mail: Alexander.Dettmann@dfki.de

F. Cordes
e-mail: Florian.Cordes@dfki.de

M. Maurus
e-mail: Michael.Maurus@dfki.de

R. Dominguez
e-mail: Raul.Dominguez@dfki.de

F. Kirchner
e-mail: frank.kirchner@dfki.de

X. T. Yan and G. Visentin (eds.), *Space Robotics*,
https://doi.org/10.1007/978-3-031-39214-6_21

first has to cover long distances, e.g., from the landing site to a scientifically interesting region, and then wants to explore a scientifically interesting but hard-to-access site in that region, we need a system that combines both capabilities. A possible solution is a system of systems, or a multi-robot exploration system, where units with different capabilities are solving the exploration problem collaboratively. This concept is demonstrated with a modular system in which a wheeled rover and a smaller legged rover form a very efficient robot team. To explore extremely hard-to-reach places, such as crater walls, canyons, or confined spaces like the lava tubes discovered on both Moon and Mars, even more than two collaborative robots are required. We present another EU-funded project that has the objective to demonstrate, in a Lunar analogue mission, how a heterogeneous team of three autonomous robots can explore a lava tube on the Moon. Finally, the assembly of infrastructure is an important aspect of future robotic missions. This includes the set-up of scientific infrastructures, such as telescopes, on the Moon, as well as the preparation of habitats prior to human arrival. These tasks are highly complex and do not only require all the robotic capabilities described above, but also the capability of robots to cooperate with human astronauts. An example for a robotic system that fulfils these requirements is given with a project that investigated human–robot cooperation to set up infrastructure components on the Moon.

21.1 Introduction

It is evident that future planetary exploration missions need robotic systems, and even more so, AI-based robotic systems. Despite the extremely important research that has been done and is on-going in the areas of propulsion systems, live support and habitats, radiation shielding and counterbalancing effects of weightlessness, as well as psychological and social studies, to support human exploration missions over very long distances and times, to deliver healthy human subjects to their destinations, and to support sustainable human teams in extreme and extremely remote paces, it is obvious that future human astronauts will need all the (robotic) help they can get. Therefore, robotic and human space exploration are fundamentally interconnected and two sides of the same medal.

However, despite years of research in terrestrial robotics and AI, we are just scratching the surface of systems that could provide the kind of help and assistance to human explorers needed to accomplish their tasks and to face the challenges they will inevitably run into. This may sound like a showstopper, but this is not the way we see it because at least we are now able to truly say that we are scratching the surface.

There are two fundamental issues with AI-based robotics: One is AI, and the other is robotics. To make that clearer it will help to give some examples of the kind of robots we are talking about if we want to go beyond scratching the surface. An assistance robot for planetary exploration missions, for example, could be of great value to forage unknown terrain and to look for shelter or certain resources. The collection of soil samples and pathfinding would also be useful tasks this robot could support. Furthermore, carrying equipment and help to load and unload equipment from exploration vehicles comes to mind as being helpful. We can even look further ahead and think of repair and maintenance tasks in and outside of human shelters and habitats. During a vehicle-based exploration it would be nice to have a robot that can change a wheel on the exploration vehicle, or at least lift the vehicle to enable a human astronaut to change the wheel. We may go on and investigate even more futuristic applications like first aid and maybe even medical surgery. After all, the next hospital will be millions of kilometres away and even a broken leg can be fatal. Finally, we may even be so bold and make the case that future human explorers will need psychological assistance which may not be delivered by the mission's psychiatrist, as he or she is the one with the broken leg. But to close this list with a more positive note: even astronauts want to play! So why not having robots that can join in to be the third player in a 3 against 3 low gravity basketball match, or in a poker or backgammon game?

The examples given above are clearly beyond the scope of performance of any single robot we know of today. That brings us back to the two problems with AI and robotics mentioned previously. For a robot, just to be physically able to provide some of the mobility and manipulation tasks that are associated with the above examples, it is necessary to have a highly complex kinematic structure and a sensorics disposition that significantly surpasses current systems. Fine and dexterous manipulation of objects with multi-fingered hands is a good example of surface scratching. It is certainly not solved but we are making great progress [1].

Another example is bi-manual manipulation in combination with multi-fingered dexterous manipulation and/or high mobility in rough environments. Again, no single system today would be fit to be packed and shipped to do the job, but progress is made here as well [2].

Some of the difficulties with mobility come from the fact that current actuators are not well suited for highly dynamic motions like jumping or other applications that deliver immediate momentum, like soccer or basketball for that matter. Of course, we are aware of systems like ATLAS [3] or SPOT [4], but hydraulic or pneumatic actuators come with too many downsides to be useful for everyday activities on earth, left alone for use in low density atmospheres. Electric actuators like in SPOT need to build up speed constantly over time before they are fully operational.

But this is just the hardware side of the story. The other problem is control and the inevitable headache one gets trying to program a humanoid robot to land on its feet after a backflip from a table, or to stay in the space domain, to run over a series of big randomly shaped rocks without losing speed. But even here we are making progress. Approaches like Whole Body Control (WBC) that interprets the problem of limb coordination as a hierarchy of prioritized optimisation problems—which unfortunately are computationally too demanding to be solved in due time before the next rock is approaching—are now moved towards predictive control methods based on models (Model Predictive Control, or MPC). These differ from previous and more traditional approaches in two significant ways: 1- they predict control instead of ramping up reaction times and 2–they give up on precision. What?! Yes, MPC combines control with fast online reactive behaviours to compensate for imprecision, or to formulate it in another way, for the impossibility to exactly compute the ballistic trajectory of a mass in motion.

But how can that be? Don't we know how to compute ballistic trajectories since Newton? This is correct, but only if we exactly know all initial parameters, and this is not the case for a robot speeding over a rocky plain. With each step, the feet will always be positioned in a slightly different manner, the traction when lifting a leg will always vary, the positions and dynamics of its extremities are always in a different state, and so on... That's why we have to say goodbye to precision and welcome the integration of model-based control, probabilistic (imprecise) predictions, and reactive (hard wired) behaviours. Before we see robots implementing this new control paradigm on planetary surfaces the biggest challenge may be to convince the top-level managers of the space agencies to accept the fact that precision is not always the best or possible way to solve a problem. Sometimes we must accept the fact that we can only manage the problem instead of fully solving it.

As hard as it is to accept the possibility of residual errors in the control of future space robotic system, it is obvious that everything possible has to be done to reduce these residuals to an absolute minimum. As a step forward in this direction, we propose an approach that can be described as 'model-based design and programming' of systems. This approach is based on the idea to create a hierarchical database of all the objects that robots are made of and link them with the code that is used to operate these objects. The hierarchy from bottom to top comprises the smallest individual components like motors, sensors, electronic circuity such as H-Bridges etc. as well as more complex objects, that can be created by combining the lower-level objects, on the next level. So, the 2'nd lowest level could be an actuator that comprises a motor, a gear, and some electronic driver board from the lowest level and controlled by a driver software to move in a desired way. Moving up the hierarchy in this fashion, more and more complex components, sub systems and finally full systems can be created out of the parts that are stored in the hierarchical database. AI tools can help to integrate and assemble objects of increasing complexity, thereby ensuring design and integration standards. This is the first step towards error minimisation.

The second step comes from the fact that the hierarchy does not just store the objects and the physical design, but also the software code that is used to control these objects. As in a classical hierarchy, the objects on a level l of the hierarchy inherit all the properties of the objects they are made of on level l-1. The same is true for the software: A motor driver for a given motor on level l-1 will be inherited by the actuator on level l, which uses this piece of code to create e.g., a classical PID controller. Therefore, if the driver from level l-1 is verified, this verified property will also be inherited by level l and all along the hierarchy to top levels, given of course that only verified code is stored on all levels in the database. How to ensure this cannot be discussed here, but to make it short, we propose that the community takes care of this, which leads us to step 3 of error minimisation: have as many people as possible to check. Therefore, we think that the design tool should be an open-source tool that can be used by anyone in the community if he/she is willing to leave the designs they made as a new object in the database. That means new users can profit from the designs of previous users and improve their functionality, but they are asked to leave the new designs in the database. Basically, this is the same principle that was used to create the largest knowledge base that planet has ever created, Wikipedia.

One could ask the question why anyone would want to start such a community-based approach in the space sector? A sector that is known for its safety, precaution and proprietary software. The answer is twofold: 1) as described above with the kind of systems we are looking at total precision is a dream that we must give up on and replace it with intelligent ways to minimize and manage risk of failure, 2) space exploration is a humanity effort which no nation or company can achieve alone, and for which all the talent available on the planet is needed. The approach for creating space systems described above can be seen as an 'open-source space system design tool' that allows everybody on the planet to provide his or her little contribution of know-how, knowledge, and creativity to the greater project.

In the next sections, we describe selected approaches, systems, and projects that were realized by the space research community over the past decade in robotic space exploration. Much of the experience gathered in these projects has finally led to the ideas described above.

21.2 Long Range Exploration

In robotic planetary exploration, the scientific output of a mission depends very much on the mobility of the robotic system. The more mobile a system is, the more scientific output it can potentially generate. In this context, mobility is defined as the radius of operation of a robotic system, i.e. the area the system can explore during the (limited) duration of a mission.

The least mobile, and thus capable, system is a robotic arm or actuator on a stationary lander. It can only make measurements and collect soil samples within a very limited workspace, i.e. directly around the lander. An example is the ESA mission Rosetta [5].

Somewhat better are tethered mobile systems, such as the Sojourner rover [6], which were able to cover a few tens of meters.

The first rover system ever really roam on a planetary surface was the Soviet-built Lunochod 1 rover [7]. This machine was launched in 1970 and was able to stay operational for 11 months, in which it covered a total of 10.54 km on the moon.

Although the engineering of the rovers improved over the years and their lifetime and operational distance increased, Lunochod 2, the successor of the first Lunochod rover, with 42 km travelled distance was able to hold the record for the longest distance covered by an extraterrestrial mobile vehicle for many years.

In the last decade, impressive results were achieved with NASAs twin MER rovers, which covered several (tens of) kilometres. By 2018, Opportunity for example had covered a distance of 45.16 km on Mars [8]. Yet, this success was to a large extend the result of the exceptional durability of the systems, which was much longer than the initially predicted 90 days.

Despite all progress, currently deployed systems still do not move more than a few tens of meters per day [9, 10, 11]. The latest of the US rovers, Perseverance, landed on Mars in February 2021 and has been active already for over a year in which it covered a mere 4.39 km [12].

In summary, planetary exploration today still resembles an ant trying to explore a football stadium and the exploration of significant amounts of the Martian (almost 145 Mio km^2) or even the lunar (14.6 Mio km2) surface is completely out of the range of these rovers. This is due not only to the challenging environmental conditions that limit the lifetime of any technical system, no matter how excellent the engineering. Fact is our current exploration rovers are awfully slow. The fastest of them, Lunochod 2, made no more than 2 km/h. More recent rovers like the Mars rovers Spirit and Opportunity travelled at 0.18 km/h, the Chinese lunar rover Yutu [13] reached 0.2 km/h.

The reason for this snail-pace lies partly in the limitations of the rover locomotion system and its power storage capabilities. However, a much more significant shortcoming is the fact, that the capabilities for decisional autonomy in current rover systems are so limited, that a human operator has to remain in the driver seat—and that this driver seat is potentially more than 54 Mio km away.

Steering a rover over a planetary surface is a complicated and elaborate process, which requires the cooperation of many engineers, scientists, and technicians. Every move is simulated in the Command-and-Control Center, and carefully evaluated by experts for potential outcomes and risks before the respective control commands are sent to the rover. Then, with growing tension and hope to have done the right thing, the human pilots in the Control room wait for the rover to execute the command and to send telemetry data and pictures to monitor rover action. For a rover on Mars, the delay between command and execution can be up to 24 min.

It is obvious that this sort of remote control will never be able to speed-up rover movements. What is needed are rovers that can perform certain tasks autonomously. This includes perceiving the environment, for example to spot obstacles and impassable terrain, and to make quick decisions based on these observations. First steps in that direction have already been taken. In July 2021, Perseverance was able to travel more than 100 m autonomously in one go [14].

However, even a task as simple as traversing one or two kilometres of more or less flat terrain without human intervention is still out of the reach of the systems that today or in the near future operate on Moon and Mars. Nevertheless, the research community has picked up the challenge and is making efforts to boost robotic decisional autonomy. Significant progress has been made in this area in the last decade.

Increased decision making on-board the space craft through increasing artificial intelligence and employing modern systems for robotic navigation, guidance and control, has the potential to overcome the aforementioned limitations. Systems that can autonomously steer a rover over long distances (as in FACILITATORS) or even adapt a plan to unforeseen situations, perform autonomous decisions and science "on-the-go" (as in ADE) were developed to demonstrate the advantages of autonomous rovers for planetary exploration.

21.2.1 FACILITATORS—Long Distance Traverse in Mars Analogue in the Moroccan Desert

The FACILITATORS project was funded by the EU in the Horizon 2020 program [15] within the Strategic Research Cluster (SRC) "Space Robotics Technologies" [16]. The idea behind the SRC was to create better and more sustainable outcomes by defining a concrete research agenda that is implemented with a sequence of interdependent "Operational Grants". The thematic focus of the SRC Space Robotic Technologies was on smart and autonomous robots for space, both in on-orbit and in planetary applications. The wave of five OGs developed basic technological building blocks, ranging from a new operation system for space robots (project ESROCOS) [17], hardware- and software modules for robotic perception (projects 3ISDS [18] and InFuse [19]) standardized interfaces between modular robot components (project SIROM) [20], to software for decisional autonomy of robots (project ERGO) [21]. The new building blocks were tested and evaluated in the FACILITATORS project, partly in the lab and partly in a large-scale Mars Analogue Mission in Morocco.

In this mission, the testing of the ERGO software stack responsible for rover autonomy was in the focus. During three weeks in November 2018, a team of up to 30 researchers from various European member states gathered in the desert near the city of Erfoud in South-Eastern Morocco to prove that ERGO and the other SRC basic building blocks are indeed able to support rover autonomy, not only in the lab, but also under the harsh conditions of a real outdoor environment. The rover platform used for the tests was SherpaTT [22], a 4-wheeled space exploration rover developed at the DFKI Robotics Innovation Center in Bremen (Fig. 21.1). SherpaTT implements a novel locomotion concept in which the 4 wheels are mounted to "legs", each of which can be controlled individually and independent from the other "wheeled-legs". This concept enables SherpaTT not only to traverse rocky plains and operate on slopes, but also to change its physical profile to negotiate narrow openings and other challenging topographies.

The mission goal of the Morocco Mars Analogue Mission was to demonstrate a long-distance traverse of at least 1 km, conducted autonomously and without human intervention. The location that was selected for the mission featured flat but gently sloped sandy areas covered with small- and medium sized rocks and interspersed by small canyons created by water runoff during the rainy season. These flats were framed by small hills with steep slopes on one side and an almost perpendicular rocky massif on the other side.

Driven by ERGO and the other SRC sub-systems, SherpaTT managed to even exceed the targeted 1.5 km by a few 100 m. However, it took the rover more than 10 h to cover this distance and ironically it was not so much the existence but rather the lack of obstacles that slowed the rover down: In some parts, the chosen track was so void of outstanding features and landmarks that the self-localization and navigation algorithms used by the autonomy module were at a loss and needed a long time to re-orient themselves.

Fig. 21.1 The SherpaTT rover in the morocco field trial 2018

This and other experiences made during the field tests proved once more that, although the terrestrial desert analogue does only partially reflect true Martian conditions, it is important to test both the hardware and software of an exploration rover in the field. The field combines numerous environmental conditions (e.g. changing lighting, fine dust, harsh terrain, significant changes in temperature, long distances to be traversed) which in the lab can only be simulated individually. This, together with the many unforeseen challenges and obstacles that have to be overcome during an analogue mission, is what makes field trials a not only valuable, but necessary extension of pure lab-testing of systems or sub-systems.

21.2.2 ADE—Long-Range Exploration with Opportunistic Science

Like the above-mentioned FACILITATOR project, the project ADE (Autonomous DEcision making in Very Long Traverses) was funded by the European Commission through the H2020 SRC on Space Robotics Technologies [23]. The ambition of this project was to design, develop, and test a fully autonomous rover system in a Mars-relevant environment.

The ADE system is capable of autonomously making all decisions necessary to achieve its mission objectives: It can control the collection of data and scientific output, perform autonomous long traverses for surface exploration, and support the fast reaction and adaptation of the rover to unforeseen situations. The feasibility of this approach was successfully demonstrated during a field test in a sand mine in Northern Germany in the spring of 2021, with the SherpaTT rover [24] as the central mobile platform. The rover was able to cover nearly 500 m in full autonomy mode within less than 3 h. The resulting mean velocity of 4.72 cm/s already accounts for stops required for map building and local path planning based on stereoscopic camera images. Furthermore, the autonomous selection of scientifically interesting locations was demonstrated with the system.

The overall architecture of the ADE demonstrator consists of three high-level subsystems as follows: (i) the rover system SherpaTT as "space segment", (ii) a ground segment capable of commanding and validating the rover operations, and (iii) testing facilities for executing and validating the system. A detailed description of the project and its outcomes is presented in [25, 26].

For autonomous operation, the SherpaTT platform was equipped with a so-called AvionicsBox, (Fig. 21.2). This payload was mechanically mounted to the front of the rover, supplied with power from the rover main power source, and connected via ethernet for communication with the rover's on-board computer. The on-board computer took care of the locomotion subsystem and the manipulator arm control, while the sensors and computational units of the AvioncsBox controlled the autonomous mission execution. The AvionicsBox features a stereo camera pair (NavCams) on a pan-tilt unit at ~ 1.9 m height, as well as a fixed stereo camera pair (LocCam) for localization and visual odometry. Furthermore, an infrared camera and a high-resolution camera (HiRes Cam) were mounted with fixed angles on the mast of the AvionicsBox.

A non-trivial challenge for the project management was the coordination of many project partners producing different software modules in parallel in a short time. To mitigate these challenges and to support software development and integration,

Fig. 21.2 Rover SherpaTT equipped with avionics box during the Field Testing of the ADE project 2021

a simulation and validation toolset was provided at an early stage of the project. The simulation allowed the partners to develop and integrate software modules even before the hardware was ready to use, based on a detailed model of the rover and of the parts to be newly integrated. Furthermore, to better plan the field tests and to anticipate issues pertaining to specific locations, simulation environments representing the areas selected for the tests were used during development, offline testing, and later as Ground Truth for the performance analysis of the results. The simulation environments were based on images and data collected with a drone. Finally, the validation toolset aimed at performing continuous testing of the integrated software and was used to ensure the functioning of the simulator while continuing its development [27].

Due to the Covid-19 travel restrictions, all European partners were connected to the team on site via teleconference systems using mobile internet (4G router) during the field trials. This inconvenience, however, revealed one of the advantages of a system with autonomous capabilities: The limited bandwidth forced the operators across Europe to only check system behaviour sporadically. This forced the team to work under even more realistic conditions concerning ground control operations (i.e. offline planning, uploading of plans etc.).

The level of autonomy of the system is an internal variable that can be pre-set by ground control in the mission plan, or changed automatically, depending on the perceived mission context and terrain directly ahead of the rover:

- In E1 mode, only direct telecommands are executed by the agent. This can be seen as a fall-back control mode.
- In E2 mode, time-tagged commands are executed at their designated time of the mission plan.
- In E3 mode, plans elaborated at the ground control centre are uploaded for execution. The scheduling of the commands is done by the agent.
- Finally, working under the highest level of autonomy (E4 mode), commanding is issued via high-level goals. These high-level commands (traverse to a given point, pick a sample at a given position, drop a sample in a given area, move to a given position to take an image, or scan for a given pattern in an area) are sent to a mission planner, which generates a plan to fulfil the high-level goals. The agent's controller then executes the plan.

A scientific detector extracting patterns from the HiRes Cam is part of the ADE autonomous agent. The scientific detector uses AI techniques for pattern recognition and can detect both, known classified classes and novel targets from images. The scientific detection is performed in seconds, to allow the rover to change its course dynamically to gather science when an opportunity arises. Once a scientific target is detected, a new goal is injected into the planner, which computes a new mission plan that considers the need to include the observation of the newly discovered target. As the scientific detector may identify more novel targets than the rover can observe, the mission planner takes an oversubscription approach: each target has a reward (determined by its novelty) and a plan is computed that schedules observation actions with the goal to maximize the total reward. Hence, novel features override better known feature in the plan, or, in other words, the robot "loses interest" in common features of the environment.

Overall, ADE is a complex system, designed to address the challenges for planetary rover exploration by providing a fully autonomous solution for planetary rovers. The complexity of the system increased when compared with previous developments like ERGO [28], which was used in the FACILITATORS-field test (see above). ADE builds on ERGO, but some new elements were added: a mobile manipulation component, a complete perception and localization system based on visual navigation, a ground station with another instance of the planner to be used on the ground for mixed-initiative, an adapted simulator with a validation toolset [27], as well as dedicated assets for ground-truth and soil traversability analysis [29].

Due to Covid-19, the field tests had to be re-scheduled from a site on Fuerteventura to a sand quarry in Northern Germany, which introduced significant challenges related to the need for remote testing and the local weather conditions. Nevertheless, the results for localization and perception met all expectations, with a long traverse of 484 m in less than 3 h and a remarkably small positioning error of only 0.4%. Other characteristics of locomotion, like hazard avoidance and the combined mobility of the rover and the robotic arm were also successfully tested. The change of the test site forced the scientific detector to be tested without adequate training of its neural network, but the interaction between the scientific detector and the rest of the system to perform serendipitous science could be fully tested and verified.

The following autonomous capabilities have been implemented and tested with the ADE system:

- An autonomous guidance system, combined with a novel perception and localization subsystem,
- a mobile manipulation subsystem deciding autonomously on the deployment of the robotic arm, and controlling the whole rover during sample approach,
- a mission planning/scheduling component allowing both, E3 and E4 modes of commanding,
- while running in full autonomy mode, the combination of the scientific detector together with the mission planner allows performing serendipitous science, modifying the plan dynamically to include new requests, without the need for human intervention. This on-board planning capability of ADE represents a step ahead for future space missions, and this technology is not only applicable to planetary exploration but also to orbital missions.
- a traversability analysis activity was conducted to develop terrain classifiers based on the use of proprioceptive features.

21.3 Traversing Extremely Rough and Inclined Terrains

Travel speed and range of an exploration rover are not the only parameters that impact the scientific results of a planetary exploration mission. Most scientifically interesting spots are found in hard-to-reach areas, such as within or near the rim of craters, on-top of exposed peaks, or at the walls of cliffs and steep slopes. A thorough exploration of these topographic features can provide much more interesting insights in the history of the celestial body under investigation than the exploration of flat areas.

Unfortunately, the exploration rovers used up to present or planned for the immediate future, such as MER, MSL, Perseverance, or ExoMars, have only limited locomotion capabilities when it comes to negotiate rough terrain, steep slopes, or even vertical cliffs. The limitations of these rovers are a result of having to find a balance between often contradictory requirements, such as mission objectives, mass and size constraints, power requirements, and risk considerations.

However, while the rovers currently roaming Mars and Moon use rather traditional locomotion systems, alternative suspension and locomotion concepts for the exploration of extremely rough and inclined terrains are developed and tested in the academic domain. Among them are walking systems, which have a long development history and currently experience a revival due to rapid developments in the field of dynamic walking.

Walking systems do offer probably the most capable concept when it comes to difficult terrains. After all, there are numerous biological examples (including human bi-pedal walking) that prove the excellent all-terrain capabilities of legged locomotion. With respect to robotic space exploration, however, walking systems have the significant drawback that they tend to be technologically complex and energy hungry. In addition, fully actuated walking systems incorporate a complex kinematic and mechanic design that needs advanced and very performant locomotion controllers. To overcome the drawbacks of technical solutions for legged locomotion, there is a need to reduce system complexity, for example by developing solutions based on hybrid legged-wheels.

21.3.1 ANT—a Generic Navigation and Locomotion Control System for Legged Space Exploration Rovers

Recent advances in legged robotics have shown that walking systems can provide a very mobile locomotion system that is able to cope with steep inclines and very rough terrain. However, to fully exploit the potential of this highly flexible locomotion concept, the navigation system for legged exploration robots needs to be closely coupled with the robotic motion control.

A navigation and locomotion control system specifically designed for legged robots was developed in the ESA-funded project ANT [30]. This system is able to perceive the terrain, to plan a path to a desired goal, and to control the path execution while traversing unconsolidated, inclined, and rugged terrain. The approach followed in this project introduced a suitable level of abstraction resulting in a modular and generic software that can support different types of walking robots. This will allow, depending on specific requirements of future space missions, to use different types of legged rovers, such as energy-efficient four-legged or the more stable six-legged robots.

The architecture of the ANT controller is shown in Fig. 21.3. A Guidance and Navigation (NAV) layer generates a path to a given target, based on the current location of the robot and the sensed environmental information that is represented in a sophisticated navigation map. The thus generated path is then transformed by the Motion Control System (MCS) into precisely planned footholds that are reactively adapted to compensate for undetected terrain irregularities. The footholds are used to plan the motion of the robot, generating kinematic references for all actuated joints. A low-level torque loop enables to implement active compliance on joint level, increasing robustness to unperceived impacts. A Hardware Abstraction Layer (HAL) is introduced to define a generic interface for different types of torque-controlled legged robots. The connection between the layers is realized through the open source "Robot Remote Control" (RRC) library [31], a lightweight, framework-independent communication library that allows the implementation of the individual layers using different frameworks such as ROCK or ROS. Additionally, the use of Docker images allows a smooth and parallel development and installation, which drastically reduces dependency issues typically introduced when combining large software systems from different institutions in the deployment target.

This generic approach was tested with two robotic systems, the hexapod CREX [32] and the quadruped AlienGo [33], both having a completely different morphologies. In general, the ANT navigation and control system supports 4-legged as well as 6-legged robots with three degrees of freedom (DoF) per leg. The approach relies on

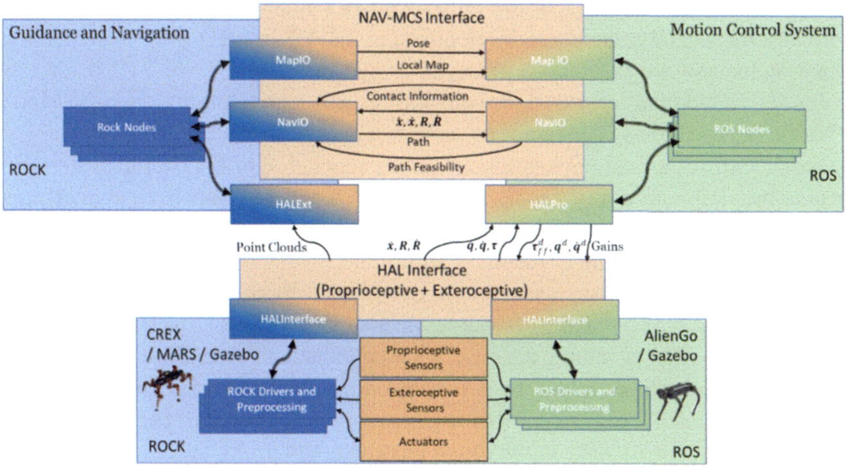

Fig. 21.3 Generic control architecture—the guidance and navigation layer, the motion generation and control layer, and the hardware abstraction layer are each interfaced through the lightweight RRC library with standardized interfaces to allow framework-independent access (orange). In this example, the approach allows the parallel use of ROS (green) and ROCK (blue)

Fig. 21.4 CREX walking on a 20° incline (left) and over a rubble field (right)

torque-controlled joints and sensors that provide dense point-clouds for the localization, mapping, path generation, and foothold selection. The omnidirectional movement of the robots on inclined planes and rubble fields could already be successfully demonstrated (Fig. 21.4). In the future, sloped rubble fields with load bearing issues (e.g. crater walls) will be targeted as well.

21.3.2 Coyote—a Hybrid Light-Weight Scout Rover

Coyote is a micro rover designed for autonomous operations in rough and unstructured terrain. The robot's design is the result of over a decade of experience with highly mobile rover systems at the DFKI Robotics Innovation Center in Bremen, Germany. Coyote III is the latest development of the ASGUARD and Coyote family, which includes more than seven predecessor platforms. The Coyote III rover was designed as a modular terrestrial testbed, with a focus on exploration tasks on Moon or Mars. The modular system concept, as shown in Fig. 21.5, allows the extension of the rover with additional payload items and an adaptation of the overall systems to specific mission- and/or test needs. Two standardized electro-mechanical interfaces (EMI) allow the online docking of additional payloads, such as a manipulator or a tether management system, to the rover at runtime.

The high mobility of the system is accomplished by a combination of special wheels mounted on a passive suspension system and a compact rover design. These hybrid "legged-wheels" mimic a simple walking behaviour to achieve an extremely high mobility on steep slopes with a wide variety of surface textures, ranging from soft sandy soils to rocky unstructured terrains as shown in Fig. 21.6. Especially in extremely soft soils and in rough unstructured terrain, the "legged-wheels" outperform standard wheels significantly.

Because they mimic a walking pattern, the wheels do not suffer from the build-up of material in front of the wheel and, as the feet are stepping on the soil based on the rotational motion of the wheel, can basically not get stuck in soft soils. This stepping motion also provides high climbing capabilities on rocks and boulders. Due

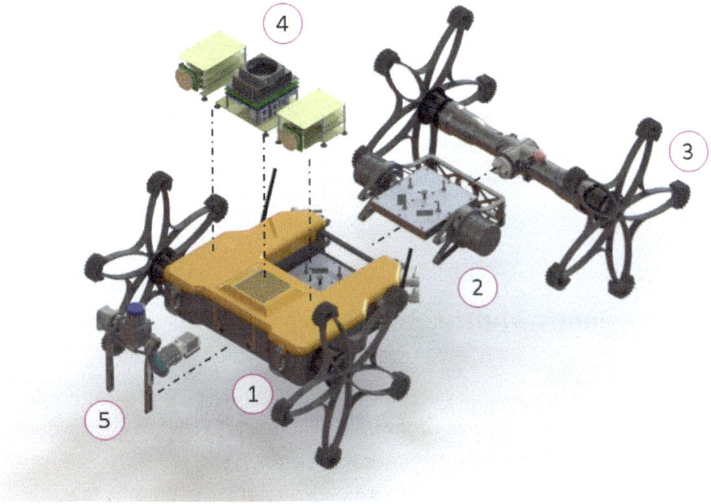

Fig. 21.5 Exploration view of coyote III showing its main elements: 1: Front Body with two directly driven hybrid legged-wheels and additional payload bay, 2: Centre Body with EMI for payload connection, 3: rear body with two directly driven hybrid legged-wheels, acting as passive boogie of the suspension system, 4: sub-system compartments, 5: sensor bench for navigation

Fig. 21.6 Coyote III navigating through boulders and showing its mobility capabilities

to the special wheel geometry, even obstacles with heights extending the leg length or wheel radius can be overcome: The leading "food" can step onto the boulder, thus anchor the wheel and pull the rest of the wheel and the vehicle upward.

The locomotion capabilities of the rover are further enhanced by the suspension system, which always ensures ground contact of all four wheels. Coyote III, for example, can traverse slopes of more than 45°, which could be demonstrated during various laboratory tests as well as extensive analogue field tests, among others in the desert of Utah (USA).

21.4 Heterogenous Multi-robot Systems

Individuals may have extraordinary capabilities, but nature provides numerous examples of animals that profit from organizing themselves in swarms, flocks, and schools – in short, by cooperating with each other. Ants are only one prominent example of cooperating individuals that together achieve goals not feasible for the individual alone.

In robotic planetary exploration, only individualistic systems were deployed up to now. These systems carry all instrumentation needed and have all locomotive capabilities required for a specific mission. However, scenarios more complex than the current ones, such as soil sampling in the wall of a canyon on Mars, or exploring a lava tube on the Moon, will require a wide range of capabilities that cannot be integrated into one single rover system. Instead, multiple specialists will have to work together as a team, each contributing its individual capabilities to the success of the mission.

Considering the example of soil sampling in a vertical cliff, a fast wheeled system might be optimal to cover the distance from the landing spot to the canyon itself, while a lightweight climbing robot may be required to climb down the canyon walls. Very likely, some kind of anchoring or tethering subsystem would be required as well. It is obvious that the two mission phases create requirements for the locomotive system that are contradictory and cannot be combined in one system. In this case, cooperating heterogeneous systems can play out their individual strengths and together achieve the goal of such a complex mission.

With a modular architecture that allows to add or change payloads, tools and sensors, such systems can be gradually extended and improved with subsequent missions. Eventually, a heterogeneous multi-robot system with modular extendibility, different locomotive systems, and immobile components provides a high degree of scalability and robustness for planetary exploration [34].

21.4.1 LUNARES, RIMRES, and TransTerrA, Consecutive Studies on Heterogeneous Robot Teams

The objective of the three projects LUNARES [35], RIMRES [34], and TransTerrA [36] was to test the feasibility of heterogeneous and modular systems for planetary exploration. All three projects implemented this concept in physical systems that were demonstrated in laboratory and/or field tests in relevant environments.

In the LUNARES project, the general feasibility of a heterogenous robotic system for planetary exploration was shown by combining legacy systems and testing them in an artificial lunar crater environment in the lab. The individual systems were initially not designed to be cooperating, hence certain compromises had to be made in terms of the electro-mechanical coupling of the rovers and inter system communication.

With the RIMRES system, however, a small legged crater explorer and a large wheeled-leg transportation rover were designed in a co-design process as part of an integrated and modular multi-robot system architecture that also supports additional modular payload units.

Both systems target the use-case of a lunar polar crater exploration mission, where the wheeled system transports the legged climbing robot to the crater rim, decouples, and sends it climbing down into the crater.

Figure 21.7 shows the wheeled-leg rover Sherpa and the six-legged climbing robot CREX. The modular payloads developed for these systems include additional battery packs, sensor modules, and a soil sampling module. A comprehensive summary of the modular capabilities is provided in [37].

The TransTerrA project used the lessons learned in RIMRES to introduce a paradigm shift with respect to the composition, task assignment and objectives for the multi-robot system. Firstly, the TransTerrA mission concept was based on landing a

Fig. 21.7 RIMRES systems. rover deploys six-legged scout from lander mock-up

mission within a lunar polar crater and deploying the systems on the crater bottom [36]. The crater selected for the reference mission provided areas with high illumination for solar power generation as well as permanently shaded regions as candidate regions for in-situ evidence of volatile materials. Secondly, the establishment of a logistics chain was used to reduce the need for the sampling rover to run multiple paths and to reduce travel distances [38].

Finally, the task of sampling was reassigned from the scout rover to the exploration rover, with the scout rover operating as a transport shuttle that fetches the samples from the exploration rover or from dedicated storage points, the so-called BaseCamps (Fig. 21.8) and transports the sample containers to the return stage at the lander. In addition, the shuttle transported energy packs between the exploration rover and several charging locations, hence implementing a logistics chain in the exploration system. To match the new requirements for the shuttle rover (i.e., speed and mechanical simplicity), the walking and climbing robot CREX was replaced by the hybrid legged-wheel robot Coyote III [39]. The TransTerrA approach of a heterogeneous, modular exploration system was tested in a field test in the desert of Utah [40, 41].

Fig. 21.8 TransTerrA systems, the exploration rover has a soil sampling module mounted to the manipulation arm. In the foreground, a stationary BaseCamp with another soil sampling module attached is visible

21.4.2 Multi-robot System Within VaMEx Initiative

The focus of the "VaMEx-Valles Marineris Exploration" research initiative launched by DLR Space Management was the exploration of Valles Marineris on Mars. The exploration of the Valles Marineris is to be carried out with the help of a heterogeneous, autonomous robot swarm consisting of rovers, legged robots and aerial vehicles. Valles Marineris is one of the largest rift valley systems in our solar system, with a width of 700 km, a length of 4000 km, and a depth of up to 10 km. Due to indications of water deposits, former volcanic activity, and shading from UV radiation, the rift valley system is particularly exciting, as it provides the conditions for the existence of extraterrestrial life.

However, exploring Valles Marineris places high navigational demands on a swarm of robots. To cope with these demands, it is necessary to develop and validate a fault-tolerant and automated joint mapping capability, with sensors and algorithms that are optimal for the Mars surface.

In precursor projects, an overall concept for an autonomous and cooperative robotic swarm on Mars has been outlined. Swarms of ground-based and aerial exploration units with local navigation and communication have been realized and tested, navigation networks, e.g., have been developed, and initial fundamental work regarding UAVs for the Martian atmosphere has led to the mars helicopter Ingenuity. Before the Russian-Ukrainian war, the option of a swarm-based Mars mission in the 2030s using the Russian nuclear-powered INPPS Flagship and its enormous payload capacity was discussed.

21.5 Exploration of Confined Spaces

Exploration rovers that can traverse long stretches of flat terrain autonomously and without human intervention are a significant step forward for planetary exploration (see Sect. 21.2). If they are able to operate on sandy or rocky terrain as well as steep slopes, their value for planetary science increases significantly (see Sect. 21.3). Working in teams and/or swarms can further improve the scientific output of a mission (see Sect. 21.4). However, what exo-geologists and exo-biologists really like to explore are cliffs with outcrops of ancient rocks that provide insight in planetary history, deep shady craters that promise traces of water or organic matter, or even caves formed by lava tubes that could not only hide all of the above but also provide shelter for astronauts in future manned missions. Caves and other confined areas have in common that they are extremely difficult to access, at least with the current robotic rovers.

Engineers at NASA as well as researchers in the space community in Europe have therefore started to develop ideas for innovative rover designs with which extremely hard-to-reach areas could be accessed. None of these systems has flown yet, but first feasibility studies have been performed in Mars and Moon analogue environments.

Lava tubes are among the most interesting, but also the most challenging environments to explore, both on the Moon and on Mars [42]. In recent years, there have been a number of feasibility studies with the goal to find ways to explore lava tubes. Obviously, as there are only limited communications with the ground control station possible from inside a lava tube, the mayor challenge is to develop an autonomous rover that is able to explore the lava tube without human intervention. An outstanding example for a new rover concept designed to explore hard-to reach areas, such as steep cliffs, crater walls or even lava tubes, is the AXEL rover developed by JPL [43] this two-wheeled tethered vehicle was designed to rappel down highly sloped terrain typical for crater walls, pits, gullies, canyons, and crevasses. It is suited even for the exploration of caves formed by lava tubes. The core module of AXEL consists of two wheels connected by a fixed boom. The wheels are actuated, and the boom contains a spool that carries hundreds of meters of tether. Using this tether, the rover can rappel down vertical slopes and return to the surface again. AXEL implements a modular concept. In the DuAxel rover configuration [44], two AXEL modules are connected by a central module to form a four-wheeled rover. Other configurations are also possible, so that this very innovative concept offers many possibilities for the exploration of hard-to-reach areas.

21.5.1 ENTERN—Environment Modelling and Navigation for Caves

The exploration of lava tubes was also addressed by the German-funded ENTERN project [45]. In ENTERN, an autonomous scout rover was tested in a real terrestrial lava tube on the Canary Islands. ENTERN focused on the development of three technological enablers that are crucial for the exploration of caves and other extreme environments in space: environment modelling, operations & control, and autonomy. Within the project, these technologies were implemented and deployed in two robotic platforms, Asguard v4 and CREX. Different space analogue locations were chosen to test the systems: a moon crater reconstruction which is part of the space exploration facility at DFKI-RIC, and a real cave, *Cueva del Viento*. This cave is part of the largest lava tube system in Europe, located on the Canary Island of Tenerife. Images of the lab- and field-tests are shown in Fig. 21.9.

On environment modelling aspects, the set of libraries for environment representation Envire were developed. In its core, Envire provides the means for allocating in a geometrical graph any C + + object and interrelate it through time-stamped transformations to any other C + + objects. This is a recurrent requirement in environment modelling software (e.g. Simultaneous Localization and Mapping, Visual Odometry, Simulation) for which an efficient implementation was needed. The two main advantages of using a common library for implementing the core environmental representations functions, are the reduction of dependencies and the ease of

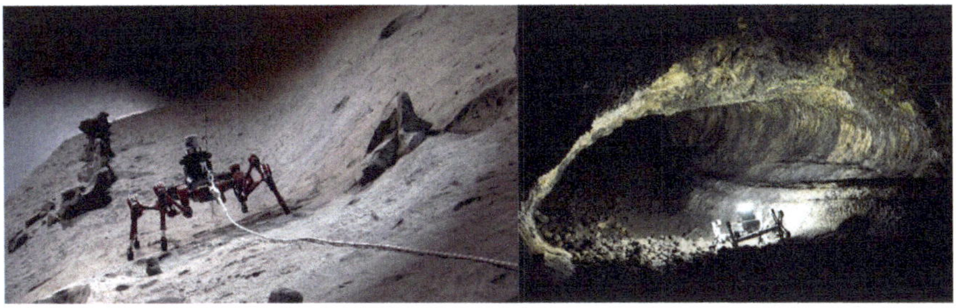

Fig. 21.9 Crex robot climbing a reproduction of a Moon crater (left) and Asguard v4 exploring a lava tube during the final field test of the project ENTERN (right)

compatibility between components, since type conversions and correspondent potential losses of information and resources are minimized. During Entern, in addition to the Envire Core a set of extensions which integrated it in the navigation architecture were developed: Map formats according to the IEEE standards [46]. Simulation based on these maps and the integration with Georeferenced Information Systems (GIS) technologies. All these libraries were provided open source [47].

On the Operations & Control aspects, the implementation and test of a reliable communication layer and control interface were addressed. The implemented software facilitated the management of robotic operations over communication delays and blackouts, enabling the exploration mission to continue despite these inconveniencies. The interface for commanding operations included a dashboard where the onboard generated maps, candidate target locations, obstacles and known explorable areas where displayed. The following commands to direct the exploration: explore volume, waypoint navigation, validation on simulation and go to waypoint where available. As soon as communication was available the update state of the commands lists was displayed, environment representations and rover position were updated. When direct communication was not possible, the operations continued without interruption on the rover.

The main contributions of the project were on the autonomy technology. Since the remote operations on a lava tube were assumed to be severely limited due to the communication limitations, efforts were focused on guaranteeing the safe exploration with minimal operator supervision. The challenges posed by autonomous navigation in underground contexts are larger than those on surfaces both from the path planning and the collision avoidance perspective: Paths crossing on different elevation levels and collisions with ceilings need to be addressed. To represent these features and generate paths that account for them the Multi-level Surface Maps [48] were used. To reduce the risk of the navigation operations and internal physics simulator was introduced to guarantee the safeness of the planned paths before its traverse. The challenge of simulating the environment where the rover is currently located onboard was eased by the common environment representation tools used by SLAM and simulation [49] [dominguez2018internal]. The Fig. 21.10 shows the lava tube

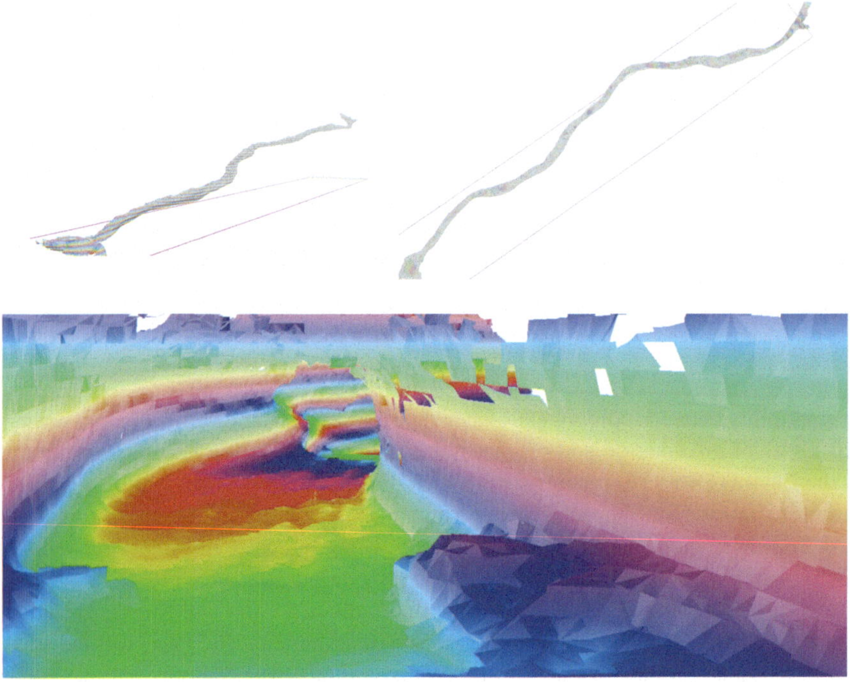

Fig. 21.10 SLAM reconstruction of the explored 244 m long lava tube. between each color repetition the elevation corresponds to one meter

reconstruction generated from LiDAR and state estimation data by the SLAM module integrated on the rover Asguard v4.

21.5.2 CoRob-X—Exploration of Cliffs and Lava Tubes

Once the means for autonomous inspection of the lava tube are available, the second challenge is to find a way to deploy the rover. Lava tubes are accessible through so-called "skylights", i.e., holes in the ceiling of the lava tube that developed when part of the ceiling collapsed. In the EU-funded CoRob-X project [50], a team of seven European research organizations and space companies studied the feasibility of using a team of several autonomous robots to enter and explore a lava tube.

The robot team consisted of three different rovers, each with a distinct portfolio of capabilities and respective tasks. The large and heavy (150 kg) SherpaTT rover had the task to transport equipment and sensor payloads from the landing site to the lava-tube skylight. More importantly, this 4-wheeled "Space-Mule" acted as an anchor to rappel a much smaller and lighter scout rover down the steep walls of the skylight and into the lava tube tunnel.

The Coyote III scout rover had already proven its value in the ENTERN project described above. In CoRob-X, it was equipped with additional scientific sensors (a ground penetrating radar) and, more impotently, a newly developed Tether Management and Docking System (TMDS). Coyote used this TDMS to rappel, spider-like, from the lunar surface into the lava tube. Once it reached the bottom, Coyote uncoupled from the tether and started exploring the lava tube tunnel. To recharge, deliver scientific data, and eventually get back to the surface, the scout rover can re-couple to the docking station.

To provide mission-critical information about dimensions, depth, and other characteristics of the skylight, the Sherpa-Coyote duo was assisted by a third rover, LUVMI-X [51]. This mid-sized, 4 wheeled lunar rover not only watched the rappelling action from another perspective, but was able collect data from inside the skylight and lava tube using a specific sensor payload. This "payload cube" equipped with HD cameras and other sensors was shot into a ballistic trajectory to enter the skylight from above. On the moon, due to the low gravity and the resulting low drop velocity, the cube will have enough time to collect data on its way down to the bottom of the lava tube. Once there, it can act as a communication relay between the scout rover and the robotic systems on the surface.

The CoRob-X project built on robotic hardware provided by the project consortium and the software building blocks that were developed within the framework of the SRC Space Robotics Technologies. These building blocks were reused and enhanced to support a multi-agent exploration team of robots (Fig. 21.11).

The feasibility of the CoRob-X concept, and the capabilities of the software building blocks developed in the SRC program to enable such a concept, are to be tested in a Lunar Analogue Mission on the Canary Islands. The mission is planned

Fig. 21.11 The four consecutive phases of the CoRob-X Lunar Analogue Mission

for early 2023 in the northeastern part of the island of Lanzarote. Here, ancient eruptions of the Corona volcano have created a system of lava tubes that is, albeit not in size, in many ways similar to the lava tubes expected on the moon. The analogue mission was not intended to simulate all aspects of a robotic exploration mission to the moon but concentrates on those aspects that are most critical for the performance of a team of autonomous robotic systems. In four consecutive mission phases, the analogue demonstrates the joint exploration and mapping of the lunar surface near the skylight (MP1), the remote exploration of the skylight using the payload cube (MP2), the rappelling of the scout rover into the skylight (MP3), and the exploration and 3-D mapping of the lava tube with the scout rover (MP4).

21.6 Mobile Manipulation and Assembly of Infrastructure

Due to its proximity to Earth, the Moon could be used as a so-called intermediate station for research on more distant celestial bodies. A research infrastructure on the Moon would also be of great value to pursue scientific and ecological interests in the long term. To achieve the ambitious goal of maintaining permanent research facilities and even habitats on the Moon, the use of existing lunar resources is essential. In-Situ Resource Utilization (ISRU) enables sustainable space exploration by using the resources available in space to produce products and services for robotic exploration [52], manned exploration [53], and commercial purposes. Studies have been conducted on the economic viability [54] of utilizing space resources in various relevant scenarios. This includes, among others, the market of LEO (Low Earth Orbit) to GEO (Geostationary Earth Orbit) transport based on hydrogen peroxide propellants, taking advantage of the fact that LEO is cheaper to access from Moon than from the Earth's surface in terms of Delta-V budget, or use of propellant. The fact that such propellant can also be used by government agencies for robotic and manned missions to explore the solar system, if provided at strategic propellant depots in cis-lunar space or on the lunar surface, creates a strong basis for public–private partnerships aimed at commercial services to sustain lunar operations.

ESA is preparing a mission to demonstrate the feasibility of ISRU on the Moon by developing a payload that will perform excavation, processing, and hydrogen reduction of regolith ore to produce a few grams of water. Advances in autonomous robotics are needed to install, build, and operate such facilities (especially if a site of high scientific interest such as the South Pole-Aitken Basin on the far side of the Moon is chosen). The use of construction technologies has been a research focus, particularly the use of regolith for civil construction [55] (e.g., preparation of launchpad berms to mitigate lander dust, construction of shelters from composing single elements, and consolidation of ground to form roads). Such ISRU plants could serve as a source of water and oxygen to enable human outposts and to deliver hydrogen and oxygen fuel to various locations in cis-lunar space. Civil construction systems could then be used to create radiation shelters for habitats delivered from Earth.

To operate an ISRU device on the Moon, research is now moving towards robotic systems to avoid that humans are permanently exposed to the harmful environment. Robotic systems can be used not only for support, but also as independently acting systems to build the necessary infrastructure. Multi-robot teams are well suited for such tasks, as multiple robotic systems with different characteristics can cooperate with each other and work together to perform a wide range of tasks. However, the key technology for any system is mobile copying, i.e., the ability to move across the difficult, unstructured terrain and perform manipulation tasks.

21.6.1 ProAct—Cooperative Robots for Infrastructure Development

The EU funded PRO-ACT project (OG-11 of the H2020 Strategic Research Cluster for space robotics technologies [56]) was aimed at developing and demonstrating cooperation and manipulation capabilities between three robots for assembling an ISRU plant [57].

One of the involved robotic systems in PRO-ACT was Mantis, a biologically inspired six-legged walking system [58]. It provides stable locomotion capabilities for exploration tasks in rough terrain, but also has the ability to raise its upper body and, by assuming a four-legged posture, to use its two upper-body limbs for dual-arm manipulation.

The other system is Veles, a mobile rover based on the IBIS [59] mobile manipulator that features a six-wheeled passive suspension system and 6-DoF mobile arm equipped with a standard HOTDOCK connector at its end-effector [60].

The robotic team is completed by a mobile gantry crane that was developed within the PRO-ACT project [61]. This crane is a cable robot consisting of two well differentiated systems, a deployable structure on four separately steerable wheels, and the cable-mounted robotic manipulator itself. The latter can be configured with different end-effectors such as a 3D printer or a gripper for payload manipulation within the work envelope.

In PRO-ACT, the three robotic systems cooperated in different configurations to fulfill following tasks:

1. Cooperative exploration and mapping
2. Terrain preparation and tool exchange
3. Cooperative unloading and assembly
4. Cooperative manipulation and transport

In the first scenario, Mantis and Veles explored unknown terrain to build a map of the environment. A cooperative localization, mapping, and planning approach was used that provided motion commands for both systems. The robot-specific motion control systems transformed these in respective actuator commands, exploiting different locomotion capabilities. Building a map cooperatively is extremely important. It benefits from the data generated by multiple robots and from additional information derived from the ability of the robots to detect each other and to localize themselves in the map, which serves as a Loop Closure.

After the cooperative mapping, the wheeled robot Veles graded the terrain and demonstrated the ability to exchange the end effector tool of its robotic arm using the HOTDOCK standard interface. The goal of the activity was to present the ability of the robot to use multiple, dedicated end-effectors on a single robotic arm, as well as to be able to effectively prepare a flat working area in a lunar environment.

In the third scenario, Mantis and Veles were supposed to unload the mobile gantry, which involves the capability to cooperatively manipulate objects. However, due Covid-19 restrictions, the cooperative tests had to be replaced by single robot actions or cooperative tasks performed on a reference system. Thus coordinated manipulation was demonstrated with two fixed Panda robotics arms manipulating a rigid bar. The experiments showed the benefit of the admittance controller, which is able to cope with both internal and external perturbations, i.e., small errors in arm trajectory tracking and significant forces manually applied to the manipulated object.

The transport of the gantry crane and its deployment could also not be demonstrated. Instead, the cooperative transport was virtually demonstrated with Mantis and Veles being active but located in their respective laboratories in Germany and Poland. However, being connected to a single VPN, the planning and simultaneous execution of actions could be successfully tested. Future work needs to pick up at this stage to also test the physical interaction.

21.6.2 TransFIT—Human–robot Cooperation on Planetary Surfaces

Future space missions will not only be confined to sending scientific equipment close to or onto astronomical objects, such as planets, moons, or asteroids, but will also involve human astronauts. Therefore, setting up local infrastructure like stationary camp sites, laboratories, or even more complex and bigger modular structures is required. Since robots only have a limited capability of solving complex tasks and only have a limited flexibility in their behaviour, (semi)-autonomous robots that support a direct and intuitive robot control by astronauts are a good starting point. In such a set-up, the robot on the one hand is able to interact directly with the human, for example by holding one module while the human is assembling a second module, and on the other hand, another astronaut can teleoperate the robot from the base camp or a space station in the orbit.

In the project TransFIT [62], the focus was on the development of basic skills needed for the assembly and installation of infrastructure, such as grabbing, holding, and connecting prefabricated components autonomously, or in collaboration with a human operator or astronaut. Such basic skills require hierarchical task planning and execution, a semantic environment representation that includes reasoning capabilities, an object and pose detection, a grasp pose estimation, motion planning including collision avoidance, as well as grasp planning able to handle the approach, grasp, and retrieval of an object.

Additionally, a multimodal human–machine-interaction approach was implemented using speech, gestures, eye tracking and EEG as inputs. With speech as the main interaction modality, the human was able to give the robot basic commands. As multiple objects of the same type can be present it is not always clear which object is meant by the human. Further, the location of the object might be unknown because the astronaut changed it. For these cases, it is helpful to get more information by using additional modalities like gestures or eye tracking to further specify the meant objects or reduce the search space. In TransFIT, IMU-based gesture recognition and eye tracking in combination with a human localization was used for that purpose.

During the project, two robotic systems were used: the RECUPERA exoskeleton [63] and the newly developed humanoid robot RH5 Manus [64] (Fig. 21.12). RH5 Manus has 3 DOFs in the neck, 7 DOFs per arm and 3 DOFs in the torso. Additionally, it has a two-finger gripper on the right and a four-finger gripper on the left (Fig. 21.13). The fingers are equipped with additional touch and near field distance sensors which are especially useful for grasping and manipulation.

If the robot does not know a command or needs to execute a skill which is not implemented yet, a direct interaction approach like teleoperation can be used. Utilizing the exoskeleton as a steering device, the operator can control the arms of the humanoid robot. The end effectors of the exoskeleton (hands and elbows) are mapped in cartesian space to the corresponding end effectors of RH5 Manus. As both systems are not kinematically equivalent, the cartesian mapping is preferable to a direct joint mapping. On the other hand, as the segment lengths of the robots like upper arm length and shoulder width are different, this must be considered for a cartesian mapping.

In the project TransFIT, a new hand interface especially designed for teleoperation was developed for the RECUPERA exoskeleton. It has three trigger pads at the thumb,

Fig. 21.12 RH5 Manus grasping a tool on a table (left). It can also be teleoperated using an active exoskeleton (right)

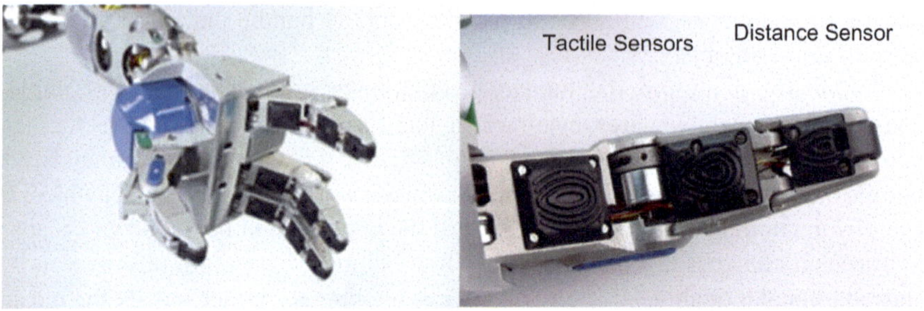

Fig. 21.13 The four-finger gripper (left) of RH5 Manus developed in the project TransFIT has tactile and distance senors integrated in the finger segments (right), which can be exploited for manipulation tasks

the index finger and the remaining fingers, a small analogue stick and two buttons. The trigger pads are mapped directly to specific fingers of the grippers facilitating a precise control of the teleoperated grippers as needed, e.g., using a pinch grasp or a power grasp. Here, the joints were mapped directly.

Manipulating objects is highly depending on motion planning, hand eye coordination and sensory feedback especially in the hands. In a teleoperated manipulation task, force feedback in combination with visual input will give the operator a good feeling of the situation at hand. In TransFIT, the forces and torques measured at the wrist sensors of RH5 Manus were mapped to the virtual wrist joints of the human operator using the exoskeleton (see Fig. 21.12). With this, the operator can feel if the gripper is in contact with an object or something is grasped.

As the autonomous and teleoperated manipulation of objects is quite the challenging task, future work must look further into combining perception, planning and execution to be able to grasp arbitrary objects repeatably and precisely while being able to adapt to dynamic changes in the environment. Dual arm manipulation tasks have even more challenges like a kinematic loop between the two arms of the robot.

Acknowledgements The authors would like to acknowledge and express their thanks to

The project ANT has been funded by ESA under contract AO/1-10289/20/NL/RA.

The project CoRob-X that was funded by the EU under GA101004130.

The Entern project (grant No. 50RA1406, 50RA1407) which was funded by the German Federal Ministry of Economics and Technology (BMWi).

The project FACILITATORS funded by the EU under GA730068.

The project LUNARES that had been funded by Federal Ministry for Economic Affairs and Technology (grant no. 50RA0706) and BIG Bremen (grant no. INNO1036A).

The project ProAct funded by the EU under GA821903

The project RIMRES, funded by the Federal Ministry for Economic Affairs and Technology (grant no. 50RA0904).

The TransFIT project was funded by the Federal Ministry for Economic Affairs and Climate Action (grant No. 50RA1701, 50RA1702 and 50RA1703).

References

1. Zhu, H., Gupta, A., Rajeswaran, A., Levine, S., & Kumar, V. (2018). Dexterous manipulation with deep reinforcement learning: efficient, general, and low-cost. arXiv, 2018.
2. Villani, L., Ficuciello, F., Lippiello, V., Palli, G., Ruggiero F., & Siciliano, B. (2012). Grasping and control of multi-fingered hands. *Advanced Bimanual Manipulation: Results from the DEXMART Project*. B, 219–266. Siciliano, Hrsg., Berlin, Heidelberg. Springer Berlin Heidelberg.
3. Atlas system page. https://www.bostondynamics.com/atlas. Last called 2022-07.
4. Spot system page. https://www.bostondynamics.com/products/spot. Last called 2022-07.
5. Schwehm, G., & Hechler, M. (1994). 'Rosetta'-ESA's planetary cornerstone mission. ESA bulletin, Bd. 77, pp. 7.
6. Bickler, D. (1998). Roving over Mars–Sojourner, *Mechancial Engineering*.
7. Dovgan, V. G., & Moisheev, A. A. (2021). First space robotic systems (On the 50th Anniversary of Lunokhod 1). *Solar System Research, 55*(7), 772–780.
8. Callas, J. L., Golombek, M. P. & Fraeman, A. A. (2019). Mars exploration rover opportunity end of mission report.
9. NASA JPL. (2022). Mars 2020 mission. Preserverance Rover. [Online]. Available: https://mars.nasa.gov/mars2020/. Zugriff am 24 02 2022.
10. European space agency (ESA). Robotic exploration of mars. [Online]. Available: https://exploration.esa.int/web/mars/. Zugriff am 24 02 2022.
11. NASA JPL. MARS Exploration Rovers. [Online]. Available: http://marsrovers.jpl.nasa.gov/home/index.html. Zugriff am 24 02 2022.
12. NASA. Where is perseverance?. Mars 2020 mission perseverance rover, March 9, 2022. [Online]. Available. https://mars.nasa.gov/mars2020/mission/where-is-the-rover/.
13. Chen, Y., Tang, Y., Fan, Y., Yan, J., Wang, C., & Zou, Y. (2022). Overview of the latest scientific achievements of chang'E-4 mission of china's lunar exploration project. *Chinese Journal of Space Science, 42*(4), 1–18.
14. NASA. Perseverance looks back after a long autonomous drive, March 1, 2022. [Online]. Available. https://mars.nasa.gov/resources/26050/perseverance-looks-back-after-a-long-autonomous-drive
15. Brinkmann, W., Cordes, F., Koch, C. E. S., Wirkus, M., Dominguez, R., Dettmann, A., Vögele, T. & Kirchner, F. (2019). Space robotic systems and artificial intelligence in the context of the European space technology roadmap. *Proceedings of Space Tech Conferences*.
16. Visentin, G. (2020). ESA AI and robotics at i-SAIRAS 2020, in Virtual Conference.
17. Arancón, M. M., Wirkus, M., Hoeflinger, K., Tsiogkas, N., Bensalem, S., Rantanen, O., Silveira, D., Hugues, J., Shilton, M. and Bruyninckx, H. (2019). ESROCOS: Development and validation of a space robotics framework. *Proceedings of the 15th Symposium on Advanced Space Technologies in Robotics and Automation (ASTRA 2019)*.
18. Dubanchet, V., & Andiappane, S. (2018). Development of I3DS: An integrated sensors suite for orbital rendezvous and planetary exploration. *Proceedings of'The 14th International Symposium on Artificial Intelligence, Robotics and Automation in Space-iSAIRAS*.
19. Dominguez, R., Govindaraj, S., Gancet, J., Post, M., Michalec, R., Oumer, N., Wehbe, B., Bianco, A., Fabisch, A., Lacroix, S., & Others. (2018). A common data fusion framework for space robotics: architecture and data fusion methods. *International Symposium on Artificial Intelligence, Robotics and Automation in Space Symposia*.
20. Vinals, J., Gala, J., & Guerra, G. (2020). Standard interface for robotic manipulation (SIROM): SRC H2020 OG5 final results-future upgrades and applications. *i-SAIRAS* 2020.
21. Ocón, J., Colmenero, F., Estremera, J., Buckley, K., Alonso, M., Heredia, E., Garcia, J., Coles, A.I., Coles, A.J., Munoz, M.M., & Others. (2018). The ERGO framework and its use in planetary/orbital scenarios. In *Proceedings of the 69th International Astronautical Congress (IAC)*.

22. Cordes, F., & Babu, A. (2016). SherpaTT: A versatile hybrid wheeled-leg rover. *Proceedings of the 13th International Symposium on Artificial Intelligence, Robotics and Automation in Space (iSAIRAS 2016).*
23. European commission, PERASPERA: Space robotics technologies. 24 02 2022. [Online]. Available. http://www.h2020-peraspera.eu/
24. Cordes, F., Kirchner, F., & Babu, A. (2018). Design and field testing of a rover with an actively articulated suspension system in a Mars analog terrain. *Journal of Field Robotics.*
25. Ocón, J., Dragomir, I., Coles, A., Green, A., Kunze, L., Marc, R., Perez, C. J., Germa, T., Bissonnette, V., Scalise, G. and Foughali, M., Kapellos, K., Dominguez, R., Cordes, F., Paar, G., Reina G., & Kisdi, A. (2020). ADE: Autonomous decision making in very long traverses. In *15th International Symposium on Artificial Intelligence, Robotics and Automation in Space (i-SAIRAS'20).* online, 2020.
26. Ocón, J., Dragomir, I., Cordes, F., Dominguez, R., Marc, R., Bissonnette, V., Viards, R., Berthet, A.C., Reina, G., Ugenti, A., Coles, A., Coles, A., Green, A., Howard, R., & Kunze, L. (2021). ADE: Enhancing autonomy for future planetary robotic exploration. In *72nd International Astronautical Congress*, Dubai, United Arab Emirates.
27. Domínguez, R., Cordes, F., Wirkus M., & Vögele, T. (2020). Space robotics software testing with the hybrid locomotion rover SherpaTT and a validation toolset. In *15th International Symposium on Artificial Intelligence, Robotics and Automation in Space (i-SAIRAS '20)*, online, 2020.
28. Ocón, J., Colmenero, F., Dragomir, I., Heredia, E., Alonso, M., Estremera, J., Marc, R., Weclewski, P., Keller, T., Woods, M., & Karachalios, S. (2019). Testing autonomous robots: A discussion on performances obtained during the ergo field tests. In *Proceedings of the 15th Symposium on Advanced Space Technologies in Robotics and Automation (ASTRA),* Leiden, The Netherlands.
29. Ugenti, A., Vulpi, F., Domínguez, R., Cordes, F., Milella A., & Reina, G. (2021). On the role of feature and signal selection for terrain learning in planetary exploration robots. *Journal of Field Robotics.*
30. Dettmann, A., Planthaber, S., Bargsten, V., Dominguez, R., Cerilli, G., Marchitto, M., Focchi, M., Barasuol, V., Semini, C. & Marc, R. (2022). Towards a generic navigation and locomotion control system For legged space exploration. *In Proc. of the Symp. on Advanced Space Technologies in Robotics and Automation.*
31. Danter, L., Planthaber, S., Dettmann, A., Brinkmann W., & Kirchner, F. (2020). Lightweight and framework-independent communication Li-Brary to support cross-platform robotic applications and high-latency connections. In *Proceedings of International Symposium on Artificial Intelligence, Robotics and Automation in Space.*
32. Machowinski, J., Böckmann, A., Arnold, S. Hertzberg C. & Planthaber, S. (2017). Climbing steep inclines with a six-legged robot using locomotion planning. In *International Conference on Robotics and Automation.*
33. Robotics, U. (2018). Aliengo. Oct 18, 2020. [Online]. Available. https://www.unitree.com/products/aliengo/
34. Roehr, T. M., Cordes, F., & Kirchner, F. (2013). Reconfigurable integrated multirobot exploration system (RIMRES): Heterogeneous modular reconfigurable robots for space exploration. *Journal of Field Robotics, 3–34*, 2013.
35. Cordes, F., Ahrns, I., Bartsch, S., Birnschein, T., Dettmann, A., Estable, S., Haase, S., Hi lljegerdes, J., Koebel, D., Planthaber, S., Roehr, T., Scheper M., & Kirchner, F. (2010). LUNARES: lunar crater exploration with heterogeneous multi robot systems. *Intelligent Service Robotics,* 1–29.
36. Sonsalla, R. U., Cordes, F., Christensen, L., Planthaber, S., Albiez, J., Scholz, I. and Kirchner, F. (2014). Towards a heterogeneous modular robotic team in a logistic chain for extraterrestrial exploration. In *12th International Symposium on Artificial Intelligence, Robotics and Automation in Space (i-SAIRAS'14).* Montreal, Canada.
37. Brinkmann, W., Cordes, F., Roehr, T. M., Christensen, L., Stark, T., Sonsalla, R. U., Szczuka, R., Mulsow, N. A., Bernhard, F. & Kuehn, D. (2018). Modular Payload-items for payload-assembly

and system enhancement for future planetary missions. In *IEEE Aerospace Conference (AeroConf'18)*, Big Sky, MT, USA.
38. Klein, E., Nilsen, E., Nicholas, A., Whetsel, C., Parrish, J., Mattingly, R. and May, L. (2014). The mobile MAV concept for mars sample return. In *IEEE Aerospace Conference (IAC'14).*
39. Sonsalla, R. U., Akpo, J. B., & Kirchner, F. (2015). Coyote III: Development of a modular and highly mobile micro rover. In *13th Symposium on Advanced Space Technologies in Robotics and Automation (ASTRA'15),* Leiden, The Netherlands.
40. Sonsalla, R., Cordes, F., Christensen, L., Roehr, T.M., Stark, T., Planthaber, S., Maurus, M., Mallwitz, M. & Kirchner, E.A. (2017). Field testing of a cooperative multi-robot sample return mission in mars analogue environment. In *14th Symposium on Advanced Space Technologies in Robotics and Automation (ASTRA'17).*
41. Planthaber, S., Maurus, M., Bongardt, B., Mallwitz, M., Vaca Benitnez, L. M., Christensen, L., Cordes, F., Sonsalla, R., Stark, T. and Roehr, T. (2017). Controlling a semi-autonomous robot team from a virtual environment. *Human Robot Interaction Conference (HRI'2017).*
42. Sauro, F., Pozzobon, R., Massironi, M. D., Berardinis, P., Santagata, T., & De Waele, J. (2020). Lava tubes on Earth, Moon and Mars: A review on their size and morphology revealed by comparative planetology. *Earth-Science Reviews, 209*(103288), 2020.
43. Nesnas, I. A. D., Abad-Manterola, P., Edlund J. A., & Burdick, J. W. (2008). Axel mobility platform for steep terrain excursions and sampling on planetary surfaces. In *2008 IEEE Aerospace Conference.*
44. Nesnas, I. A., Matthews, J. B., Abad-Manterola, P., Burdick, J. W., Edlund, J. A., Morrison, J. C., Peters, R. D., Tanner, M. M., Miyake, R. N., Solish, B. S. & Anderson, R. C. (2012). Axel and DuAxel rovers for the sustainable exploration of extreme terrains. *Journal of Field Robotics*, Bd. 29, 663–685.
45. Schwendner, J., Hidalgo-Carrio, J., Domınguez, R., Planthaber, S., Yoo, Y.-H., Asadi, B., Machowinski, J., Rauch C., & Kirchner, F. (2015). Entern–environment modelling and navigation for robotic space-exploration. *ESA/Estec Symposium on Advanced Space Technologies in Robotics and Automation (ASTRA)*, 13th, May.
46. IEEE standard for robot map data representation for navigation, 1873–2015. *IEEE Standard for Robot Map Data Representation for Navigation*, pp. 1–54.
47. Envire library for environment representation and plugins. https://github.com/envire
48. Triebel, R., Pfaff, P., & Burgard, W. (2006). Multi-level surface maps for outdoor terrain mapping and loop closing. In *2006 IEEE/RSJ International Conference on Intelligent Robots and Systems.*
49. Domınguez, R., Arnold, S., Hertzberg C., & Böckmann, A. (2018). Internal simulation for autonomous robot exploration of lava tubes. In *Proceedings of the 15th International Conference on Informatics in Control, Automation and Robotics.*
50. Dettmann, A., Voegele, T., Ocón, J., Dragomir, I., Govindaraj, S., De Benedetti, M., Ciarletti, V., Hassen-Khodja, R., Germa, T., Viards, R., Paz-Delgado, G. J., & Mantoani, L., & Mantoani, L. M. (2022). Corob-X: A cooperative robot team for the exploration of lunar skylights. In *Proc of the Symp. on Advanced Space Technologies in Robotics and Automation.*
51. Losekamm, M. J., Barber, S., Biswas, J., Chupin, T., Evagora, A., Fau, G., Fodorcan, D., Gancet, J., Kubitza, S., Madakashira, H. K., & Murray, N. (2021). LUVMI-X: A versatile platform for resource prospecting on the moon. *In Earth and Space, 2021*(289–299), 2021.
52. Simon, T., & Sacksteder, K. (2007). NASA in-situ resource utilization (ISRU) development and incorporation plans. In *Presentation at Technology Exchange Conference,* Galveston, TX, 2007.
53. David, L. (2016). Inside ULA's plan to have 1000 people working in space by 2045. *Space.com*, June, Bd. 29, 2016.
54. Blair, B. R., Diaz, J., Duke, M., Lamassoure, E., Easter, R., Oderman M. & Vaucher, M. (2002). Space resource economic analysis toolkit: The case for commercial lunar ice mining. *Final report to the NASA Exploration Team.*
55. Urbina, D., Madakashira, H.K., Salini, J., Govindaraj, S., Bjoerstad, R., Gancet, J., Sperl, M., Meurisse, A., Fateri, M., Imhof, B. & Hoheneder, W. (2017). Robotic prototypes for the

solar sintering of regolith on the lunar surface developed within the Regolight project. In *68th International Astronautical Congress (IAC)*, Adelaide, Australia.
56. https://www.h2020-peraspera.eu
57. Govindaraj, S., Gancet, J., Urbina, D., Brinkmann, W., Aouf, N., Lacroix, S., Wolski, M., Colmenero, F., Walshe, M., Ortega, C., & Balazs, B. (2019). PRO-ACT: Planetary robots deployed for assembly and construction of future lunar ISRU and supporting infrastructures. In *Proceedings of the 15th Symposium on Advanced Space Technologies in Robotics and Automation (ASTRA-2019).*
58. Brinkmann, W., Danter, L. C., Stark, T., Dettmann, A., De Benedetti, M., Govindaraj, S., Sans Nieto, I., But, A., Colmenero, F. J., Heredia E., & Alonso, M. (2022). Mantis with in a multi robot team for lunar exploration and construction tasks. *IEEE Aerospace Conference.*
59. https://piap.lukasiewicz.gov.pl/en/2018/01/26/an-ibis-robot-on-watch-for-safety-of-polish-airports/
60. Letier, P., Siedel, T., Deremetz, M., Pavlovskis, E., Lietaer, B., Nottensteiner, K., Roa Garzon, M. A., Sánchez Garcia, J., Corella, L., & Gancet, J. (2020). HOTDOCK: Design and Validation of a New Generation of Standard Robotic Interface for On-Orbit Servicing. *In International Astronautical Congress, IAC, 2020*, 2020.
61. Govindaraj, S., Brinkmann, W., Colmenero, F. J., Nieto, I. S., But, A., de Benedetti, M., Danter, L. C., Alonso, M., Heredia, E., Lacroix, S., Kleszczynski, D., Purnell, J., Picton, K., Aouf, N., & Lopes, L. (2021). Building a lunar infrastructure with the help of a heterogeneous (Semi) Autonomous multi-robot-team. In *proceedings of the 72nd International Astronautical Congress.*
62. DFKI, TransFIT, April 1, 2022. [Online]. Available. https://robotik.dfki-bremen.de/en/research/projects/transfit/
63. Kumar, S., Wöhrle, H., Trampler, M., Simnofske, M., Peters, H., Mallwitz, M., Kirchner, E.A. and Kirchner, F. (2019). Modular design and decentralized control of the recupera exoskeleton for stroke rehabilitation, *Applied Sciences*, Bd. 9.
64. Boukheddimi, M., Kumar, S., Peters, H., Mronga, D., Budhiraja, R., & Kirchner, F. (2022). Introducing RH5 Manus: A powerful humanoid upper body design for dynamic movements. In *2022 IEEE International Conference on Robotics and Automation (ICRA).*

Chapter 22
HOTDOCK and the MOSAR Walking Manipulator, Technologies for Modular Spacecraft Assembly and Reconfiguration

Pierre Letier, Mathieu Deremetz, Jeremi Gancet, Gerhard Grunwald, Maximo A. Roa, Bernhard Brunner, and Xiu Tian Yan

Abstract The concept of modular and reconfigurable spacecraft is a possible game-changer solution to answer the future challenges of the space industry and exploration, that includes, space sustainability, economical profitability, system reliability and large structure on-orbit assemblies. This paper introduces the EC H2020 MOSAR project, aiming at refining and demonstrating the concept of modular spacecraft. More specifically, the chapter will focus on the design, validation and future perspectives of the two robotic system developments, which are the standard interconnect (SI) HOTDOCK and the MOSAR Walking Manipulator (MOSAR-WM). The exploitation of standard interconnects, providing mechanical /data/power/thermal transfers between connected entities, as well as more advanced and capable robotic systems, for performing the operations, open new horizon of possibilities for future space missions, ranging from On-Orbit Servicing of existing satellites to large On-Orbit Assembly and reconfiguration of modular spacecraft.

22.1 Introduction

A major trend in space is the emergence of on-orbit servicing, maintenance, refuelling and lifetime extension capabilities [1], which have been demonstrated by companies in North America, as for example with the recent re-fuelling operation performed

P. Letier (✉) · M. Deremetz · J. Gancet · M. A. Roa · B. Brunner
Space Applications Services, Zaventem, SA, Belgium
e-mail: pierre.letier@spaceapplications.com

M. Deremetz · J. Gancet · G. Grunwald · M. A. Roa · B. Brunner
Institute of Robotics and Mechatronics, German Aerospace Center (DLR), Cologne, Germany

M. Deremetz · J. Gancet · M. A. Roa · B. Brunner · X. T. Yan
RAS Group, DMEM, University of Strathclyde, Glasgow, Scotland
e-mail: x.yan@strath.ac.uk

X. T. Yan and G. Visentin (eds.), *Space Robotics*,
https://doi.org/10.1007/978-3-031-39214-6_22

by Northrop Grumman's MEV-1 servicer [2]. A further trend is the huge investments made by NASA to promote the commercialization of low Earth orbit space assets, which will continue with the development of commercial space stations and platforms. The interest for space sustainability is also growing very fast, with the main focus on the removal of the space debris and satellite end-of-life management, as illustrated by the ESA Clean Space initiative [3] and the mission CLearSpace-1, planned for 2025 [4], targeting the first debris removal from orbit.

Offering cost effective, performant, reliable, scalable and flexible solutions will be essential to make the European space industry spearheading strategic segments of the commercial space. The development of novel, European technologies that would allow standardizing satellites components, facilitating their assembly, reducing time between customer's orders and commissioning in space, repairing and upgrading components directly in orbit instead of replacing entirely a failed or damaged satellite, would provide extremely valuable competitive edges to establish European leadership in a number of key commercial space segments. In that context, Standard Interconnects (SI) and small-medium scale versatile orbital robots are considered presently as promising assets for opening new horizons of possibilities and configurations for such future space missions [5].

MOSAR is a space oriented research project of the European Commission proposing a new paradigm for the on-orbit assembly, operations and servicing of spacecraft. With a consortium of nine industrial and academic partners, MOSAR shall elaborate and refine the concept of Modular Spacecraft, identify and recommend standards and design the ground and space infrastructure elements that will enable its realization and sustain its progressive deployment and economical exploitation. This is achieved by the development of a ground demonstrator of on-orbit modular satellite reconfiguration, and by the analysis and elicitation of recommendations for the exploitation and maturation of the associated technologies.

The first section will provide a general overview of the MOSAR project, its objectives, as well introduce, in the perspective of the mission, the ground demonstrator aiming at illustrating the concepts of modular spacecraft. The two next sections describe respectively the design of the HOTDOCK SI and robotic walking manipulator. Then the main results of the robotics demonstration are summarized, before providing the conclusions, with a focus on the future perspectives of each device.

22.2 MOSAR Project Overview

22.2.1 MOSAR Mission Concept

The technologies developed in MOSAR aim to support the development of fully modular and on-orbit reconfigurable spacecraft. This ambitious objective is derived into a demonstrative mission concept illustrated in Fig. 22.1, with an artistic view and spacecraft design.

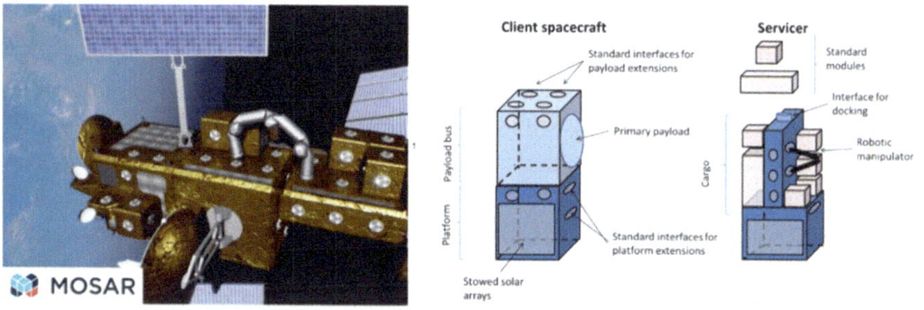

Fig. 22.1 MOSAR mission concept artistic view and derived spacecraft design (client and servicer)

The baseline scenario is the one of a Servicer Spacecraft (SVC) transporting a cargo of Spacecraft Modules (SM) and a dedicated Walking Manipulator (WM). This last enables a number of operations with the transfer of SM from and to the Client Spacecraft (CLT), with the purpose of adding, replacing or enhancing client platform or payload functionalities.

This mission concept aims at making use of MOSAR technologies to introduce enhanced functionalities on space missions, mainly on two topics:

Standardization of design thanks to modularity approach: a common design of the product line allows multiple configurations to meet different mission objectives thanks to the standardized interfaces that could be used to plug additional payloads and service modules. In this way, the customization needed to meet specific mission objectives do not need a redesign of the main elements of the spacecraft.

The spacecraft, equipped with multiple SI and some modules, performs its original missions for some years with the option of introducing new elements or reconfiguring the mission by replacing or adding new modules brought by a servicer satellite equipped with a robotic manipulator.

The client satellite is built around a standardized common platform, with all the required main functionalities. The platform is designed to cover an envelope of payloads and functions. It can be initially specialized, on-ground, with a primary payload and services modules. Both the platform and the payload are equipped with SI enabling the connection of the standard modules, for initial configuration on-ground or for addition or replacement of units during on-orbit mission.

Spacecraft Modules are individual structures associated with a specific function of the spacecraft, that once assembled, will ensure the spacecraft operations. They can cover system or payload functions. The SM are equipped with SI, enabling Modules to Modules and Modules to spacecraft connections, as well their manipulation with the robotic manipulator, equipped with a SI as end-effector. Each SI provides mechanical, data, power and thermal transfer capabilities, allowing a full configuration of the CLT functions. That includes mechanical structural integrity, data transmission, power routing and thermal management along the elements.

The autonomous transfer and configuration of the SM follow an execution plan prepared and validated off-line, in the Monitoring and Control Centre (MCC), on the ground segment. The MCC includes a satellite design, modelling and validation tool, specifically targeting modular satellites applications. It also allows the automatic planning of the assembly or reconfiguration sequence that can be verified with a multi-physics simulator (FES). All these elements are working iteratively together to prepare a valid execution plan that is finally uploaded to the spacecraft for execution. Based on the monitoring and feedback information received from the spacecraft during the operations (e.g. detected failed module), the MCC can update the execution plan. The MCC finally includes visualisation front-end to support the design, verification and monitoring activities during sequence execution.

22.2.2 MOSAR Project Objectives

The first objective of MOSAR is to develop and mature key technologies and product ecosystem that will support in the future the deployment and operations of modular spacecraft:

- Extension and integration of the ESROCOS, ERGO, INFUSE, I3DS and HOTDOCK common building blocks, the outcomes from the previous Operational Grants (OG), belonging to the same European Research Cluster. The aim is to encourage reusability, and initiate the standardization in the European space robotic endeavour.
- Development of a repositionable robotic manipulator system (MOSAR- WM) allowing the manipulation and relocation of the Spacecraft Modules Development of a modular Spacecraft Modules ecosystem, consisting of: Standardized functional modules; Suitable power and data buses technologies allowing flexible and hot-reconfiguration; Functional Engineering Simulation (FES) environment offering assistance for the design of modular satellites, for the verification of obtained designs (multi-physics simulation), and for the motion planning and rehearsal of assembly / reconfiguration operations.

The second objective of MOSAR is to develop a sound technology ground demonstrator of on-orbit modular satellite reconfiguration relying on robotic capabilities, exploiting the technologies described above, in order to illustrate the operative mission described in the previous section.

22.2.3 From Mission to Demonstration

MOSAR demonstrator aims at reproducing the main functionalities needed for on-orbit module assembly and reconfiguration of spacecraft, as described in the mission section above. Figure 22.2 illustrates the concept for the spatial mission, involving a client and a servicer satellite, already docked. From this concept, a physical ground demonstrator has been integrated, including representative components for

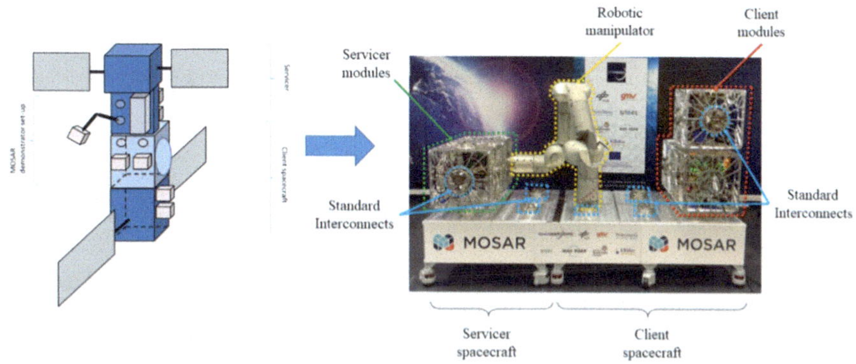

Fig. 22.2 MOSAR ground demonstrator derivate from MOSAR mission concept

the servicer and client spacecraft buses, several functional modules, some of them being already installed on the client, and other ones brought by the servicer, and the walking manipulator. The MOSAR ground demonstrator has been built in order to provide the most realistic vision compared to the proposed space design approach. It had however to take into account practical constraints, mainly imposed by the presence of the gravity, reducing the capability to perform modular operations with big assets. The nominal module size in the demonstrator has been selected in accordance to the 1U size unit, in order to limit the modules weight to the lifting capabilities of the MOSAR-WM. The operational concept is however fully extendable to other shape, sizes and structures of elements.

The ground demonstrator has been focusing on the validation of the sequence of operations needed for the preparation of the operations with the ground support tools (design, simulation and planning), the operational concept of manipulator and spacecraft module relocation, resources re-allocation and further client satellite operations. It addresses the following topics:

- Hardware: with the possibility to re-configure the physical arrangement of the spacecraft and/or providing means to add/replace/upgrade specific functions.
- Software: with the possibility to re-configure components responsibilities and support the re-configuration operations
- Data: with the possibility to re-route TM/TC and data transmission along the different elements
- Power: with the possibility to re-route and control the power transmission along the elements.
- Thermal: possibility to manage thermal loads, to increase dissipation capabilities or maintain thermal stability of modules.
- Visual: possibility to support operations through external visual detection system, targeting the inspection of spacecraft configuration and detection of anomalies.

The next sections will focus on the description of the HOTDOCK SI and MOSAR-WM components, as well as their validation during the MOSAR demonstration. More details on the other parts of the system and the related demonstration results are provided in [6].

22.3 HOTDOCK: Space Standard Interconnect

22.3.1 Motivation

Most robotic interfaces with space heritage today such as the Power and Data Grapple Fixture device (PDGF) and the Wedge Mating Interface (WMI) have been designed for robotic operations on the ISS using the Space Station Remote Manipulator System (SSRMS, a.k.a Canadarm2) and the Special Purpose Dexterous Manipulator (SPDM, a.k.a Dextre). These interfaces are generally too bulky and massive for the emerging LEO/GEO robotic servicing market and future exploration missions to the Moon or Mars.

In this context, Space Applications Services has initiated the development of HOTDOCK, an innovative standard interconnect building block for commercial exploitation in space (Fig. 22.3). It is a mating device providing androgynous coupling to transfer mechanical loads, electrical power, data and (optionally) fluids (for thermal or refuelling applications) through a single interface. Integrated to spacecraft and payload structures, it provides mounting points for components assembly and reconfiguration. Mounted at the tip of a robot arm, it works as an end-effector for quick connect/release of spacecraft modules during their manipulation or as a quick tool adaptor. HOTDOCK eases the replacement of failed modules (ORU) as well as swapping of payload elements, and provides chainable data interfaces for building spacecraft modules assemblies.

22.3.2 HOTDOCK General Overview

HOTDOCK is a compact mating device that provides the following four interface functions (Fig. 22.4):

- The mechanical interface that provides the alignment, connection and mechanical load transfer capabilities.
- The electrical power interface that offers the possibility to control the transfer of power through the component.
- The bi-directional data interface that ensures high-rate communication between systems.

Fig. 22.3 HOTDOCK passive (left) and active (right) standard interconnect

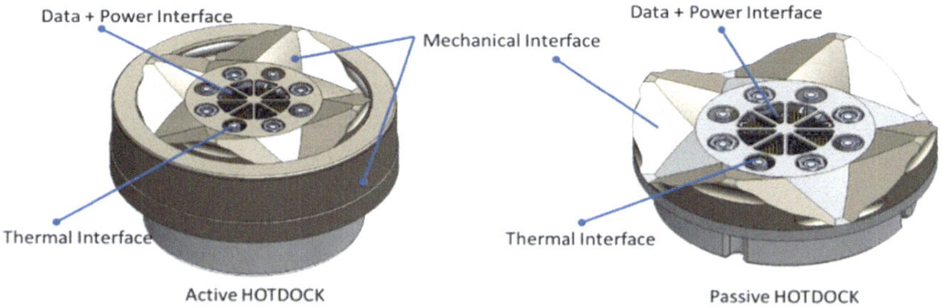

Fig. 22.4 Active and passive version of HOTDOCK with highlight of the four main interface functions

- The fluidic interface that can optionally be integrated to allow fluid (e.g. refuelling) and thermal energy transfer.

Based on the same design and shape, different declinations of HOTDOCK have been developed to offer mass and cost optimization as function of the application. While the *Active* interface has all the above features, the *Passive* interface is a simplified version of the *Active*, which provides the form-fit geometry and a fixed power/data transfer interface, without any active capture mechanism. As a third variation, the *Mechanical* version offers only the form-fit geometry. An *Active* interface is able to connect independently to another *Active*, *Passive* or *Mechanical* version. The two simpler declinations of HOTDOCK present a reduced volume and mass, do not include moving parts and can conveniently and affordably be integrated on spacecraft buses and payloads, in prevision of future potential connection in space with an *Active* HOTDOCK.

Table 22.1 summarizes the main characteristics of the HOTDOCK ground model. More details on each of the key features are provided in the next sections.

Table 22.1 HOTDOCK characteristics

Dimensions (diameter×height)	Active: 148 × 70 mm Passive: 120 × 35 mm Mechanical: 120 × 25 mm
Mass	Active (Fluidic): 1.4 kg(1.55 kg) Passive (Fluidic): 0.35 kg(0.5 kg) Mechanical: 0.25 kg
General features	Androgynous design with 90° symmetry (for all interfaces)
Misalignment tolerances	± 15 mm in translation 10° in rotation
Capture range (without contact)	2 mm
Diagonal engagement	130° approach aperture
Load transfer (non-destructive)	3000 N in traction 350 Nm in bending moment
Coupling sequence	20 s
Power transfer	2.5 kW @ 120 V (with specific pin configuration), with switch relay option and power measurement
Data transfer	CAN, SpW, Ethernet, TTE
Thermal transfer	up to 2.5KW (conduction up to 50W)
Powering	24 V, 0.15A (average consumption)
Control, TM/TC	CAN

22.3.3 HOTDOCK Mechanical Interface

The HOTDOCK mechanical interface provides the following features (Fig. 22.5):

- Androgynous design: both on the mechanical and electrical connections, such that one HOTDOCK can mate any other ones, with similar design.
- 90-degree symmetry: on the mechanical and electrical side, that provides additional redundancy and increases the range of possible position/orientation for mating, for easier robotic manipulation or compatibility with parallelepiped spacecraft modules

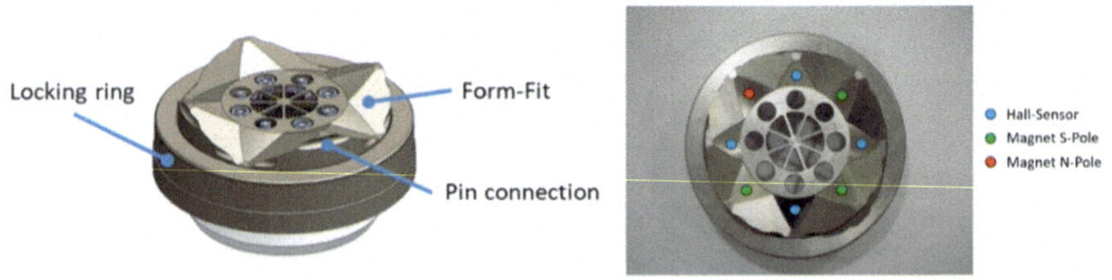

Fig. 22.5 Left: HOTDOCK mechanical interface elements; Right: Hall sensor configuration

- Locking mechanism: the active version of HOTDOCK is equipped with a patented locking mechanism to enable the mating and fixation to another HOTDOCK device. The coupling mechanism relies on four locking pins, embedded in the locking ring structure (Fig. 22.5-Left). During the mating process, they are pressed with the mated partner, by the motion of the locking ring. This integrated compact solution, without external deployed mechanical parts, offer a high level of mechanical robustness during the approach phase. The locking mechanism is compatible with a remaining misalignment of up to 2 mm. The use of the full circumference of the device allows high levels of load transfer. HOTDOCK has been tested with loads of 3kN in traction and 350Nm/700Nm in bending moment (for active/passive or active/active configuration). The mechanism is driven by a BLDC sensor-less motor connected to an internal gearbox and barrel cam mechanism. An absolute position sensor measures the motion, enabling the control of the connection states of the interface. The same mechanism is also used to drive the motion of the central power and data connector plate and the fluidic interface.
- Form-fit guidance: tooth like geometries on the circumference enable self-guidance positioning during the final approach process and compensation for a range of transversal and orientation misalignments. The form-fit is also designed to support straight and diagonal engagement. The form fit has also a main role in the mechanical load transfer capabilities of the interface.

Figure 22.6 illustrates the tests and verifications that have been performed to characterize the range of attraction (misalignment tolerance) and the diagonal engagement capabilities of the HOTDOCK form-fit geometry. The form fit geometry allows guiding the final approach of the interface before starting the mating process, with the support of a compliant robotic manipulator. The geometry allows 15 mm of initial misalignment in all directions, as well as 10° in rotation. This enables mating alignment without necessarily requiring external visual servoing capability. This is furthermore supported by Hall-effect sensors embedded in the mechanical structure, that are used for proximity measurement as well as verification of alignment and relative orientation of the two mated interfaces Fig. 22.5-Right. HOTDOCK offers a 130° approach aperture, enabling the simultaneous mating between a number of interfaces exposed in "rectangular" and "hexagonal" configurations.

22.3.4 HOTDOCK Power and Data Interface

The HOTDOCK power and data interfaces are integrated in the inner section of the device, through a PCB connector plate (Fig. 22.7-Left). Both electrical power and data are transferred via a set of spring-loaded connectors also known as "POGO" connectors. They are particularly tolerant to misalignment and prevent accumulation of dirt or dust. This keeps the power and data transfer interface simple and subsequently improves its reliability.

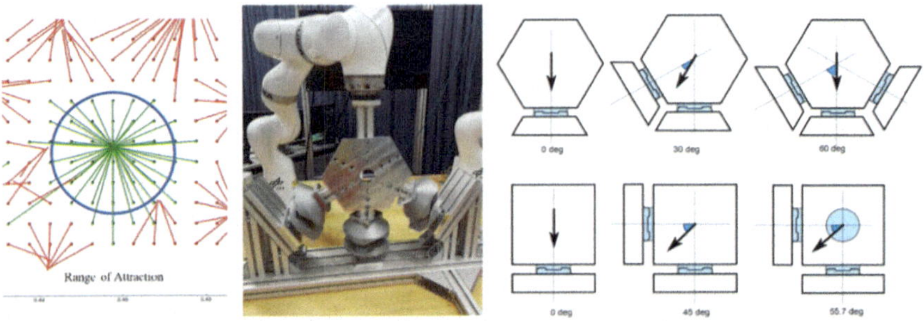

Fig. 22.6 Left: Experimental result of HOTDOCK form-fit range of attraction; Centre: Test setup to validate HOTDOCK angular engagement; Right: Compatibility of HOTDOCK engagement with standard module shapes

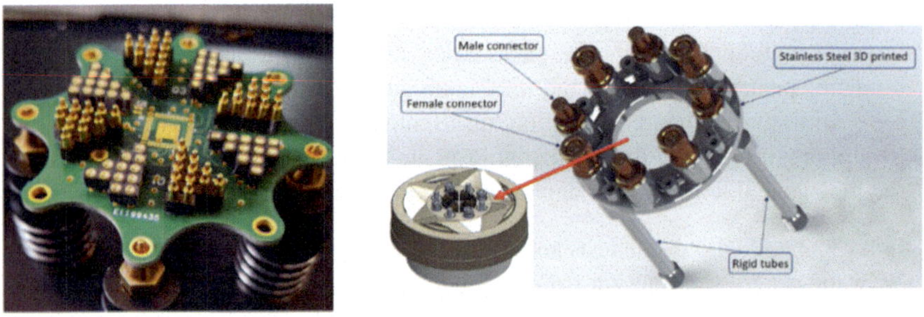

Fig. 22.7 Left: HOTDOCK central power and data connector plate, equipped with POGO connectors; Right: HOTDOCK fluidic interface structure

The PCB is split in four sections, with 24 independent POGO pins/pads, arranged in mirror to ensure the androgynous and 90° symmetry characteristics of the interface, offering also the option of redundant data transmission. The flexibility of the connector layout makes it easy to integrate the POGOs in various patterns, as function of the application and the required power/data transfers needs.

In the MOSAR configuration of the power interface (used in the demonstration projects described afterwards), it can transfer up to 10A @ 120 V. It can also be equipped with an internal power switch relay or control an external one, while also measuring the current and voltage of the power line. The data interface has been currently validated for Ethernet, CAN and SpaceWire bi-directional transmission. It is equipped with an LVDS switch to enable signal data routing through the different quadrant, as function of the interface orientation, while keeping required track impedance for high-speed signals.

22.3.5 HOTDOCK Fluidic Interface

Thermal management is an important topic in space environment. HOTDOCK can be equipped with an internal fluidic interface to enable thermal fluid transfer between two connected HOTDOCK units. Developed in collaboration with MAG SOAR, the HOTDOCK thermal interface consists of 8 hydraulic Stäubli connectors (four male and four female) integrated in a 3D printed titanium structure (Fig. 22.7-Right). The fluidic interface deployment relies on the regular HOTDOCK actuation mechanism. The additional weight for the interface is 150 g. With a maximum fluid pressure of 16 bar, it can transfer up to 2500W of heat power.

22.3.6 HOTDOCK Controller and Operations

The HOTDOCK controller is implemented as a round-shaped custom PCB to seamlessly fit at the back of the HOTDOCK mechanism (Fig. 22.8). As illustrated in the avionics architecture diagram, the controller is interfaced on the outside with the HOTDOCK CAN control TM/TC line, the HOTDOCK power line (24 V) and the power/data transfer lines. On the other side, it interfaces with the central POGO connector plate. The PCB controller is also equipped with an external interface to support update of the firmware and for debugging purpose. The HOTDOCK controller manages the following tasks:

- Motor control: field-oriented control of the brushless motor (Sect. 3.2), by interfacing the front-end chain (H-bridge, gate drive and buffer) through PMW signal.

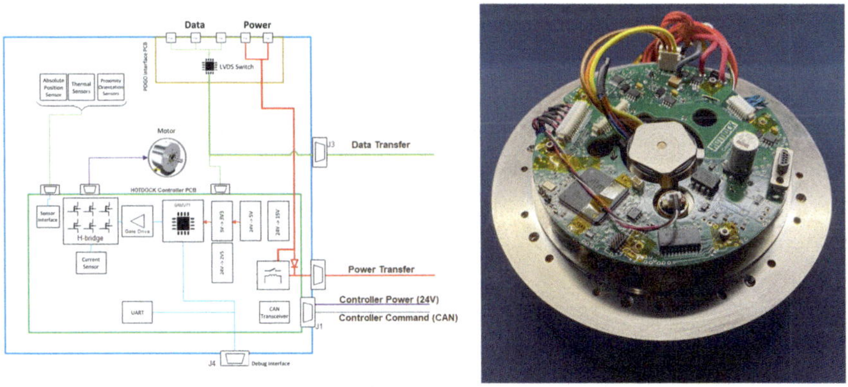

Fig. 22.8 HOTDOCK avionics architecture and integrated PCB controller board

- Sensors interface: processing of the analogue and digital internal signals of HOTDOCK including the Hall effects proximity/orientation sensors, the absolute encoder locking sensor, the voltage/current of the interface power transfer and the 4 internal temperature sensors, used to monitor the state of the device.
- Local management: with the implementation of a Finite State Machine and FDIR management.
- Local electrical power conversion: from the 24 V powering bus, for the PCB components.
- TM/TC: to enable telemetry and telecommands with an external OBC/EGSE through CAN bus.
- LVDS switch: control of the connector plate data routing.
- Power relay: latching relay control to enable the bi-directional control of the power flow through the interface. It can also be used to control external relays for higher current applications.

22.3.7 HOTDOCK Controller and Operations

In MOSAR, HOTDOCK plays a central role in all the project applications and demonstration. As illustrated in Fig. 22.9, HOTDOCK has been successfully implemented for the following applications:

- As end-effectors of the walking manipulator (on each extremity) for the operations of manipulation of the spacecraft modules and the motion of the robot along the spacecraft structures and on top of the modules
- As SI for the spacecraft modules, each of them being equipped with active, passive and mechanical versions, for their interconnection and connection to the spacecraft bus and MOSAR-WM, including data (SpaceWire) and electrical power transmission

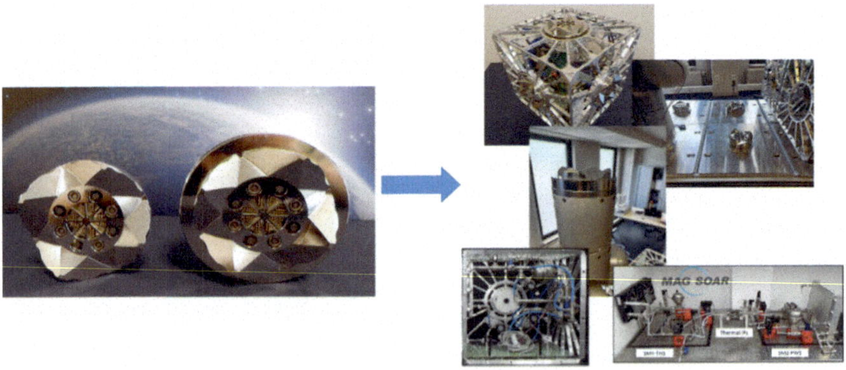

Fig. 22.9 Integration of active and passive HOTDOCK SI in MOSAR components

- As interface point for the spacecraft buses, for the connection of the SM and the WM
- As key component for the test and validation of thermal power between two modules

22.4 MOSAR Walking Manipulator

22.4.1 Motivation

For the purpose of the MOSAR project specificities, ISS space manipulators (Canadarm, ERA), capable to relocate themselves, have been identified as an excellence baseline to be transposed to an autonomous small scale case. ESA R&D projects, DEXARM and EUROBOT, have been also identified as excellent baseline to be adapted in MOSAR, especially in term of selected technology toward space qualification and development perspectives (multi-am robot).

From this analysis, the development of the walking manipulator for MOSAR has been initiated by Space Applications Services. The principal key target was to propose a new compact and fully integrated robotic manipulator with high lifting capabilities and the possibility to re-locate itself along structures, through the connection of standard connectors equipped on each end-effector.

22.4.2 MOSAR-WM General Overview

The result is the MOSAR-WM (Fig. 22.10), a 7-DOF redundant and symmetrical anthropomorphic manipulator, with HOTDOCK based end-effector, embedded intelligence and internal harnessing (RCU, SCUs, data and power buses), an autonomous impedance based control (vision system is not needed), and a lifting capability to carry, simultaneously, itself and a substantial payload (10 kg at 1 g in the entire workspace), all in a compact form factor and light structure (1.6 m/30 kg). Thank to these features, this is making the MOSAR-WM a precursor and unique robotic system with regard to the existing SoA in the category of autonomous anthropomorphic small relocatable robotic manipulator, currently developed at TRL4, and targeting future on-orbit servicing applications. Table 22.2 summarizes the main characteristics of the MOSAR-WM ground model.

Fig. 22.10 MOSAR-WM relocatable manipulator (connected on the side of a MOSAR module)

22.4.3 MOSAR-WM Mechanical Design

The MOSAR-WM comprises the following elements:

- Structure (Limbs): The walking manipulator is composed by eight structural elements. The overall configuration of the manipulator is based on a human-like arm with asymmetric joints (Fig. 22.11).
- Motorization (Robotic joints): The walking manipulator has seven revolute joints according to the following symmetric configuration $R{\perp}R{\perp}R{\perp}R{\perp}R{\perp}R{\perp}R$, where R indicates a revolute joint and $\perp$ the orthogonality between two successive joint axes (see Fig. 22.11). Each joint is equipped with a specific developed rotary hollow shaft BLDC based actuator. Two different sized have been designed to optimize the torque-to weight ratio. They are all equipped with two position sensors (incremental and absolute) and a torque sensor (Fig. 22.12).
- End-effector: the two tips are each equipped with an active HOTDOCK SI, as depicted in previous section. It guarantees the mechanical connection of the arm to its base structure, as well the possible connection to payload or tool changers.

Table 22.2 MOSAR-WM characteristics

Length/Degrees of Freedom	1.6 m/7-DOF
Maximum reach / joints limits	1.2 m/−175 to 175 (45) degrees
Weight	32.5 kg
Payload	10 kg (at the end-effector in all workspace)
Joint torque capabilities (permanent)	Joint 1-3-4-5-7: 175 Nm Joint 2–6: 260 Nm
Joint maximum rotational speed	0.15 rad/s
Joint sensors	Position/torque/temperature
Mechanical Structure/design	Aluminium based Symmetrical design Hollow shaft joint and structure design (fully internal harnessing)
End-effector	Each end tips equipped with an active HOTDOCK featuring mechanical/data/power transfer
Avionics	Embedded control unit (Intel NUC), USB interface Individual servo controllers at each joint (etherCAT) CAN, WiFi and SpW communication interface
Safety	Safety brakes on each joint Software joint position, velocity and torques limits
Power and consumption	Input: 48 V Stationary (with holding brakes): 35W Operational: [80–200]W as function of speed and payload mass

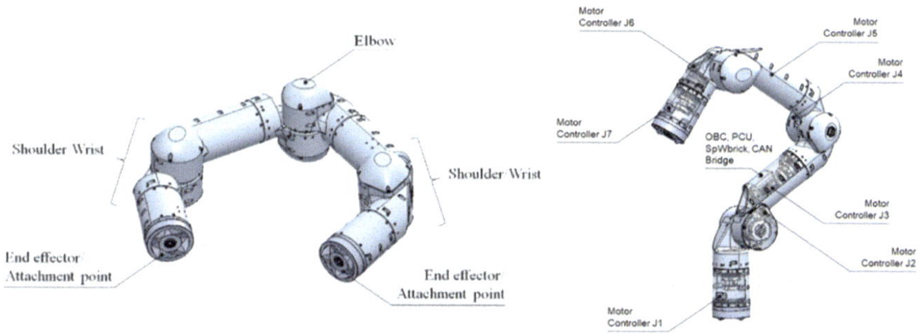

Fig. 22.11 MOSAR-WM design and avionic integration

It provides data and power transmission (both for the self-powering of the arm and the power transmission to the other extremity).

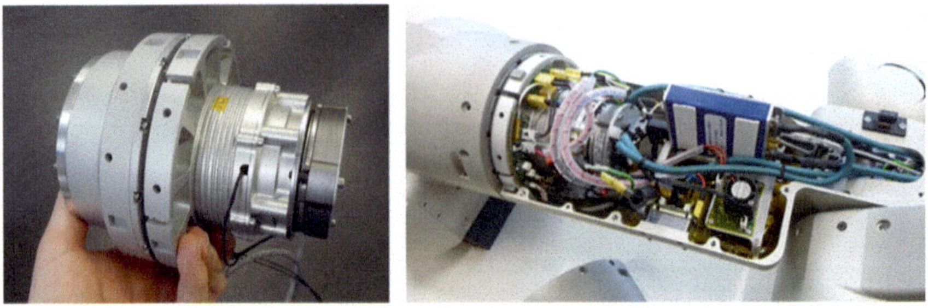

Fig. 22.12 MOSAR-WM actuator and avionics integration

22.4.4 *MOSAR-WM Avionics*

The MOSAR-WM avionics architecture is depicted in Fig. 22.13. The WM Controller (WMC) is the central avionics component. It manages the internal control of the arm, of the two HOTDOCK at the WM tips, and the communication with the satellite on-board computer (OBC). The WMC uses an Intel NUC board, running Ubuntu 18.04 LTS 64-bit. This solution, offers the required compactness while providing a standard platform to integrate the different software applications.

The WMC interfaces the three main data buses running along the robotic arm:

- SpaceWire bus for "high level" non-deterministic telemetry and telecommand (TM/TC) communication with the servicer OBC through the RMAP protocol.

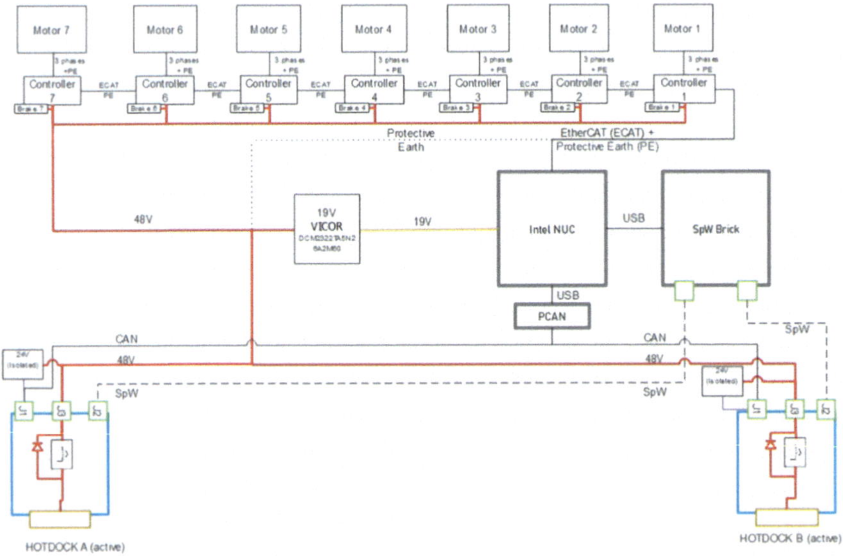

Fig. 22.13 MOSAR-WM avionics architecture

This relies on a Star-Dundee SpaceWire Brick Mk3, connected by USB to the WMC. The Brick also ensures the transmission of the SpaceWire signal between the two HOTDOCK extremities.
- CAN bus for local control of the HOTDOCK interfaces. This is done with a USB-CAN interface.
- EtherCAT bus for low-level deterministic communication with the joint drivers. This is done by the implementation of an EtherCAT Master application on the WMC and replacement of the Ethernet drivers by EtherCAT driver, to allow the connection directly though the WMC Ethernet port.

Each joint of the WM is equipped with a local controller for the closed loop position/current control of the actuator and the measurements of the joint sensors. All joint controllers are interfaced through the EtherCAT bus, managed by the WMC, which ensures real-time exchange of information required for the control algorithm of the arm at 2 kHz.

The robotic arm is powered by the servicer 48 V power bus, through the power interface connector of HOTDOCK. The 48 V bus is directly interfaced with the joint drivers that will manage the power interfaces to the motors and sensors of the joints. Local DC/DC convertors provide the required 24 V to power the two HOTDOCKs and 19 V for the WMC. The arm is able to control the power transfer, passing through the interfaces, thanks to power relays embedded in HOTDOCK. This allows to power-on a module connected at the end-effector and communicate with it through SpaceWire.

22.4.5 MOSAR-WM Software and Control

The WM comprises a series of software units to control the different operations that should be performed on the WMC (see Fig. 22.14):

- The robot monitor and control, split in two main parts:
  - A low-level control software for controlling the robotic joints over Ethercat, either through the servo control units to move in position mode, or through the WMC in torque mode.
  - A high-level control software to enable/disable the arm motions and set the position or torque commands to each robotic joint at high rate, based on advanced control methods (e.g. Cartesian Impedance Controller).
- The HOTDOCK monitor and control, a low-level software that manages the CAN bus communications between the WMC and the HOTDOCK.
- The TM/TC Engine, a Spacewire communication handler to convert high-level information from the spacecraft OBC into low-level information for HOTDOCKs and robot control.

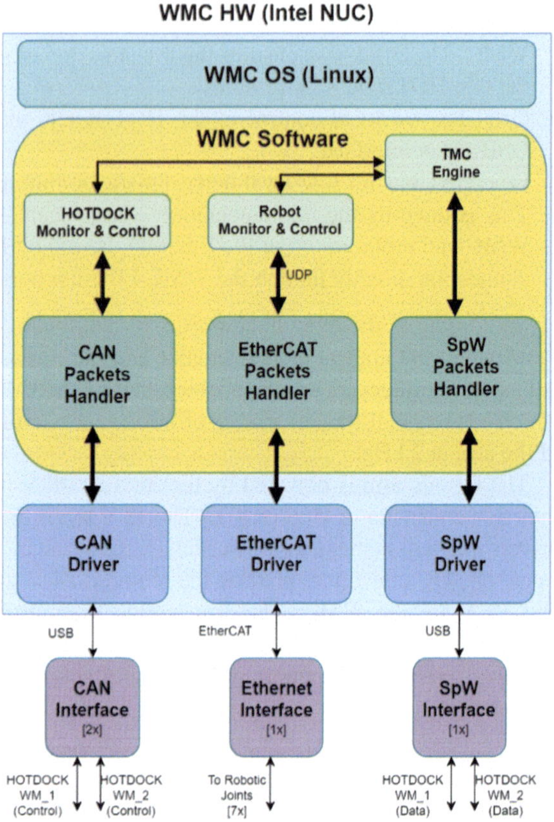

Fig. 22.14 WM software architecture

The WM controller communicates with the WM via an encapsulated data layer, which hides the specific communication details (i.e. the Master I/F and the CAN Driver). That means, that data packets are interfaced to the appropriated communication bus using an internal API (IP-based) to local SW processes running close to the OS/Driver layer and in charge of the encoding of the high-level information in the appropriated low-level format. The data communication between the WM controller and the EtherCAT Master must fulfil the real-time requirements, given by the control rate of the WM. As both software modules are running on the WMC, the implemented interface is based on the UDP/IP protocol.

Each joint of the WM is equipped with a local controller for the closed loop position/current control of the actuator and the measurements of the joint sensors. All joint controllers are interfaced through an EtherCAT bus (master/slave), which ensures the real-time exchange of information required for the control algorithm of the arm at 2 kHz.

The control of the MOSAR-WM is based on impedance controllers, which allow a modulation of the mechanical impedance at the joint or arm level while interacting with the environment. In this way, the system successfully completes interaction tasks such as docking or grasping objects, while making it very robust to external perturbations [7].

The Impedance Controller has two modes of operation. The first one is in Cartesian space, which sets the desired impedance at the tool centre point (TCP). The second one is at joint level, for setting individual desired joint impedances. In both cases, stiffness and damping matrices are predefined, i.e., they are controller parameters that define the compliant behaviour [8].

22.5 MOSAR Demonstration

As part of the final MOSAR validation and demonstration, several scenarios have been illustrated, representative of the concept of assembly and reconfiguration of modular satellite, as described in the first section of this chapter.

Figure 22.15 depicts the nominal sequence of operation for the transfer of modules between the client and the servicer spacecraft mock-ups, through the use of the MOSAR-WM and the HOTDOCK SI. It includes different type of operation with the (dis)connection and transfer of 10 kg modules under 1 g, relocation of the MOSAR-WM on the spacecraft structure and reconfiguration of the data and power network through the HOTDOCK devices. During the scenarios, both position and impedance based control have been demonstrated, highlighting valid different control strategies approach, not requiring external systems like visual based tracking.

The relocation results obtained during the test campaign validate the ability of the WM to perform successful relocation operations from one SI to another in planar and 3D configurations. Figures 22.16 and 22.17 illustrate in particular walking sequences on the demonstrator structure, respectively on parallel levels and on orthogonal surfaces. In Fig. 22.16, the WM managed to reach an SI located at 80 cm horizontally and 40 cm vertically from its attachment point, while in Fig. 22.17 it reaches the vertical panel SI located at (80 cm,20 cm,20 cm) from its initial attachment point and then the top panel SI located at (20 cm, 0, -20 cm) from the side panel SI.

Finally, the triple HOTDOCK orthogonal approach and simultaneous connection has been also validated (required when a module needs to be inserted in existing slot). The sequence of operation is illustrated in Fig. 22.18. The WM is approaching a module in a corner limited by the Client bus and two other modules. Once all the HOTDOCK are aligned, the command can be sent simultaneously to the three interfaces to perform the three connections (bottom and two sides). The experiment was also validated with the connections in sequence.

Fig. 22.15 MOSAR manipulation sequence of spacecraft module from client to servicer spacecraft

Fig. 22.16 Walking on top of spacecraft modules

Fig. 22.17 Walking on side of spacecraft modules

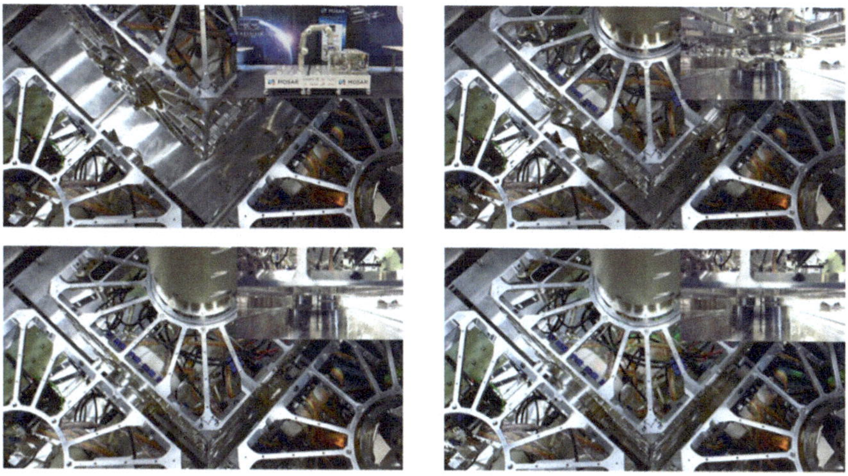

Fig. 22.18 HOTDOCK triple orthogonal alignment and simultaneous connection

22.6 Conclusions and Perspectives

This chapter described the innovative HOTDOCK SI and MOSAR-WM technologies, in perspective of their integration and demonstration in the scope of the H2020 MOSAR project.

Experimental results obtained during the MOSAR tests and validation campaign are very promising for the future use of these technologies for on-orbit operations:

- Fully integrated manipulator with all embedded components, with a full kinematics symmetry and high load capabilities.
- Smooth and repeatable operations of the MOSAR-WM, during motion, modules manipulation and HOTDOCK connection, highlighting the relevance of the kinematic configuration and the support of HOTDOCK mechanical features (as the form-fit).
- The possibility to control the arm both in position and impedance without the need of external visual based control.

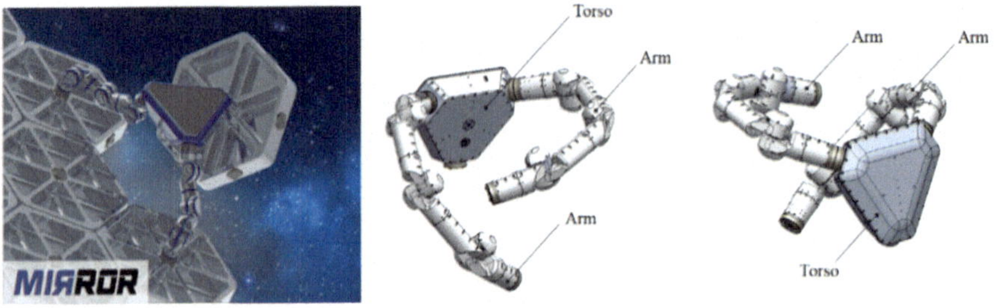

Fig. 22.19 Artisitc view and detailed design of the MIRROR Multi-Arm Robot concept

- The capability for HOTDOCK to support diagonal approach, large misalignments and simultaneous connections in different planes, as well as to sustain the high loads imposed at the MOSAR-WM base, including during the disconnection phase (when switching robot base). This is highly relevant for the use of HOTDOCK in future on-orbit applications.
- The capability for HOTDOCK to support data and power transfer with 90 degrees symmetry (mandatory for MOSAR like applications), including reconfiguration of SpW buses, for the full control of the arm.
- The mechanical robustness of HOTDOCK against collisions during operations.
- The capability for HOTDOCK to provide live telemetry of proximity and connection status to support autonomous operations.

Future activities on both robotics development will focus on the maturation of the technologies, including the TRL increase. This is done through follow-up implementation and maturation of these concepts in ESA projects, as the ESA MIRROR activity [9] with the implementation of a dual arm robotic system for assembly of large segmented telescopes, and the ESA OSIP SKYBEAM study, an application to cooperative robotics for building very large assembly in space.

HOTDOCK SI will also be part of the next EC projects targeting in-orbit demonstration, based on the previous SRC developments. Activities will also cover design evolution of HOTDOCK for qualification campaign, with testing in relevant environment (Fig. 22.19).

Acknowledgements The authors would like to acknowledge and express their thanks to all the supporting staff and partners of the MOSAR Project: Space Applications Services, DLR, GMV, Thales Alenia Space, SITAEL, Mag Soar, University of Strathclyde and Ellidiss and the financial support for MOASR as part of the project "PERASPERA" on Space Robotics Technologies from the European Union's Horizon 2020 research and innovation programme under grant agreement No 821966.

References

1. Davis, J. P., Mayberry, J. P., & Penn, J. P. (2019). *(2019) On-orbit Servicing: Inspection, repair*. Upgrade, and Assembly of Satellites in Space, The Aerospace Corporation.
2. SPACE NEWS: Northrop Grumman's MEV-1 servicer docks with Intelsat satellite, Caleb Henry, 2020. https://spacenews.com/northrop-grummans-mev-1-servicer-docks-with-intelsat-satellite/
3. ESA Website. http://www.esa.int/Safety_Security/Clean_Space
4. ESA Website. https://www.esa.int/Safety_Security/Clean_Space/ESA_commissions_world_s_first_space_debris_removal
5. Post, M. A., et al. (2021). Modularity for the future in space robotics: A review. Acta Astronautica, 2021, vol. 189, p. 530–547.
6. Pierre Letier et al. (2021). MOSAR: Demonstration of modular spacecraft assembly and reconfiguration, In *72nd International Astronautical Congress (IAC)*, Dubai 25–29 October 2021.
7. Lange, F., et al. (2016). Robotic simulation of on-orbit servicing including hard impacts. *In IEEE/RSJ Int. Conf. Intelligent Robots and Systems – IROS 2016*, pp. 2992–2997.
8. Albu-Schäffer, A., et al. (2007). A unified passivity-based control framework for position torque and impedance control of flexible joint robots, *Int. Journal of Robotics Research*, vol. 26, pp. 23–39, 2007.
9. M. Deremetz et al. (2021). Concept of Operations and Preliminary Flight Model Design of a Modular Multi-Arm Robot using Standard Interconnects for On-Orbit Large Assembly, In *72nd Intenational Astraunotical Congress (iAC)*, Dubai, 2021

Author Index

A
Alonso, Mercedes, 395
Aouf, Nabil, 395
Arancón, Miguel Muñoz, 121
Arm, Philip, 135
Azkarate, Martin, 231

B
Balachandran, Ribin, 281
Bartsch, Sebastian, 191, 209
Bazerque, Clément, 49
Benedetti De, Matteo, 395
Bensalem, Saddek, 121
Bianco, Alessandro, 433
Bissonnette, Vincent, 355
Brinkmann, Wiebke, 395
Brunner, Bernhard, 281, 515
Bruyninckx, Herman, 121
But, Alexandru, 395

C
Colmenero, Francisco Javier, 395
Cordes, Florian, 191, 483

D
Danter, Leon C., 395
Deremetz, Mathieu, 515
De Stefano, Marco, 281
Dettmann, Alexander, 483
Doermer, Manfred, 209
Domínguez, Raúl, 49, 483

F
Filippetto, Daniele, 375

G
Gancet, Jeremi, 49, 515
Germa, Thierry, 49, 355
Govindaraj, Shashank, 49, 395
Grunwald, Gerhard, 331, 355, 515

H
Heredia, Enrique, 395
Hoeflinger, Kilian, 121
Hugues, Jérôme, 121
Hutter, Marco, 135

K
Kirchner, Frank, 191, 483
Kisdi, Aron, 433
Kleszczynski, Dominik, 395
Kolvenbach, Hendrik, 135, 157
Kruse, Gerard, 105

L
Lacroix, Simon, 49, 395
Lampariello, Roberto, 281
Letier, Pierre, 355, 375, 515
Lingenauber, Martin, 281
Li, Youhua, 433
Lopes, Luis, 395
Lopez Negro, Pablo, 355
Lu, Ying, 433

M
Magnani, Piergiovanni, 97
Maurus, Michael, 483
M.Giordano, Alessandro, 281

X. T. Yan and G. Visentin (eds.), *Space Robotics*,
https://doi.org/10.1007/978-3-031-39214-6

Michaud, Stéphane, 105
Mishra, Hrishk, 281

N
Negro, Pablo Lopez, 375
Nieto, Irene Sanz, 395
Nottensteiner, Korbinian, 331, 355

O
Oumer, Nassir, 281

P
Palazzetti, Roberto, 433
Paz-Delgado, G. J., 231
Pérez-del-Pulgar, C. J., 231
Picton, Kevin, 395
Post, Mark A., 433
Purnell, Joseph, 395

R
Rantanen, Olli, 121
Roa, Máximo A., 281, 331, 355, 515
Roßmann, Jürgen, 305
Rodríguez, Ismael, 331, 355
Roehr, T. M., 191
Rognant, Mathieu, 355
Romeo-Manrique, P., 231
Rouvinet, Julien, 355
Rudin, Nikita, 135
Rusconi, Andrea, 97
Rybus, Tomasz, 157

S
Sánchez-Ibáñez, J. R., 231
Shilton, Mark, 121
Siegfried Koch, Christian Ernst, 355
Silveira, Daniel, 121
Smisek, Michal, 49
Sonsalla, R. U., 191, 483
Stelzer, Martin, 281

T
Tanzariello, Claudio, 375
Thieling, Jörn, 305
Trinh, Souriya, 355
Tsiogkas, Nikolaos, 121
Tubby, Wayne, 433
Turetta, Alessio, 355

U
Ummel, Antoine, 355

V
Valsecchi, Giorgio, 135
Viards, Raphael, 49
Vinals, Javier, 209, 455
Visentin, Gianfranco, 1, 13
Vögele, Thomas, 483

W
Winnendael van, Michel, 105
Winter, Matthias, 249
Wirkus, Malte, 121

Y
Yan, Xiu-Tian, 1, 13, 49, 157, 209, 375, 433, 515

Topics Index

A
Accelerometers, 53
Actuation, 39, 143, 457
Air bearing, 158, 161, 163
Arms, 98, 101
Autonomy, 38, 39, 70, 122, 249, 263, 404, 405

C
Calibration, 63
Case studies, 129
Components, 127, 253, 348, 404
Control, 70, 90, 92, 122, 128, 131, 163, 166, 167, 170–172, 174, 202, 235, 239, 256, 264, 266, 290, 291, 297, 324, 417, 418, 441, 443, 473, 495
  adaptive, 174, 202
  learning-based control, 147
  model-based control, 147
Control interface, 407

D
Data, 62, 66, 67, 69, 76, 77, 79, 81, 125, 127, 129, 131, 173, 222, 277, 409, 458, 473, 479
  data fusion, 66, 67, 79, 81, 173
Debris, 214
  debris removal, 6, 37, 160, 179, 180, 214, 281, 283
Design methodology, 13, 215, 437, 438
Digital Twin, 312, 320–323, 328

E
Energy, 40, 110
Environment, 14–17, 19, 20, 32, 35, 38–40, 50, 51, 56–58, 60–64, 66, 67, 70–72, 75, 77–79, 82, 85, 89, 91, 92, 97, 98, 103, 122, 127, 128, 136, 141–143, 147–151, 157, 158, 163, 165, 166, 172, 175, 179, 180, 213, 215, 216, 218, 231, 232, 234, 238, 239, 245, 252–254, 258, 260, 261, 272, 306–313, 315, 317–321, 323–325, 327, 328, 396, 409, 415, 423, 425, 430, 434, 438, 442, 444, 445, 447, 450–452
European Space Agency (ESA), 88, 92, 99, 100, 102, 103, 110, 122, 125, 140, 150, 151, 162, 163, 214, 233, 234, 236, 276, 277, 326, 331, 340, 465, 510

F
Field, 149, 202, 252, 272, 273, 277, 448, 449, 491
Flexibility, 35, 68, 72, 76, 100, 103, 166, 194, 195, 201, 215, 223, 227, 228, 261, 308, 404, 437, 475
Framework, 62, 69, 121, 123, 124, 409
Fusion, 38, 40, 50, 61–63, 66, 67
  fusion algorithm, 77, 89, 417
  fusion framework, 50, 62, 76, 78, 79, 90
  information fusion, 68
  multi-scale fusion, 67
  sensor fusion, 67, 68
Future work, 133, 451, 452

G
Gyroscopes, 55

X. T. Yan and G. Visentin (eds.), *Space Robotics*,
https://doi.org/10.1007/978-3-031-39214-6

H

Hardware in, 7, 261, 367
HOTDOCK, 515, 522, 523, 525, 526, 528, 529, 531, 533, 536
Hypervisor, 127

I

Inertial measurement units, 53
Infrastructure, 31, 32, 38, 125, 193, 197, 202, 203, 211, 213, 214, 227, 261, 298, 299, 331, 356, 375, 377, 396, 420, 507
In-space assembly, 50, 332–335, 371
Intelligence, 70, 138, 170, 307, 328, 475
Interface, 125, 195, 196, 209, 210, 216, 273, 275, 320, 417, 455, 466, 475

K

Kinematics, 126

L

Learning, 147
Leg, 144
Legged systems, 135, 138
  space legged robots, 138
  terrestrial legged robots, 140
Lifetime, 37, 192, 214, 282, 383, 388
Lightweight, 101
Localisation, 38, 69–71, 78, 84, 85, 87, 251, 255, 256, 264, 266, 273, 412, 419–423
  direct localisation, 87
  indirect localisation, 87
Locomotion, 39, 106, 145, 150, 239

M

Magnetometers, 57
Manipulation, 199, 200, 209, 210, 404, 421, 425, 455, 466
Manipulators, 97, 166
Manoeuvrability, 109
Mars, 3, 5–7, 20, 38, 40, 51, 78, 88, 91, 101, 102, 105–107, 129, 136–138, 150, 196, 231–233, 236, 249, 252, 261–264, 266, 271, 272, 274, 276, 305, 307, 312, 326–328, 331, 397, 489
Martian, 150, 211, 232, 249, 252, 254, 255, 264, 273
Mechatronic, 13, 16, 51, 331, 433, 437, 438, 443
Mission, 19, 35, 85, 193, 198, 201, 211, 228, 381, 397, 404, 406–408, 505
  mission planning, 201, 404, 406–408
Modelling
  sensing modelling, 62
Modular, 191, 198–200, 204, 213, 226–228, 385, 386, 436
Modularization, 195
Modular Spacecraft Assembly and Reconfiguration (MOSAR), 11, 133, 340, 376, 385–393, 515–520, 524, 526–536
Moon, 40, 78, 105–107, 136–139, 150, 331, 396, 397, 503
Multiple robot, 86
Multi-robot systems, 191, 192, 197, 200–202, 204, 498

N

Navigation, 17, 38, 40, 50–52, 57, 61, 68–70, 75, 78–80, 85, 86, 89–92, 111, 143, 171, 173, 194, 211, 232, 234, 237–240, 245, 246, 251, 252, 271–273, 375, 377, 396, 433, 435, 436, 445, 449, 450, 452, 494, 497
Nuclear scenario, 131

O

Orbital, 3, 4, 6, 31, 37–39, 50, 67, 72, 75, 76, 79, 81, 83, 86, 91, 92, 129, 157, 162, 164, 166, 167, 173, 179, 192, 194, 211, 212, 214, 216–219, 224, 227, 228, 275, 276, 281, 286, 291, 296, 298, 300, 331, 332, 338, 355, 356, 380, 383, 390, 422, 466
  orbital robot, 31, 162, 164, 281

P

Payload, 4, 31, 32, 97, 98, 101, 103, 150, 151, 164, 166, 194, 197–199, 203, 204, 209–211, 214, 218, 219, 222, 227, 237, 298, 300, 333–336, 338, 356, 384, 387, 388, 391, 392, 396, 399, 400, 402, 421, 441, 443, 455, 459, 475, 497
Perception, 15, 38, 49, 50, 58, 62, 66, 69, 70, 72, 73, 75, 76, 78, 81, 86–90, 271, 307, 322, 324, 397, 423, 428, 493

perception processing, 72
sensors, 51, 52, 63, 305, 413, 459
Planetary, 38, 70, 129, 211, 213, 219, 227, 237, 395, 396, 412, 483
planetary robot, 3, 20, 31, 66, 79, 149, 152, 213, 214
planetary scenario, 129
Planning, 15, 38, 49–52, 70, 75, 76, 78–80, 84–87, 90, 91, 127, 132, 149, 157, 166, 167, 170, 171, 179, 180, 192, 195, 201–204, 212, 231–236, 238, 240, 242–246, 256, 259, 271–273, 320, 326, 327, 348, 397, 406–409, 412, 417, 421, 422, 425, 433, 438, 440, 443, 444, 447–449, 452, 493

R

Reconfigurable, 86, 191–204, 209, 210
Reconfiguration, 50, 86, 191, 194–204, 210, 211, 214, 229, 376, 381, 383, 385, 387, 388, 392, 393, 465, 466
Reliability, 62, 89, 122, 123, 307, 397, 443
Resources, 71, 82, 98, 101, 149, 173, 201, 236, 252, 310, 376, 384, 396, 406

S

Sample Return, 7, 38, 232, 305, 307, 326–328, 397
Scenario
orbital scenario, 130, 212
Self-alignment, 217
Simulation, 114, 215, 236, 249, 252, 254, 313
Soil, 79, 80, 82, 111, 150, 434, 438, 443
Space environment, 310, 338
lighting, 19, 50, 52, 79, 83, 84, 149, 253, 261, 292, 297
radiation, 19, 51, 89, 136
thermal, 4, 18, 33, 52, 152, 195, 209, 210, 215, 218, 219, 223, 224, 226–228, 291, 299, 337, 338, 356, 384, 386–388, 391, 420, 455, 474
vacuum, 53, 291, 337, 464, 477
Space qualification, 89
Standard interconnect, 359, 364, 370, 417, 520, 521
HOTDOCK, 515, 516, 518, 520–523, 525–527, 530, 531, 533, 535
Status, 133
Stereo-cameras, 57
Subterranean, 149
Support tools, 127
System Interconnect for Robotic Manipulation (SIROM), 204, 209–212, 214–229, 455–462, 464–481

T

Teleoperation, 7, 129, 130, 281, 283, 287, 296–298, 300, 336, 341, 423
Terrain, 79, 80, 116, 250, 254, 507
Terrainability, 108
Testbeds, 7, 305, 307, 310, 327, 397
Testing, 112, 130, 131, 164, 165, 226, 236, 249, 252, 254, 260, 263, 271, 273, 276, 464, 491
The ASSERT Set of Tools for Engineering (TASTE), 124–129, 131, 132
Thermal, 223, 227, 388, 474, 480
heat exchanger, 223
Time-of-Flight, 58
TOF sensors, 58, 60
Trafficability, 106

V

Validation, 15, 63, 123, 124, 129, 132, 138, 152, 158, 167, 171, 173, 215, 219, 227, 245, 287, 297, 305–310, 312, 324, 326, 328, 396, 397, 420, 427, 461, 474
functional validation, 326
on-ground validation, 281, 287, 298
Verification, 90, 103, 121, 123, 124, 126, 128, 132, 165, 166, 170, 215, 218, 227, 245, 281, 305–307, 326, 355, 360, 369, 396, 397, 466
Virtual, 215, 425
virtual environment, 311, 343
virtual graspings, 425
virtual integration, 308
virtual laboratory, 320, 321
virtual manipulation, 430
virtual network, 418, 427
virtual robot, 231, 238
virtual rover, 245, 259
virtual scene, 238
virtual terrain, 237, 239
virtual test, 310
virtual validation, 326

W

Walking, 38, 98, 135, 138, 140–143, 147, 151, 198, 236, 385, 386, 388, 396, 465, 496
Wheel, 108, 111, 112, 441

The manufacturer's authorised representative in the EU is Springer Nature Customer Service Centre GmbH, Europaplatz 3, 69115 Heidelberg, Germany. If you have any concerns regarding our products, please contact ProductSafety@springernature.com

Printed and bound by CPI Group (UK) Ltd, Croydon, CR0 4YY
01/10/2026
02225962-0017